Political and Legal Dimensions

INDUS WATERS TREATY

Political and Legal Dimensions

INDUS WATERS TREATY

Ijaz Hussain

OXFORD

UNIVERSITY PRESS

OXFORD
UNIVERSITY PRESS

Oxford University Press is a department of the University of Oxford.
It furthers the University's objective of excellence in research, scholarship,
and education by publishing worldwide. Oxford is a registered trade mark of
Oxford University Press in the UK and in certain other countries

Published in Pakistan by
Ameena Saiyid, Oxford University Press
No.38, Sector 15, Korangi Industrial Area,
PO Box 8214, Karachi-74900, Pakistan

ISBN 978-0-19-940354-7

Typeset in Baskerville
Printed on 80gsm Local Offset Paper

Printed by The Times Press Pvt. Ltd., Karachi

To
Jahangir Ahmad and 'Deputy' Naseem Ahmad

One can pay back the loan of gold,
but one dies forever in debt to those who are kind.
                                    —Malayan Proverb

Contents

Preface   xiii

Introduction   1

Chapter 1

Indus Rivers and Plains, Canal Building, and   13
    Sindh-Punjab Water Dispute
    A) The Indus Rivers   14
    B) The Indus Plain   20
    C) Canal Building before the British   22
    D) British and the Canal System   24
    E) Origins of Sindh-Punjab Water Dispute   26

Chapter 2

Genesis of the Dispute   39
    A) Radcliffe Award and its Background   39
    B) Motives behind the Award   45
    C) Standstill Agreement and the Aftermath   61
    D) May 1948 Joint Statement and its Validity 71

Chapter 3

World Bank: The Honest Broker?   100
    A) The Lilienthal Proposal and its Background 100
    B) Parties' Interests in the Negotiations   103
    C) Bank's Initiative and Start of Negotiations 108
    D) Bank Cuts the Gordian Knot   116
    E) Analysis of Pakistan's Claim to the Eastern Rivers 125
    F) Bank Dictates: Smoking Gun Evidence   133
    G) Crossing the Rubicon   137

H) Towards a Water Treaty   149
I) Appraisal of Ayub Khan's Role   158
J) Bank's Motive to Side with India   163

Chapter 4

Terms of the Treaty   186

A) Uniqueness of the Treaty   186
B) Safeguards for Pakistan and India   188
C) Permanent Indus Commission   193
D) Conflict Resolution Procedure   194
E) Territorial Scope of the Treaty   202
F) Miscellaneous Provisions   207

Chapter 5

Treaty in Action-I   215

A) Salal Dam   215
    a) History of the Dispute   215
    b) Pakistan's Objections and India's Rebuttal   217
    c) The Settlement   220
B) Wullar Barrage/Tulbul Navigational Project   221
    a) History of the Dispute   222
    b) Arguments of the Parties and their Critique   225
C) Baglihar Dam   229
    a) History of the Dispute   230
    b) Did Pakistan Seize the World Bank Belatedly?   237
    c) Appointment of the Neutral Expert   240
    d) Pakistan's Objections   242
    e) Point Counter Point   245
        i)   India's Arguments   245
        ii) Pakistan's Arguments   247
    f) Neutral Expert's Determination and its Critique   250
    g) Why Did Pakistan Lose the Case?   257
    h) Allegation of Stealing Pakistan's Water   270

Chapter 6

Treaty in Action-II   281
  A) Kishenganga Dam   281
    a) History of the Dispute   281
    b) Point Counter Point   283
      i) Pakistan's Arguments   283
      ii) India's Arguments   288
    c) Partial Award and Court's Reasoning 294
    d) Final Award and Court's Reasoning   300
    e) Critique of the Award   304
  B) Nimoo Bazgo and Chutak Dams   311
    a) Basic Facts   311
    b) India Claims Carbon Credits   312
    c) Controversy about Jamaat Ali Shah   313
  C) Restructuring Governmental Water-Related Machinery   318

Chapter 7

Treaty and Climate Change   333
  A) Is Climate Change Real?   334
  B) South Asia and Climate Change   344
  C) Climate Change and Indus Rivers   347

Chapter 8

Treaty in Trouble   355
  A) Terrorism as Grounds for Termination   358
  B) *Rebus Sic Stantibus* as Grounds for Termination   364
  C) Kashmiri Concerns as Grounds for Termination   369
  D) Unfairness as Grounds for Renegotiations   371
  E) India's Motive for Revisiting the Treaty   375
  F) India's Water Diplomacy with its South Asian Neighbours 381
    a) Nepal   382
      i) Kosi Agreement   382
      ii) Gandak Agreement   384

iii) Tanakpur Agreement   385
iv) Mahakali Treaty   386
b) Bangladesh   388
i)   Farakka Barrage Agreement   388
ii) Teesta River Dispute   396

Conclusion   410

Bibliography   438
A) Official Documents   438
a) General   438
b) World Bank-Related Documents   441
B) Books and Reports   443
C) Periodical Literature   452
D) Newspaper Articles   457
E) Case Law   464
a) General   464
b) Baglihar and Kishenganga   465
F) Pakistan's TV Talk Shows (Available on the Internet)   466
G) Interviews or Correspondence with Water Experts   466

Appendix 1: Inter-Dominion Agreement   468

Appendix 2: Text of Indus Waters Treaty   470

Appendix 3: Details of Projects by India on Western Rivers   516
A) River Chenab   516
a) Commissioned Projects   516
b) Location   518
c) Under Active Consideration and Power Potential   519
d) Locations of Possible Diversions   520
B) River Jhelum   521
a) Commissioned Projects   521
b) Location   522
c) Power Potential   523

C) River Indus 531
    a) Commissioned/Under Construction Projects 531
    b) Location 532
    c) Power Potential 533
D) Rivers Common with Afghanistan 535
    a) Common Rivers in Khyber Pakhtunkhwa 535
    b) Common Rivers in Balochistan 536
    c) Average Annual Inflows 537
    d) Location Map of Ongoing/Proposed Projects 538
        by Pakistan
    e) Projects in Afghanistan 539
E) Wullar Barrage and Storage Project/Tulbul 540
    Navigation Project
    a) Location 540
F) Baglihar HEP 541
    a) Location 541
    b) Salient Features 542
G) Kishenganga Hydroelectric Plant 543
    a) Kishenganga Hydroelectric Plant on River Neelum 543
    b) Composite Map of Kishenganga, Wullar, LJHP, 544
        URI-I, and URI-II

Appendix 4: Western Rivers Inflow at Rim-Stations (MAF) 545

Index 551

List of Maps, Figures, and Graphs

MAPS:

Map 1.1:   Indus Basin   13
Map 5.1:   Wullar Barrage   221
Map 5.2:   Baglihar Dam   229
Map 6.1:   KHEP and the NJHEP Kishenganga Dam (I)   282
Map 6.2:   KHEP and the NJHEP Kishenganga Dam (II)   283

FIGURES:

Figure 3.1: Shoaib's Draft and Iliff's Alterations (I)   135
Figure 3.2: Shoaib's Draft and Iliff's Alterations (II)   136
Figure 5.1: Salal Dam   216
Figure 5.2: Wullar Barrage/Tulbul Navigational Project   222
Figure 5.3: Baglihar Dam   230
Figure 6.1: Nimoo Bazgo Dam   311

GRAPHS:

Graph 7.1: Rise in Global Mean Sea Level (1901–2010)   341
Graph 7.2: Energy Accumulation within the Earth's   342
           Climate System
Graph 7.3: Globally averaged greenhouse gas concentrations   343
Graph 7.4: Radiative forcing in 2011 relative to 1750   344
Graph 7.5: Western Rivers Inflow at RIM-Stations   349
           (Average MAF)
Graph 7.6: Western Rivers Inflow at RIM-Stations Pre-   349
           Independence vs. Post Tarbela (Average MAF)

Preface

I EMBARKED ON THIS STUDY IN EARLY 2010 WHILE I WAS A MEMBER OF the faculty at the National Institute of Pakistan Studies (NIPS), Quaid-i-Azam University, Islamabad. I had imagined that I would be able to bring this project to completion fast, ideally around the fiftieth anniversary of the signing of the Indus Waters Treaty (IWT), signed on 19 September 2010. Being an academic, I had in-depth knowledge of the subject, and had written on the IWT in the *Daily Times* as a regular columnist. However, I soon realized that the publication date that I had fixed was too ambitious. This was so for two reasons: firstly, my duties at the NIPS did not leave enough time for me to conduct serious research; and secondly, I did not have sufficient funds to undertake a visit to the Library of Congress in Washington DC—an indispensable resource housing a treasure trove of material on the Treaty. I had hoped that the Quaid-i-Azam University, the Higher Education Commission, or the Fulbright Educational Foundation would provide the funds for the proposed visit, however, that did not happen.

Given the importance of the water issue in Pakistan, I decided to utilize my personal resources. I had enough funds to pay for my air travel to Washington but I did not have the financial resources for my lodging. Hence I decided to contact my friend, Jahangir Ahmad, who lived in Washington; he generously offered to be my host. Graciously accepting his offer, in the summer of 2010 I embarked on my journey to Washington.

While I was working at the Library of Congress, an American friend of Pakistani origin, Dr Suleiman Wasty, who worked on World Bank projects, indicated that the Bank had recently declassified documents relating to the negotiations held on the Indus waters dispute under its auspices. Consequently, I requested that he arrange a visit for me to the archives. Dr Wasty fixed an appointment for me with Ms April Miller, the Bank's archivist, who provided a comprehensive list of documents and asked me to identify those that I was interested in. These documents were stored at the University of Pennsylvania and the Bank arranged for them to be brought from there.

I went through some of them but did not have the time to go through them all. Hence I decided to go back to Washington in 2012 to go over the remaining documents.

A mention regarding the working conditions at the archival section of the Bank is merited here. It has a small reading room with limited number of seats for researchers. The atmosphere is quite intimidating. There is a raised platform at the head of the room from where an archivist monitors the activities of the researchers. A closed circuit camera is also installed. Researchers are never left alone and have to enter their time of arrival and departure in a log book that is placed at the entrance. They have to log in every time they leave the room and re-enter. There are dos and don'ts that they have to observe in the reading room which have to be followed diligently. If they fail to do so, they are severely reprimanded. This is meant to deter researchers from pilfering or damaging the sensitive documents.

Coming to the Treaty-related documents. There are two categories of them. Those documents that belong to the Bank fall in the first category, and the documents which are the property of the Governments of Pakistan and India fall in the second category. The Bank was at liberty to declassify the documents belonging to the first category without seeking permission from India or Pakistan. As for the documents in the second category, it could declassify those only after seeking permission from the Governments of Pakistan and India. They allowed declassification of only a few documents. This becomes obvious when one sifts through them. One suddenly comes across a tag which indicates that the concerned government had not allowed the Bank to release it. The tag, however, provides the title of the missing document, the date of its preparation, the names of its author, and its addressee. It also indicates that the researcher is entitled to request the archival section to seek permission from the concerned government. And although I submitted a request to the archival section to allow me to go through several documents, nothing came out of it. It is disconcerting that even after the passage of more than a half century, the governments of India and Pakistan are not prepared to reveal the facts involved in the negotiations of the Indus waters dispute. The problem is further compounded by the fact that both governments have not declassified the Treaty-related documents which are in their possession. However, since I was keen to consult these documents, I contacted the relevant authorities.

I started with my own country and wrote to Mr Chaudhry Rauf who was Establishment Secretary in the Government of Pakistan. Simultaneously, I also approached the Vice Chancellor of Quaid-i-Azam University, Dr Masoom Yasinzai, to write to the Mr Rauf to assist my claim, and he did. However, at the end of the day, the entire demarche was an exercise in futility. In theory, the Pakistan Government is required to declassify documents if twenty-five years have elapsed since the time of the documented event. India too is supposed to follow the same protocol. Yet this does not happen.

When I was not able to go through the documents that were in the possession of both the governments, I decided to use, for this book, the declassified documents that I had been able to research at the World Bank. My study of them had revealed the kind of diplomacy that was at play.

The negotiations under the Bank's auspices started in 1952 but in due course reached an impasse because India and Pakistan would not agree to a conflict resolution. To curb this impasse, in 1954, the Bank proposed the division of the six rivers of the Indus Basin by allocating exclusive use of three Eastern Rivers to India and three Western Rivers to Pakistan. This was not acceptable to Pakistan hence the deadlock. The Indian government expressed reluctance in continuing negotiations unless the Bank's proposal was accepted by Pakistan. Eventually Pakistan government relented. While rummaging through the archives, I came across a two-page document which unequivocally revealed that the Bank had forced Pakistan to do so. The document in question is Pakistan's draft reply written by its Executive Director at the Bank, Muhammad Shoaib. He tried to meet the Indian demand (acceptance of the Bank Proposal) by scribbling a text on the document which is feeble in character. Thereafter, one can view World Bank's Vice President Sir William Iliff's correction of the text. It states that the Government of Pakistan accepts the Bank's proposal. The fact that the Pakistan government accepted the corrected version is proven by the fact that soon after this episode, an official letter was written to the Bank accepting the proposal suggested by Iliff. The significance of this discovery was that at that time it was rumoured that the US had forced Pakistan to surrender the three Eastern Rivers to India in perpetuity, but there was no hard evidence to support this theory. This document confirmed this conspiracy theory and brought it to the realm of hard political reality (for details see Chapter 5).

***

In bringing this book to fruition, I would like to acknowledge the help and support that was extended to me by individuals and institutions.

I am most grateful to my friend Jahangir Ahmad and his wife 'Deputy' Naseem Begum and for their warmth and hospitality during my stay in Washington. Jahangir would drop me every morning to the Bank or the Library of Congress and would also pick me up later in the day. I must confess that without their help this study would not have been possible.

I would like to thank Jamaat Ali Shah, Pakistan's former Indus Commissioner. He was always willing to help me. He not only provided data and documents relating to various aspects of the study but also offered his candid views on various water-related issues. I would also like to thank the incumbent Indus Commissioner and Vice President, Water Resources Division, NESPAK, Mirza Asif

Baig, for providing me with the documents related to the *Baglihar* and *Kishenganga* cases. He clarified many points that I would find obscure. My debt of gratitude to Bashir Ahmad, Advisor to the Pakistan Indus Commission, Vice President, Water and Agricultural Division, NESPAK, and member of the Pakistan delegation to the *Baglihar* case, for taking time out to read my manuscript, make useful suggestions, and answering my queries that related to the *Baglihar* case.

I am equally indebted to engineer Shams-ul-Mulk, former Chairman, WAPDA, and current President, Society for Promotion of Engineering Sciences and Technology (SOPREST), Pakistan. He enlightened me with his views on various aspects of the Treaty. Thanks to Professor Raymond Lafitte, Neutral Expert in the *Baglihar* case for answering my queries, and also to M. A. Salman, former lead counsel for the World Bank. He provided me with his critical evaluation on Pakistan's strategy relating to the *Baglihar* and *Kishenganga* cases. Thanks also to Shafqat Kakakhel, former Ambassador of Pakistan and ex-Deputy Secretary General, UNEP; Javed Hafiz, former Ambassador of Pakistan; Asif Kazi, Honorary Vice President, International Commission on Large Dams (ICOLD); Sardar Muhammad Tariq, Executive Director, Global Water Partnership, South Asia; M. H. Siddiqui, consultant, Punjab Irrigation Department; Dr Rafique Afzal, former Dean, Faculty of Social Sciences, QAU; and Dr Naeem Qureshi, former Chairman, Department of History, QAU.

My profound appreciation for the personnel of the World Bank, especially archivists Ms April Miller and Ms Vlada Alekankina; personnel of the Library of Congress, Washington DC; Mr Tahir Naqvi, Librarian, NIPS, QAU; and Mr Manzoor Anjum, Chief Librarian, University of Engineering and Technology, Lahore.

Last but not the least, particular thanks to my wife, Ismat Hussain, who was always there to take care of internet related glitches; to my sons, Aun Hussain and Nabeel Hussain, who, from their abode in Canada, kept providing me with documents which were not available in Pakistan, and were a source of help in a variety of other ways; and to my brother-in-law Munir Zaidi for the graphics in the book.

Finally, the views expressed in this book are strictly mine—
I alone am responsible for any act of omission or commission in the
writing of the book—hence my views should not be attributed to my
abovementioned benefactors.

Ijaz Hussain
Islamabad, January 2016

Introduction

THE GREAT AMERICAN WRITER AND WIT, MARK TWAIN, FAMOUSLY quipped, 'Whiskey is for drinking, but water is for fighting over.'[1] His statement makes sense when we realise that water is an enormously precious commodity as out of the 2.97 per cent of the fresh water found on planet earth, only 0.03 per cent is available for use by a population of over 7 billion people.[2] The problem is further compounded by the fact that 80 million people are born every year which puts additional pressure on the existing fresh water resources.

With this backdrop of galloping human population and static fresh water resources, Ismail Serageldin, the founding chairman of the Global Water Partnership and the former chairman of the World Commission for Water in the 21st Century, gave an ominous warning in 1995, '[i]f wars of this century were fought over oil, the wars of the next century will be fought over water, unless we change our approach to managing this precious and vital resource.'[3] He is not alone in making such a dire prediction. Three successive secretaries-general of the United Nations, starting with Boutros Boutros Ghali, are on record having made similar statements.[4] For example, the current Secretary-General Ban Ki-moon, while addressing the World Economic Forum at Davos in 2008, stated:[5] 'A shortage of water resources could spell increased conflicts in the future. Population growth will make the problem worse. Same is true about the climate change. As the global economy grows so will its thirst. Many more conflicts lie just over the horizon.' His predecessor, Kofi Annan, observed in 2001 that '[f]ierce competition for fresh water may well become a source of conflict and wars in the future'.[6] According to some commentators, there are possibilities of armed conflicts between states in the future over sharing of water resources. No two countries are more serious candidates for the first water war in the twenty-first century than Pakistan and India. To comprehend this phenomenon we need to understand the nature of the water issue between them.

The water dispute between Pakistan and India owes its origin to the Punjab Boundary Commission Award headed by Cyril Radcliffe. In violation of its mandate to demarcate the boundaries on the basis of geographical contiguity, headworks in Madhopur and Ferozpur (originally Muslim majority areas) were allocated to India. As a result of this unfair award, Pakistan, which should have been an upper riparian vis-à-vis India, unfortunately became a lower riparian. In December 1947, the two countries concluded an accord called the Standstill Agreement by virtue of which they decided to continue with the existing arrangement to share the waters of the Indus system of rivers till 31 March 1948, after which they were to draw up a new agreement. Around the time the Standstill Agreement expired, Pakistan approached India for its renewal. However, instead of engaging Pakistan in negotiations, India responded by stopping the flow of water to eleven West Punjab canals with the result that this area, which was the breadbasket of Pakistan, was simply devastated. Following this blow, a high-powered Pakistani delegation rushed to New Delhi in order to get the flow of water to these canals restored. India refused to accept Pakistan's plea for restoration by invoking the Harmon doctrine according to which India was entitled to do what it liked with rivers that flowed through its territory. Following Pakistan's entreaties, India finally agreed to restore the flow of water but in return for a price. It asked for the proprietary rights over the Eastern Rivers to be surrendered to India, an agreement that came to be variously known as the Delhi Agreement, the Inter-Dominion Agreement, the Document, or the Joint Statement.

This agreement was, from Pakistan's perspective, not only a dictated document but also an ad hoc arrangement which was to lapse with the passage of time. The two countries then entered into negotiations to find a permanent and equitable solution to the water dispute. The negotiations were protracted, continuing till 1952 without a settlement. Relations between the two countries became very tense and at times it appeared as if they would go to war. Around this time, David E. Lilienthal, the former chairman of the Tennessee Valley Authority (TVA), wrote an article in the *Collier's* magazine in which he drew attention to the grave threat that the water dispute between Pakistan and India posed to international peace. He advocated the intervention of the international community for resolution of the water dispute between the two adversaries. This was the period of the Cold War as the Korean War was raging. It was in this backdrop that the president of the World Bank, Eugene Black, who was a friend of Lilienthal, happened to read this article and offered the services of his institution to mediate between the two countries to bring about a settlement between them. The leadership of Pakistan and India accepted the offer, and started a protracted round of trilateral parleys involving the two countries and the World Bank which lasted for about eight years. On 19 September 1960, the two countries, with the active assistance of the World Bank, succeeded in drawing up the Indus Waters Treaty which settled the water dispute between them by allocating proprietary rights over the Eastern Rivers in perpetuity to India, and almost the same rights over the Western Rivers to Pakistan. What the Treaty meant in practical terms to the two countries is revealed by the following World Bank note:[7]

> The allocation to India of the whole flow of the Eastern Rivers calls for a diversion of water in Pakistan on a gigantic scale. It envisages the cutting off of about 6 million acres of land in Pakistan from its historic source of supply (the Eastern Rivers) and making this area dependent, through a complicated system of works, on a source of supply (the Western Rivers) which, in the case of the Pakistan Sutlej Valley Canals, is from 300 to 400 miles away from the nearest point of use. An idea of the magnitude of the operation can be gained from the fact that the area in Pakistan historically irrigated by the Eastern Rivers (and now proposed to be fed from the Western Rivers) is roughly the equivalent of the total irrigated area of Egypt. We are not aware of any precedent for a diversion on this scale from actual historic uses. India will actually divert from Pak cultivators nearly twice as much water as Egypt hoped to gain by building the High Aswan Dam.

The Treaty has been in operation now for the last fifty odd years. During this period, it has kept peace between the two countries on water-related matters. One can say that it has, so far, been a success story. However, that does not mean that there has been no water-related dispute between the two countries. In fact, there are a number of cases that have arisen after its conclusion.

The first one relates to the Salal Dam which India planned in the 1970s on the Jhelum River. Pakistan had objections against its design which the parties, after some hiccups, successfully resolved through bilateral talks. The second one relates to the Wullar Barrage/Tulbul Navigational Project which India planned to construct on the Jhelum River reportedly for navigational purposes. Here too Pakistan raised serious objections against the project as a result of which India decided to stop work on it. The work remains suspended as it is currently one of eight issues in the Composite Dialogue under the Pakistan-India peace process which awaits resolution.

The third dispute pertains to the Baglihar Dam which India has built on the Chenab River. The parties initially tried to resolve it through bilateral negotiations but failed to do so. The matter was ultimately referred to a neutral expert appointed by the World Bank who favoured India on the critical question of gated spillway. The next case relates to the Kishenganga Dam which India is building on the Jhelum River. The two countries tried to resolve it bilaterally but failed to do so. Pakistan then referred it to the International Court of Arbitration which ruled that India was entitled to divert water from the Kishenganga/Neelum River. However, it pointed that the latter was not entitled to reduce the water level in the reservoirs of the run-of-river plants below the dead storage level except in case of emergency.

Another case relates to Nimoo Bazgo Dam which India has constructed on the Jhelum River. Pakistan had serious objections against its design which, in its opinion, is contrary to the Treaty and to which it claims it never gave its consent. An ancillary matter related to the Nimoo Bazgo issue is the question of carbon credits that the United Nations, under the Framework Convention on Climate Change (UNFCCC), has awarded to India on the ground that Pakistan had issued an environmental impact assessment report in favour of the latter. Pakistan denies it and was at one time planning to file a suit against India on the issue before the International Court of Arbitration but subsequently decided against it.

In the course of implementing the Treaty over the last fifty-six years, new issues have cropped up which the two countries never envisaged at the time they negotiated the Treaty. The paramount issue in this regard is that of environment regarding which the Treaty carries no provision; although it did anticipate yearly fluctuations in weather patterns, no large scale and long term environmental changes were considered. It is common knowledge that most of the water in the Western Rivers has been principally fed by glaciers. However, in recent decades the climate change resulting from greenhouse gases are impacting the flow of water in the Indus system of rivers. In other words, the rise in temperature is accelerating the melting of glaciers in the Himalayas. It has been reported that some glaciers in Kashmir have simply disappeared whereas others are fast shrinking. Premature melting leads to massive discharge in rivers resulting in flash floods, while at the time of agricultural activity there is a lack of adequate water supply. Similarly, it has been reported that forests that once covered 37 per cent of the total area of Kashmir have been reduced to a mere 11 per cent. This has reportedly happened due to the presence of Indian troops who, in order to flush out militants during the Kashmir insurgency, have chopped them down. Deforestation has led to a decline in the water flow in, among others, the Western Rivers. There is a fear that climate change, by affecting changes in hydrology or atmospheric conditions of the region, may render the Treaty obsolete.

Though the Treaty has been in operation during the last fifty odd years, India is terribly unhappy because it deems it a hindrance in its thirst for more water. Therefore, it would like to get rid of it, but has never officially expressed any intention to jettison it except for a brief period when, following the terrorist attack on the Indian Parliament in 2002, the Indian officials including Prime Minister Atal Bihari Vajpayee threatened to scrap it as one of the measures to punish Pakistan. Members from unofficial circles have occasionally expressed the desire to do away with it.

There are two schools of thought in this regard. One is of the view that the Treaty is a partitioning and negative document, and is as such beyond redemption. In their opinion, it needs to be replaced by a new treaty in its entirety. The other does not reject the Treaty as such, but favours major changes in it in order to bring it in line with current needs. The Indian government never officially denounced the Treaty nor advocated its replacement with a new one or sought its amendment. However, taking advantage of the provisions in the Treaty for the construction of the run-of-river hydropower plants, it has embarked on a strategy to build a series of small and big dams on the Western Rivers which Pakistan regards as an existential threat. Given the seriousness of the matter, individuals and institutions have warned that Indian designs can lead to armed hostilities between the two countries.

Despite the seriousness of this threat, and the passage of more than half a century since the Treaty became operational, it is amazing that there is a poverty of literature on the subject. There are in total only seven full-fledged studies which have been undertaken so far. The first one is entitled *The Indus Rivers: A Study of the Effects of Partition* by the American writer Alloys Arthur Michel. Published in 1967, it traces the story of Pakistan's water problems and their solution. Since the author is a geographer, he approaches the subject from the geographical viewpoint though he mixes it with history. He makes comments on the political or legal aspects of the topic but only *en passant*, and they suffer from lack of sound scholarship. For example, while dealing with the question of the delineation of the Punjab Boundary by Radcliffe, he absolves him of any wrongdoing. It is true that the testimony of Christopher Beaumont, which exposed the role that Cyril Radcliffe played in this affair, came out after the publication of his book, but there was still considerable incriminating material against the latter at the time Michel wrote the book. However, he does not give it much credence and, strangely, puts his stamp of approval on Radcliffe's neutrality and fairness. The second study is by Niranjan D. Gulhati entitled *Indus Waters Treaty: An Exercise in International Mediation* which was released in 1973. The author was initially a member and, from 1954 onwards, head of the Indian delegation which participated in the World Bank sponsored negotiations on the Indus waters dispute. He covers the history of negotiations and related events as an insider. He too makes comments on the political and legal aspects of the dispute, but on the margin. He claims to have made 'a studied attempt to be objective, as far as it is possible for one who was at the time identified with one of the parties'.[8] Notwithstanding this claim, his book is considered to be written from a quintessentially Indian viewpoint.

The third study is entitled *Water Rationality: Mediating the Indus Waters Treaty* by Undala Z. Alam. The writer is a British citizen of Indian origin. She undertook this study as a doctoral dissertation which she submitted to the Department of Geography at the University of Durham in September 1998. She has not published it as a book but has made it available to the public on the internet. She focuses on the role of the world as a mediator and, like the previous writers, mostly leaves out the political and legal aspects. She has tried to be even-handed and has succeeded to a considerable extent.

The fourth study is entitled *Indus Waters Treaty in Retrospect* by Bashir Malik and was published in 2005. The author was an employee of the Water and Power Development Authority (WAPDA) of Pakistan. He worked on different water-related projects in Pakistan and abroad, and was the director of the Tarbela Dam project built under the Treaty. He evokes conspiracy theories to explain Pakistan's negotiating behaviour before and during the Indus waters discussions. He is highly emotional in his approach. Since he is an engineer by profession, he does not focus on the political and legal dimensions of the subject. The fifth study is entitled *The Indus Waters Treaty Regime* by Rakesh

Kumar Arora, published in 2007. It deals with different aspects of the Treaty such as various options for resolving the conflict, water and human security, water rationality, and Pakistan's perspective on the Treaty, but does not focus on its political and legal dimensions. Arora is not neutral; he presents a purely Indian point of view.

The sixth study is in Urdu entitled *Teen Darya Kaisay Khoe, Sutlej, Beas aur Ravi* (How did we Lose the Three Rivers?) by Ramiz Ahmed Malik which came out in 2008. The writer is the former chief engineer of the Department of Canals, Punjab. He too relies on conspiracy theories to explain why Pakistan took the step it did. Like the book by Bashir Malik, it is an exercise in raw emotion and does not add much to the existing literature on the subject. The seventh and the last study is titled *Indus Waters and Social Change: The Evolution and Transition of Agrarian Society in Pakistan* by Saiyid Ali Naqvi. The writer is a former senior official of WAPDA in which capacity he oversaw the construction of the Tarbela and Mangla Dams. The book was published in 2013 and, as the title suggests, focuses on social change in Pakistan's agrarian society as a result of the Treaty. The book presents Pakistan's perspective. It totally shies away from touching the political or legal questions of the subject under study. It follows from the foregoing that there has been no full-fledged study by a Pakistani or a foreigner, exclusively focusing on political and legal aspects of the Treaty.[9]

The present study is an attempt to fill this huge gap by dealing with political and legal aspects of the Treaty in a systematic, professional, and dispassionate way. This study strives to address a whole gamut of questions relating to the Treaty and make the study as exhaustive as possible. We have also tried to shatter certain myths that have grown around the Treaty and which most Pakistanis, including the intelligentsia, have come to believe as the gospel truth. For example, there is a common belief in Pakistan that the Treaty was unfair and a huge mistake; that Ayub 'sold' the three Eastern Rivers to India in exchange for money; that it was a military dictator who 'sold' the Eastern Rivers, implying thereby that a civilian government would have never done it; that India has no right to build dams and reservoirs on the Western Rivers; and that the ones it is in the process of building are in utter violation of the Treaty. We have tried to be as objective as possible in this study. However, in all candour, it is a very difficult task to be so because no matter how hard one may try, one's biases and preconceived ideas imperceptibly shape the interpretation of facts with which one is confronted. Hence, we would suggest with humility and without bravado that this is a study from the Pakistani perspective.

The book is divided into eight chapters. The first chapter focuses on the general background of the Indus Basin system. It poses a number of questions. What is the topography of the Indus Rivers? What was the state of the canal system during antiquity and the Mughal periods? What motivated the British to develop the canal system in India? Where did the water dispute between Sindh and Punjab originate? The second chapter deals with the genesis of the water dispute between Pakistan and India. It addresses a host of questions most important of which are as follows: Why did Pakistan and India fix 31 March 1948 as the termination date of the Standstill Agreement? Why did Pakistan fail to protect its water interests beyond the Standstill Agreement? Why did the West Punjab government approach its counterpart in East Punjab for the renewal of the Standstill Agreement just two days before its expiry? Why did Radcliffe award Madhopur and Ferozpur headworks, which were located in Muslim majority areas, to India? Why did Jinnah accept Radcliffe as chairman of the Boundary Commission? Was the Inter-Dominion Agreement or the Joint Statement of 4 May 1948 signed under duress? The third chapter looks into the role of the World Bank as an 'honest broker' in finding a solution to the water dispute between Pakistan and India. It raises the following questions: Why did Lilienthal take an interest in this dispute? What motivated the Bank, which is not a dispute settlement agency, to get into the convoluted business of mediating between two highly quarrelsome neighbours? Why were Lilienthal and the president of the Bank, Eugene Black, interested in developing the Indus Basin as an integrated management project by Pakistan and India when they knew, or should have known, that the former would never agree to it? Why was the proposal, which the Bank floated in 1954 to break the deadlock in the negotiations, almost a rehash of the Indian plan? Were the Bank and India hand in glove with each other? Why did India concede, almost in its entirety, the proprietary rights over the Western Rivers to Pakistan? Why did the US and other Western countries agree to give Pakistan almost a billion dollars for replacement and development works on the Western Rivers? Did Ayub Khan 'sell' the waters of the three Eastern Rivers in exchange for money?

The fourth chapter deals with the terms of the Treaty. It tries to bring out the principal features of the Treaty in some detail. It also elaborates on the shortcomings of the Treaty that the two countries have discovered in the course of its implementation during the last half century. The fifth and sixth chapters focus on how the Treaty has worked in practice. They deal with projects which Pakistan and India have either bilaterally settled, are bilaterally negotiating, have settled, or are in the process of settling through the involvement of a third party, namely Wullar Barrage/Tulbul Navigational Project, Salal, Baglihar, Kishenganga, and Nimoo Bazgo Dams. They raise the following questions: Why did Pakistan lose the *Baglihar* case? Has the determination by the Neutral Expert in the *Baglihar* case effected any change in the terms of the Treaty? What has happened to Manmohan Singh's promise to compensate Pakistan for the 200,000 cusecs of water[10] that the latter lost when India filled the Baglihar Dam in violation of the Treaty? The seventh chapter deals with climate change and the Treaty. It focuses on the following questions: What is climate change and is it real? If it is real, does it affect the flow of waters of the Indus system of rivers? If the answer is in the affirmative, can India invoke the principle of *rebus sic stantibus* to revoke the Treaty? The eighth chapter deals with the controversy which has arisen in India regarding the future of the Treaty. It raises the following questions: Should the Treaty be revised or replaced altogether by a new Treaty? Can India repudiate the Treaty on the grounds of *rebus sic stantibus* or the change of circumstances? Can India terminate the Treaty on the grounds of 'cross-border terrorism' by the Lashkar-e-Taiba (LeT) in proper India or in the Indian-administered Jammu and Kashmir? Can India terminate the Treaty on the grounds that it failed to consult the Kashmiris at the time it was negotiated? Is the Treaty unfair to India? Is there a case for reverting to Lilienthal's original proposal for a joint management of the Indus Basin by Pakistan and India for its optimal utilization?

REFERENCES AND NOTES

1. Notwithstanding the attribution of this quote to Mark Twain, his biographer, Joe Fulton, has denied it in these words: 'It is a great quote, but I don't believe that Twain ever said that.' See <www.mcclatchydc.com/news/politics-government24609343.html>.

2. The US Census Bureau put the figure at 7.023 billion on 12 March 2012 (US Census Bureau—World POP Clock Project), whereas the UN Population Fund reached that figure on 31 October 2011. 'Population seven billion: UN sets out challenges', *BBC* (26 October 2011).

3. See <www.serageldin.com/water.htm>.

4. Boutros Boutros Ghali stated in 1991 that the 'next war will be fought over water, not politics'. 'I Support the Algerian Government', *Middle East Quarterly* (September 2007).

5. T. Deen, 'Climate Change Deepening World Water Crisis' (19 March 2008). See <www.globalpolicy.org/socecon/environment/climate/2008/0319deepwater.htm>.

6. Aaron T. Wolf, Alexander Carius, and Geoffrey D. Dabelko, 'Water Conflict and Cooperation', *ECSP Report 10* (2004), cited by Per Steineide Refseth, *The Indus River Basin, 1999–2008: An Intellectual History in Hydropolitics*, Master's thesis (Institute of Archaeology, Conservation and History, University of Oslo, Spring, 2013), 26.

7. Paper to guide Black in his discussion with the Indian Prime Minister during the October Annual Meeting in Delhi, dated 18 September 1958. See File no. 1787282, *Indus Basin Dispute—General Negotiations—Correspondence* 07, Start Date 8/1/57.

8. N. D. Gulhati, *Indus Waters Treaty: An Exercise in International Mediation* (Bombay, Calcutta, etc.: Allied Publishers, 1973), 5.

9. Noting that the Indian chief negotiator during the Bank-sponsored Indus waters negotiations, N. D. Gulhati, wrote a superb book on the Treaty from

the Indian perspective, John Briscoe has lamented that no similar attempt was made by a Pakistani. John Briscoe, 'Troubled Waters: Can a Bridge be Built over the Indus?', *Economic and Political Weekly* (11 December 2010), vol. XLV, no. 50, 32, note 1.

10. A cusec is a flow of 1 cubic foot of water per second while one cusec flowing 24 hours is equal to 2 acre feet.

1

Indus Rivers and Plain, Canal Building, and Sindh-Punjab Water Dispute

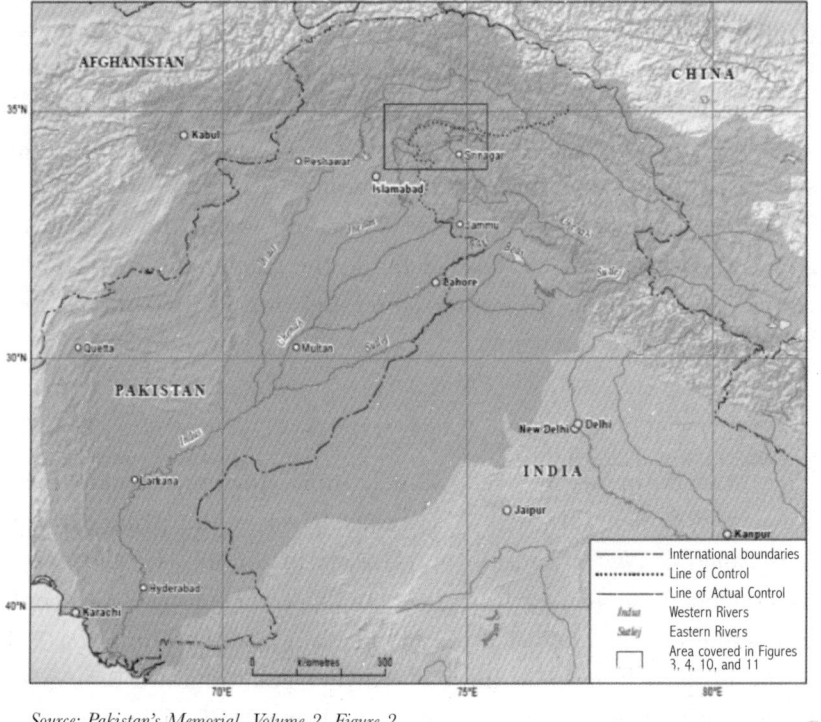

Source: Pakistan's Memorial, Volume 2, Figure 2.

Map 1.1: Indus Basin

THE INDUS RIVERS CONSTITUTE THE LARGEST CONTIGUOUS IRRIGATION system of the world. They command an area of 20 million hectares and an annual irrigation capacity of over 12 million hectares. The Indus Basin has four riparians, namely Pakistan, India, Afghanistan, and China. In terms of its distribution, Pakistan covers 52 per cent, India 33.51 per cent, and Afghanistan and China about 13 per cent. However, only the former two are parties to the Treaty. Afghanistan and China were not invited to the negotiations, which were held under the auspices of the World Bank, to resolve the Indus water dispute between Pakistan and India. Since Afghanistan and China are contemplating embarking on water-related projects in the Indus Basin located in their territories, it is increasingly feared that this may affect the water rights of Pakistan and India under the Treaty and pose a threat to the stability of the Treaty in the future. The alluvial plains of the Indus Basin cover about 25 per cent of the land area of Pakistan and about 9.8 per cent of India. Pakistan's dependence on external water resources is 76 per cent whereas that of India is 34 per cent. The quantum of water that flows through the Indus system of rivers does not remain constant. It varies from year to year, depending mostly on melted snow and ice in the Himalayan and Karakoram Ranges, and rainfall in the catchment areas. It also changes from season to season, increasing during the summer and decreasing during the winter.

A) THE INDUS RIVERS

The Indus Basin has six rivers, namely Indus, Jhelum, Chenab, Beas, Ravi, and Sutlej. Of these, the Indus River is the principal one. It is called *Sindhu* in Sanskrit and *Mehran* in Sindhi.[1] In Tibet, it is called 'the River issuing from the mouth of the Lion'. The people inhabiting the area between the Karakoram and the Himalayas call it 'the Eastern River' and those of Khyber-Pakhtunkhwa have given it the name of *Abasin* or 'the Father of Rivers'. In ancient times, it was known as *Sindhu* or divider, keeper, or defender. In the lower reaches, it was known as the 'ocean' and at its delta 'as the fresh water sea'.[2] The Rig Veda, the religious book of Hindus, composed in about 1500 BCE mentions it. It is 1,980 miles long (3,080 kilometres) and has a drainage area of roughly 450,000 square miles (1,165,000 square kilometres) of which about 175,000 square miles (453,000 square kilometres) lie in the Himalayan Range and the remainder in the semi-arid plains of Pakistan. It is the twelfth largest river in the world. It has about 207 billion cubic metres of annual flow of water, which is twice that of the Nile and three times that of the Tigris and Euphrates put together. It passes through the middle of Pakistan.

It originates in a spring near Mansarovar Lake on the northern side of the great Himalayan Range in Kailas Parbat, in Tibet, at an altitude of about 19,000 feet. In the uppermost reaches, it flows in a northwesterly direction for about 200 miles before crossing the southeastern boundary of the Kashmir region. The Zaskar River joins it at a short distance from Leh in Ladakh. It flows for about 500 miles in Gilgit Baltistan (formerly the Northern Areas) where the Shyok River joins it. Mighty glaciers feed it on the slopes of the Karakoram Range, the Nanga Parbat massif, and the Kohistan highlands. Apart from Shyok, Shigar, and Gilgit, other streams also join it in Gilgit Baltistan. It has thirty-two tributaries, all of which lie in Pakistan except for a part of the Eastern tributaries which partly lie in India. The Jhelum, Chenab, Beas, Ravi, and Sutlej join it in the plains of Punjab. The Kabul River that originates from Afghanistan joins it at Attock, in Pakistan.

There are seven barrages on the Indus River from where a large number of canals take off. The first one is the Ghazi Barrage which, unlike the other six barrages that are meant for irrigation purposes, is merely dedicated to power generation. It produces 1450 MW of electricity near the Barotha village (hence the commonly used appellation Ghazi-Barotha Hydropower Project). The second one is the Jinnah Barrage located near the town of Kalabagh. It feeds the Thal Canal which serves a region of the western Punjab called Sindh Sagar. The third one is the Chashma Barrage which feeds two canals, namely Chashma Jhelum Link Canal and Chashma Right Bank Link Canal. The fourth one is the Taunsa Barrage. It feeds three canals, namely the Muzzafargarh Canal, the Taunsa Panjnad Link Canal, and the Dera Ghazi Khan Canal. The fifth one is the Guddu Barrage. Guddu feeds three canals: the Begari Canal, the

Desert Canal, and the Ghotki Canal. It may be mentioned here that a canal called Pat Feeder takes off from the Desert Canal which serves Balochistan.

The sixth barrage is the Sukkur Barrage (previously called Lloyd Barrage). It feeds seven canals, namely Rohri Canal, Nara Canal, two Khairpur Feeders, Northwest Canal, Rice Canal, and Dadu Canal. The seventh and the last one is the Kotri Barrage or the Ghulam Muhammad Barrage. It feeds four canals: Pinyari Canal, Fuleli Canal, Lined Canal, and Kotri Beghar Feeder. Pakistan has built the world's second largest earth-filled dam on the Indus, the Tarbela Dam, which initially held 11 MAF of water but the Mangla Dam, following the raising of its height, has since snatched this distinction. This is because Tarbela's capacity has come down considerably due to the silting of its reservoir over the years. In one go, it serves the purpose of irrigation and hydropower. Its power generation capacity stands at 3,478 MW.

The Jhelum River is called *Vitasta* in the Rig Veda and *Vyeth* in Sanskrit. The ancient Greeks called it *Hydaspes* and the historian, Alberuni, *Biyatta* or *Jailam*.[3] It has a total length of 505 miles (813 kilometres). The Verinag spring located 80 kilometres from the capital Srinagar, in the Indian-administered Jammu and Kashmir, is believed to be its principal source. It flows past Srinagar and through Dal and the Wullar Lakes. The Neelum/Kishenganga River and the Kunhar River flowing from the Kaghan Valley join it at Domel in Muzaffarabad, the capital of Pakistan-administered Jammu and Kashmir. Another tributary, the Poonch River, joins it near the Kohala Bridge which joins Jammu and Kashmir with Pakistan. It then flows into the Mangla Dam reservoir in district Mirpur to which it brings on an average about 2,388 MAF of water. With the recent increase of the height by about 30 per cent, the storage capacity of the dam has increased by 2.88 MAF, from 4.51 MAF to 7.39 MAF, making it the largest water reservoir in Pakistan. The increase in height has also augmented the power generation capacity from 1,000 MW to 1,100 MW. The Jhelum River emerges from Mangla Dam and enters the plains of Pakistan's Punjab. At the Rasul Barrage, which is located on the Jhelum River, two canals, namely the Lower Jhelum Canal and the Rasul Qadirabad Link Canal, take off. The Jhelum River joins the Chenab River at Trimmu in the district, Jhang. It has historical importance as the forces of Alexander and Porus engaged in a pitched battle on its eastern bank near the present-day city of Jhelum.

The Chenab River is called *Asikini* in the Vedas, *Akesines* by ancient Greeks, and *Candrabagga* or *Candraha* by Alberuni.[4] It originates in the upper Himalayas in the Lahaul and Spiti district of Himachal Pradesh in India. The Chandra River and the Bhagga River come together in a village called Tandi to form the Chandrabhagga River. It then becomes the Chenab River when it joins the Marau River at Bhandera Kot, which is 12 kilometres from a town called Kishtwar in the Indian-administered Kashmir; from there it enters the plains of Pakistan's Punjab. The Jhelum River joins it at Trimmu and the Ravi River at Ahmedpur Sial. It merges with the Sutlej River near Uch Sharif, and the Beas River connects with the Sutlej River near Ferozpur in India. It then joins the Indus River at Mithankot in Pakistan. Measured from its source to its confluence, it is 772 miles (1,235 kilometres) long. It has five barrages located at Merala, Khanki, Qadirabad, Trimmu, and Panjnad from where a number of canals take-off. Their details are as follows: Upper Chenab Canal and Merala Ravi Link Canal at Merala Barrage; Lower Chenab Canal at Khanki; Qadirabad Balloki Link Canal, which transfers the water of the Chenab River to the Ravi River at Qadirabad; Haveli Canal, Trimmu-Sidhnai Link Canal, and Rangpur Canal at Trimmu; and Panjnad Canal and Abbasia Canal at Panjnad Barrage. It has the same place in the folklore of Punjab as the Rhine has for the Germans, and the Danube for the Austrians and the Hungarians. It owes this iconic status to the fact that the Punjabi romantic epic, *Heer Ranja*, and the legend of *Sohni Mahiwal* are woven around it. India has constructed two hydropower dams on the Chenab River, Salal Dam, and Baglihar Dam.

The Beas River was known as *Arjikiya* in the Vedas, *Hyphasis* by the ancient Greeks, and *Biyah* by Alberuni.[5] Many consider the present name a corruption of the Sanskrit name, *Vipasha*, which it was known as at one time. It constituted the frontier of Alexander's conquest. It was here that Alexander's troops revolted by refusing to proceed further, having been away from home for eight years. It originates on the southern face of the Rohtang Pass in Kullu, some 13,326 feet above sea level. It joins the Sutlej River at the southwestern boundary of the Kapurthala district in the Indian Punjab and is the shortest of the Punjab Rivers. Its principal tributaries are Bain, Bangana, Luni, and Uhal. It is roughly 450 kilometres long and its drainage basin is about 20,303 square kilometres large.

The Ravi River was known as *Iravati* or *Parushni* in the Vedas and *Hydraotes* by the ancient Greeks.[6] It originates in Himachal Pradesh. It is about 725 kilometres long and drains an area of approximately 14,442 square kilometres. Two of its major tributaries, the Budhil and Nai or Dhona, join it 40 miles from its source. Another major tributary, the Seul River, joins it just below Bharmour, the capital of Chambra. Still another tributary, Siawa, joins it near Bissoli. The Ravi River flows through the base of Dalhousie Hill, past the town Chambra, and then cuts a gorge in the Dhauladhar Range before entering the Punjab plains near Madhopur. It then flows along the Pakistan-India border for about 80 kilometres before entering Pakistan and joining the Chenab River. The Ujh River, which originates in the Kailash Mountains at a height of 14,100 feet, close to the Bhaderwah Mountains, joins it at Nainkot in Pakistan. The Ravi River flows past Lahore, takes a turn at Kamalia, and then debouches into the Chenab River, south of the town of Ahmedpur Sial.

The Bambanwala-Ravi-Bedian-Dipalpur Link Canal (BRBD), crosses it upstream in Lahore. It has a number of canals whose details are as follows: Lower Bari Doab Canal and the two Balloki-Suleimanke Link Canals that take off from the Balloki Barrage, and Sidhnai Mailsi Link Canal and Sidhnai Feeder Canal that take off from the Sidhnai Barrage. India has constructed a dam called Ranjit-

Sagar Dam at Thein near Madhopur. The Harappa civilization, which flourished between 2600 and 1800 BCE, was located some 70 miles above the Ravi and Chenab confluence. It has been claimed that a village was located at the Harappa site in 3300 BCE or some 700 years before the advent of the Harappan civilization around 2600 BCE.

The Sutlej River was known as *Sutudri* in the Vedas, *Zaradros* to the ancient Greeks, and *Sataruda* or *Shataldar* by the Alberuni.[7] It is the longest of the five rivers of Punjab and has eight tributaries. All of its tributaries, except for Rohi Nala which is the smallest, join it in India. It originates in western Tibet in the Kailash Mountain Range near the sources of the Indus, the Ganges, and the Brahmaputra Rivers. It travels through the Pir Panjal and Siwalik Mountain Ranges before entering the plains of the Indian Punjab. It continues in the southwestern direction where it is joined by the Beas River in Harike-Paton at Amritsar, in the Indian Punjab, and forms 65 miles (105 kilometres) of the Pakistan-India border before entering Pakistan. After flowing another 220 miles (350 kilometres), it joins the Chenab River, west of Bahawalpur. The two then join the Panjnad River.

The Sutlej River has two barrages in Pakistan from where a number of canals take-off. The first one is located at Suleimanke from where three canals emerge: Pakpattan Canal, Fordwah Canal, and Eastern Sadiqia Canal. The second barrage is located at Islam which feeds three canals: Mailsi Canal, Qaimpur Canal, and Bhawal Canal. Downstream to the Sidhnai-Mailsi-Bhawal section of the Trimmu-Sidhnai-Mailsi-Bhawal Canal, it is taken across the Sutlej through a siphon called Mailsi Syphon. The Sutlej River is about 946 miles long and its catchment area is about 47,000 square kilometres. India has constructed a 742 feet high multipurpose dam called Bhakra with a storage capacity above 9 MAF and, a few miles downstream of it, the Nangal Dam for the purpose of power generation.

Apart from the six Punjab Rivers, the Kabul River which originates from Afghanistan is another major tributary of the Indus Rivers. It originates at the foot of the Unnar Pass in the Hindu Kush Mountains. It drains into eastern Afghanistan over the fractured edge

of the Iranian plateau before entering Pakistan just upstream of the
Warsak Dam in the Khyber Pakhtunkhwa province. The Chitral
River, which originates north of Tirich Mir, flows for about 200
miles in Pakistan and then enters Afghanistan under a different name,
Kunhar River, and joins the Kabul River about 50 miles upstream
of the Pak-Afghan border. The Kabul River then re-enters Pakistan
where the Swat River joins it from the left side and the Bara River
from the right side below the Warsak Dam; and the Kalpani River
3 miles downstream of Nowshera city before it joins the Indus River
near the Attock fort.

B) THE INDUS PLAIN

The Indus plain covers about 200,000 square miles (518,000 square
kilometres) from the Himalayan piedmont to the Arabian Sea. It has
a gentle slope with an average gradient of about 1 foot per mile.
It can be divided into upper and lower regions, each with different
physiographic features. The Indus and its five tributaries constitute
the upper region whereas the Indus alone makes up the lower region.
The Suleiman Range joins the Indus plain and the Indus merges with
the Panjnad River (which is a confluence of the five Punjab Rivers)
at Mithankot. The Indus plain is bounded on the west by the Kirthar
and Suleiman Ranges, on the northwest by the Salt Range, and on
the north by the foothills of the Himalayas. It extends into Kashmir
and India on the northeast and the east. In the southeast, the Thar
Desert narrows its width whereas its southernmost region is located
in the Indus Delta. The entire Indus plain is topographically more
or less consistent, except for the Kirana Hills and limestone ridges
at Sukkur and Hyderabad which interrupt the flow of rivers. There
are four *doabs* or areas between the Indus Rivers: Sindh Sagar or the
Thal Doab between the Indus and the Jhelum Rivers, Chaj Doab
between the Jhelum and the Chenab Rivers, Rechna Doab between
the Chenab and the Ravi, and the Bari Doab between the Ravi
and the Sutlej Rivers. Beyond the Sutlej River lies Bahawalpur and
southwest of Bahawalpur is the province of Sindh. Out of about 50

million acres of the Indus plain, river channels occupy about 2 million acres, salt or tidal flats of the Indus Delta another 2 million acres, and the alluvial sand and silt occupy the remaining 46 million acres.[8]

The Indus Rivers have formed four types of land, namely active floodplains, meander floodplains, cover floodplains, and scalloped interfluves. The active floodplains, which cover about 5.2 million acres, are adjacent to the rivers. The Indus Rivers inundate them during floods which renders the soil suitable for cultivation during the winter season. Next to the active floodplains are located the meander floodplains which have resulted from the change of course by the major rivers. They constitute natural drains and are extensively waterlogged. The cover floodplains were formed by sheet flooding which deposited a flat mantle of fine grained alluvium. At some places, they lie next to the active floodplains while at others they are bordered by relatively high meander floodplains. The scalloped interfluves or bars are the elongated and extremely level plains which exist in the centres of Bari, Rechna, and Chaj Doabs. Covering about 4.5 million acres of land, they were formed by ancient rivers. They are covered by highly uniform soils of sandy silts and constitute the most fertile regions of Punjab. The Thal Doab, compared to the three foregoing doabs, has sand flats and irregular low hills covered with shifting sands of the Thal Desert. It is noteworthy that the wind, and not running water, has shaped the topography of the Thal Desert, the smaller areas of other Doabs, and the Cholistan and Tharparker deserts. This has resulted in the sand dunes, hills, and wind-eroded remnants dominating the landscape. At the North end of Thal, Chaj, and Rechna Doabs, near the rim of the plains, the slope of the land steepens somewhat and the alluvial deposits become thinner.[9]

These sloppy plains are generally beyond the canal system and are formed by deposits carried by the water of melted glaciers from the Himalayan foothills rather than by river floods. Similar sloppy plains cover large areas at the foot of the Suleiman and Kirthar Mountains on the right bank of the Indus. North of the rim lies the Potwar region which is irregular, hilly, and eroded with small alluvial basins. From Mithankot to the sea, the Indus flows between natural embankments and is, therefore, above the general level of the floodplain. The lower Indus is marked by natural embankments which are reinforced by artificial bunds built at some distance from the principal low water channels in order to contain the river when it is in spate. It is noteworthy that the British focused more on controlling the flood than promoting irrigation in Sindh. As a result, by the turn of the twentieth century, they had dyked 330 miles of the channel from Kashmore to the Indus Delta on the right bank and two-thirds of the left bank.[10]

Finally, a word about rainfall in the Indus plain is in order here. Jacobabad, which is located to the northwest of Sukkur, has the lowest annual rainfall. It gets from 3 to 4 inches in summer, though winter rainfall is not unknown. Hyderabad and Multan each get an average of 6 to 7 inches of rainfall annually whereas Faisalabad gets about 14, Lahore about 20, and Sialkot 35 inches of rain.[11]

C) CANAL BUILDING BEFORE THE BRITISH

Despite the existence of these rivers, there is no trace of canals during antiquity and part of the medieval period, perhaps because of the former's shifting character. Experts, however, are of the view that irrigation in the Indus Basin is almost as old as the civilization itself. It is claimed that some five thousand years ago, the Mohenjo-Daro and Harappa civilizations depended on irrigation from the Indus and Ravi Rivers respectively for their sustenance. The famous archaeologist, Sir John Marshall, states in his book *Mohenjo-Daro and the Indus Civilisation*:[12] 'They [the discoveries of Mohenjo-Daro and Harappa] exhibit the Indus peoples of the fourth and third millennia BCE in possession of a highly developed culture in which no vestige of Indo-Aryan influence is to be found .... Their society is organized in cities, their wealth derived mainly from agriculture and trade .... They cultivate wheat and barley as well as the date palm.' Dr Wheeler, in his book *India and Pakistan*, writes that the Indus people grew food crops and cotton with the help of some sort of irrigation system.[13] Human beings started with inundation canals but then moved to perennial canals with permanent headworks. There existed an extensive system of inundation canals in Bahawalpur State which took off from Sutlej and Panjnad. The Amirs of Sindh dammed a side channel of the Indus for an inundation canal which subsequently became the perennial canal called Fuleli.

It is believed that perennial irrigation in the subcontinent began circa 300 AD when the Grand Anicut of Madras was dammed. As far as Punjab is concerned,[14] it was Feroz Shah Tuglaq who, in the fourteenth century, built a canal drawing water from the Jumna River. He did not do so to irrigate land for agricultural purposes but to supply water to a hunting estate called Hissar, some 100 miles northwest of Delhi. The Mughal Emperor,[15] Akbar the Great, restored it circa 1568. He, however, used its water not only for pleasure but agriculture purposes as well. Shah Jahan,[16] the grandson of Akbar, extended this canal in order to expand irrigation to lands northwest of Delhi. It was this canal which later became the precursor of the Western Jumna Canal. He also brought a branch of this canal into Delhi in order to supply water to the Red Fort which was the imperial palace. Jahangir,[17] the son of Akbar, was the one who built the first perennial canal. He built a fortress and a hunting lodge across the Ravi River as well as a pleasure garden around a large, walled-in reservoir in the vicinity called Hiran Minar at Sheikhupura, near the city of Lahore. To supply water to this reservoir, he got a canal dug some 50 miles away on the Ravi River. In 1633, his son, Shah Jahan, followed his footsteps and built a canal known as 'Huslie' (which later came to be known as Bari Doab System) on the left bank of the Ravi River in order to irrigate the Shalimar Gardens in Lahore. Later, the Sikh rulers built a branch of this canal to carry water to fill the tank at the sacred Golden Temple in Amritsar, in addition to providing water for crop irrigation. Shah Jahan was also responsible for improving the Jumna Canal by damming the western branch of the Jumna River.

He also constructed a new channel to carry water to Delhi with an improved design.

D) BRITISH AND THE CANAL SYSTEM

When the British[18] came to India, it did not entail a 'clean slate' as far as the construction of perennial canals was concerned. The Mughals had left an enduring legacy in the matter. The British had no experience in the field because irrigation is not only unnecessary in Great Britain but it was also not in use in North America, South Africa, or Australia, which they had colonized before 1849 when they conquered Punjab. They were familiar with the Egyptian irrigation system which depended on inundation canals. As far as perennial irrigation is concerned, they embarked on it by restoring the Western Jumna Canal in 1821 and constructed permanent headworks on it in 1836. They dispatched teams of engineers to Italy, France, Spain, and Northwest Africa to learn about their experience in perennial irrigation. They borrowed irrigation techniques from the Middle East and the Mediterranean countries, and introduced them in India. By the time British rule came to an end in India, they had introduced the most extensive irrigation system in the world, in the Indus Basin. The extent of their contribution can be gauged from the fact that most of the techniques and formulae which the British introduced in canal construction and operation in the Indus Basin are now prevalent all over the world.

Following the conquest of Punjab, the British set about constructing a canal system in the Indus Basin. According to Michel, they were motivated by several considerations to undertake this enterprise.[19] To begin with, they were interested in augmenting the agriculture output to increase revenue. Secondly, they were spurred by the desire to 'do something for irrigation' and to demonstrate to the native population that European science could do much more for them than their predecessors had done in this field. Thirdly, they were keen to provide employment to the Sikh soldiers who had turned to brigandage or were simply sitting idle following the disbanding of the Sikh army after its defeat at the hands of the British in 1846.[20] They engaged them in public works and settled them on the lands colonized by the new canals. Fourthly, they were motivated by the desire to keep famine, which was always knocking at the door, at bay. It is noteworthy that it was the Agra famine of 1837–38 that impelled them to improve the Western Jumna Canal.

Of the foregoing considerations, Michel thinks that the need to employ the idle Sikh soldiers was the decisive one. Two scholars, Gilmartin and Whitcombe, adduce the following two additional reasons for the British canal building exercise:[21] a) increasing government control over the local population by encouraging them to take up settled agriculture and thereby minimizing the security threat that they potentially pose to the power of the state; b) creation and development of new social elite through the settlement policies that were to follow the water development schemes. In their opinion, these elite in turn helped secure British rule because they owed their new-found status to the British Empire.

Irrespective of the British motive in developing water resources of the Indus Basin, the first canal that the British built was the Upper Bari Doab Canal (UBDC) on the Ravi River. Since they had no experience in the matter, they undertook it on a trial and error basis. To begin with, they were not certain about the minimum inflow of the River. Besides, they sanctioned the project on the understanding that it would require an intake of 3,000 cusecs, be 247 miles long, provide irrigation for 180 miles while the remaining part of the Canal would be used for navigation purposes, and would rejoin the Ravi River 56 miles above Multan. However, the fact of the matter is that the project was not well-planned, with the result that the canal was shortened by 100 miles, terminating at Changa Manga rather than rejoining the Ravi River above Multan. It had other problems as well, for instance, it lacked proper headworks. It was opened in 1859 and irrigation started in 1861.

The second canal that the British built in Punjab was the Sirhind Canal. The Maharaja of Patiala encouraged the idea in 1861 by offering to pay the cost of surveys, project preparation, and have the works constructed on his territory, provided the British furnished the personnel for construction free of cost and paid for the remaining work. The latter happily accepted the offer because the Maharaja was their faithful ally as he had provided supplies and soldiers to the British during the Great Uprising of 1857. The parties signed a treaty for the construction of the Sirhind Canal and work started in 1869. Like the UBDC, it too had problems, particularly regarding the design of the headworks.

E) ORIGINS OF SINDH-PUNJAB WATER DISPUTE

The British had embarked on a canal building exercise in India but without a policy. However, they soon remedied this lacuna when, in the course of the construction of the Sirhind Canal, Captain J. H. Dyas, the director of canals in the Punjab Government, spelled out the considerations for canal construction. This is what he stated in the matter:[22]

> Looking as an Engineer merely at the question of the employment for irrigation of the water of any river, and assuming that no special reasons exist for irrigating any tract, I am of the opinion that the best line for a canal is that from which the largest extent of the country can be irrigated at the smallest cost, irrespective of the name or nature of the existing Government of the country in question. If a larger extent of Putiala than of British territory can be irrigated from the Sutlej with the same quantity of water at a smaller cost, then it would appear to me that the Sutlej was intended for the irrigation of Putiala, and I would run accordingly.

In other words, Captain Dyas emphasized the cost benefit factor rather than other considerations when determining the acceptance or rejection of an irrigation project. When he made this statement, he was not voicing his personal opinion in the matter but that of his government as is evident from the following official statement: 'The only project which should be entertained by the Government of India is the best that can be devised irrespective of the territorial boundaries.'[23] Now these two statements (that of Dyas and the official one), which have identical import, constitute a seminal event in the history of canal building in the Indus Basin as they had far-reaching political and economic consequences regarding the relations between Punjab and Sindh. These consequences were so serious that their tremors are being felt even today. The Sindhis who were on the receiving end of this policy have a different perspective on it. They view it as an attempt by the British to reward Punjab's ruling elite for its mercenary role as the following statement by a leading Sindhi nationalist, Rasul Bux Palejo, suggests:[24]

> The ruling class of the Punjab was the ally of the British since 1807 treaty of Amritsar with them. It was their junior partner in their war of subjugation of surrounding Muslim areas and Afghanistan and later on, became the most loyal swordsman of the British empire, having proved its super-loyalty to the empire, by helping it in a big way, in the suppression of the 1st Indian war of Independence in 1857, for which it had been loyally rewarded, besides other bounties, by the then greatest irrigation system of Asia, the Indus rivers irrigation system in Punjab.

Continuing this line of thought, Palejo approvingly quotes the following statement from a review of Imran Ali's book, *The Punjab Under Imperialism: 1885–1947*:[25]

> For Punjab, the ... [British] imperialists and empire-builders devised an entirely different scheme of exploitation. Punjab was not naturally fertile and rich in water resources as Bengal ....
>
> Unlike Bengal, where large landholdings of the Muslims were broken and parceled out to cultivators, in Punjab ... the imperialists bestowed their patronage, in huge parcels of canal-irrigated and thus perennially fertile land, to a new breed of landed, Muslim aristocracy they had conjured up for their convenience, a century-and-a-half hence, we in Pakistan are still contending with that imperialist legacy and paying a colossal premium in national disarray and political chaos for that 19th century 'convenience'.

The wounds of the Sepoy mutiny were still fresh. They [the British RBP] had succeeded in putting it down with the help of Punjabi mercenary soldiers recruited for them by petty Muslim middlemen. So they showered their favors in spades on these middlemen when virgin land became cultivable, thanks to the canals. Petty middlemen became aristocrats, overnight. Naturally, they were beholden to their masters and readily became their pawns in the imperial game of rapacious plunder.

As we pursue the story of the British attempt at canal building in India, a severe famine struck North India in 1878 which compelled the government in Delhi to accelerate the construction of irrigation canals. For this purpose, it asked the state governments to make proposals for new canals. The Government of Punjab suggested the construction of a canal on lower Ravi called Sidhnai and a new Chenab Canal. It also proposed the Lower Sohag and Para Canals on the Sutlej River. It is important to note here that the latter two canals (on the Sutlej River) were inundation-based in nature as opposed to the former two which later on became perennial. The Sidhnai Canal was initially mooted in 1856 but the Punjab government later revived it. It was a small project as it was to irrigate 351,000 acres of land only. The work on the project started in 1883 and the canal was opened in 1886. The Chenab Canal Project, adopted in 1891, was to irrigate a vast area of 1.1 million acres of land. Later on, with the completion of the Triple Canals Project, the Chenab Canal (which later came to be known as the Lower Chenab Canal) was to irrigate 2.9 million acres of land. It is noteworthy that the Chenab Canal which constituted both an administrative and engineering feat was the largest canal built during British rule.

Towards the early twentieth century, the British had built an extensive irrigation system by harnessing all the Punjab Rivers except the Beas. This led to a lot of colonization and settlement work on previously barren lands with the result that Punjab started producing grain surpluses and was at the threshold of becoming the 'breadbasket' of northern India. Resultantly, revenues increased enormously, leading to a demand to extend the irrigation system to more areas for greater profits. However, before it could be expanded further, there was a shortage of water in areas where it was most needed. This forced planners to review the situation in the provinces and undertake planning on a regional basis. The idea to extend the irrigation system gained impetus when a serious famine took place in India during 1899–1901 in which one million people perished. However, before undertaking any future canal building project, the British government felt the need for a stocktaking of the progress made in the matter so far. The Governor General, Lord Curzon, set up a body in 1901, headed by Sir Colin C. Scott-Moncrieff; it came to be known as the Indian Irrigation Commission and reported on the state of irrigation in India as protection against famine. The Commission toured the whole of India for that purpose at the end of which, in 1903, it submitted a report.

According to the Commission's report, there was about 4 billion cubic feet of water in the Indus Basin which was going to waste, into the sea. It proposed that out of 4 billion, at least 0.5 billion should be utilized for irrigation purposes. It would have recommended the transfer of water of the Indus Basin to an area outside of it but, since this was impracticable, it was forced to recommend its utilization within the Basin. It toyed with the idea of building storage dams but jettisoned it as that would be prohibitively expensive. It thought of barrages on the lower Indus which could supply water to perennial canals, replacing inundation canals, which was a characteristic feature of irrigation in Sindh so far, but abandoned it too on account of high costs. It came to the conclusion that the water should be used within the Basin to produce grain surpluses and then send them to the areas affected by famine. In its estimation, Punjab was the best candidate to implement this idea because '... it is here that the greatest progress has been made in irrigation works of this class; ... and it is here that there is still the greatest field for further expansion .... Thus there is no province in which, taken as a whole, the direct profits of irrigation works, ... have been so high, or in which new protective works can be proposed with the same confidence in their remunerativeness as financial investments.'[26]

To give practical shape to the idea, the Commission recommended the Triple Canals Project comprising the Upper Jhelum Canal, Upper Chenab Canal, and Lower Bari Doab Canal. It also considered projects for the irrigation of Sindh Sagar Doab but decided against them till the completion of the Triple Canals Project. While taking a decision on a major enterprise like the Triple Canals Project, it also recognized a limit to this policy. Consequently, it decided that from now onwards, no major project would be undertaken without taking into consideration the interests of all regions and without consultation with all concerned parties. In other words, it acknowledged that each riparian, whether upper or lower, was entitled to a fair share of water. This constituted a complete reversal of the earlier policy that the government in Delhi, as seen above, had adopted in which it had directed that 'the only project which should be entertained by the Government of India, is the best that can be devised irrespective of the territorial boundaries.' This was a seminal development with enormous political and economic consequences, among others, in the relations between Punjab and Sindh.

Following the submission of the Commission's report, John Benton of the Punjab Irrigation Branch prepared detailed plans and financial estimates of the Triple Canals Project. The Government of India gave its approval to the project in 1905, and the construction work on the Upper Jhelum Canal and the Upper Chenab Canal started the same year, whereas that of the Lower Bari Doab Canal started in 1907. The Punjab government was delighted with this development while the Sindh government, which was still part of the Bombay Presidency, expressed its reservations on the grounds that withdrawals in Punjab would affect existing inundation canals in Sindh. It also proposed the construction of perennial canals system in Sindh. However, these protests were completely ignored on the grounds that since storage dams were not feasible, in view of the paucity of funds and the state of machinery which was available, the water would go to waste by ending up in the sea.

Following the submission of the Commission's report, surveys along the Sutlej River started in 1906. The Government of India, the Government of Punjab, and the rulers of Bikaner and Bahawalpur states discussed the issue of new irrigation projects and reached an agreement in 1919 on the Sutlej Valley Project which envisaged the construction of four barrages and eleven canals. The Project was prepared in 1920 and approved in 1921. It is necessary to mention here that the Government of India Act 1919 gave partial autonomy to the provinces regarding, among others, the matter of irrigation. What it signified in practical terms was that now a province could undertake a large-scale irrigation project without the approval of the secretary of state which it could not do earlier. The latter also arbitrated all disputes between the provinces and their decision was final. It is noteworthy here that while irrigation became a provincial subject after 1919, it was simultaneously a reserved subject. In other words, the secretary of state had to give approval to all projects affecting two (or more than two) provinces or costing more than Rs. 5 million on the recommendation of the Government of India. On the other hand, the Governor General gave direct approval to projects costing less than Rs. 5 million and affecting one province only. It is obvious that the Indian government through the Act of 1919 affirmed the policy that the Indian Irrigation Commission had laid down according to which Sindh's consent was necessary for diversion of water from the Indus. Here too, the purpose of the exercise was obviously to protect the interests of Sindh.

The next milestone in the development of laws for the protection of water rights of the provinces was the Government of India Act 1935, which separated Sindh from the Bombay Presidency and made it a separate province. It increased the quantum of provincial autonomy by making irrigation a full provincial subject. What it stipulated was that the government in Delhi would intervene only if one province made a formal complaint against another province regarding the latter's interference with its water supplies. In such a situation, the 1935 Act entitled the Governor General to appoint a commission to investigate the complaint and submit a report. It also entitled the latter to pass final orders in the matter. However, the disputants enjoyed the right of appeal against the orders of the commission to His Majesty-in-Council for a review. Sections 130 to 134 of the 1935 Act prescribed the procedure for appeals, which incidentally was not easy to execute; as a result, the pace of the river valley development suffered. It was this complex procedure that was responsible for some of the delays that occurred in the Bhakra Project that the Punjab government had planned on the Sutlej River before Partition.

Here the question arises, why did the government in Delhi accord preference to Punjab but not to Sindh? According to Michel, there were many reasons why Sindh, unlike Punjab, did not invite the interest of the British administrators. To begin with, Punjab had much to offer in the form of through-flowing streams, beautiful hill stations, Mughal-Sikh monuments, and a hardworking peasantry. In comparison, historical cities in Sindh like Debal and Thatta lay in ruins; others like Hyderabad and Sukkur were no match to Lahore and Amritsar; and Karachi was no more than a fishing village. There were areas of Sindh that were prone to epidemics like the plague and malaria.[27] After they occupied Sindh, the British, for some time, mistakenly believed that it was an utterly unhealthy place to live.[28] Finally, Sindh was neither a famine area nor lay close to famine-prone districts like UP, Bihar, and Bengal; which Punjab, because of its 'granary' status, was in a better position to supply. Allied to this was the question of whether the settlers would take up new lands and whether sufficient workers would be available to till them. Additionally, a perennial canal needed to be large enough to be remunerative. Lastly, information regarding the area that could be irrigated and the best site for the proposed barrage was not available.

Resuming our story about the Sutlej Valley project, the Government of Bombay, which represented Sindh at the time, was not pleased with it.[29] It lodged a complaint against it with the government in Delhi, which appointed a committee in 1919, headed by Mr Cotton, to look into the matter. The Government of Punjab rejected the complaint on the grounds that it was entitled to use as much water from the rivers that pass through its territory as it needed. The Cotton Committee, which submitted its report in 1919, did not agree with the Punjab government's contention. It recommended the Sukkur Barrage Project and decided that no water should be allocated to Punjab from the Indus River till the effects of the construction of the proposed Sukkur Barrage had become evident. The Government of Punjab opposed this by presenting the Thal Canal Project. The Viceroy, Lord Chelmsford, rejected the Punjab government's proposal and the secretary of state gave its preliminary approval to the Sukkur Barrage Project in 1921 and its final approval two years later in 1923. The Punjab government protested against this move on the grounds that it amounted to partiality towards Sindh. Besides, it raised the issue of duties relating to the Sukkur Barrage Project. The Government of Bombay rejected the objection by contending that Punjab had more than its share of perennial canals in the Indus Basin whereas Sindh did not have a single one. It also raised its voice against the Thal Canal Project which it described as the greatest threat to Sindh's water rights.

The Government of India examined the competing claims of the parties, and at once rejected their objections to the Sukkur Barrage Project and the Canals Projects on the grounds that they had been adopted for the benefit of the region. It also dismissed the Punjab government's objection against the waiver of duties on the Sukkur Barrage on the grounds that scant rainfall and peculiar conditions prevailing in Sindh justified them. It therefore refused to re-open the subject. The construction work on the Sukkur Barrage Project began in 1923. It was formally dedicated in 1932 and irrigation started the same year. The central government's decision was disconcerting for the Punjab government because it posed a threat to its future projects, namely the Bhakra Dam, Thal Canal, and Trimmu Canal.

The Punjab government proposed the development of a small experimental Thal Canal involving merely 750 cusecs of water to

the Government of India in 1924, but it decided to drop the idea
the next year in favour of a bigger project. In 1926, the Government
of India announced its decision in the matter. To begin with, the
question of the quantum of water required for the Sukkur Barrage
would not be open for discussion as long as it did not come in line and
the government did not have experience with perennial irrigation in
Sindh at its disposal. Next, since the Government of Bombay would
be uncertain about the effect of withdrawals for the supply of the
Sutlej Valley Canals in Punjab, the latter would be entitled to object
to further withdrawals from the Indus Basin system till there was
definite proof that supplies for the Sukkur Barrage Project would not
be in jeopardy as a consequence. It also pointed out that the proof
in question must be based on a more accurate gauging of the Indus
Basin system.

The Government of Punjab protested against the decision of the
central government, which then referred the matter to the secretary
of state who concurred with it. The latter then constituted an eight-
member committee, headed by the chief engineer of the United
Provinces (UP), Mr Anderson, with the direction that it should not
make a recommendation for withdrawals by the upper riparian that
may adversely affect not only the existing, but also the future rights
of the other riparian over the waters of the Indus Basin system. The
Government of Punjab responded to it by proposing to build a dam
on the Sutlej River, now known as the Bhakra Dam. Invoking the
Government of India Act 1935, the Government of Sindh lodged
a complaint with the central government in 1939 concerning the
negative effects that the said project was likely to have on the water
supply to Sindh. To resolve the controversy, in 1941 the central
government constituted a three-member commission, headed by
Justice B. N. Rau, who was a judge of the Calcutta High Court. It
was variously called the Indus Commission or the Rau Commission.
Sindh and Punjab conducted themselves before the Commission as
plaintiff and defendant respectively. Sindh used the Commission as

an equity proceeding against Punjab in order to put a check on its further interference with the supply of water to Sindh.

After prolonged deliberations, the Rau Commission in 1942 submitted a report which upheld the complaint filed by the Government of Sindh. It was of the view that the projected withdrawals for the Punjab projects mentioned in the complaint, coupled with the requirements of other irrigation projects, which were already in operation or were about to be completed, would cause material injury to Sindh's inundation canals each year particularly in the month of September. It recommended the construction of two barrages, one each in upper and lower Sindh, costing a hefty sum of Rs. 16 crore. It fixed Punjab's contribution at Rs. 2 crore as a compensation for the damage that it was likely to cause to Sindh's irrigation system. It directed the Punjab government to desist from taking any action on its proposed projects up to October 1945 and instructed the two governments concerned to reconcile their differences through negotiations while maintaining a status quo on the distribution of water between them.

In compliance with the directive of the Commission, the Governments of Punjab and Sindh entered parleys through their chief engineers, under the supervision of Sir Claude Angles, the director of Central Irrigation and Hydro-Dinlock Research at Poona, Bombay Presidency. The parties concluded the Sindh-Punjab Water Agreement in September 1945, by virtue of which Punjab was to take one share of water, and Sindh three from the Indus at Ghazi Ghat. According to Article 8 of the Agreement, Punjab was forbidden from constructing any dam on the River Indus or any of its tributaries without the concurrence of the Sindh government. The Agreement fixed priorities and provided a framework for sharing waters of the rivers of the Indus Basin for canals that existed in 1945 or were to be planned in the future. It also presented a framework for projects that were to be planned in the future and for sharing the surplus supplies of water beyond the need of the existing or proposed schemes. It laid down, in detail, the schedule for sharing the water supplies during the lean period as well. Following the Agreement, the Government of Punjab asked for the postponement of the reference of its dispute with the Government of Sindh to His Majesty-in-Council for six months in order to enable it to reach a settlement on the amount of money that it was to pay to the Sindh government.

This was the state of affairs regarding the water issue between Punjab and Sindh in the Indus Basin when Partition took place in 1947. The Partition's affect on the water issue has been described by Michel below:[30]

> Partition changed the parties to the dispute, or rather put three parties where two had been, but it did not change the interests or, fundamentally the arguments. From 1947 to 1960, the Government of Pakistan had to argue Sindh's case, plus Bahawalpur's and part of the Punjab's, while the Government of India inherited the remainder of the Punjab's and all of Rajasthan's (including Bikaner's) ....
> [N]either Independence nor the one-unit rule introduced into West Pakistan in 1955 served to eliminate the dispute between the upper and lower regions (now termed Northern and Southern Zones by the engineers and planners) of the Indus Plains within that country. That Sindh and West Pakistan could make common cause against India did not eliminate the conflict of interest between them, a conflict that is inherent in the physical geography of the Indus Plains.

On the eve of Partition, a seminal event took place when the Punjab Boundary Commission, which the British Indian government had established under Cyril Radcliffe to fix the boundary line between East and West Punjab, allocated to India the headworks of the canals which irrigated West Punjab, now part of Pakistan. This award sowed the seeds of a full blown crisis between the two new dominions, which took place less than a year after their independence when India stopped the supply of water to Pakistan that it had been receiving before Partition. This gave rise to a water dispute between Pakistan and India. The nature, motive, and impact of this award is the subject of our study in the next chapter.

REFERENCES AND NOTES

1. Saiyid Ali Naqvi, *Indus Waters and Social Change: The Evolution and Transition of Agrarian Society in Pakistan* (Karachi: Oxford University Press, 2013), 3–29. We would like to acknowledge that we have drawn liberally on Naqvi's book in writing this section.

2. Jean Fairly, *The Lion River: The Indus* (London: Penguin Books, 1975), xiv.

3. Saiyid Ali Naqvi, *Indus Waters*, supra note 1, 14.

4. Ibid. 16.

5. Ibid. 18.

6. Ibid. 17.

7. Ibid. 18.

8. Ibid. 23–4.

9. Ibid. 24–5.

10. Ibid. 25.

11. Ibid.

12. Abdul Latif Mirza and Ch. Mazhar Ali, 'Canal Irrigation', in Mubashir Hasan (ed.), *Hundred Years of PWD* (Lahore: Publications Committee of PWD, October 1963, Centennial), 4–5.

13. Ibid. 5.

14. Aloys Arthur Michel, *The Indus Rivers: A Study of the Effects of Partition* (New Haven and London: Yale University Press, 1967), 49.

15. Ibid.

16. Ibid.

17. Ibid. 50.

18. Ibid. 50–1.

19. Ibid. 65–6.

20. One writer describes how the Sikh soldiers conducted themselves after their demobilization in these words: 'Sikh peasants and soldiers were suspect and given little chance for employment in the army or in the police force, both of which were largely Muslim. Under the circumstances it was not very surprising that the militant spirit of the disbanded Khalsa soldiery (over 40,000 of whom were let loose after the Anglo-Sikh wars) turned to crime. The central districts of the Punjab became infested with dacoits, almost all of whom were Sikhs. Thuggee became rampant—most of the fraternity being either Mazhabi or Sainsi Sikhs.' Khushwant Singh, *A History of the Sikhs*, vol. 2 (New Delhi: Oxford University Press, 1999), 94–5.

21. D. Gilmartin, 'Scientific Empire and Imperial Science: Colonialism and Irrigation in the Indus Basin', *The British Journal of Asian Studies*, vol. 53, no. 4 (1994), 1127–49; E. Whitcombe, 'Irrigation', in Dharma Kumar (ed.), *The Cambridge Economic History of India* (Cambridge: Cambridge University Press,

1982), 677–737; cited by Daanish Mustafa, 'Critical Hydropolitics in the Indus Basin', *Water and Geopolitics* (18 June 2010), 379.

22. 'Memorandum dated 22 October 1862, by Captain J. H. Dyas on Captain Crofton's Preliminary Report of the Sutlej Canal Project', *Government of the Punjab, Agricultural Proceeding No. 3* (16 December 1864), 82, cited by Michel, *Indus Rivers,* supra note 14, 69–70.

23. A. M. R. Montague, 'Presidential Address to the Punjab Engineering Congress, 33rd session, 1946', *Proceedings of the Punjab Engineering Congress, 33rd session, 1946,* iv, cited by Michel, *Indus Rivers,* supra note 14, 70, note 25.

24. Rasul Bux Palejo, *Sindh-Punjab Water Dispute 1859–2003,* 9, <www.panhwar.com>.

25. Ibid. 9–10.

26. *Report of the Indian Irrigation Commission,* vol. II (Provincial), chapter XIV–The Punjab (Calcutta: Office of the Superintendent of Government Printing, India, 1903), 1.

27. Michel, *Indus Rivers,* supra note 14, 100.

28. This opinion, though false, was based on the British experience. One British report explains its basis in these words: 'For some time after the British occupation the climate of Sindh was believed to be preeminently unhealthy. Our first experiences of it were unhappy. When the army for Afghanistan was passing through the Province [sic] in 1839 the sickness in the camp at Thatta was so terrible as to leave no doubt in the popular mind that the whole of the lakh and twenty five thousand [125,000] Pirs buried on the Makhli hill [at Thatta] had been stirred to vengeance by the desecration of their tombs; and in the autumn of 1843 more than two-thirds of Sir Charles Napier's forces were prostrated. But ampler experience has shown that Sindh is, upon the whole, a healthy country.' *Gazetteer of the Province of Sindh,* compiled by E. H. Aitken in two volumes, vol. A (Karachi: Government of Pakistan, 1907), 486, cited by Michel in ibid.

29. While describing the water dispute dealt with by the Cotton Commission and subsequently, we have drawn liberally on *Sindh-Punjab Water Dispute,* supra note 24, 10–14.

30. Michel, *Indus Rivers,* supra note 14, 117.

2

Genesis of the Dispute

A) RADCLIFFE AWARD AND ITS BACKGROUND

THE GENESIS OF THE WATER DISPUTE OVER THE INDUS BASIN IS FOUND in the award that the Punjab Boundary Commission rendered. The Congress party and the Muslim League leadership had instructed the latter '... to demarcate the boundaries of the two parts of the Punjab on the basis of ascertaining the contiguous majority areas of Muslims and non-Muslims. In doing so, it will also take into account other factors.' The division of the Punjab was a very tedious affair because the province had been developed as a single unit which included the common irrigation and hydroelectric system. There were large tracts of land in the Punjab that the rivers Ravi and Sutlej irrigated. The headworks of the Ravi River was located at Madhopur in the district, Gurdaspur, whereas that of Sutlej was situated at a place called Ferozpur in an area known by the same name. According to the 1941 census, district Gurdaspur was a Muslim majority area as three out of its four tehsils had a Muslim majority (Gurdaspur 52.1 per cent, Batala 55.06 per cent, and Shakargarh 51.3)[1] and only Pathankot had non-Muslim majority (77 per cent). However, Gurdaspur, Batala, and Pathankot were allocated to India and only Shakargarh came to Pakistan. Similarly, the headworks of Sutlej were also located in the Muslim majority area as its tehsils had Muslim majorities (Ferozpur 55.2 per cent, Zira 65.2 per cent, and Fazilka 75.12 per cent).[2] However, the award, in violation of the partition principle outlined above, allocated Ferozpur and Zira tehsils to India. According to Justice Muhammad Munir, the Muslim League representative on the Punjab Boundary Commission, Radcliffe, told him in no uncertain terms that since the Ferozpur, Zira, and Fazilka tehsils were to be included in Pakistan, he did not need to argue the point.[3] Similarly, Justice Din Muhammad, the other Muslim League representative on the Commission, claimed that the latter had assured him that the Ferozpur headworks and some other areas were to be given to West Punjab (Pakistan).[4] However, when the award saw the light of day, contrary to the assurances that Radcliffe had given to Justices Muhammad Munir and Din Muhammad, he had placed Ferozpur and Zira in East Punjab (India). Despite the fact that the Punjab and Bengal Boundary Commissions were composed of two representatives from the Congress party and the Muslim League each,[5] keeping in view the fact that they could not see eye to eye on the allocation of territories, the award was considered to be the handiwork of Radcliffe alone who had headed these commissions.[6]

How do we explain the award of these Muslim majority areas to India? The boundary award itself offers no explanation for this result although Radcliffe tried to give his reasoning in the report that he appended to the award. According to him, since the boundary demarcation—because of the existence of the canal, rail, and road communication systems—was a complicated affair, he decided to award, '… [the] not inconsiderable areas east of the Sutlej River and in the angle of the Beas and Sutlej Rivers in which Muslim majorities are found' to India.[7] He justified it on the grounds that '… it would be in the true interests of neither State to extend the territories of the West Punjab to a strip on the far side of the Sutlej and that there are factors such as the disruption of railway communications and water systems that ought in this instance to displace the primary claims of contiguous territories.'[8] In other words, Radcliffe justified the award of the Muslim majority areas to India on the basis of 'other factors', namely the maintenance of the integrity of water and railway communication systems. This also becomes clear in the memorandum that Radcliffe prepared for the British under-secretary of state for India, Arthur Henderson, in which he also justified the award of the Madhopur headworks to India on the grounds that the headworks of the canals which irrigate the Amritsar district lie in the Gurdaspur district, and it was important to keep as much as possible of these canals under one administration. Does Radcliffe's justification of the award of the two headworks, based on the integrity of the canals system to India, carry conviction? It utterly fails to do so when we examine the issue, keeping in view the two Boundary Commissions' terms of reference. As stated above, the commissions were instructed to demarcate the boundary on the basis of the contiguity of the Muslim and non-Muslim areas though they were at the same time entitled to take 'other factors' into consideration. The commissions did not bother defining either the meaning of the expression 'other factors' or what weight was to be given to them in comparison to the communal principal, which in hindsight appears to be a major flaw.[9] It seems as if this was done deliberately, and both the Muslim League and the Congress party were parties to it as they did not bother to challenge it.[10]

Notwithstanding this reality, the fact remains that the 'other factors' formula, whatever it may be, did not carry as much weight as the partition principle. According to Arthur Henderson, in the following statement that he made before the British House of Commons,[11] '[T]he primary basis of demarcation must be majority population. In certain cases there may be factors which justify departure from this principle.' In other words, the majority population was the general principle and 'other factors' was an exception. In case of departure from the general principle there must be a compelling reason, the force of which should be immediately apparent to any dispassionate person. In spite of this fact, Radcliffe disregarded the communal principle and decided the matter on the basis of the integrity of the canal system. Incidentally, in addition to the expression 'other factors', the commissions also left ambiguous whether a district or a tehsil would be the basic administrative unit for 'ascertaining the contiguous majority areas of Muslims and non-Muslims'. Next, even if we were to agree with Radcliffe that the integrity of the canal system was a guiding principle, he should not have awarded the Madhopur headworks to India because the canals that flowed from them mostly irrigated areas in Pakistan. Consider the following: according to the data which is available[12] for the period between 1922–23 and 1931–32, the Upper Bari Doab Canal (UBDC) irrigated only 90,000 acres in the Gurdaspur district (7 per cent), 418,000 acres in the Amritsar (32 per cent), and 792,000 acres in the Lahore district (61 per cent). According to the former chief engineer of the Punjab Irrigation Department, Ramiz A. Malik, the area that the UBDC irrigated was 75 per cent Muslim.[13] It is obvious that control of the UBDC system mattered much more to the Lahore district than to the Gurdaspur and Amritsar districts (taken separately or jointly). This was also true insofar as the municipal supply of water to the population of Lahore (632,136) was concerned, as it depended much more than Amritsar (389,581) for its needs of drinking water on the 'Lahore branch' of the UBDC. Commenting on this, Michel observes:[14] 'Thus if irrigation considerations were taken in combination with communal population, Radcliffe would have been amply justified in awarding both the Gurdaspur and Lahore districts to Pakistan, and with them control of the UBDC headworks and over two-thirds of the area served by the system.'

Similarly, using the 'integrity of the canal system' as the guiding principle for demarcation of the boundary, Radcliffe could not have awarded the Ferozpur headworks to India. Reflect on this: three canals flowed from the UBDC, namely the Dipalpur Canal which irrigated Lahore and Montgomery (present day Faisalabad) regions, the Eastern Canal which irrigated Ferozpur and Zira tehsils, and the Bikaner Canal which irrigated the state of Bikaner. While the Bikaner Canal irrigated the Hindu majority state of Bikaner, it also irrigated the Muslim majority state of Bahawalpur. These canals mostly irrigated Muslim majority areas, which should have assured that the award of the Ferozpur headworks would go to Pakistan. However, Radcliffe decided to award it to India. We conclude from this that Radcliffe was not justified in awarding the Gurdaspur and Ferozpur headworks to India. It is noteworthy that impartial Indian commentators accept this viewpoint. For example, the former Indian foreign secretary, Jagat Mehta, states:[15]

> Specifically, the biggest anomaly was that the headworks of the two big canal systems (the Upper Bari Doab Canal starting in Madhopur on the Ravi and the Dipalpur Canal starting in Ferozpur) were left in India, whereas almost all of the area irrigated by these canals was on Pakistan's side of the border. Out of the total water carried in the canals, 80 per cent or 64.4 million of acre-feet was to fall inside Pakistan whereas only 8.5 million acre-feet were earmarked for land that was to fall inside India.

Michel also recognizes the inequity of the Radcliffe award, though his take on the matter is somewhat different. He is of the opinion that since despite the fact that three out of four tehsils of the Gurdaspur district had a Muslim majority, the latter awarded the Gurdaspur headworks to India on the basis of irrigation considerations. Therefore, he should have, on the same basis, awarded the Ferozpur headworks to Pakistan. Simply put, he contends that if the Ravi was the logical boundary in the Gurdaspur district, then the Sutlej was the logical boundary in the Ferozpur district. He justifies this kind of arrangement on the grounds that it would have compelled the parties to cooperate with each other and might have obviated the need to divide the Indus system of rivers in 1960. This is how he argued this point:[16]

> Alternatively, by giving Pakistan control of *either* Ferozpur or Madhopur (and either award could have been justified on a communal-majority principle—the former on a contiguous-tehsil basis, the latter on a contiguous-district basis), each party would have had some leverage against unilateral closure by the other. This would certainly have placed Pakistan in a stronger bargaining position with respect to the Indus Waters Treaty of 1960, and would probably have shortened and simplified the negotiations.

Pakistan, on the basis of the communal principle and 'other factors' (namely irrigation considerations which favoured Pakistan as shown above), deserved the Ferozpur headworks, yet Radcliffe awarded neither of the two (Ferozpur or Madhopur) to Pakistan.[17] Despite the injustice towards the latter, he advised Pakistan (and India) to cooperate with each other in the matter:[18] 'I think it only right to express the hope that, where the drawing of a boundary line cannot avoid disrupting such unitary services as canal irrigation, railways, and electric power transmission, a solution may be found by agreement between the two States for some joint control of what has hitherto been a valuable common service.' It is interesting to note that Radcliffe tendered this advice despite the fact that the Muslim League and the Congress party leadership had rebuffed him earlier on this very point, as explained by the British author Leonard Mosley:[19]

> Radcliffe immediately contacted the Viceroy and told him that he would like to submit a proposition to Jinnah and Nehru. Whatever he decided as to the lines of demarcation, he said, would it not be a good idea if both the leaders agreed at once, before the announcement of his Award, that the Punjab Water System should be a joint venture run by both the countries. It would thus safeguard the interests of both peoples and form a basis of cooperation which might prove fruitful in the years to come.
>
> He was rewarded for his suggestion by a joint Muslim-Hindu rebuke. Jinnah told him to get on with the job and inferred that he would rather have Pakistan deserts than fertile fields watered by courtesy of Hindus. Nehru curtly informed him that what India did with India's rivers was India's affair. Both leaders were obviously furious with him and hinted that he was playing politics. It was his one and only attempt to try to make a constructive suggestion.

Considering the hostile attitude of the parties towards each other, Radcliffe's advice seemed to be impracticable. In view of the fact that the Indian leadership was utterly opposed to the creation of Pakistan and was looking for ways and means to destroy it, his suggestion created suspicion in the mind of the Muslim League leadership. In its opinion, giving India control of the Ferozpur and Madhopur headworks amounted to entrusting it with the perfect tool to destroy it (Pakistan's reservations were totally justified when we realize that India closed the canals supplying water to Pakistan on 1 April 1948). If Radcliffe was truly interested in seeing cooperation between the two countries, he should have at least awarded one of the two headworks to Pakistan, which would have simultaneously rendered the two countries' lower and upper riparians towards each other thus bringing a balance in their relationship. This would have guaranteed that no one would do any mischief to the other and might have promoted neighbourly relations instead. Had this happened, relations between the two countries might have been quite different. In any case, the Radcliffe award became the catalyst for India's assertion to the exclusive use of waters of the Eastern Rivers, which in turn gave birth to the water dispute between the two countries.

B) MOTIVES BEHIND THE AWARD

This raises the most important question: why did Radcliffe favour India at the expense of Pakistan? To answer this question, one needs to look at the politics that the Muslim League and the Congress party were conducting at the time. The former was demanding the creation of Pakistan out of the territories that constituted British India whereas the latter called it, in the words of Gandhi, the 'vivisection of mother India'. Though the Congress party finally acquiesced to the demand out of the tactical need to get rid of the British, its president, Acharya Kirpalani, made a candid admission that '[n] either the Congress nor the nation has given up its claim of a united India'.[20] It was not alone in this animus towards Pakistan. India found in the British viceroy of the time, Lord Mountbatten, who was supposed to be absolutely neutral in the Hindu-Muslim fray, a more than willing ally. This comes out from the number of disparaging titles that the latter conferred on Jinnah and Pakistan. For example, he described Jinnah as 'suffering from megalomania in its worst form',[21] 'a psychopathic case',[22] 'evil genius',[23] 'clot',[24] 'bastard',[25] 'lunatic',[26] and the new Muslim state as 'this mad Pakistan'.[27] Going beyond these belittling titles, when Jinnah thwarted him in his ambition to become the common Governor General of Pakistan and India, he threatened him with dire consequences:[28] 'I [Mountbatten] asked him [Jinnah], "Do you realise what this will cost you?" He said sadly, "It may cost me several crores of rupees in assets," to which I replied acidly, "It may well cost you the whole of your assets and the future of Pakistan".'[29] This animosity, once again, comes out in the interview that Larry Collins and Dominique Lapierre, authors of the bestselling book *Freedom at Midnight*, have referred to in their book where Mountbatten made the following confession:[30] '... had he [Mountbatten] known the Muslim leader was dying, he would have been strongly tempted to delay independence to await his death. Then, perhaps, an independent Pakistan would never have come into being.' This statement leaves no doubt about the extent of hostility that Mountbatten carried against Jinnah and Pakistan.

With this animus, Mountbatten set about hurting Pakistan. The opportunity presented itself during the working of the Boundary Commissions which were entrusted with the task of partitioning the provinces of Punjab and Bengal, and their demarcation into Muslim and non-Muslim areas. Following the decision to establish two Boundary Commissions, Mountbatten discussed the question of their composition with Jinnah and the Congress party leaders which resulted in two proposals. The first one provided for three UN experts and three experts representing each side of the partitioned province. The second foresaw two representatives, each of the Congress party and the Muslim League, headed by an independent chairman. Jinnah favoured the first proposal but according to Alan Campbell-Johnson, press attaché to Mountbatten, Nehru shot it down on the grounds that it would involve 'cumbersome procedure and unacceptable delay'.[31] Jinnah then opted for the second proposal but wanted the appointment of three law lords from Britain as neutral members. This time the British torpedoed the idea stating that in view of their age, the law lords would not be able to stand the oppressive Indian summer. Given the impasse, Jinnah had no option but to agree to the proposition of a commission composed of four members (two each representing the Congress party and the Muslim League) headed by an independent chairman.

Notwithstanding the agreement on the composition of the commissions, the issue of the appointment of the chairman had not been sorted out. On 19 June 1947, Mountbatten proposed that in case of disagreement on the choice of the chairman, the president of the International Court of Justice should be invited to nominate one. The next day, Liaquat Ali Khan told Mountbatten that the chairman who was to be nominated by each commission should be a business manager rather than an arbitrator. He made this proposal because '[h]e anticipated that agreement would not be reached by the Boundary Commissions, and, therefore, that two sets of recommendations would be submitted to the Governor General who would pass them to the Partition Council. It was not beyond the bounds of possibility that the Partition Council would be able to come to an adjustment among themselves; but if they failed to do so, reference would have to be made to the Arbitral Tribunal.'[32] The Congress party rejected Mountbatten's proposal. Realizing the impossibility of an agreement between Congress and the Muslim League and the need to move forward, on 23 June, Jinnah proposed the appointment of a distinguished member of the English bar as an independent chairman, which the Congress party accepted. Following the agreement, Mountbatten suggested the name of Sir Cyril Radcliffe who fulfilled that qualification as chairman. Jinnah knew him and concurred with the nomination. Nehru initially opposed his nomination, apparently under the impression that 'he [was] a conservative and therefore likely to favour the League'[33] and urged the acceptance of the Federal Court as the final arbitrator which Jinnah was strongly against.[34] However, he too finally accepted Radcliffe's nomination.

Here, the question as to why Jinnah accepted Radcliffe as chairman is raised. The answer is to be found in the reputation that the gentleman enjoyed, in terms of his integrity, in legal circles. This is testified by Philip Zeigler, Mountbatten's biographer, who paints him as a lawyer of 'great intelligence, probity and an intellectual toughness',[35] and someone who depicts 'monumental integrity and independence of mind'.[36] Apart from his impeccable standing, Jinnah must have believed that Radcliffe would follow the tradition of 'fair play' for which the British are known.[37] He never doubted Radcliffe's reputation and considered the much-vaunted British tradition of 'fair play' to be more than just hearsay (after all, Britain is known as the Perfidious Albion because of its alleged treachery to other nations).[38] However, this was exposed by Radcliffe's inequitable award of territories. To understand the treachery that Radcliffe, along with Mountbatten and Nehru, committed against Pakistan, we begin by looking at the 'sketch-map' story that came to light about the time of Partition.

The former prime minister of Pakistan and an important Muslim League leader, Chaudhri Muhammad Ali, had personal knowledge of the circumstances and execution of the Radcliffe award.[39] According to him, the governor of Punjab at the time of Partition, Sir Evan Jenkins, left a 'sketch-map' of the original Radcliffe boundary award which his successor, Sir Francis Mudie, discovered. Sir George Abell, the private secretary to the viceroy, had prepared it on 8 August 1947, on the basis of information that he received on the phone from Christopher Beaumont, the secretary to the Radcliffe Commission. According to Chaudhri Muhammad Ali, it showed the Ferozpur and Zira tehsils as part of Pakistan. However, when the final award saw the light of the day, it showed those areas as part of India rather than Pakistan. Chaudhri Muhammad Ali further narrates that on 9 August, he went to Delhi to consult Jinnah and Liaquat on a certain matter. He stated that when he met the latter, he shared disturbing reports according to which the Punjab Boundary Commission was preparing to award the Muslim majority areas in Gurdaspur, Amritsar, and

Jullundur to India. Liaquat instructed him to meet Lord Ismay, secretary to the viceroy, and convey to him on Jinnah's behalf that if the reports in question turned out to be true, this would seriously impact relations between Pakistan and the United Kingdom.

Continuing the story, Chaudhri Muhammad Ali says that upon returning to Delhi, he went straight to the viceroy's house to meet Lord Ismay who was closeted with Radcliffe. He waited for his turn to meet him. After Radcliffe left, Chaudhri Muhammad Ali met Ismay and conveyed Jinnah's message to him. Ismay expressed complete ignorance about Radcliffe's thinking in the matter and stated that neither the viceroy nor he ever discussed the issue of boundary demarcation with Radcliffe. Ismay tried to assure him that Radcliffe alone was to decide the matter, and that no suggestion of any kind whatsoever was ever made or would be made to him. Chaudhri Muhammad Ali tried to emphasize his point to him by pointing towards a map hanging against the wall in the room. While doing so, to his utter surprise, he discovered a pencil line drawn across the map of Punjab which exactly followed the one that the reports indicated. In this situation he did not deem it necessary to explain the point any more. According to Chaudhri Muhammad Ali, Ismay turned pale and suggested that someone had been fooling with the map. The line shown on the map differed from the final boundary which was announced on 17 August in one respect: the Ferozpur and Zira tehsils were shown as part of Pakistan.

We have noted above that George Abell had prepared the 'sketch map' on instructions from Beaumont and sent it along with a letter on 8 August to Jenkins. On 11 August, he dispatched a telegram to Jenkins which read: 'eliminate salient'. What did this cryptic message signify? The British historian, Andrew Roberts, states that it meant that 'the Sutlej salient, where Ferozpur and Zira were located, had in the intervening period been awarded to India rather than to Pakistan.'[40] To prove his point, he refers to the statement that Jenkins made in 1948 in which he stated,[41] 'I recall this clearly: Mudie and I were sitting together at the time and I understood clearly that the whole of the Ferozpur district was to be in India.' Incidentally, Mudie who succeeded Jenkins as the governor of Punjab had the same take in the matter which he expressed in his unpublished memoirs:[42]

> This meant that Ferozpur was to go to India, thus depriving the Pakistan Army of most of its weapons. No explanation of why this sudden change was made at the last moment was given, or has ever been given, but I find it difficult to believe that it was not the result of pressure put on Radcliffe by Mountbatten and his Government.

Philip Zeigler, the official biographer of Mountbatten, however, disputes this version of the story. According to him, Governor Jenkins had asked the viceroy to let him know in advance the terms of the award so that he could move his troops to the areas where trouble was most likely to take place. Mountbatten instructed his Private Secretary Abell to do the needful, who in turn approached Radcliffe's office and supplied him with a rough draft of the award with the observation that 'there will not be any great changes from this boundary'. Regarding this, Zeigler observes that the correctness of this comment depends on the definition of the expression 'great'. Continuing his version of the story, he concedes that a telegram was dispatched to Jenkins with the message 'eliminate salient' which he accepts that the latter rightly interpreted (that it referred to Ferozpur) and corrected it accordingly. He further mentions that Jenkins was not surprised with this development on the grounds that he recorded, twenty years later, that he had never viewed more than a rough guide of the map; and that it was inconceivable that Mountbatten had any hand in altering it. Zeigler also notes that Pakistan did not express outrage at the award of Ferozpur (and Gurdaspur) to India whereas India did on the award of Chittagong Hill Tracts to Pakistan, and that Justice Munir, as member of the Boundary Commission, stated that he was certain from the outset that Ferozpur would go to India.[43]

The question arises then: how did the news about the change of the Punjab map leak? The answer lies in the fact that the deputy commissioner of Ferozpur received instructions from the governor of

Punjab to select his headquarters outside of the three tehsils of the Ferozpur district, namely Ferozpur, Zira, and Fazilka, as these were to be allocated to Pakistan. For example, the chief irrigation engineer of the state of Bikaner, Kanwar Sain, states in his memoirs that the chief engineer of the Punjab Irrigation Department, Sarup Singh, learned it from the deputy commissioner himself. Subsequently, Singh sent a sealed letter to Sain who immediately went to Delhi and along with the prime minister of Bikaner, Sardar Pannikar, sought an audience with Mountbatten, which was granted. He complained to him about the allocation of the areas in question to Pakistan at the expense of the state of Bikaner. That same evening, the Government of India announced that the Radcliffe award was delayed by a few days.[44]

How did Kanwar Sain achieve this remarkable feat? He has an interesting story to tell in this regard. He claims that he persuaded the Maharaja of Bikaner to allow him to threaten Mountbatten that if the Ferozpur headworks were allocated to Pakistan, the Maharaja would have no option but to accede to the latter. According to him, the behaviour of Mountbatten during their meeting was quite interesting. He observes that when Sardar Pannikar brought up the subject, Mountbatten shouted at him:[45] 'The Viceroy has nothing to do with the Radcliffe Commission. The Commission has been appointed by His Majesty's Government. Radcliffe does not to report to me.' He further observes that Pannikar retorted that in that case, Bikaner would revise its earlier decision and join Pakistan. Mountbatten made a protestation of innocence and showed indifference but Sain states that, 'I could see a change in the colour of the face of Lord Mountbatten. He said nothing and we left His Excellency's room.'[46] No wonder Radcliffe postponed the award by a few days. According to Sain, 'when the award was announced on 17 August, we were happy to find that the Ferozpur headworks and the entire area ... were left with India.'[47]

Mountbatten's apologists, however, refute this story on the grounds that there is no record of any meeting between the Viceroy Pannikar and Sain either in Mountbatten's papers or in his diary.[48] There are others who refute the allegation against Mountbatten stating that it was based on mere hearsay; there was no hard evidence to support it.[49] Philip Zeigler, on his part, does not deny that the meeting took place but rejects the allegation. He contends that it is inconceivable that Radcliffe, a man of monumental independence and integrity, meekly submitted to someone who had no official *locus standi* in the matter, and that Mountbatten risked his reputation and all that he had achieved in India for little advantage. He also discounts the possibility that Pannikar could have intimidated Mountbatten with his bluff because the latter must have known that both economically and politically, the Maharaja of Bikaner, who was one of his closest allies among the princes, was not in a position to recant and accede to Pakistan. He concludes by asserting that it is easy to believe that Nehru pressurized him to change the award but hard to justify his succumbing to the latter's pressure.[50]

The Mountbatten version of the story more or less stood as fact in the West till February 1992, when in a strange turn of events a retired circuit judge and the former secretary to the Boundary Commission, Christopher Beaumont, exposed it as utterly fallacious. He did so by issuing a testimony which confirmed what the Pakistanis had always maintained with regard to the change of the Punjab map by Radcliffe. According to him, following the completion of the work of the Boundary Commission which had allocated Ferozpur to Pakistan, V. P. Menon, an important figure in Indian politics at the time, appeared outside the door of the vice-regal lodge where Radcliffe and Beaumont were staying on 11 August around midnight. He met the latter and told him that he wanted to see Radcliffe who 'less politely' refused his request. The next morning, he narrated to Radcliffe what had happened the night before. Later that evening (12 August), Radcliffe informed Beaumont that Lord Ismay had invited him to lunch at his place but had forbidden him from bringing Beaumont with him saying that there wasn't sufficient space at the table for an extra guest. Since Beaumont had lived in the house that Ismay occupied for six months, he knew from personal experience that that was patently untrue. He further claimed that during the course of the lunch, the map of Punjab was changed by allocating Ferozpur and Zira to India.[51]

How do we explain this treachery? Beaumont holds Mountbatten, Nehru, and Radcliffe responsible for it. At the same time, he tries to exonerate Mountbatten and Radcliffe as far as possible on the grounds of mitigating circumstances. In the case of Mountbatten, while admitting that he cheated, Beaumont tries to defend him stating that he was 'overworked and overtired'; that he was persuaded by Nehru and Menon that the award of Ferozpur to Pakistan would lead to a war between India and Pakistan; and that the Maharaja of Bikaner exploited his personal friendship with him to drive home the point that the canal headworks at Ferozpur were of utmost importance to his state. He also mentions that Mountbatten liked Nehru and disliked Jinnah but that was the last on the list of reasons that he gives for the change of the award. As far as Radcliffe is concerned, Beaumont tries to exonerate him by noting that he was new to India and had never been to East of Gibraltar; that Ismay and Mountbatten persuaded him that the award of Ferozpur to Pakistan would lead to civil war; and that he yielded because of 'overwhelming political expediency'. Although Beaumont makes a case for Mountbatten and Radcliffe, he shows absolutely no sympathy in his testimony for Nehru (and V. P. Menon).[52]

Beaumont deserves credit for showing enormous courage by telling the truth about what transpired at the lunch that took place on 12 August 1947 at the vice-regal lodge. However, one notes that in doing so, he (consciously or otherwise) seems to show partiality towards his two compatriots. He is soft on them on the grounds of mitigating circumstances but shows no sympathy for Nehru. Perhaps this is because Nehru was not an Englishman. However, to be honest, it is hard to feel any sympathy for this trio.[53] There were no mitigating circumstances to absolve Mountbatten. As shown above, he felt such a depth of ill will towards Jinnah and Pakistan that he was ready, as he himself confessed, to do anything to abort the birth of the new state. It was this obsession that drove him to pressurize Radcliffe to change the Punjab map. As far as Radcliffe was concerned, the fact that he 'destroyed all his papers before he left India' is really intriguing. This is so because people in positions of power preserve them in order 'to set the record straight' (which Radcliffe amply needed to do in the present case), or simply to leave their imprint on history. However, he did not do so. Was he trying to hide something? His yielding to Mountbatten's pressure, for whatever reasons, shows that his reputation as a lawyer of 'monumental integrity and independence of mind' was built on sand. Perhaps acting on Mark Twain's dictum, 'get yourself known as an early riser and then you can sleep till noon', he thought that he could get away with any action based on the strength of his reputation alone without realizing that the truth becomes unveiled sooner or later.

As far as Nehru is concerned, of all the characters involved in the drama surrounding the boundary award, he emerges as the most flagitious. This is not because he had a Machiavellian approach in politics (considering that there have been many others like him in human history, he does not deserve to be pilloried on this count)[54] but because while, on the one hand, he made a commitment to do something which he did not intend to fulfil, he meanwhile pretended to occupy a moral high ground. There are numerous examples which prove this point. For example, he made a pledge to let the Kashmiris decide their own future through a plebiscite under UN auspices without ever intending to carry it out, but at the same time claimed to be a votary of peace.[55] Compared to him, Gandhi stands out as a man of great moral standing when we realize that at one stage he was implacably opposed to the creation of Pakistan but once the Congress party conceded the idea in principle, he disagreed with both Patel and other Congress leaders to withhold Pakistan's share of funds. When they persisted with the idea, he protested with a fast (unto death) for their release, which ultimately led to his assassination. Incidentally, the leader who outshines all others in terms of moral superiority was indeed Jinnah who, while knowing fully well that the Congress and the British leadership had successfully conspired to deprive his new nation of the territories that rightfully belonged to it, decided not to create a fuss over the inequity of the Radcliffe award, and accepted it with grace and dignity. The following statement stands testimony to it:[56]

> The division of India is now finally and irrevocably effected. No doubt we feel that carving out of this great independent Muslim State has suffered injustices. We have been squeezed in as much as possible, and the latest blow that we have received was the Award of the Boundary Commission. It is an unjust, incomprehensible and even perverse award. It may be wrong, unjust and perverse; and it may not be a judicial but a political award, but we have agreed to abide by it and it is binding upon us. As honourable people we must abide by it. It may be our misfortune but we must bear up this one more blow with fortitude, courage and hope.

Jinnah showed magnanimity in accepting an award which Radcliffe had rendered with a *mala fide* intent. Given the fact that the Muslim League (along with the Congress party) had given their prior consent to abide by the award, was the Government of Pakistan entitled to denounce it? To answer this question, we need to examine the theory and practice of international arbitral awards. It is true that international arbitral awards are deemed final and binding, and are normally not questioned even if the parties in dispute are not happy with them. Does that mean that they cannot be challenged on any grounds whatsoever? The opinion in the matter is divided.[57] According to one view, the principle of finality of awards is absolute and the plea of nullity is not available to any party to the dispute under any circumstance whatsoever. This view derives force from Article 81 of the Hague Convention I of 1907 and the absence of an international mechanism to undertake it. According to another perspective, the previous contention is not valid because it is at variance with the traditions of arbitration. The protagonists of the second view cite the judgment by the World Court in the *Case Concerning the Arbitral Award Made by the King of Spain* which, according to them, proceeds on the assumption that an award could be nullified in certain cases. They also seek justification from the fact that in some instances the losing party raised the plea of nullity and refused to abide by the award. Despite jurisprudential divergences, the balance of opinion in international law tilts in favour of the second viewpoint.

If jurists are divided on the question of the finality of awards, they are equally divided on the grounds of their nullity.[58] To begin with, the jurisprudence in the matter is not very rich. Besides, whenever the plea of nullity was advanced, it was the losing party which did so and the winning party opposed it. Despite these difficulties, jurists have identified four grounds of nullity: 1) those relating to grounds of jurisdiction; 2) those relating to the procedure of hearing and award; 3) essential errors; and 4) fraud and corruption. Since it is the last ground which is of interest to us, we shall focus on it.[59] As far as the argument of fraud is concerned, it relates to the presentation of evidence, which is irrelevant in the case of the Radcliffe award because no such objection was ever brought up. It was an argument of corruption which the Muslim League raised, which has two aspects. One relates to the use of corrupt means in tampering with the evidence, which does not concern us because the latter never raised it. The other relates to the conduct of members of the arbitration tribunal. This is the relevant part because the Muslim League accused Radcliffe of having changed the award under the influence of Mountbatten.

On the basis of this discussion, we are of the opinion that the Pakistani government was entitled to denounce the Radcliffe award as null and void under international law. If it did not do so, there were perhaps political considerations behind it.

We have seen that Radcliffe awarded the Ferozpur (and Zira) tehsil(s) to India. He also awarded Gurdaspur to India which resulted in depriving Pakistan of the Madhopur headworks. Did he do it by gerrymandering the Punjab map? There was a rumour at the time that the Boundary Commission was deliberating that Radcliffe pulled a 'Ferozpur' on Pakistan on the issue of Gurdaspur. Beaumont dismisses this allegation in his testimony and there is no evidence available to support it. Irrespective of whether or not Radcliffe gerrymandered the map, it does not mean that he rightfully awarded Gurdaspur to India because, as seen above, with three out of four tehsils having a Muslim majority, it should have gone to Pakistan.

How do we explain the fact that it went to India? The answer lies in a possible secret understanding between the British and the Congress party leadership. The idea may seem outlandish on the face of it. However, upon close scrutiny, it is not. Take the case of the great Bengal city and port of Calcutta. Whereas the British told the Muslim League that the Bengal Boundary Commission would decide Calcutta's fate, they had reached a secret deal with the Congress party leadership that it would be awarded to India. Sardar Patel openly admitted this in a public speech delivered on 15 January 1950: 'We made a condition that we could only agree to partition if we did not lose Calcutta. If Calcutta is gone, then India is gone.'[60]

As for Gurdaspur, what is the proof of a deal between the British government and the Congress party? The Muslim League leaders involved in the transfer of power politics made claims in this regard. For example, Chaudhri Muhammad Ali points towards a possible deal in the following rhetorical question: 'If, long before Radcliffe's Award, Mountbatten was assuring the Maharaja of Kashmir that India could safeguard the security of the state as well as Pakistan could, was it not likely that he had reached an understanding with Congress leaders in respect of Gurdaspur district similar to the one regarding Calcutta?'[61] Similarly, Sir Zafrulla Khan claims in his memoirs that Radcliffe awarded Gurdaspur to India on the basis of a secret deal that Mountbatten reached with Nehru and Krishna Menon in the second week of May 1947 in Simla. In this deal, the Congress party agreed to remain in the British Commonwealth with a Dominion status—for which the British government was very keen—in return for three concessions by the British: 1) that they would transfer power in two months' time; 2) that they would give Calcutta to India; and 3) that they would include Gurdaspur and Batala which were Muslim majority areas in India.[62] He makes this claim but, like Chaudhri Muhammad Ali, fails to provide any evidence thus making it baseless.

Irrespective of the question whether or not the Muslim League leaders' claim could be taken seriously, why would the British make a secret deal in favour of India and against Pakistan? To answer this question, we begin by examining the geostrategic importance that the State of Jammu and Kashmir enjoyed in the British scheme of things, for the defence of India. According to the noted British historian, Alastair Lamb, since the middle of the nineteenth century, Whitehall had viewed the state as an important outpost to keep Russian influence out of the north-western part of the subcontinent. This perception did not change even after the transformation of Russia into the Soviet Union. In order to realize its objective, in 1935, it acquired from the Maharaja of Jammu and Kashmir, the Gilgit Lease over the Gilgit Agency for a period of sixty years with the view of observing and actively countering the Soviet influence in Sinkiang (let us not forget that the lease was acquired in the first place to keep the area in question out of the Maharaja's control). On the eve of Partition, it found it hard to transfer the defence of Gilgit to the Maharaja. With this background, the tenancy of the lease was to devolve on Pakistan because the leased areas had a Muslim majority, besides being non-contiguous with India. Now if the British were ready to accept Pakistan as the guardian of the Gilgit Agency, they could have left the Gilgit Lease untouched. However, they were not prepared to do so because they were of the view that the state needed to be put in firm hands and that India rather than Pakistan could do the job much more effectively.[63] Lamb describes the British thinking in these terms:[64]

> The possible consequences of the partition of the Indian Empire for the security of the Northern Frontier must have been apparent to foreign policy specialists in the Indian Government during the final days of the British Raj. If the State of Jammu and Kashmir asserted its independence, then India would be deprived of its main benefit from the creation of the State since the sale of the Vale of Kashmir to the Dogras in 1845; it would no longer serve as a vital guard for a difficult frontier. If the State of Jammu and Kashmir joined Pakistan whose

stability and durability appeared to many British observers in 1947 to be extremely doubtful, then the Northern Frontier might become an open door into the subcontinent for all sorts of undesirable influences which it had been British policy for generations to exclude. For better, it could well have been argued, that the guardianship of the entire Northern Frontier be entrusted to a bigger, stronger, and apparently more reliable of the two successors to the British Raj, India.

Lamb argues[65] that the British shared their thinking with Jawaharlal Nehru, the future prime minister of independent India. He draws this conclusion from the statement that Mountbatten made after a meeting that he had with the latter on 22 April 1947. In this meeting, he pointed out that princely states would be free to accede to any successor state that they liked, independent of geographical considerations. Nehru's observation regarding this point was, '... the future of Kashmir might produce a difficult problem'.[66] Lamb also cites, as evidence in this regard, the advice that Mountbatten tendered to the secretary of state for India, Lord Listowel, on 29 April 1947, in which he exhorted him to return the entire area covered by the Gilgit Lease to the State of Jammu and Kashmir before the transfer of power. According to him, both Listowel and Nehru concurred with this view. He also notes that Jinnah was possibly kept out of the consultation process. It was with this background that the British terminated the Gilgit Lease prematurely and the area covered by it reverted to the State of Jammu and Kashmir.

In the wake of this development, the next step was to get the state acceded to India. However, this presented a serious hurdle as all lines of communications from the princely state led to West Pakistan; none went to India except for an unmetalled road that passed through Gurdaspur. In this context, Gurdaspur assumed critical importance as its allocation to Pakistan would mean that Jammu and Kashmir would be constrained to accede to Pakistan which, as seen above, neither the British nor the Indian leadership wanted. The location of the Madhopur headworks in Gurdaspur must have been an added attraction for India. Taking this into account, the British appeared determined to see Gurdaspur awarded to India for which there is also overwhelming circumstantial evidence. For example, on 3 June 1947, Mountbatten made a radio broadcast on the Partition Plan in which he declared that 'the ultimate boundaries will be settled by the Boundary Commission and will almost certainly not be identical with those which have been provisionally adopted.'[67] When quizzed in a press conference held on 4 June regarding the meaning of this observation, Mountbatten explained it[68] by saying that in the district of Gurdaspur, the Muslim population was 50.4 per cent and the non-Muslim population 49.6 per cent (his figures, however, were incorrect. The Muslim population was 51.14 per cent and not 50.4 per cent as stated). He went on to say that with a difference of 0.8 per cent, the Boundary Commission was unlikely to award the whole of the district to Pakistan. As observed by Chaudhri Muhammad Ali, this statement shows that Mountbatten had made a minute study of the population figures of the Gurdaspur district. If he was as disinterested in what Radcliffe was doing as he pretended to be, the intriguing question arises: why did he give so much attention to the population profile of Gurdaspur?[69]

The answer to this question lies in the strategic importance that Gurdaspur enjoyed. This emerges clearly from the observation that Mountbatten made to the Nawab of Bhopal on 4 August in which he stated that Kashmir was '... so placed geographically that it could join either Dominion, provided part of Gurdaspur were put into East Punjab by the Boundary Commission'.[70] In addition to this, there is another incident which shows a possible collusion between the Congress party leadership and Mountbatten on the Gurdaspur issue. In the middle of July, Radcliffe was in Lahore for an aerial view of the Punjab terrain for which purpose Mountbatten had supplied his own plane. However, he could not fly because of a storm. Justice Din Muhammad, one of the Muslim League nominees for the Punjab Boundary Commission, revealed that he saw the flight plan he got from the pilot which, in the words of Justice Munir, '... was exactly identical with the boundary line [regarding Gurdaspur] drawn on the map attached to the award. These facts show conclusively that the boundary line had been determined before the award was made.'[71]

C) STANDSTILL AGREEMENT AND THE AFTERMATH

Well before the Radcliffe award, the British government had constituted two bodies for the division of Punjab's physical assets, namely the Punjab Partition Committee and the Arbitral Tribunal. The Partition Committee was to deal with matters relating to the division of the Punjab province whereas the Tribunal was to arbitrate on matters which the Partition Committee referred to it. The latter had constituted various subcommittees to deal with different subjects. Of these subcommittees, Committee B on 'Division of Physical Assets' dealt with matters relating to the future management of joint assets, the division of other physical assets, and the evaluation of these assets for accounting between Pakistan and India. The purpose of this exercise was to make Pakistan account for the amount by which the assets of West Punjab exceeded 60 per cent of the assets of the pre-Partition Punjab. Thus, Committee B was to determine the nature of assets, their worth, and the basis of their division. Of these assets, the principal one was the canal systems and the Crown lands that they irrigated. Committee B consisted of eight members, four each from the Muslim League and the Congress party. It rendered a unanimous report in July 1947 which, in paragraph 15, stated that '... there is no question of varying the authorized shares of water to which the Zones [meaning East and West Punjab] and various canals are authorized.'[72] Even though the Committee agreed on the maintenance of the pre-Partition division of the water resources, it could not agree on the valuation of the canal systems through which these resources were distributed or the value of the irrigated Crown lands. The Committee B report was put before the powerful Punjab Partition Committee (composed of ministerial representatives of the East and West Punjab and presided over by the governor) which adopted it *in toto*. While doing so, it appointed two of its members, Messrs Sarup Singh and S. A. Hamid, the chief engineers designate of the East and West Punjab respectively, to submit a joint proposal for the implementation of paragraph 15.

Both the Punjab Partition Committee and the Arbitration Tribunal accepted—as agreed and settled—that the authorized shares of water of the East and West Punjab would not be varied. This comes out from the awards of the Arbitration Tribunal on various financial issues where it proceeded on the assumption that the parties would continue to receive supplies of water allocated to them before Partition.[73] In any case, the two chief engineers subsequently submitted a joint proposal which recommended the continuation of the *status quo*. On 20 December 1947, the two countries signed a Standstill Agreement which provided that until the end of the *rabi* crop, i.e. 31 March 1948, the *status quo* would be maintained.[74] The joint proposal also recommended that the two countries conclude 'a further agreement' before the expiry of the Standstill Agreement. On 29 March, East Punjab notified West Punjab that the Standstill Agreement would expire on 31 March. The latter requested its extension and proposed a meeting of their chief engineers. The Indian response to Pakistan's request came on 1 April 1948 in the shape of stopping water supplies from the UBDC to eleven canals in West Punjab (which included Central Bari Doab Canal system, the Dipalpur Canal system, and the Bahawalpur State distributary) affecting some 1.6 million acres of land. India did so on the grounds that the Standstill Agreement had expired and that it had not been renewed. It is noteworthy that the Indian authorities did so without any warning[75] which showed their *mala fide* intent. This is so because even if we consider for a moment that India was not under any legal obligation to continue the supply of water to Pakistan, it was at least under a moral obligation to send a warning that it intended to cut-off water supplies.

So, we are compelled to ask, why did India choose 1 April to cut-off the water supplies to Pakistan's canals? Was it a mere coincidence or was it playing some kind of prank on Pakistan or is there some other explanation for it? Such an important step being taken can neither be a coincidence nor an April fool's prank; rather a decision consciously made. India chose the date in question because, as stated above, the term of the Arbitration Tribunal was to expire on 31 March 1948, and it knew that after the latter had ceased to exist, Pakistan would not be able to seize it and thus get a favourable verdict on its water rights.

The Indian version of the episode is different. Gulhati rejects this allegation and claims that East Punjab took this step temporarily in order to establish its exclusive ownership over the UBDC. In its view, in the absence of a formal agreement, West Punjab would have acquired some kind of a legal right to it which they did not want. He does not adjudge East Punjab's action but concedes that under the prevailing conditions, Pakistan and the West Punjab authorities were justified in regarding it as provocative.[76] His explanation is good but not the real one. The circumstantial evidence clearly points to the fact that East Punjab cut-off the water supplies on that fateful date in order to stop Pakistan from approaching the Arbitration Tribunal and getting a favourable award.

The next question is why didn't Pakistan seek an extension of the Standstill Agreement well before its expiry instead of waiting till the eleventh hour to request its extension? The answer to this question partly lies in the fact that the East Punjab ministers and officials gave oral assurances to their counterparts in West Punjab that the existing arrangement would continue even after the termination of the Standstill Agreement. The testimony of Chaudhri Muhammad Ali in this regard is quite revealing:[77]

Despite the fact that the Radcliffe Award had placed the control of headworks vital for Pakistan in the hands of India, the West Punjab government remained content because of the agreement reached by ... Committee B and the Punjab Partition Committee, that the prepartition shares of water would not be varied. No formal document specifying the precise shares of East and West Punjab in irrigation waters was drawn up and signed. The West Punjab ministers and officials felt assured by the repeated declarations of their counterparts in East Punjab that there was no question of any

change in the prepartition arrangements for canal waters. The same
declarations were also made by the East Punjab representatives before
the Arbitration Tribunal, when the disputed question of valuation of
the canal system came up for a hearing.

S. I. Mahbub, the chief engineer of the Punjab Irrigation Department,
who took part in the meetings of the Partition Committee, narrated
a story[78] which is very telling about the cunning Indian leadership on
the one hand and gullibility of the Pakistani leadership on the other.
According to him, at the end of a meeting of the Partition Committee
in Lahore, Shaukat Hayat, Punjab Revenue Minister and member
of the Partition Committee, asked his counterpart from East Punjab,
Sardar Swaran Singh, whether his government intended to stop the
supply of water to Pakistan's canals. To this, the Sardar replied: 'What
are you talking about! Do brothers ever deprive brothers of water?'
The answer seemed to have mollified the Pakistani representative
and the Punjab government did not bother to seek the renewal of
the Agreement until much later. Had it sought an extension earlier
and had India refused, Pakistan had the option to seize the Arbitral
Tribunal. This was the last thing that the Indian leadership wanted.
Therefore, it kept on falsely assuring the Pakistani leadership of the
continued supply of water till such time as the Arbitration Tribunal
had ceased to exist. Since Pakistan failed to seize the Tribunal, it
missed the opportunity to get an award because two days after its
request for continuation of the Standstill Agreement, it ceased to
exist. Sir Patrick Spens, the chairman of the Arbitral Tribunal and
the former Chief Justice of India, made a statement before the joint
meeting of the East India Association and the Overseas League on
23 February 1955, on the issue which tells a lot about the gullibility
of Pakistan's leadership in believing verbal promises:[79]

> I remember very well suggesting whether it was not desirable that
> some order should be made about the continued flow of water ....
> But we were invited by *both* the Attorney Generals [of India and
> Pakistan] to come to our decision on the basis that there would be no

interference whatsoever with the then existing flow of water, and the award which my colleagues made in which I had no part, they made on that basis. Our awards were published at the end of March 1948. I am going to say nothing more about it except that I was very much upset that almost within a day or two there was a grave interference with the flow of water on the basis of which our awards had been made.

Was the East Punjab government, following the expiry of the Standstill Agreement, under any obligation to continue the water sharing arrangement that previously existed? The answer to this question is in the affirmative for two reasons. Firstly, the Arbitration Tribunal, as pointed out by Chaudhri Muhammad Ali while dealing with the question of valuation of the canal system, accepted the Indian contention that Pakistan should be accountable for a higher valuation on the grounds that '... existing allocation of water would be respected, for, without water, canals are dry ditches, a liability and not an asset'.[80] In other words, the Arbitration Tribunal made Pakistan accountable for a higher valuation of the canals that irrigated its territory on the assumption that India would respect the existing allocation of water. Thus it was under legal obligation to respect the existing arrangement even in the absence of the Standstill Agreement. It was equally bound to do so on the grounds, as noted above by Sir Patrick Spens, that its Attorney General (along with that of Pakistan) made a commitment before the Tribunal that there would be no interference whatsoever with the existing supplies of water. Secondly, India was not entitled to disturb the existing water-sharing arrangement with Pakistan because the latter had acquired it on the basis of historic uses. It signifies that even in the absence of the Standstill Agreement, Pakistan could claim the preexisting share of water on the basis of customary international law.[81]

Before proceeding further, we propose to examine another matter pertaining to the foregoing issue. We have seen above that the East Punjab ministers and officials made continuous declarations that there would be no question of any change in the pre-partition arrangements for canal waters. The East Punjab representatives also repeated them before the Arbitration Tribunal. Since most of these declarations were oral in character, the question of their legal value arises. More importantly, we are interested to know whether or not they can be equated with a treaty. To answer this question, we begin with the definition of 'treaty' as defined in the Vienna Convention on the Law of Treaties 1969. According to it, a 'treaty' means 'an international agreement concluded between States in written form and governed by international law, whether embodied in a single instrument or in two or more related instruments and whatever its particular designation'. Since the Vienna Convention defines a 'treaty' as an international agreement in written form, does it signify that oral agreements are bereft of any legal validity?

To elaborate on this, we seek help from the report of the International Law Commission which codified the convention. According to its commentary, the definition is not intended to deny the legal force of oral agreements under international law or to imply that some of the principles contained in the Commission's draft articles may not have relevance in regard to oral agreements but only to promote clarity and simplicity.[82] Article 3 further clarifies that because the convention does not apply, among others, to international agreements that are not written down, it shall not affect the legal force of such agreements.[83] It may be mentioned here that the World Court even went beyond the convention in the *Legal Status of Eastern Greenland* case where the question arose whether an oral declaration containing a promise made by the foreign minister of one country to that of another, irrespective of whether or not the former acted within the limits of his constitutional authority, would be binding on the state whose minister made the declaration. Concerning the oral statement incorporated in the Ihlen Declaration, which it came to be known as, the Court declared that '[it] considers it beyond all dispute that a reply of this nature given by the Minister for Foreign Affairs on behalf of his Government in response to a request by the diplomatic representative of a foreign power, is binding upon the country to which the Minister belongs'.[84] In light of these observations, India was bound to respect any oral commitments that the East Punjab ministers and other relevant officials made but failed to do so.

When the East Punjab authorities cut-off supplies, Nehru gave the impression that the latter acted of their own accord and without taking the central government into confidence. According to press reports, he admonished them for their actions.[85] However, Nehru's reprimand seemed to be no more than a public relations exercise. Given the Machiavellian character attributed to him by Sheikh Abdullah,[86] it could have been a coordinated move between the government in Delhi and the East Punjab authorities in order to dupe the others. There were different explanations for the cessation of water supplies. According to one interpretation, 'it might have well have been a calculated effort to wreck Pakistan's very existence and force her back into unity with India—as many Pakistanis believed.'[87] According to another, the following combination of factors have been advanced:[88] 'The first is legalistic—that of an upper riparian establishing its sovereign rights. Others include an Indian maneuver to pressure Pakistan on the volatile Kashmir issue, to demonstrate Pakistan's dependence on India in the hope of forcing reconciliation, or to retaliate against a Pakistani levy of an export duty on raw jute leaving East Bengal. Another interpretation is that the action was taken by the provincial government of East Punjab without the approval of the central government.'

Irrespective of India's motives and whether or not the central and provincial governments were hand in glove with each other, Pakistan's leadership not only continued to believe in the Indian promises but also gave no serious thought to seize the Arbitration Tribunal. Shaukat Hayat reveals that the matter of seizing the Tribunal was discussed in the party meeting but was rejected on the grounds that even if we '... got a favourable decision, how were we going to get it implemented when the headworks had been unfairly allotted to India, in the so-called Radcliffe Award'.[89] Firoz Khan Noon, another important Muslim League leader who was subsequently Pakistan's prime minister, also expressed similar views on the matter:[90] 'I always resisted any suggestion of ours to go to the International Court of Justice at the Hague, because ... what is the use of winning such suits in courts, if such a decision cannot be backed up by physical power to enforce it ....' The Pakistani leadership's thinking in the matter seemed to be muddled because it demonstrated its lack of knowledge about the moral and legal value of an award. This attitude came from a party to the dispute that did not have a military option to resort to and whose only option was to put moral and legal pressure on India through the Arbitration Tribunal's award. Commenting on this, Chaudhri Muhammad Ali was spot on in observing:[91] 'On the side of East Punjab there was Machiavellian duplicity. On the part of West Punjab there was neglect of duty, competency, and lack of common prudence ....'

The truth of the matter is that the malaise ran much deeper than that. The Muslim League leadership had neither the vision nor a strategy to deal with Partition-related issues. For example, the Muslim League's proposal to invest the rulers rather than the people of princely states, with the right of accession, led to the loss of Jammu and Kashmir for Pakistan as Maharaja Hari Singh decided to take advantage of this principle proposal and accede to India.[92] It was no different in the case of the supply of water.

Following the stoppage of the water supplies, the chief engineers of the two Punjab provinces met on 18 April in Simla and concluded two agreements of limited duration for restoration of the water supplies to Pakistan's canals. One agreement was to terminate on 30 September 1948, and the other on 15 October 1948. These agreements were to come into force from the date of ratification by the governments of Pakistan and India. In the wake of this development, on 3 May 1948, a three-member Pakistani delegation headed by the federal finance minister, Ghulam Muhammad (other members of the delegation were Shaukat Hayat and West Punjab finance minister, Mian Mumtaz Daultana), went to Delhi for further discussion of the Simla agreements. The Indian delegation was led by the Indian prime minister, Pundit Jawaharlal Nehru (other members of the delegation were East Punjab irrigation minister, Swaran Singh, and the union minister for mines and power, N. V. Gadgil). It is worth noting that the Pakistani delegation was not headed by its prime minister whereas the Indian delegation was. This perhaps indicates the kind of importance that the Pakistani leadership attached to the water issue.

One gets the same impression from the memoirs of Shaukat Hayat, a member of the Pakistani delegation, who out of 376 pages of his book barely devotes half a page to this issue. In any case, the Pakistani delegation termed the stoppage of the water supplies a violation of the terms of the Punjab Partition Committee and of the assurances held out by the Indian officials in that regard. It appealed to its counterpart for restoration of the *status quo ante* but the East Punjab government representative present at the meeting rejected the demand. They argued that 'they would not restore the flow of water to the canals unless West Punjab acknowledged that it had no right to the water'[93] and that the restoration was contingent on the payment of seigniorage charges. In a cable dated 21 April 1948, sent to its counterpart, it unambiguously stated, 'we shall immediately open the canals on hearing your acceptance of these terms'. The same message was sent again in a cable dated 24 April.

Pakistan rejected the Indian demand out of hand and invited India to submit the dispute to the International Court of Justice for adjudication. The latter turned it down, which was understandable because it knew that its position on the issue was untenable under international law and that it would, in all probability, lose the case. It is ironic to note that whereas the Pakistani delegation took part in the negotiations held in Jullundur, as revealed by Shaukat Hayat, it did not deem it fit to submit the dispute to the Arbitration Tribunal on the grounds that it would be hard to get it implemented. That same delegation was now keen to get it adjudicated by the World Court, which incidentally did not enjoy jurisdiction over it because of the Indian Declaration.[94] This demonstrates the confusion from which the Pakistan delegation suffered over the issue. In the face of the Indian refusal to get the issue adjudicated by the World Court, Ghulam Muhammad then appealed to Mountbatten, who was then the Governor General of India, to use his good offices for an amicable and fair settlement of the matter. The lack of experience of the Pakistani delegation is noticeable, since Mountbatten was ostensibly responsible for creating the situation in the first place.

On 4 May, the Indian government submitted to the Pakistani delegation the text of a statement in which they asked, in the words of Chaudhri Muhammad Ali, '... to sign without changing a word or a comma'[95] if it wanted to have water supplies restored. Given the dire water situation in West Punjab, which posed a serious threat not only to crops but to millions of humans especially in the city of Lahore, it had no choice but to sign the document. In the wake of this development, India let the water flow in the CBDC and the Dipalpur Canal systems (though not in the Bahawalpur State Distributary and the tail reaches of nine minor distributaries and forty-one water courses taking-off from the UBDC and extending into Pakistan till November 1953).

The preceding analysis brings us to question why Pakistan buckled under India's pressure so meekly and signed the document in question without protest. In our judgment, the right course of action for its delegation should have been to return home without signing the document and to call India's bluff. This might have forced the latter to reach a compromise with Pakistan. This was absolutely possible because, according to irrigation experts, though the critical period for sowing the summer crops in Punjab starts towards the second half of April, most of them are sown from May through July.[96] Had this strategy failed and had India persisted with its blackmail by continuing to block Pakistan's share of water, resulting in the loss of life and crops, Pakistan could have drawn the international community's attention to the terrible death and destruction that India's brutal action was wreaking on Pakistan. This could have brought the UN Security Council into the equation which could have forced India to restore water supplies to Pakistan's canals.

Additionally, Pakistan could have monopolized its membership of the British Commonwealth by threatening to leave it. This might have forced Britain, which was responsible for this situation in the first place and had the moral responsibility to take care of it, to defuse the situation.

Had this strategy failed and there was no favourable response from the international community including Britain, Pakistan could have resisted India's onslaught unaided. Nations take every possible step to safeguard their vital interests as, for example, Russia did by putting up stiff resistance against the French troops when Napoleon invaded it or during World War II against the marauding Nazis. Pakistan could have stoutly resisted India's water terrorism as Pakistanis were fired with a new-born nationalistic fervour and were prepared to make any sacrifice at that time including that of their lives. However, Pakistan's leadership failed to rise to the occasion and relinquished its historic rights in the Eastern Rivers to India at the first hint of pressure. The explanation for this abject surrender to India lies in its failure to galvanize the nation and devise a proper plan to defend its water rights. This was a repeat of its inability to properly draft the Standstill Agreement and seek its renewal in time, as seen above.

D) MAY 1948 JOINT STATEMENT AND ITS VALIDITY

The piece of paper which the Pakistani delegation had signed for the restoration of water supplies has come to be known in various ways as Document, Joint Statement, Delhi Agreement, or Inter-Dominion Agreement.[97] As to its contents, it begins by stating the position of the two Punjabs. Hence, it declares that the East Punjab government was of the view that under the Punjab Partition (Appropriation of Assets and Liabilities) Order, 1947 and the Arbitral Award, the proprietary rights in the waters of the rivers flowing through East Punjab wholly belonged to it and that the West Punjab government had no claim whatsoever on any part of them as a matter of right.

The West Punjab government, however, contested this viewpoint on the grounds 'that the Arbitral Award had conclusively decided the issue, by implication, in its favour. It also contended that on the basis of international law and equity, it had rights to waters flowing though the East Punjab Rivers. Of the conditions that led to the restoration of water through Pakistani canals, the West Punjab government disputed two. One was the right to the levy of seigniorage charges for water, and the other was the capital cost of the Madhopur headworks and carrier canals to be taken into account. Without prejudice to its legal rights, the East Punjab government had assured the West Punjab government that it had no intention of suddenly withholding water without giving it time to tap alternative sources. The West Punjab government on its part recognized the concern of the East Punjab government to develop areas where water was scarce and which were underdeveloped in relation to parts of West Punjab. Without taking into account the question of law involved, the East Punjab government was to progressively diminish the supply of water to provide reasonable time to the West Punjab government to look for other options. The West Punjab government agreed to deposit a certain amount of money, fixed by the Prime Minister of India in the Reserve Bank of India, and to immediately transfer to the East Punjab government an amount over which there was no dispute. The two governments also agreed that their representatives would meet again after examination by each side of: the legal issues involved, the method of estimation of the cost of water by the East Punjab government, the technical survey of water resources, and the means of using them to supply water to the canals.

Subsequently, a meeting of the representatives of the two countries took place on 21 July 1948, in Lahore. The Pakistani side was led by Foreign Minister Sir Zafrulla Khan, and the Indian by their railways minister, Gopalaswamy Ayyangar. The purpose of the exercise was to give final shape to the May 1948 arrangement. The Indian delegation desperately wanted the Pakistani delegation to accept that the proprietary rights in the waters of the Eastern Rivers were wholly vested in the East Punjab government, and get it incorporated in a final agreement. The Pakistani delegation wanted the said arrange-

ment to be rewritten. For want of a meeting of minds on the issue,
a deadlock ensued between the parties. Subsequently, India informed
Pakistan through a wire message dated 18 October 1948 that it
considered the Joint Statement a permanent agreement and that any
further negotiations would be held on that basis in these words:[98]

'We deny that [the] Delhi Agreement provides for continuance of
supplies to West Punjab until a final agreement is arrived at between
the two dominions. We consider that if a party refuses to come to a
final settlement of the dispute without any reservation or if there is
unreasonable delay on the part of a party in concluding such final
settlement, it is open to the other party to put an end to the agreement
by giving reasonable notice.'

It is obvious that the negotiations floundered because Pakistan
considered the Joint Statement temporary whereas India proceeded
on the assumption that it was a permanent agreement. Was Pakistan
justified in its stance? In our assessment, it was because the Joint
Statement unambiguously stated in its opening paragraph that the
parties disagreed on the issue of the proprietary rights to the waters
of the Eastern Rivers. In other words, they concluded the statement
without prejudice to their legal position in the matter. It signifies
that Pakistan at least signed it believing that the parties would
reconcile their viewpoints through a new dispensation. The fact
that Pakistan wasn't agreeable to the levy of seigniorage charges for
water or the capital cost of reviving the flow of water to Pakistan's
canals also demonstrates that it regarded the Joint Statement as a
provisional agreement.

Pakistan claimed that the two countries did not envisage that the
Joint Statement would last as long as it did. It also affirmed that Nehru
did not think that it would last more than eight weeks and that it could
be terminated at any time on reasonable notice.[99] Sir Zafrulla Khan
sums up Pakistan's position in these words: 'The terms and conditions
of this agreement made it obvious that it was designed as a temporary
measure of relief while the parties would seek to adjust the difference

between them over their legal rights through negotiations and, failing that, through third-party determination either by way of good offices or mediation or by way of arbitration or judicial decision.'[100] In light of this statement, Pakistan's claim that the Joint Statement was of temporary nature seems to be justified. However, the stipulation in the Joint Statement that the East Punjab government would progressively diminish supply of the waters to Pakistan's canals to give reasonable time to the West Punjab government to tap alternative sources seems to weaken Pakistan's stance on the matter.

On 1 September 1948, the East Punjab engineers informed their West Punjab counterparts that the Joint Statement for the continuation of the water supplies to the CBD and Dipalpur Canal systems would expire on 30 September. Pakistan's Foreign Minister, through a wire dated 15 September (repeated on 26 September), sought assurances from Delhi for their continuation till the conclusion of the final agreement. Nehru obliged by stating that, 'you may rest assured that the supply of water would be continued ... until the matter can be further discussed at an Inter-Dominion Conference.' However, in a wire dated 24 October, he went back on the promises made to Pakistan on the grounds that '[i]f a party refuses to come to a final settlement without reservation, or if there is an unreasonable delay on the part of the party in concluding a final settlement, it is open to the other party to put an end to the agreement by giving reasonable notice.'[101] We have already noted that through another wire, he had also let Pakistan's government know that the 'Delhi Agreement is based on recognition by West Punjab government of the right of the East Punjab government to progressively diminish supply of water to West Punjab in order to give reasonable time to enable West Punjab government to tap alternative sources.'[102] He had also informed Pakistan's government that future meetings between the two parties would take place on that basis.[103]

Pakistan's government reminded Nehru that when the Indian representative had joined the Punjab Partition Committee, he had accepted that the authorized shares of the two countries in the waters common between the two countries would not be varied. In a letter dated 24 November 1950, addressed to the Prime Minister of Pakistan,[104] Nehru rejected Pakistan's assertion by contending that the Indian representative never made such a commitment which was, in Churchill's words, nothing but a 'terminological inexactitude' on his part. Again, in a letter dated 11 December 1950, addressed to the Prime Minister of Pakistan,[105] Nehru repeated his argument. He also claimed that 'India only proposes to utilize the waters to which she is entitled, but before doing so she has generously agreed not to prejudice any existing irrigation in Pakistan with such waters until Pakistan has had reasonable time to tap alternative sources …'.

While these talks were going on, India embarked on the construction of the Bhakra Dam to take increasing quantities of water from the Eastern Rivers which had so far catered to Pakistan. Pakistan's government claimed, on the basis of India's first five-year plan, that India intended to gobble up the entire flow of these rivers which had and was irrigating 5 million acres of land in Pakistan. It opposed the Indian move on the grounds that India enjoyed more rainfall than Pakistan. It also argued that unlike Pakistan, the latter had many large rivers such as the Jumna and the Ganges Rivers which, with a mean annual flow of 400,000,000 acre-feet of water, had more than double the waters of all the rivers in the Indus Basin; and that whereas the flow of the Jumna was being used to some extent, that of the Ganges was scarcely used for irrigation purposes. It referred to the recommendations that A. N. Khosla, a leading Indian authority on irrigation and subsequently leader of the Indian delegation to the World Bank-sponsored talks on the Indus water dispute, made in 1943 in which he proposed that India should irrigate East Punjab with waters from the Jumna and the Ganges Rivers which, in his view, were logical sources of water. Towards this end, he proposed three dams each on the Jumna and the Ganges Rivers.[106]

In order to break the deadlock, on 5 October 1949, India proposed the establishment of a committee to negotiate the matter which Pakistan's government accepted. The committee established consisted of three members from each side.[107] It met during 27–29 March 1950, in Karachi. Pakistan's delegation, led by Chaudhri Muhammad Ali, proposed that existing uses be met from the available sources and that new supplies be provided from flood waters by constructing dams on the Beas, Ravi, Sutlej, and Chenab Rivers. He proposed sharing the costs in proportion to the benefits derived, and an equitable division of the new supplies on the basis of relevant considerations.

Khosla, who was leading the Indian delegation, disagreed with the proposal. Instead, he suggested that the entire flow of the Sutlej, on which India was building the Bhakra Dam, be reserved for India's exclusive use. The waters of the Beas, Ravi, and Chenab, he stated, would be used for catering to the existing uses of Pakistan, except that India 1) would have the use of the surplus, if any, in the Beas over and above the requirements of the Sutlej Valley Canals; and 2) would have the supplies of the CBD Canals for irrigation in the Amritsar and Gurdaspur districts. He was of the view that supplies for the CBD Canals and deficiencies in the Sutlej Valley Canals could be met from the construction of a link canal on the Chenab, the entire flow of which would be available for Pakistan. If surplus from the Chenab was inadequate to meet these requirements, a dam should be built on the Chenab to meet the shortfalls in addition to providing extensions for irrigation in Pakistan.

India proposed that the matter be referred to engineers and if they did not complete their assignment within six months then their mandate could be further extended. Pakistan had reservations about the proposal because, in its opinion, sufficient data was already available and the issue could be taken care of rather quickly, but accepted it lest an objection on its part delay its resolution. Following Pakistan's acceptance of the Indian proposal, the two countries then decided to refer the two plans to their engineers for consideration and take a decision on the basis of their recommendations in the next meeting of the Negotiating Committee.

However, when the Committee met in Delhi in May 1950, India went back on the plan that it had submitted earlier at the Karachi meeting. Instead, it presented a new plan according to which it proposed to not only appropriate the entire flow of the Sutlej for its exclusive use (which it had indicated in the earlier meeting) but also that of the Beas and Ravi. Additionally, it wanted 10,000 cusecs from the Chenab through a tunnel built at a site called Marhu in Jammu and Kashmir. The Pakistani delegation outrightly rejected the new Indian proposal. This created a deadlock because India refused to budge from its position.[108]

In a Note addressed to India dated 23 August 1950, Pakistan denounced the Joint Statement as a dictated one in these terms:[109]

> ... With millions of people facing the loss of their herds, the ruin of their crops and eventual starvation from lack of water, Pakistan was under compulsion to accept whatever India proposed ....
>
> The so-called Delhi Agreement, if ever it was binding on Pakistan, has since long expired. The Government of Pakistan [has] on numerous occasions given to the Government of India notice to this effect. If, however, in view of the Government of India any further action or notice is needed to render the so-called Agreement of 4 May 1948, without present effect, it will be appreciated if this communication is accepted as such action or such notice.

It also sent a Note to the UN Secretary General in which it took the position that the agreement in question which India had registered with the UN Secretariat under Article 102 of the Charter '... if ever binding upon Pakistan had ... ceased to be effective'.[110] The West Punjab and Pakistan's governments took the position that 'the withholding of water essential in an arid region to the survival of millions of its inhabitants is ... an international wrong and a peculiarly compelling use of force contrary to the obligations of membership in the United Nations. Any concession obtained by any such means cannot confer upon the offender any enforceable rights.'[111] India blatantly rejected Pakistan's contention in the matter. In a letter addressed to Pakistan's Prime Minister in September 1950, Nehru reacted to Pakistan's position by stating that he participated in the discussions and helped in the drafting of the agreement in cooperation with the representatives of Pakistan present in the meeting, and that it was reached unanimously and in a friendly atmosphere. He concluded the letter with these words:[112] 'It has been stated in your government's communication that the agreement of 4 May 1948 was accepted by Pakistan under "compulsion". This has surprised and distressed me greatly ... I cannot imagine how any question of compulsion could possibly have arisen in these circumstances. There was then no kind of threat or even suggestion about stopping the flow of the water.' India also responded to Pakistan's Note to the UN Secretary General by rejecting it.[113]

This brings us to the question whether Pakistan really signed the Joint Statement under duress as it contended and, if so, whether it was justified to denounce it under international law. The question of coercion is a statement of fact which can be determined by looking at the circumstances under which Pakistan signed the Statement. However, we first need to focus on the principle of international law that governs the issue. Article 52 of the Vienna Convention on the Law of Treaties 1969 covers it in the following terms: 'A treaty is void if its conclusion has been procured by the threat or use of force in violation of the principles of international law embodied in the Charter of the United Nations.' Before undertaking a content analysis of this Article, it may not be out of place to point out here that before 1919, treaties concluded under duress were considered absolutely valid and binding. Things started to change following the establishment of the League of Nations whose covenant prohibited the use of force as a means of settling international disputes and the conclusion of the Kellogg Briand Pact (or the Pact of Paris) in 1928. This pact forbade the recourse from using force for solutions of international controversies and prescribed peaceful means for their settlement (the Pact was titled International Treaty for the Renunciation of War as an Instrument of National Policy). By the time the United Nations was established in 1945, the new principle was firmly enshrined in the corpus of international law. Given the fact that the issue of validity or otherwise, of the statement in question, arose in 1948—it is this principle that regulated it. In other words, if it is proved that Pakistan was made to sign it under coercion then it would be deemed null and void.

An examination of Article 52 rests on the meaning and scope of the expression 'threat or use of force'. Does this refer to 'armed force' only or does it extend to 'economic or political pressure' as well? To answer this question, we need to look back at the *travaux préparatoires* of the conference which codified the convention and where the issue of coercion was debated. The conference was divided on the issue between the Western countries and the 'new states' of Africa, Asia, and Latin America. The Western countries favoured the narrow interpretation embracing it exclusively in terms of 'armed force', whereas the 'new states' supported the wider interpretation embracing 'economic or political pressure' as well. The latter group of states submitted an amendment for the incorporation of their viewpoint to the International Law Commission's draft of the convention which enshrined the position of the Western countries on this issue and which was the basis of discussion. After a lengthy and at times acrimonious debate between the delegates, the conference decided to accept the new states' amendment but with a difference. It resolved to incorporate it as an annexure to the convention under the title Declaration on the Prohibition of the Threat or Use of Economic or Political Coercion in Concluding a Treaty rather than in the body of Article 52 itself. It is also noteworthy that the Declaration condemns the threat or use of pressure in any form by a state against another but does not declare a treaty void if concluded through the threat or use of political or economic pressure.[114]

The resulting text of Article 52 is indeed unsatisfactory when viewed from the perspective of the ground realities of international relations which tell us that strong countries frequently use coercive measures (short of military force) such as political or economic pressure or both to achieve their foreign policy objectives. The Mexican judge, Padilla Nervo, admirably described this stark reality in his dissenting opinion in the *Fisheries Jurisdiction* case which deserves to be reproduced verbatim:[115]

> A big power can use force and pressure against a small nation in many ways, even by the very fact of diplomatically insisting on having its views recognized and accepted. The Royal Navy did not need to use armed force, its mere presence on the seas inside the fishery limits of the coastal State could be enough pressure. It is well known by professors, jurists and diplomats acquainted with international relations and foreign policies, that certain 'Notes' delivered by the government of a strong power to the government of a small nation, may have the same purpose and the same effect as the use of threat of force.
>
> There are moral and political pressures which cannot be proved by the so-called documentary evidence, but which are in fact indisputably real and which have, in history, given rise to treaties and conventions claimed to be freely concluded and subjected to the principle of *pacta sunt servanda.*

It is true that India did not resort to the threat or use military force to get Pakistan to sign the Joint Statement[116] but it is equally undeniable that it used political and economic coercion to achieve its objectives as we propose to demonstrate.

Nothing can be more devastating than depriving an individual or a nation of water because it is possible to survive in the face of other deprivations, but not without this commodity. This is what India did on 1 April 1948 when it cut-off the water supplies to canals that crossed from India into Pakistan. This was nothing short of water aggression or, in today's parlance, 'water terrorism'. It ravaged an estimated 1.66 million acres of land. Commenting on the situation, one Pakistani observer noted the impact of the Indian action as follows:[117] 'There was acute distress which, with every day that passed, became more and more intolerable. In large areas where the subsoil is brackish there was no drinking water. Millions of people faced the ruin of their crops, the loss of their herds, and eventual starvation due to lack of water.' David Lilienthal, the head of the Tennessee Valley Authority (TVA) and the catalyst for Pakistan-India negotiations on the water dispute under the World Bank's auspices, made a similar observation when he said, 'In the spring of 1948 ... India cut-off most of the supply of water to Pakistan for a month causing distress, loss of crops, and general disruption.'[118]

Incidentally, India also indirectly concurred with this assessment though it tried to downplay it. In one communication to Pakistan, India admitted the 'hardship and distress occasioned by stoppage of water'.[119] Similarly, in another communication, it hinted at 'the hardship involved, if water supplies were stopped for any length of time'.[120] When the Pakistan government protested against India's unjustified actions, the latter refused to entertain Pakistan's request 'unless West Punjab acknowledged that it had no right to the water'.

India did not stop at the economic strangulation of (West) Pakistan by cutting off the water supplies. It also resorted to exerting political pressure by arm-twisting the Pakistan delegation in Delhi to sign the text of a statement 'without changing a word or a comma' in return for restoration of the water supplies. Pakistan did the Indian bidding because it had no other option and thus the Joint Statement was born whereby Pakistan was made to surrender its rights to the waters of the Eastern Rivers.

Pakistan signed the Joint Statement under economic and political coercion and not out of its free will. It may be pointed out here that the World Bank also shared this view as evidenced by the following observation that William Iliff, the vice-president of the Bank, made in a confidential note that he penned for Eugene Black after his meeting with Nehru:[121]

I mentioned the attitude which Pakistan took about the 1948 Agreement, one of her points being that it had been entered into by Pakistan 'under duress'. I reminded the [Indian] PM that he had once asked me if I could imagine a little man like himself being 'duressed' by two giants like Ghulam Mohammed (sic) and Zafrulla Khan. I said I was quite prepared to agree; but I said, this was not the sort of duress that Pakistan had in mind. She had, however, been up against the situation where the EP [East Punjab] had actually cut-off the supplies of water to the C.B.D.C. channels at a critical crop period; the Agreement of 1948 had had to be negotiated against that background, and Pakistan's vital objective had to be that supplies of water to these channels should be assured. *To that extent Pakistan had a point that there was 'duress'.*

India also indirectly agreed that the Joint Statement was not obtained through free and fair means. Hence, F. S. Nariman, a senior member of the Indian delegation in the *Baglihar* case, accepted it when in the course of the proceedings he made the following statement:[122] 'It was insisted on these safeguards [i.e. detailed design criteria for India's hydropower projects on the Western Rivers] because it [Pakistan] did not trust India. Entirely understandable, given the history of conflict and violence between the two countries, *and India's decision in 1948 to abuse—mark my words abuse—its control over the shared water resources.'*

This brings us to the timing of the denunciation of the Joint Statement, which though academic is an important one. Pakistan signed the Joint Statement in May 1948, whereas it denounced it for the first time in August 1950 when, in a Note addressed to India, it declared that, '... with millions of people facing the loss of herds, the ruin of crops, and eventual starvation from lack of water, Pakistan was under compulsion to accept whatever India proposed .... The so-called Delhi Agreement, if ever it was binding upon Pakistan, has since long expired.'

If Pakistan considered the Joint Statement to be dictated, why did it take so long to denounce it? India raised the same query in its Note dated 15 September 1950. They asked Pakistan to explain why it did not denounce the Joint Statement in the joint communiqué issued on 7 May 1948 which related to the same subject, nor in the Note submitted three weeks later by the West Punjab government to the Punjab Partition Committee, nor still in the Inter-Dominion conference held in Lahore three months later. It also sought explanation for more than two years' delay in raising the issue of compulsion.[123] The answer to these questions is to be found in Pakistan's fear of India stopping the supply of water to its canals again as happened in April 1948. In any case, in a Note dated 18 September 1951 (repeated on 13 March 1953) addressed to Pakistan, India offered to refer the question of the Joint Statement's continuing validity to an impartial tribunal for adjudication.[124] Pakistan rejected the Indian offer and instead proposed the submission of the entire water dispute to adjudication. India felt confident of its position on the question of validity of the Joint Statement whereas Pakistan felt vulnerable on the issue.

The two countries continued bilateral negotiations to resolve the water issue but failed to make any progress. Frustrated by the unending stalemate, Pakistan again invited India to submit the dispute to the World Court for adjudication. India, on its part, rejected Pakistan's overtures again. However, in September 1950, Nehru replied positively to Pakistan's offer, but with a difference. To begin with, he sought an investigation of the facts. He also linked the issue with the question of evacuee property that arose from the transfer of population from each country as a result of Partition. He then made a counter offer to refer the issue to a court composed of two Pakistani and two Indian judges. Pakistan showed willingness to accept Nehru's offer on the condition that the court was headed by an impartial judge. Alternatively, it proposed to refer the issues on which the court was deadlocked to another court headed by an impartial judge. This was indeed a fair proposal because without a neutral head, there was no possibility of breaking the deadlock that would have obviously arisen between the partisan judges of the two sides. Nehru rejected the proposal on the grounds that[125] 'to think *ab initio* of a third party will lessen the sense of responsibility of the [Indian and Pakistani] judges and will also be ... a confession of our continued dependence on others. That would be hardly becoming for proud and self-respecting independent nations.' He insisted that as proposed by him, the court would render a resolution of conflicts easier and more expeditiously than 'a tribunal sitting thousands of miles away'. Continuing this argument, he stated: 'You seem to think ... only outsiders can decide for us. I confess that I am unable to appreciate the force of this argument, which ... reduces us to a dependent status relying upon the pleasure of others; this is wholly repugnant to me and, in my view, incompatible with the dignity of both India and Pakistan.'

Responding to Nehru, the Pakistani Prime Minister contended that the World Court '... is our Court. India and Pakistan by accepting the statute of the Court and agreeing to the jurisdiction, far from impairing their sovereignty, [would exercise] it in aligning themselves with those nations that have freely chosen to live under the rule of law.'[126] He maintained that a court of undoubted standing, 'sitting thousands of miles away' would adjudicate 'the question dispassionately and without getting entangled into the barbed wire of political controversy'.[127] However, his advocacy in favour of the World Court was not amenable to Nehru. While Nehru was utterly opposed to a third party's role in disputes involving Pakistan and India, he was very keen to submit India's disputes with China, particularly the Sino-Indian border dispute, to a neutral forum. For example, on 10 December 1962, he stated in the Indian Parliament:[128] 'I am prepared ... even to refer the basic dispute of the claims on the frontier to a body like the International Court of Justice at the Hague.' Similarly, he repeated the same proposal again in these words:[129] 'We would be perfectly prepared to refer the matter [of differences between India and China] to the International Court of Justice or to arbitration if it was agreed to.' Again, he reiterated this proposal on 5 March 1963, by stating that:[130] 'We have suggested that we are prepared to refer the frontier disputes to the International Court of Justice at the Hague or to arbitration.'

Before concluding this chapter, we propose to deal with the philosophical peg on which India hung its rejection of Pakistan's proposal for referral of the water dispute to the World Court. Nehru justified it on the grounds that it was not a justiciable issue.[131] What did he mean by this objection and was it well-founded?

Disputes are divided into two categories, namely justiciable or legal and non-justiciable or political. Political or non-justiciable disputes are not disposed to being settled through generally recognized rules of international law for reasons of material insufficiency and inadequate development of the law of nations. The litmus test to determine which disputes are justiciable and which are non-justiciable is not the limitation of judicial function based on the applicability of legal rules, but rather the relative importance of the subject matter of the controversy. Concretely put, those disputes are political or non-justiciable and affect the vital interests of a state. Proceeding from this distinction, states often contend that only disputes which are of minor importance are open to obligatory judicial settlement and those which touch their vital interests are excluded.[132] It is also contended that the UN Charter and the Statute of the World Court also make a distinction between legal and political disputes.[133]

While the issue is being debated, we note that each dispute has a political and a legal aspect to it. It is therefore open to being settled both politically or legally. With this perspective, it is not fair to contend that the political significance of the subject can be an obstacle to its being settled by legal means. It is the refusal of a state to submit a dispute to judicial settlement rather than its intrinsic nature which makes it political. It is true that the UN Charter and the Statute of the World Court maintain the distinction between legal and political disputes. However, the latter has, for all practical purposes, refused to uphold this distinction as attested by its jurisprudence.

For example, the World Court, in the *Certain Expenses of the United Nations* case, rejected the argument that it should refuse to give an opinion because the question was intertwined with political considerations on the grounds that '... most interpretations of the Charter will have political significance, great or small. In the nature of things it could not be otherwise. The Court, however, cannot attribute a political character to a request which invites it to undertake an essentially judicial task, namely the interpretation of a treaty provision.'[134] Similarly, in the case concerning *Legality of the Use by a State of Nuclear Weapons in Armed Conflict*, the World Court rejected the plea that the question posed was essentially a political one:[135]

The fact that this question also has political aspects, as, in the nature of things, is the case with so many questions which arise in international life, does not suffice to deprive it of its character as a 'legal question' and to 'deprive the Court of a competence expressly conferred on it by its Statute' (*Application for Review of Judgment No. 158 of the United Nations Administrative Tribunal Advisory Opinion. I.C.J. Reports* 1973, p. 172, para. 14). Whatever its political aspects, the Court cannot refuse to admit the legal character of a question which invites it to discharge an essentially judicial task, namely, an assessment of the legality of the possible conduct of States with regard to the obligations imposed upon them by international law .... The Court also finds that the political nature of the motives which may be said to have inspired the request and the political implications that the opinion given might have are of no relevance in the establishment of its jurisdiction to give such an opinion.

The foregoing discussion shows that the distinction between justiciable and non-justiciable or legal and political disputes is for all practical purposes of no consequence. Therefore, Nehru's refusal to submit the water dispute to the World Court, or to any other adjudicatory body for that matter, on the grounds that it was non-justiciable was meant to avoid a third party's involvement. We can assume that he invoked this argument because he feared that the Court would render a judgment adverse to the Indian cause.

Pakistan and India continued to discuss the water issue but without any breakthrough and without the possibility of a resolution. By 1951, the parties had virtually reached a cul-de-sac in their relations. It was at this stage that the World Bank emerged on the scene as a kind of *deus ex machina* to mediate in the dispute between the two countries, which they accepted. Consequently, a new phase in the history of conflict resolution of the water issue began. This is what we propose to discuss in the next chapter.

REFERENCES AND NOTES

1. Sharifuddin Pirzada, 'Radcliffe Award', in Mian Muhammad Sadullah, Sharif-al-Mujahid et al. (eds.), *The Partition of the Punjab 1947: A Compilation of Official Documents,* vol. I. (Lahore: National Documentation Centre, 1983), xiii.

2. Ibid.

3. Muhammad Munir, 'Days to Remember', *The Pakistan Times* (22 June 1980), cited in ibid. xviii.

4. *The Civil and Military Gazette* (25 April 1958), cited in ibid. xvii.

5. The Punjab Boundary Commission was composed of Justices Muhammad Munir and Din Muhammad representing the Muslim League; and Justices Mehr Chand Mahajan and Teja Singh representing the Congress party. The Bengal Boundary Commission was composed of Justices S. M. Akram and S. A. Rehman representing the Muslim League and Justices C. C. Biswas and O. K. Mukerjee representing the Congress party.

6. Mountbatten explained to Radcliffe that in theory a panel of four judges would assist him by submitting a joint recommendation on the boundary between the two dominions. However, skeptical of the fact that the judges, who represented hostile parties with mutually divergent interests, would ever reach an agreement, he told the latter in no uncertain terms that he alone would have to take responsibility for making all the decisions. Larry Collins and Dominique Lapierre, *Freedom at Midnight* (Pan Books, 1977), 218.

7. 'Award of Sir Cyril Radcliffe', in Mian Muhammad Sadullah, Sharif-al-Mujahid et al. (eds.), *Partition of the Punjab 1947,* vol. III, supra note 1, 284.

8. Ibid.

9. Lord Mountbatten was quizzed as to the meaning of the expression 'other factors' in his press conference of 4 June 1947. He was vague in his response. It has been said that he was vague because otherwise it would 'have led Nehru and Jinnah into another of their unending arguments'. It has also been said that, convinced that the relations between the two dominions would remain friendly, the commander-in-chief of the Indian army, Field Marshall Sir Claude Auchinleck, authorized Radcliffe 'to ignore the elements that were usually the first concern of a nation in setting its frontiers, defense considerations' (Larry Collins and Dominique Lapierre, *Freedom at Midnight,* supra note 6, 219). According to Andrew Roberts, Lord Listowel, the secretary of state for India, in a letter addressed to Mountbatten stated that 'other factors' could be 'assumed to have had security, as well as geographic, religious and economic aspects'. See 'Lord Mountbatten and the Perils of Adrenalin', in Andrew Roberts, *Eminent Churchillians* (London: Weidenfeld & Nicholson, 1994), 92.

10. The Muslim League apparently accepted the 'other factors' formulation in the hope of getting Calcutta without ever suspecting that the British and the

Congress leadership would reach a secret understanding to award the famed city to India.

11. *Partition of the Punjab 1947*, vol. III, supra note 1, 161. Emphasis added.

12. Aloys Arthur Michel, *The Indus Rivers: A Study of the Effects of the Partition* (New Haven and London: Yale University Press, 1967), 188–9.

13. Ramiz Ahmed Malik, *Teen Darya Kaisay Khoe, Sutlej, Beas aur Ravi: Muhaida-e-Sindh Taas Ke Andruni Kahani* (in Urdu), (Lahore: Takhleqat, 2008), 45.

14. Michel, *Indus Rivers*, supra note 12, 189.

15. Jagat Mehta, 'The Indus Waters Treaty: A Case Study in the Resolution of International River Basin Conflict', *Natural Resources Forum*, vol. 12, no. 1 (1988), 71.

16. Michel, *Indus Rivers*, supra note 12, 193. Emphasis original.

17. It is noteworthy that as far as Punjab was concerned, Radcliffe used 'other factors' to award no less than nine Muslim majority tehsils or parts of tehsils to India—namely Gurdaspur, Batala, Ajnala, Nakodar, Jullundur, Zira, Ferozpur, Shakargarh (part), and Kasur (part)—but did not use the same factor to award even a single tehsil or part of it with non-Muslim majority to Pakistan. He, however, used 'other factors' in Bengal to award Chittagong Hill Tracts which was a non-Muslim majority area to Pakistan. This was possibly done to compensate Pakistan for the loss of Calcutta which he awarded to India. It is important to remember that even though the Chittagong Hill Tracts was a non-Muslim majority area, it was not Hindu in character. He apparently did this to convey the impression that he was being even-handed which seems doubtful for the simple reason that whereas Gurdaspur and Ferozpur had enormous strategic value, the Chittagong Hill Tracts had none whatsoever.

18. 'Award of Radcliffe', supra note 7, 284.

19. Leonard Mosley, *The Last Days of the British Raj* (London: Weidenfeld and Nicholson, 1961), 195.

20. *Amrita Bazar Patrika* (Calcutta, 18 August 1947), cited in Mohammad Ayub Khan, *Friends not Masters: A Political Autobiography* (Lahore, Karachi, Dacca: Oxford University Press, 1967), 115.

21. M. P. File no. 62, cited in Latif Ahmed Sherwani, *The Partition of India and Mountbatten* (Karachi: Council for Pakistan Studies, 1986), 94–5; see also H. V. Hodson, *The Great Divide* (Karachi: Oxford University Press, 1985), 218.

22. M. P. File no. 207, cited in Sherwani, *Partition of India*, supra note 21, 15.

23. Andrew Roberts, 'Lord Mountbatten', supra note 9, 82.

24. Ibid.

25. Ibid.

26. Ibid.

27. M. P. File no 192, cited in Sherwani, *Partition of India*, supra note 21, 21; see also Hodson, *Great Divide*, supra note 21, 523.

28. M. P. File no. 62, cited in Sherwani, *Partition of India*, supra note 21, 94–5; see also Hodson, *Great Divide*, supra note 21, 218.

29. Mountbatten explained the background of this statement by observing that he felt that 'Pakistan was relatively weak in numbers, in military force, in economic strength, in administrative experience; if its future depended upon its bargaining muscle in face of the much more powerful Indian government, in a settlement of disputed issues of partition, then that future was indeed in jeopardy.' He then went on to claim that he, as the Governor General of India, was responsible for a settlement of the vitally important question of Rs. 55 crores of cash assets and the canal waters in favour of Pakistan; see also Muhammad Yusuf Saraf, *Kashmiris Fight for Freedom*, vol. II (Lahore: Ferozsons Ltd., 1979), 1393. This explanation shows that Mountbatten was the Baron Munchhausen of politics because we know that by his own admission he did what he could to stop the emergence of Pakistan as an independent state; that he was responsible for the award of the Ferozpur headworks and possibly Gurdaspur to India; that Gandhi, who made a sacrifice of his life, and not Mountbatten, was responsible for the release of the cash balances to Pakistan; and that if India restored water to Pakistan's canals, it got its pound of flesh in the shape of the May 1948 Joint Statement which adversely affected Pakistan's water rights. Notwithstanding the harm that he did to Pakistan, he had claimed that he favoured Pakistan regarding cash assets and the water issue. The former British prime minister, Anthony Eden, described him as a 'congenital liar' and Sir Gerald Templar famously remarked: 'You're so crooked, Dickie, if you swallowed a nail you'd shit a corkscrew.' Andrew Roberts, 'Lord Mountbatten', supra note 9, 133.

30. See the Golden Jubilee issue of *Time* magazine, Asia Edition (11 August 1997), 32; see also Larry Collins and Dominique Lapierre, *Freedom at Midnight*, supra note 6, 131–2, footnote. It is indeed a measure of the intensity of Mountbatten's animus towards Jinnah and Pakistan that even after a lapse of more than half a century following the emergence of Pakistan as an independent state (during which instead of mellowing down, which is what normally happens to individuals, and be contrite for the harm that he did to the new state), he unabashedly remarked that had he known Jinnah's illness, he would have delayed partition in order to abort it.

31. Alan Campbell-Johnson, *Mission with Mountbatten* (London: Robert Hale, 1953), 124.

32. Sharifuddin Pirzada, 'Radcliffe Award', supra note 1, ix.

33. Philip Zeigler, *Mountbatten: The Official Biography* (London: Book Club Associates, 1985), 397.

34. Stanley Wolpert, *Jinnah of Pakistan* (Rawalpindi: RAF Book Club, 1995), 332.

35. Zeigler, *Mountbatten*, supra note 33, 402.

36. Ibid. 421.

37. Sir Zafrulla Khan recounts in his memoirs that when he heard of the appointment of Radcliffe as chairman of the Boundary Commission, he was

very upset because, being a practicing lawyer and a member of the British parliament, he was susceptible to pressure. He reveals that previously the British government had proposed his name as president of the Arbitration Tribunal, set up to deal with the Partition matters but on second thought rejected it in favour of Sir Patrick Spens. In his opinion, given the circumstances of Partition and the extensive judicial powers which were conferred on the chairman, it was inappropriate on the part of Prime Minister Attlee to appoint Radcliffe, especially when he was not considered fit for appointment on an equally important judicial position. Sir Muhammad Zafrulla Khan, *Tehdis-e-Naimat*, 500, <www.slideshare.net/muzaffertahir9/tehdis-enaimat>.

38. The British boast that the tradition of 'fair play' is quintessentially British. This is understandable because at one time, while ruling a large part of the globe, they tended to present everything British—be it literature, diplomacy, architecture, system of government, judicial process, or anything else—as the best and exemplary in the world. Thus the term 'fair play' came to be regarded as being synonymous with Britain. On closer scrutiny we discover that it is a false notion and the world at large tends to disagree with them. For example, writing at the end of the nineteenth century, this is what the *New York Times* had to say in the matter: '"British fair play" has been celebrated extensively but it is worthy of note that it has been celebrated exclusively by Britons. Foreigners are so far from being impressed with the British quality for which the natives so much admire themselves that upon the continent of Europe Albion has for some generations been known as perfidious.' *The New York Times* (5 March 1890).

39. Chaudhri Muhammad Ali, *The Emergence of Pakistan* (Lahore: Research Society of Pakistan, 1973), 217–19.

40. Andrew Roberts, 'Lord Mountbatten', supra note 9, 94.

41. Ibid.

42. Ibid.

43. Zeiglar, supra note 33, 419–20.

44. Kanwar Sain, *Reminiscences of an Engineer* (New Delhi: Young Asia Publications, 1978), 117. It is noteworthy that it was A. N. Khosla, chairman of the Central Waterways, Irrigation and Navigation Commission (now the Central Water and Power Commission) who informed Nehru of the award, among others, of Ferozpur to Pakistan. In the letter that he sent to Nehru, he argued that '[b]oth from the strategic and irrigation point of view it will be most dangerous to let Ferozpur go to Pakistan. Whatever may be the decision about area west of Sutlej, no area east of Sutlej must on any account go to Pakistan.' Nehru in turn sent that letter to Mountbatten for transmission to Radcliffe. Nicholas Mansergh and Penderel Moon (eds.), *The Transfer of Power, The Mountbatten Viceroyalty: Princes, Partition and Independence* (London: Her Majesty's Stationery Office, 1983), vol. XII, 618–20.

45. Kanwar Sain, *Reminiscences*, supra note 44, 121–2.

46. Ibid.

47. Ibid.

48. Andrew Roberts, 'Lord Mountbatten', supra note 9, 101. Commenting on Mountbatten's apologists' attitude, Roberts has the following observation to make: 'As has already been seen, however, in matters concerning the Boundary Awards he purposefully did not keep records of conversations which he considered "off the record"' (ibid.). It is interesting to note that, as indicated in the main body of this chapter, Radcliffe also destroyed his papers and notes relating to the boundary awards. The two characters who were privy to the boundary awards did not keep their papers and notes but decided to destroy them which makes their actions seem questionable and suspicious.

49. See for example, Robert G. Wirsing, *India, Pakistan and the Kashmir Dispute* (New York: St. Martin's Press, 1994), 20. Contrary to writers like Wirsing who are more loyal than the king, there were others who admitted that Radcliffe was guilty of the wrong he was accused of. For example, Beaumont claims that following his return to the United Kingdom, he visited Radcliffe in his chambers to learn the truth about the change of the Punjab map. He states that, 'he [Radcliffe] was very sheepish and never denied it. He didn't welcome my visit, said he was very busy and shuffled me off' (Andrew Roberts, 'Lord Mountbatten', supra note 9, 97). Even more damning evidence in the matter emerged from Philip Noel-Baker, the British secretary of state for commonwealth relations, who had been asked by Prime Minister Attlee to investigate the allegation that Sir Zafrulla Khan had made regarding the change of the Punjab map. According to the report that he submitted to Attlee, Radcliffe admitted that, 'he showed the first draft of the proposed Award to the authorities in Delhi and that, on further consideration, he made the Award in terms which departed from the first draft'. Andrew Roberts, 'Lord Mountbatten', supra note 9, 97.

50. Zeigler, *Mountbatten*, supra note 33, 421.

51. Ijaz Hussain, *Kashmir Dispute: An International Law Perspective* (Islamabad: Quaid-i-Azam Chair, National Institute of Pakistan Studies, Quaid-i-Azam University, 1998), Appendix XII, 271–4. It is noteworthy that Radcliffe had recorded the testimony in 1989 and he wanted it to see the light of the day after his death. However, more than forty years after the Radcliffe award, he was so appalled by the pro-Mountbatten line being fed to his grandson who had taken up the 'transfer of power in India' as a special subject in part II of Tripos in history at Cambridge University that he decided, after getting permission from the British Foreign Office, to make the testimony public prematurely. Andrew Roberts, 'Lord Mountbatten', supra note 9, 93; see also *The Telegraph* (16 May 2002).

52. Ijaz Hussain, *Kashmir Dispute*, supra note 51, 271–4.

53. Recently, a British theatre company presented a play entitled *Drawing the Line* on the Radcliffe award which was a huge success on the London stage. Whereas

it shows Jinnah in a sympathetic light, it presents Nehru as a manipulative character. According to the storyline, the latter uses Edwina Mountbatten to get the Viceroy to bully and browbeat Radcliffe into gerrymandering the Punjab map to award Ferozpur to India. Radcliffe, on the other hand, emerges as a hapless and sympathetic character in this drama of high passions and calculated cunning. His moral dilemma reaches a crisis stage when Mountbatten confronts him over Ferozpur and Calcutta both of which were to go to Pakistan but did not. Irfan Hussain, 'Edwina Mountbatten, Nehru and Radcliffe', *Dawn.com*, (20 January 2014).

54. Sir Zafrulla Khan narrates three incidents in his memoirs that reflect poorly on Nehru because of his Machiavellian character. According to the first incident, during the mid-1950s, the Indian ambassador to the Netherlands hosted a reception in honour of the Indian prime minister, Jawaharlal Nehru, who was transiting through The Hague to India after attending the Commonwealth Prime Ministers' Conference. He also invited, among others, the judges of the World Court to the function. Sir Zafrulla, who was serving as a judge at the World Court at that time, could not attend it stating a prior engagement. He recalls that President Klaestad from Norway also declined the invitation on the grounds that 'it was given in honour of a man who had no respect for his own word' (Sir Zafrulla Khan, *Tehdis-e-Naimat*, supra note 37, 622). He narrates another incident in which Joseph·Korbel, President of the UNCIP— set up by the UN Security Council to get the UNCIP Resolutions on Kashmir implemented—after leaving office told the latter that he and his colleagues in the Commission always wondered why he (Zafrulla) and the Pakistan prime minister were reluctant to accept a verbal promise by Nehru, who was a highly respected personality of international standing, and always insisted on a written commitment by him. He confided in Zafrulla that after interacting with him, all the members of the Commission were unanimously of the opinion that '[a]ll your suspicions of Pundit Nehru were more than fully justified' (Ibid. 544). Finally, Sir Zafrulla recounts that Sir Owen Dixon, the UN Representative on Kashmir, made a proposal to Nehru and the Pakistan prime minister for the purpose of holding a plebiscite in Jammu and Kashmir. He told them that he would elaborate it if they would be prepared to discuss it with him and with each other, to which they gave their consent. He revealed that later on Nehru backed out of his commitment, arguing that no proposal was ever made to him and that he never discussed it with the latter (Sir Owen). He states that when he told Jha Shankar Bajpai that he could understand if Nehru rejected the proposal on the grounds that it was not in India's national interest but certainly not if he contended that no proposal was ever made to him and that they never discussed it, the latter politely said: 'Sir Owen I conceive the Prime Minister must have suffered an attack of temporary amnesia.' According to Sir

Zafrulla, Sir Owen reacted to it in his presence by making comments on Nehru which he did not deem appropriate to reproduce. Ibid. 546–7.

55. On 12 January 1949, or a week after the adoption of the UNCIP Resolution in favour of plebiscite in Kashmir, Nehru shared his secret thoughts with Sheikh Abdullah in these words: 'You know well that this business of plebiscite is still far away and there is a possibility of the plebiscite not taking place at all (I would suggest however that this should not be said in public, as our bona fides will then be challenged)' (S. Gopal (ed.), *Selected Works of Jawaharlal Nehru*, Nehru Memorial Fund (Oxford University Press), second series, vol. 9, 198, cited in A. G. Noorani, 'How and Why Nehru and Abdullah Fell Out', *Economic and Political Weekly*, vol. XXXIV, no. 5 (30 January 1999), 270; see also A. G. Noorani, *Article 370: A Constitutional History of Jammu and Kashmir* (Karachi: Oxford University Press, 2011), 198–205). Despite his hypocrisy on the question of a plebiscite in Kashmir, Nehru pretended to be a moral crusader. President Kennedy, a one-time admirer of Nehru, upon hearing of India's invasion of Goa wryly commented: 'The priest has been caught in the brothel.' See also Sir Zafrulla's *Tehdis-e-Naimat* cited in Muhammad Yusuf Saraf, *Kashmiris Fight for Freedom*, supra note 29, 1094.

56. *Quaid-i-Azam Muhammad Ali Jinnah: Speeches as Governor General* (Karachi: Pakistan Publications, 1963), 32–3.

57. B. S. Murty, 'Settlement of Disputes' in Max Sorenson (ed.), *Manual of International Law* (London, Melbourne, Toronto: Macmillan; New York: St. Martin's Press, 1968), 692–3.

58. Ibid. 693–6.

59. Ibid. 695–6.

60. *The Hindu* (16 January 1950), cited by Sharifuddin Pirzada, 'Radcliffe Award', supra note 1, XXXIII; see also Muhammad Ali, *Emergence of Pakistan*, supra note 39, 216; and Andrew Roberts, 'Lord Mountbatten', supra note 9, 105.

61. Muhammad Ali, *Emergence of Pakistan*, supra note 39, 216.

62. Sir Zafrulla Khan, *Tehdis-e-Naimat*, supra note 37, 509–14.

63. Alastair Lamb, *Kashmir: A Disputed Legacy 1846–1990* (Lahore, Karachi, Islamabad: Oxford University Press, 1992), chapters 3, 4, and 6.

64. Ibid. 74. It is noteworthy that the Indian writer H. L. Saxena puts forward a totally opposite narrative. According to him, Mountbatten deliberately aborted the restoration of the Gilgit Lease to the Maharaja and ensured its transfer to Pakistan so that the Anglo-Americans could have a base there. H. L. Saxena, *The Tragedy of Kashmir* (New Delhi: 1975), Preface, cited in ibid. 17.

65. Lamb, *Kashmir*, supra note 63, 106.

66. N. Mansergh and P. Moon (eds.), *The Transfer of Power, The Mountbatten Viceroyalty: Formulation of a Plan, 22 March–30 May 1947*, vol. X (London: Her Majesty's Stationary Office, 1981), no. 194, 365.

67. Muhammad Ali, *Emergence of Pakistan*, supra note 39, 215.

68. Ibid. 215–16.

69. Mountbatten explained his position on this point in these words:

> My mention of Gurdaspur at my press conference was meant as an example of the sort of problem that might arise, on which I had been briefed by my staff, whose business was to look up details which otherwise I would not have known. Like everyone else concerned, I was well aware that controversy over the boundary would probably be focused on two areas, apart from Calcutta, the northern Punjab, in which Gurdaspur was obviously a critical district, and the border between Assam and East Bengal. I might have quoted the latter example as readily as the former. Incidentally, the Assam award went in favour of Pakistan.

(Muhammad Yusuf Saraf, *Kashmiris Fight for Freedom*, supra note 29, 1392). Mountbatten's explanation is hardly convincing when we juxtapose it with his confession to the Nawab of Bhopal in which he stated that Kashmir was 'so placed geographically that it could join either dominion, provided part of Gurdaspur were put into East Punjab by the Boundary Commission'.

70. N. Mansergh and P. Moon (eds.), *Transfer of Power*, vol. XII, supra note 44, no. 335, 509.

71. Muhammad Munir, *From Jinnah to Zia*. (Lahore: Vanguard Books Ltd., 1979), 16.

72. Muhammad Ali, *Emergence of Pakistan*, supra note 39, 318.

73. The Arbitration Tribunal's awards in question were as follows: Reference no. 1 (valuation of canals); no. 2 (valuation of crown wastelands brought under irrigation); no. 3 (valuation of irrigated forests); and no. 5 (apportionment of income from seigniorage levied by Punjab on non-riparian states on account of irrigation water supplied from the Sutlej).

74. Though the two countries signed two Standstill Agreements, we are referring to them as if they were one for the sake of simplicity.

75. Jean Fairley, *The Lion River: The Indus* (London: Penguin Books, 1975), XIV.

76. N. D. Gulhati, *Indus Waters Treaty: An Exercise in Mediation* (Bombay, Calcutta, Delhi, etc.: Allied Publishers, 1973), 63–4 and 66, footnote.

77. Muhammad Ali, *Emergence of Pakistan*, supra note 39, 319.

78. Bashir A. Malik, *Indus Waters Treaty in Retrospect* (Lahore: Brite Books, 2005), 91. The conversation took place in Punjabi which, according to Malik on the authority of Mr Mahbub, took place as follows. Shaukat Hayat: *O, Sardara tusi sada paani taan nehi bund kar deo gey?* Swaran Singh: *Bhai jee kee kendeo kadee koee bharawaan da paani wee band karda aeh!*

79. Muhammad Ali, *Emergence of Pakistan*, supra note 39, 320. Emphasis added.

80. Ibid.

81. See chapter 3, 'World Bank: The Honest Broker', under E 'Analysis of Pakistan's Claim to the Eastern Rivers'.

82. 'Draft Articles on the Law of Treaties with Commentaries', *Yearbook of International Law Commission* (1966), vol. II, 189, para. 7.

83. Oppenheim is also of the opinion that, though the Vienna Convention envisages treaties in written form, 'this does not affect the legal force of oral agreements.' Sir Robert Jennings and Sir Arthur Watts (eds.), *Oppenheim's International Law: Peace*, vol. 1, parts 2 to 4 (Longman, 1992), 1201, para. 6.

84. *PCIJ*, Series A/B (1933), no. 53, 71. It is noteworthy that in the *United States v. Gonzalez* case, the court held a conversation by telephone with another government to constitute an 'arrangement'. Sir Robert Jennings and Sir Arthur Watts, *Oppenheim's International Law*, supra note 83, 1201, note 13.

85. Gulhati claims that seventeen months after the incident, in the course of a discussion in which he (Gulhati) was present, Nehru rebuked the East Punjab government 'for having taken the law into their own hands'. Gulhati, *Indus Waters Treaty*, supra note 76, 64, footnote.

86. Sheikh Abdullah, the Kashmiri leader who knew Nehru intimately, described his politics as follows: 'His [Nehru's] idealism was tainted with politicking and conjurations of Machiavelli. He was an admirer of the famous ancient political thinker Chanakya and kept the *Arthashastra* by his bedside. Nehru was influenced by Machiavelli's doctrines and practiced them with us in Kashmir.' See the autobiography of Sheikh Abdullah (translated from Urdu by Khushwant Singh) entitled *The Flames of Chinar* (Viking Penguin India, 1993), 74–5; see also the autobiography in Urdu, *Aatish-e-Chinar* (Lahore: Chaudhry Academy, Urdu Bazar, undated), 351.

87. Jean Fairley, *Lion River*, supra note 75, 212.

88. Aaron T. Wolf and Joshua T. Newton, 'Case Study of Transboundary Dispute Resolution: Indus Water Treaty', 3, <www.transboundarwaters.orst.edu/research/case_studies/Indus-New.htm>.

89. Shaukat Hayat Khan, *The Nation that Lost its Soul: Memoirs* (Lahore: Jang Publishers, 1995), 203.

90. Firoz Khan Noon, *From Memory* (National Book Foundation, 1993), 278.

91. Muhammad Ali, *Emergence of Pakistan*, supra note 39, 319.

92. On the insistence of the Muslim League, rulers of the Princely States were invested with the right to decide the future of their state. In contrast, the Congress party wanted the people to enjoy this right.

93. Muhammad Ali, *Emergence of Pakistan*, supra note 39, 320.

94. The World Court exercises its jurisdiction vis-à-vis states on a voluntary basis and thus cannot force a country to submit a case before it against its wishes. As far as India is concerned, the British Indian government made a declaration on 19 November 1929, accepting the compulsory jurisdiction of the Permanent Court of International Justice, the predecessor of the International Court of Justice, which it subsequently substituted with another one on 7 March 1940. The new court set up in 1945, following the establishment of the

United Nations, accepted this declaration as valid as India had inherited the mantle of original membership of the United Nations after Partition. In other words, India was deemed to have accepted the compulsory jurisdiction of the World Court because the declaration was still in force as India had not served a termination notice. An examination of the terms of the 1940 declaration reveals that India did not confer blanket jurisdiction on the World Court in all cases in which it was involved with other countries. It excluded, among others, cases involving a UN member which was also a member of the British Commonwealth. Simply put, India refused to let the Court enjoy jurisdiction over disputes involving Pakistan as it was a member of the British Commonwealth. Hence Pakistan was not in a position to take India to the World Court even if it wanted to do so. The foregoing analysis shows that the Pakistan delegation was totally ignorant about the realities of international justice. See 'Pakistani and Indian Attitudes towards the World Court' in Ijaz Hussain, *Issues in Pakistan's Foreign Policy: An International Law Perspective* (Lahore: Progressive Publishers, 1988), 204–5. If it is argued that Pakistan was merely trying to bring India under pressure by proposing to seize the World Court, then it contradicts itself because it did not believe the force of public opinion or morality to influence India as evidenced by its refusal to seize the Arbitration Tribunal which could have given a binding verdict in its favour.

95. Muhammad Ali, *Emergence of Pakistan*, supra note 39, 321.

96. Gulhati, *Indus Waters Treaty*, supra note 76, 66.

97. For the text of the May 1948 Joint Statement, see *United Nations Treaty Series*, vol. 54, 45 (Appendix 1 of the book). India called the May 1948 Joint Statement 'An Agreement' whereas Pakistan described it as a 'Document'. Iliff claims that Pakistan's foreign minister in a conversation with him disagreed with the contention of his government that it was not an agreement, and held the view that though it was not a treaty, it was 'certainly entitled to be described as an international agreement'. Memo by Iliff on a Talk with Pakistan's Foreign Minister, 3 May 1960. File no. 1787284, Title: *Indus Basin Dispute-General Negotiations-Correspondence* 09, Start Date 1/1/60.

98. Telegram no. 1681 PRIMIN dated 18 October 1948, from the Government of India to the Government of Pakistan, 'A Review of the Efforts Made to Settle the Irrigation Water Dispute between Pakistan and India', an undated document of the Government of Pakistan. File no. 1787276, Title: *Indus Basin Dispute-General Negotiations-Correspondence* 01, Start Date 4/1/49.

99. Telegram dated 8 May 1948, from the Prime Minister of India to the Prime Minister of Pakistan.

100. Sir Muhammad Zafrulla Khan, *The Agony of Pakistan*, 1974, 73.

101. 'Pakistan: The Struggle for Irrigation Water and Existence' (Washington DC: Embassy of Pakistan Publication, November 1953), 10. File no. 1787279, Title: *Indus Basin Dispute-General Negotiations Correspondence* 04, Start Date 11/1/53.

102. Telegram no. 1681, supra note 98.

103. Ibid.

104. Letter dated 24 November 1950, from the Prime Minister of India to the Prime Minister of Pakistan.

105. Letter dated 11 December 1950, from the Prime Minister of India to the Prime Minister of Pakistan.

106. Pakistan: The Struggle for Irrigation Water and Existence, supra note 101, 11.

107. Pakistani members of the Committee were Chaudhri Muhammad Ali, the Secretary General, Government of Pakistan; M. A. Majid, chief secretary, West Punjab; and Pir Ibrahim, chief engineer, irrigation, West Punjab. Whereas the Indian members were B. K Gokhale, secretary, Ministry of Works, Mines and Power, Government of India; A. N. Khosla, chairman, Central Waterways, Irrigation and Navigation Commission; and M. R. Sachdev, chief secretary, East Punjab.

108. Gulhati gives an entirely different account of the negotiations. According to him, Pakistan was prepared to concede proprietary rights to all the waters of the Ravi and the Beas which it was using before Partition. India was agreeable to the idea provided Pakistan agreed to let it divert and use waters of the Chenab through a tunnel at Marhu, and Pakistan could make up any deficiency by constructing a dam in India at Dhiangarh. He argues that the negotiations failed because Pakistan would not agree to let India construct any dam on the Chenab. Besides, he also contends that the entire exercise was an attempt to find merely an ad hoc settlement (Gulhati, *Indus Waters Treaty*, supra note 76, 81–2).

109. Ibid. 79–80.

110. *United Nations Treaty Series*, vol. 85, 356.

111. See 'Indus Water Dispute', *Encyclopaedia of Public International Law*, vol. 2 (Amsterdam, etc.: North-Holland, Elsevier, 1995), 963.

112. Letter dated 12 September 1950, from the Prime Minister of India to the Prime Minister of Pakistan.

113. *United Nations Treaty Series*, vol. 128, 300.

114. Annexure 1 covers the Declaration which forms part of the Final Act of the Convention.

115. Fisheries Jurisdiction (United Kingdom v. Iceland), Jurisdiction of the Court, Judgment, *ICJ Reports* 1973, 47; and, Fisheries Jurisdiction (Federal Republic of Germany v. Iceland), Jurisdiction of the Court, Judgment, *ICJ Reports* 1973, 91.

116. To those who may think that article 52 cannot apply to the Joint Statement because whereas it was concluded in 1948, the Convention came into force much later, and it applies retroactively and not from the date of the coming into force of the Convention because it represents *lex lata*. For details, see 'Draft Articles', supra note 82, 247, para. 8.

117. Muhammad Ali, *Emergence of Pakistan*, supra note 39, 321.

118. David E. Lilienthal, 'Another "Korea" in the Making?', *Collier's* (4 Aug. 1951).

119. Wire no. 1006 dated 21 April 1948, from Chief Secretary, East Punjab, Simla, to Chief Secretary, West Punjab, Lahore. See 'A Review of the Efforts', supra note 98.

120. Express Postal Telegram dated 26 April 1948, from Chief Secretary, East Punjab, to Chief Secretary, West Punjab, Lahore. Ibid.

121. Iliff's Conversation with Nehru, dated 9 August 1960, File no. 1787284, Title: *Indus Basin Dispute-General Negotiations-Correspondence* 09, Start Date 1/1/60. Emphasis added.

122. *Baglihar Hydro-Electric Plant* (Pakistan v. India), Transcript of Meeting no. 5, vol. 2, 8 November 2006, 285. Emphasis added.

123. Gulhati, *Indus Waters Treaty*, supra note 76, 80.

124. By virtue of the two 'Notes' dated 18 September 1951 and 13 March 1953, addressed to the Pakistan government, the Indian government offered to refer the question of the 'validity' of the May 1948 Joint Statement to an impartial tribunal. Taking advantage of this offer, the Pakistan government conveyed to the Bank's president its willingness to submit for adjudication to the World Court or any other effective and impartial tribunal, all the issues in the water dispute and to work out under the good offices of the Bank a joint declaration with India towards this end. Black did not accede to this request from Pakistan on the grounds that 'any attempt at this time to participate in working out a procedure for adjudication could not help the Indus Basin Party and might even hinder it by detracting from the main problem of getting along with the preparation of a comprehensive plan.' Letter of 11 August 1953 from President Black to the Government of Pakistan, *Indus Basin Dispute-General Negotiations-Correspondence* 04, supra note 101. *The Indus Basin Irrigation Water Dispute* (Government of Pakistan Publication), 61.

125. Letter dated 8 October 1950 from the Prime Minister of India to the Prime Minister of Pakistan, 'A Review of the Efforts', supra note 98; see also S. M. Burke, *Pakistan's Foreign Policy: A Historical Analysis* (Karachi: Oxford University Press, 1973), 52, notes 2–4.

126. S. M. Burke, *Pakistan's Foreign Policy*, supra note 125, 52, note 2.

127. Ibid.

128. Ibid. 52.

129. Ibid.

130. Ibid.

131. Gulhati, *Indus Waters Treaty*, supra note 76, 89.

132. See Hersch Lauterpacht, *The Function of Law in the International Community* (Hamden, Connecticut: Archon Books, 1966), 139–201; Hans Kelsen, *Peace Through Law* (Chapel Hill: Univ. of North Carolina Press, 1944), 23–32; B. S. Murty, 'Settlement of Disputes', in Max Sorenson (ed.), *Manual of Public International Law* (London, etc.: Macmillan; New York: St. Martin's Press,

1968), 677–8; Sette-Camara, 'Methods of Obligatory Settlement of Dispute', in Mohammad Bedjaoui (ed.), *International Law: Achievements and Prospects* (Paris-Unesco; Dordrecht, etc.: Martinus Nijhoff, 1991), 524–6; Goodrich, Hambro, and Simmons, *Charter of the United Nations: Commentary and Documents* (New York, London: Columbia University Press, 1969), 281–3.

133. See article 96 of the UN Charter and Article 36, para. 2 of the Statute of the World Court.

134. *ICJ Reports* 1962, 155.

135. *ICJ Reports* 1996, 73–4; see also the following cases: United States Diplomatic and Consular Staff in Tehran (*ICJ Reports* 1980, 21, para. 40); Military and Paramilitary Activities in and against Nicaragua (Jurisdiction and Admissibility) (*ICJ Reports* 1984, 432, para. 90).

3

World Bank: The Honest Broker?

A) THE LILIENTHAL PROPOSAL AND ITS BACKGROUND

GIVEN THE IMPASSE IN THE NEGOTIATIONS, PAKISTAN SERIOUSLY contemplated taking the matter to the UN Security Council. In the meantime a momentous development with far reaching consequences for the water dispute took place. Nehru invited David E. Lilienthal, the former head of the US Tennessee Valley Authority (TVA) and the American Atomic Energy Commission to visit India as his personal guest.[1]

While planning his visit, Lilienthal decided to include Pakistan in the itinerary on the suggestion of the well-known American journalist, Walter Lipmann, who believed that failure to do so would 'be almost the proportions of an international incident' as Pakistanis were very sensitive on the issue; that the real reason behind the dispute over Kashmir was the struggle for the control of waters which flowed from there into West Pakistan; and that he could help resolve the water issue by applying his TVA experience.[2] Before proceeding to the subcontinent, Lilienthal met Sir Zafrulla Khan, the foreign minister of Pakistan and Chaudhri Muhammad Ali, Secretary General of the Pakistan delegation to the UN who, according to him, 'broke out a wonderful map of Pakistan and talked about rivers of the Punjab'.[3] During his sojourn in the subcontinent, he met, among others, Liaquat Ali Khan and Nehru, but discussed different subjects with them. With the latter he held intensive discussions concerning Indo-US relations whereas with the former he broached the issues revolving around the use of common rivers between the two countries. Liaquat Ali Khan told him that '[u]nless the Kashmir issue is settled it is unreal to try the issues about water or about evacuees.'[4] It is intriguing to note that Lilienthal did not, as such, broach the subject of Indus waters with the two prime ministers.[5]

Upon his return to the US, Lilienthal published two articles in *Collier's* magazine, one of which was entitled 'Another "Korea" in the Making?'[6] which subsequently assumed seminal importance on the water dispute between Pakistan and India. In this article, he warned that a population of more than 360 million people, or one-sixth of humanity, inhabiting the two countries were on the verge of war on account of Kashmir. He did not deem this assessment either speculative or alarmist. He was of the opinion that if war broke out between the two antagonists, since the UN was already involved in the Kashmir dispute, it would 'undoubtedly put the USA into another and bigger Korea'. He observed that the prevention of war and promotion of peace and a sense of community between Pakistan and India lay neither in a plebiscite in Kashmir nor adjudication of the water issue by the World Court or an arbitration tribunal (both of which Pakistan was seeking). He was so convinced of his viewpoint that he suggested, *'even if Kashmir plebiscite could be held*, peace would not come'; and that even if the World Court adjudicated the water issue in favour of Pakistan (provided of course India consented to submit the case before it), it would antagonize India.[7]

If resolution of the Kashmir dispute was not the way out of this impasse, then what was? He noted that Pakistan possessed some of the best food growing areas in the world in western Punjab and Sindh. However, without water for irrigation, there was a danger that 20 million acres would dry up in a week and tens of millions of people would starve, thus turning these areas into a desert. He warned that, '[n]o army, with bombs and shellfire, could devastate a land as thoroughly as Pakistan could be devastated by the simple expedient of India's permanently shutting off the sources of water that keeps the fields and the people of Pakistan alive. India has never threatened such a drastic step, and indeed denies any such intention, but the power is there nonetheless. ... Life depends upon irrigation canals, some of them hundreds of years old. Not only farms depend on irrigation; as great a city as Lahore takes her drinking water from a canal.'[8]

Starting with the premise that a solution of the water issue could create an atmosphere conducive for the eventual settlement of the Kashmir dispute, Lilienthal proposed that India should start putting Pakistan's fears at rest by confirming its present use of water on the condition that the latter worked with it in a joint use of the Indus Basin. He drew attention to the fact that barely 20 per cent of the water of the Indus Basin was used for irrigation purposes while the remaining flowed to the Arabian Sea unused. He believed that if this wasted water or even a large fraction of it could be put to use, it could take care of the needs of both countries. Conscious of the national sensitivities of the two adversaries, he pleaded that the matter be approached on a functional rather than political basis. Arguing that it was not a religious or a political issue, he proposed that, '[t]he whole Indus system must be developed as a unit—designed, built, and operated as a unit, as a seven-state TVA system back in the US.'[9] He held that towards this end, the two countries, in association with the World Bank, should jointly finance the building of storage dams for the diversion and distribution of water; and establish an authority composed of technical men from the three parties to operate it. He proposed that once the scheme was designed, an India-Pakistan agency or a supranational body like the one provided under the Schumann Plan in Europe or a special corporation like the Port of New York Authority should take over its operation.[10]

Lilienthal's proposal showed the way out of the impasse. Eugene Black, the president of the World Bank who happened to be a friend of Lilienthal, read the article, commented that the proposal 'makes good sense all round', and decided to react favourably to the suggestions that the latter had made.[11] Lilienthal told Black that, in his opinion, Pakistan would accept his proposal as he had already shown an advance copy of the article to its ambassador to the US and that John Laylin of the law firm, Covington and Burling, which represented Pakistan's interests in the US, had reacted favourably to it.[12] However, he was not certain about India's reaction. He therefore wanted him to convince Nehru of its benefits. Subsequently, when both the governments agreed to consider the water issue separately from the 'territorial' dispute over Kashmir, he claimed that they had done so unknown to each other.[13] At this stage, the question of how Black was going to approach the two governments, i.e. through a speech in the Bank, or through a conference, arose. In September 1951, he decided to do so by addressing separate letters to the prime ministers of Pakistan and India in which he drew their attention to Lilienthal's proposal and tentatively offered them the good offices of the Bank for a discussion and settlement of the dispute strictly on the basis of technical and engineering grounds. It is interesting to note here that it was the Bank which took the initiative in the matter rather than the two governments. The Pakistan government showed much enthusiasm for the idea whereas the Indian government's response was reserved. Nehru was of the view that '[a]ny settlement would have to take into account the May 1948 agreement', of which he sent a copy with his letter.[14]

B) PARTIES' INTERESTS IN THE NEGOTIATIONS

Before proceeding further with the negotiations on the Indus water dispute, we need to answer three important questions. First, why did Pakistan accept the Bank's offer when it knew full well that the Lilienthal proposal envisaged the development of the Indus Basin as an integrated unit, which was an anathema to Pakistan as it viewed any such proposal as a device to undo Partition? Secondly, why did India, which—with the exception of the referral of the Kashmir dispute to the UN Security Council in 1948 and that of the Rann of Kutch dispute to an arbitration tribunal in 1966[15]—has always refused to accept third party involvement in its disputes with Pakistan (and other South Asian neighbours), consent to the Bank's good offices in the water dispute?[16] Thirdly, why did the Bank, which was conceived at Bretton Woods as a body dealing with economic and financial issues confronting the world rather than political matters, suddenly consent to play the role of a good officer between Pakistan and India?[17]

Even before receiving formal letters from the World Bank to act as good officer, the two countries started manifesting their reaction towards Lilienthal's proposal. Pakistan entertained certain misgivings towards it as revealed in the following entry by Lilienthal in his journal: 'Laylin said his client would ask him (actually, I am sure they already have) whether I have any connections with Indians of any kind. I gave him a categorical answer. I thought this interesting; indicates that they thought my visit to Nehru, being there at his invitation, etc., meant that I was hired as an advisor or something.'[18] It must have been thought that Pakistan would have strong reservations towards it on account of the fact that it proposed the development of the Indus Basin as a unit. However, contrary to all expectations, Pakistan's legal counsel, John Laylin, reacted favourably to the article even before the Pakistan government received the entire text as evidenced from the following statement by Lilienthal in his journal:[19]

> The Pakistan Government is prepared to accept 'The Lilienthal Proposal' as it stands. This is the remarkable news that this quite hectic day has produced. This word I got from Laylin, ... counsel for Pakistan in this country. The only question he has is: What is the best way to express that acceptance so as to bring Nehru to the same conclusion?
>
> Well, for a magazine article this is quite an initial impact. And the curious thing about it is that it was the Paks, I felt, would regard it with suspicion, dismay, and explosive disagreement.
>
> Laylin said that his initial reaction to a reading of the piece yesterday (we sent him one of our advance copies) was to urge an immediate acceptance without qualification, publicly stated in Karachi tomorrow ....
> Then he was concerned about whether such an immediate answer would not lead Nehru to reject it, on the ground, generally, that what Pakistan wanted could not be in the interests of India and Nehru.

How do we explain Pakistan's sudden and, if we may say, unduly hasty acceptance? The explanation is very simple. For it, the Bank's offer was a godsend because bilateral negotiations on the water issue with

India had broken down and the latter was not agreeable to settling the dispute in the World Court under any conditions. In fact, it was not amenable to third party involvement in its disputes with Pakistan, including the water dispute. Pakistan had threatened to take the issue to the UN Security Council but it knew that its threat was 'full of sound and fury signifying nothing' because the Council was impotent; it had miserably failed to get its resolutions on Kashmir envisaging the right of self-determination for Kashmiris implemented. In this context, the Lilienthal proposal provided the only way out of the cul-de-sac in which it found itself on the water issue. It is quite possible that Pakistan's top leadership might have calculated that with Pakistan being closer to the US than India, the Bank, which was led by an American, would favour Pakistan during the negotiations, though as demonstrated by subsequent developments this was no more than wishful thinking on its part. As we shall see below, the Bank, on the most critical issue of allocating the waters of the six rivers to the two countries, betrayed Pakistan by unabashedly favouring India.

As for India, the publication of Lilienthal's article created general disappointment and disagreement in the country which Lilienthal attributed to 'a few inadequate paragraphs in the press'.[20] The matter became complicated after *Dawn,* the Pakistani daily newspaper, reprinted it and, in the words of Khosla, 'garbled' it.[21] Lilienthal notes that for the Indians, the mere fact that the article was published in Pakistan meant that 'it could not be fair to India'.[22] In the face of these misgivings, Lilienthal claims that it was Khosla who, convinced that 'here was a great idea, one that will live in history', persuaded Nehru to accept the proposal.[23] In any case, Black provided Nehru, who was opposed to any third party involvement in India's disputes with Pakistan, an incentive to accept the Bank's offer when he specified in his letter that the Bank's involvement would be restricted to 'the development of the Indus water resources ... on an engineering basis...', and 'a common project that is functional, and not political, in nature and that could therefore be undertaken separately from the political issues'.[24] B. K. Nehru also contends that Nehru agreed to the

Bank's interventionary role because, in his opinion, 'this was not a political question. He did not want unnecessary tension with Pakistan and he had faith in the impartiality of the World Bank.'[25]

In our judgment, this explanation is far from convincing because what could be more political than controlling the water flowing into Pakistan whose control could invest India with enormous leverage over the latter? The real reason for accepting the Bank's offer of good offices (which subsequently turned into much more than that) lies in several advantages that Nehru saw India could accrue. First, he saw in it an opportunity to get 'proprietary rights' over the Eastern Rivers that he had coerced Pakistan into accepting through the May 1948 Joint Statement and which it was now reneging on. This is testified by the fact that Nehru's response to the Lilienthal proposal was that '[a]ny settlement would have to take into account the May 1948 agreement.' Similarly, when at the start of negotiations under the Bank's auspices, Pakistan sought necessary assurances from India for the continuation of the existing water supplies, Nehru, instead of doing the needful, demanded the Bank make Pakistan accept the May 1948 Joint Statement.

Secondly, India accepted the Bank's offer because it knew that without a settlement of the waters dispute with Pakistan it could not hope to get any loan from the Bank, which had a policy not to finance a project under dispute (it had adopted this policy because, in its opinion, in such a case not only was the investment risky but a completed project could also exacerbate the existing dispute). India got a taste of this policy in 1949 when it approached the Bank for the financing of the Bhakra-Nangal multi-purpose project on the Sutlej River and for the Damodar Valley project in the state of Bihar. The Bank refused the loan request for the Bhakra-Nangal project on the grounds that it was a disputed matter. India subsequently objected to Pakistan's request for a loan for the construction of a barrage at Kotri, on the Indus River, on the same grounds.[26]

Thirdly, Nehru perhaps realized that without a settlement of the water dispute, the two countries could go to war. It is undeniable that since India was the upper riparian vis-à-vis Pakistan, Nehru could have achieved his objective of enforcing the May 1948 Joint Statement to Pakistan without involving the Bank, as he did with Jammu and Kashmir by occupying it and by refusing to implement the UN resolutions, but that would have been at a terrible and probably unacceptable price. It could have led to a full-fledged war with Pakistan because, unlike the situation in 1948 when India temporarily stopped water flowing into Pakistan's canals, in order to assert India's 'proprietary rights' over waters of the Eastern Rivers, it was now planning to permanently close these waters to Pakistan. It was obvious that Pakistan would have never accepted this as it would have transformed Pakistan into a desert. The Pakistanis at that time believed that instead of dying of thirst and hunger, which would result from the stoppage of water by India, it was preferable to go down fighting the Indians. A war with Pakistan over the water issue did not suit Nehru as images of suffering and starving Pakistanis in the international media would have hurt his ambition of global leadership which he was harbouring. He subsequently realized this ambition by founding, along with other like-minded Third World leaders, the Non-Aligned Movement (NAM).

However, these explanations for Nehru's motives are far from satisfactory. Nehru would not have easily agreed to abandon the principle of exclusion of third parties from India's disputes with Pakistan without some assurance from the Bank that it would support India's claim to 'proprietary rights' over the waters of the Eastern Rivers. Do we have any evidence in support of this contention? Honestly speaking, we have none whatsoever. Such deals in international diplomacy tend to take place clandestinely. Notwithstanding the lack of evidence, the question of Nehru's motives, in our opinion, is legitimate and needs to be explored. Since it is inextricably linked to why the Bank got involved in the water dispute in the first place, we propose to defer it for the time being and take up both the questions subsequently.

After Lilienthal's return from a trip to the subcontinent, there was lot of discussion within the Bank over the latter's purported role during the negotiations. Lilienthal thought that it was imperative for the prime ministers of the two countries to reach a political agreement on the principles that would guide the negotiators. He was of the view that since the parties had already agreed to the content of his article in writing, the Bank engineers in their capacity as negotiators should be instructed to follow the principles outlined there as the basis for discussions. He wanted the principle of the development of the Indus Basin and its utilization for the benefit of the region (as a unit) to be emphasized. He was opposed to the idea floated by General Wheeler, the Bank's chief engineer, that the Bank representative should act as the arbitrator of disputes. He instead asked that his role be restricted to assisting the parties in working out their own solutions; that the Bank should appoint an engineer who was not an American; and that meetings should not take place in the US, the UK, Pakistan, or India, but preferably in Scandinavia or Switzerland.[27]

The Bank did not pay much heed to his recommendations. It held almost all of its meetings at the venues he had advised against; appointed an American as an engineer; and despite declaring that it would 'assist in solving problems without being in the position of a mediator', got itself fully involved in the negotiations going even beyond the mediatory function that it had assigned for itself in the beginning. It is noteworthy that Lilienthal's fears that negotiations would suffer if the Bank did not heed his advice proved utterly unfounded. In fact, it was the Bank's unstinting and deep involvement in the negotiations that guaranteed their success and the eventual conclusion of the Treaty.

C) BANK'S INITIATIVE AND START OF NEGOTIATIONS

Following a brief discussion within the confines of the Bank, Black addressed two identical letters[28] to the prime ministers of Pakistan and India. In said letters, he asked them to agree to the following three points which were based on the principles that Lilienthal had spelled out in his article and which the two governments had already accepted in principle; a) Indus waters are sufficient for the present and future needs of the two countries; b) these resources should be cooperatively developed and used most effectively to promote the economic development of the Indus Basin as a unit; and c) the issue of development and use of the Indus Basin water resources should be solved on a functional and not political plane, independent of past negotiations, claims, and political issues. He proposed the two countries depute qualified engineers, of high standing, to jointly prepare a comprehensive plan for the most effective utilization of the water resources of the Indus Basin. He also suggested the nomination of a Bank engineer to work with the designees of the two countries and participate 'as an impartial advisor, free to express his views on any aspects of the matter and available to perform such other services as might be mutually determined to be appropriate. He could thus assist in solving problems without being in the position of an arbitrator.' Subsequently, he clarified[29] that the principles mentioned in his previous communication only provided the broad basis of discussion and were not intended to be rigidly fixed terms of reference. He also stated that '[e]xcept as the two sides may hereafter agree, legal rights will not be affected and each side will be free to withdraw at any time ...'.

Both the countries welcomed the Bank's initiative. However, they had certain reservations. Nehru agreed with proposition (a) but had reservations with regard to (b) and (c), particularly with the caveat 'without regard to past negotiations and past claims ...'. This had a reference to the May 1948 Joint Statement which, we have seen, Nehru was not only reluctant to jettison but also wanted it to be the starting point of any negotiations. Pakistan too was unhappy with the contents of the letter but for entirely different reasons. In its opinion, Black's letter had failed to mention that it was entitled to the existing uses of water which Lilienthal had mentioned in his article and without which Pakistan would not have consented to enter into negotiations.

When Lilienthal saw the letter, he was upset with it as evidenced by the following entry in his journal: 'But it did not include the specific thing I most wanted, and which was needed to reassure Pakistanis, something explicitly in the text of my *Collier's* article, and subsequently agreed to, in effect, by both sides. This was an understanding that Pakistan water-users would have a right to no less than the amounts of Indus water presently used by Pakistan. What was in the letter, actually, was simply the statement that there was enough water in the river to continue present uses, which is, or could be, quite a different thing.'[30] Pakistan's legal counsel, John Laylin, who had drawn Lilienthal's attention to this development feared that the vague language of Black's letter to the prime ministers would arouse the fears of Pakistanis—that the water supply that they were currently using could be diminished while the negotiations were underway. He said that he would be satisfied if the exact words of Lilienthal's article were used.[31] It is a mystery as to why Davidson Sommers, the Bank's legal counsel, who had drafted the letter, made that omission. Interestingly, Lilienthal too was intrigued by it.[32]

Khawaja Nazimuddin, then prime minister of Pakistan, expressed concern over the Bank's failure to guarantee Pakistan's existing uses. Lilienthal then wrote to Khosla, chairman of India's Central Water and Power Commission and advisor to the Indian prime minister on engineering matters, and sought his help in making Nehru provide the necessary guarantee to Pakistan about the existing uses. However, Nehru wrote a letter to Black in which he insisted on getting Pakistan to accept the May 1948 Joint Statement as a treaty, recognizing India's exclusive rights to the waters of Beas, Sutlej, and Ravi.[33] This created a deadlock in the negotiations. In order to break it, Lilienthal recommended Black discontinue the correspondence and instead seek personal audience with the two prime ministers. Acting upon that advice, Black who was already planning a visit to the subcontinent did so during January and February 1952, and met the two prime ministers. This resulted in the two countries formally reaffirming to accept the Bank as a 'good officer' and proceed with the negotiations as outlined in Black's letter. India also made a commitment to keep supplying Pakistan the quantum of water that it was already receiving as long as the negotiations continued, as revealed by Black's letter:[34] '[L]egal rights will not be affected and each side will be free to withdraw at any time; but while the co-operative work continues with participation of the Bank neither side will take any action to diminish the supplies available to the other side with the existing uses.' All the parties including the Bank also agreed to the ultimate objective of the working party which it described as 'carry[ing] out specific engineering measures by which supplies effectively available to each country will be increased substantially ...'. They also agreed to accept not only the three main principles of Black's letter of 8 November 1951, but also give the negotiators the freedom '[w]ithin the broad outline of the basic framework ... to put forward or consider proposals in pursuance of the general objective'.

It is clear that whereas Black's letter confirmed Pakistan's existing use of water (though the language is couched in terms of reciprocity, obliging Pakistan to do the same to India, the obligation was one-sided because, since India was not receiving any water supplies from Pakistan)[35] it did not confirm Pakistan's acceptance of the May 1948 Joint Statement as a binding treaty. However, Gulhati claims that Black did not 'confirm' Pakistan's 'present use of water' as had been proposed by Lilienthal and by Pakistan's lawyers.[36] What he meant was that Black did not accept Pakistan's claim of uninterrupted flow of water as a legal right. He is right in this assertion but wrong when he suggests that Lilienthal favoured Pakistan's claim as a legal right. On the contrary, without emphasizing the merits of Pakistan's claim, he opposed it on the grounds that even if the World Court decided the legal issue in its favour, it would not bring peace between the two countries.[37] He merely wanted India to 'confirm' Pakistan's present use of water as a functional proposition in order to move things forward. Gulhati also claims that India made the 'concession' in favour of Pakistan because it did not at that time fully appreciate its significance.[38] This interpretation fails to carry conviction because both India and the Bank knew full well that without this 'concession' the negotiations would have been a non-starter. Pakistan would have neither agreed to reduction in the water supplies that it was receiving at that time nor accepted the May 1948 Joint Statement as a binding treaty.

Following the agreement on this thorny issue, the delegations of Pakistan, India, and the Bank met in Washington in May 1952, for the first round of negotiations.[39] The brief opening session took place on 6 May. Lilienthal, who also attended it, believed that the solution of the water problem lay in the construction of reservoirs in India because the best dam sites were located there. He favoured their construction, with the concurrence of Pakistan, and their operation as an integrated system with the latter sharing the expenditure incurred on them. When the Pakistani,[40] the Indian[41], and the Bank engineers[42] met subsequently, they entrusted themselves the task of, 'prepar[ing] an outline of program[s] and lists of studies for possible technical measures to increase the supplies of water available from the Indus system of rivers for purposes of economic development'. Towards this end, they decided to form a working party which consisted of Hamid, Khosla, and Wheeler. Black explained its purpose and function in these words:[43]

> The working party would hold an initial meeting for the purpose of determining the procedure to be followed in working out the plan, the steps needed to be taken, the order and manner in which those steps would be undertaken and the persons by whom they would be undertaken, and would set target dates for completion of the various steps. On reaching agreement on these matters, the working party would promptly, without the need of any further authorization, put the agreed procedure into effect and begin work on the plan.

The opening session was followed by a substantive session which lasted from 7 May to 18 June. It was tumultuous, which indicated the difficulties that lay ahead. The three delegations that participated in the meeting were composed exclusively of technical people. The Bank officials were of the opinion that the task before them was purely of an engineering nature and that they would take care of it rather easily. Accordingly, Black insisted, during his address at the preliminary meeting, that 'the Indus problem is an engineering problem and should be dealt with by engineers. One of the strengths of the engineering profession is that, all over the world, engineers speak the same language and approach problems with common standards of judgement. We all look to the working party to approach its duties in that spirit— as a group of experts dealing co-operatively and objectively as a technical problem.'[44]

Lilienthal, on his part, expressed the view that, 'the engineers from India and Pakistan, who have been sent by their Governments, can by their work lift the hearts of millions of people and find accord where dispute now exists: a great responsibility rests on them.'[45] As for Wheeler, he stated: 'The objective of the Working Party here assembled is to develop a comprehensive long-range plan for the most effective utilization of the total water resources of the Indus River System in the development of the region. This is peculiarly the province of engineers, and I share with you the high satisfaction that the task has been thus relegated for a factual solution.'[46] This approach made sense because this was a period of innocence—the Bank was not fully cognizant of the complexities involved in Pakistan-India relations.

Pakistan and India were totally opposed to the Lilienthal proposal which fervently advocated the development and management of the Indus Basin resources on an integrated basis. We ask why, especially when the proposal could unquestionably guarantee the most optimal and efficient results. The question assumes all the more significance because of voices that have risen, in recent years, in favour of an integrated approach.[47] Pakistan opposed the Lilienthal proposal because of the enormous trust deficit that it had developed vis-à-vis India on account of the Kashmir issue, India's stoppage of water to Pakistani canals on 1 April 1948, and the imposition of the May 1948 Joint Statement. It would not be wrong to say that Pakistan opposed anything and everything that indicated the undoing of Partition. The Lilienthal proposal was the ideal candidate for it, despite the fact that Lilienthal had tried to take care of Pakistan's misgivings by proposing a joint mechanism in the shape of an Indo-Pakistan agency, a supranational international agency on the model of the Schumann Plan in Europe or some special corporation like the Port of the New York authority to exploit and manage the water resources of the Indus Basin.[48] India, on the other hand, opposed the proposal because, 'she could go ahead, at considerably less ultimate cost to herself, with the Bhakra-Beas-Rajasthan Project.' Nor did many Indians want to accept Partition 'as a final and permanent thing'.[49] Perhaps, it also rejected it because it feared that the integrated development approach, instead of undoing Partition, which it fervently and unabashedly favoured, was likely to give Pakistan a virtual veto over all of its engineering projects relating to the Indus Basin, including the Bhakra-Beas-Rajasthan Canal Project.

A former Indian foreign secretary, Jagat Mehta, writing on the subject in the 1980s, opined[50] that the Bank abandoned the Lilienthal thinking of integrated development too easily which, in his view, it should not have done. In his opinion, it could have forcefully lobbied Western governments for this proposal and got commitments of a billion dollars for replacement works and a substantial bonus of advantages for Pakistan. He cited the example of the Nile agreement of 1929 between Sudan and Egypt where Great Britain, by dint of its imperial position, forced the upper and lower riparians to resolve their differences. He also believed that, whenever necessary, the Bank could also politely threaten to withdraw from the negotiations if Pakistan did not accept its independent proposals. The latter, in his opinion, being a lower riparian, would not have risked such a breakdown. In our judgment, Mehta did not seem to realize that, unlike Great Britain, which enjoyed the status of an imperial power over Sudan and Egypt, and hence was able to impose its will on them, the Bank, or for that matter the Western countries, did not enjoy a similar position vis-à-vis Pakistan and India. Hence, they were not in a position to force Pakistan to do their bidding. Besides, Mehta tends to forget that both Pakistan and India opposed the idea of integrated development. With this backdrop, there was no question of imposing it on Pakistan.

The working party found it difficult to proceed because the Indian and Pakistani designees could not agree to a common basis of discussion.[51] Pakistan wanted the working party to fix the surplus waters to be divided between the two countries after deducting that portion which was to cater to the 'existing uses' from the Eastern Rivers and to exclude the three Western Rivers from its deliberations. India wanted it to determine the total quantum of water available in the Indus Basin and allocate it to the two countries on the basis of their requirements. The two countries also differed with regard to the boundary of the Indus Basin. Clearly, though the negotiators were engineers, lawyers pulled their strings from behind the scenes.

At the end of the substantive session, that lasted three weeks, not much was achieved except for an 'outline of programme'. Pakistan accepted the Indian demand to proceed on the basis of the entire waters of the Indus Rivers system and requirements of the each party. India, in return, consented to allow Pakistan to submit studies relating to 'existing uses'. Later on, when each side put forward its list of studies for the preparation of the comprehensive plan, there were sharp differences between them on the interpretation of the 'outline of programme'. The session ended with the decision by the working party to abandon any idea of a joint study and instead agreed to let the parties conduct further studies separately and meet in Karachi on 1 November 1952 for the purpose of exchanging completed studies and forms, visiting sites, and determining further action.

Michel has attributed the failure of the working party to resolve the issue to the fact that the two governments had not already taken a political decision regarding the shares or ultimate uses of the increased supplies, a decision the engineers were not competent to take.[52]

D) BANK CUTS THE GORDIAN KNOT

At the Karachi session, which lasted from 1–8 December, the working party exchanged some technical studies and decided to undertake, along with the advisors, a visit to the works, irrigated and un-irrigated areas, and dam sites. It covered about 9,000 miles of the Indus Basin over a period of six weeks in the course of which it collected a lot of useful data. It held its next session on 23 January 1953 in Delhi where the Pakistani and Indian engineers exchanged more technical studies. However, they failed to agree on an integrated development plan. Consequently, the Bank asked the two parties to prepare comprehensive but separate plans, focusing on preliminary cost estimates and construction schedules for the engineering works involved.

When they submitted their respective plans at the next session held on 6 October 1953, they were utterly at cross purposes with each other. The Pakistani plan proposed to allocate to Pakistan the entire flow of the waters of the Indus Rivers except for that which India was actually using. In concrete terms, the Pakistani plan staked claim to all the waters of the Western Rivers (namely Chenab, Jhelum, and Indus) and 70 per cent of the Eastern Rivers (namely Beas, Sutlej, and Ravi). The Indian plan, on the other hand, proposed to allocate to India all the waters of the Eastern and 7 per cent of the Western Rivers. The differences between the two could be summarized in terms of 'existing uses from existing sources' versus 'existing uses from alternative sources'.

There was more or less agreement between the two countries on the estimated quantum of water available in the Indus Rivers. India put the total usable water at 119 million acre feet whereas Pakistan calculated it to be 118 million acre feet. However, they differed on the allocation that each one claimed, particularly on account of the meaning of the term 'existing uses'. Pakistan took the view that 'existing uses' connoted at once the actual historic withdrawals and those envisaged by the Anderson Committee, the Rau Commission, and the Punjab-Sindh Chief Engineers Agreement of 1945. India construed the expression in terms of water that Pakistan withdrew by virtue of prior allocation. In other words, it rejected Pakistan's definition of 'existing uses' in terms of pre-Partition irrigation plans except for that of the Anderson Committee (Trimmu-Haveli and Thal) on the grounds that they had not been formalized. It was also of the view that Pakistan was entitled to water from the Western Rivers only and not from both the Eastern and Western Rivers. In short, out of the 119 million acre feet of usable water that Indian experts had calculated in the Indus Rivers, the Indian plan allocated 90 million acre feet to Pakistan and 29 million to India. Out of the 118 million acre feet of water which the Pakistani experts had calculated, the Pakistani plan allocated 102.5 million acre feet to Pakistan and 15.5 to India. In the words of the Bank, the two plans 'differed widely in concept and substance'.[53] The following table brings out a comparison between them in quantitative terms:[54]

Total Uses Excluding Losses and Unusable Supplies (in millions of acre feet)

Plan	For India	For Pakistan	Total Usable
Indian	29	90	119
Pakistani	15.5	102.5	118

It is clear that Pakistan and India sharply differed in their claims. They could not bridge them, creating a complete deadlock in the negotiations which, according to Michel, resulted from 'the lack of a preliminary political agreement between India and Pakistan regarding sharing of the waters, costs of development, and operation of an irrigated system'.[55] He also attributed it to 'the physical inability of the Indus Rivers system ... to supply the amount of water required to fulfill the hopes and plans of the two nations'.[56] To break the deadlock, the Bank, which claimed to be 'neither judge nor jury but a mere spectator with goodwill', decided to go beyond its role of a 'good officer'. On 5 February 1954, it submitted a memorandum in which it put forward its own proposal for consideration of the parties whose basic thrust was as follows:[57]

The entire flow of the Western rivers (Indus, Jhelum and Chenab) would be available for the exclusive use and benefit of Pakistan, and for development by Pakistan, except for the insignificant volume of Jhelum flow presently used in Kashmir.

The entire flow of the Eastern rivers (Ravi, Beas and Sutlej) would be available for the exclusive use and benefit of India, and for development by India, except that for a specified transition period India would continue to supply from these rivers, in accordance with an agreed schedule, the historic withdrawals from these rivers in Pakistan.

The transition period would be calculated on the basis of the time estimated to be required to complete the link canals needed in Pakistan to make transfers for the purpose of replacing supplies from India. A temporary cooperative administration would be needed to supervise the carrying out of the transitional arrangements.

Each country would construct the works located on its own territories which are planned for the development of the supplies. The costs of such works would be borne by the country to be benefited thereby. Although no works are planned for joint construction by the two countries, certain link canals in Pakistan will, as stated above, be needed to replace supplies from India. India would bear the costs of such works to the extent of the benefits to be received by her therefrom. An appropriate procedure would be established for adjudicating or arbitrating disputes concerning the allocation of costs under this principle.

Simply put, the Bank allocated the entire flow of waters of the Western Rivers for the exclusive use and development of Pakistan except for an 'insignificant' volume for use in Kashmir.[58] It assigned the entire flow of waters of the Eastern Rivers for the exclusive use and development of India. In concrete terms, the Bank proposal put Pakistan's share of water at 97 million acre feet and that of India at 22 million. The following table provides a comparative quantitative division of waters by the Pakistani, Indian, and Bank plans:[59]

Total Uses Excluding Losses and Unusable Supplies (in millions of acre feet)

Plan	For India	For Pakistan	Total Usable
Indian	29	90	119
Pakistan	15.5	102.5	118
Bank	22	97	119

The Bank proposal specified that for a transitional period India would let Pakistan make the historic withdrawals from the Eastern Rivers according to an agreed schedule. It did not spell out the duration of the transition period but stated that it would be calculated on the basis of the time required to complete the link canals needed in Pakistan to make transfers of water for the purpose of replacing supplies from India. Each country was to construct the works on its territory that it planned for development of supplies and was to bear its own costs. India was to bear the costs of link canals in Pakistan which the latter would construct, to replace supplies from India, to the extent of the benefits that it was to receive from them. The Bank claimed that its proposal, 'provides a fair division of waters. It protects existing uses from disturbance and allocates surplus supplies, … in accordance with the principle of equitable apportionment.'[60] It justified it on the grounds that it made the two countries independent from each other to develop their own resources and remove the element of friction between them in these words:[61]

> The Bank proposal also embodies the principle that, in view of existing circumstances, allocation of supplies to the two countries should be such as to afford the greatest possible freedom of action by each country in the operation, maintenance and future development of its irrigation facilities. It is desirable, as far as practicable, to avoid control by India over waters on which Pakistan would be dependent, and to enable each country to control the works supplying the water allocated to it and determine in its own interests the apportionment of waters within its own territories. This principle has not merely the negative advantage of minimizing friction between the two countries … and of

avoiding the necessity of costly and perhaps ineffective permanent joint
administration. It also has a positive advantage. There is every reason
to believe that leaving each country free to develop its own resources,
and without having to obtain the agreement of the other at each point,
will in the long run most effectively promote the efficient development
of the whole system ....

A further advantage of the Bank proposal lies in the fact that, after
works are completed, each country will be independent of the other in
operation of its supplies. Each country will be responsible for planning,
constructing and administering its own facilities in its own territories
as it sees fit. This should provide strong incentives to each country to
make the most effective use of water, since any efficiency accompanied
by works undertaken by either country for storage, transfer reduction
of losses and the like will accrue directly to the benefit of that country.
The same will be true of the efficiency achieved in operations.

The Bank proposal envisaging division of the six Indus Rivers
between Pakistan and India was a departure from the idea of an
integrated development that Lilienthal had proposed in his famous
article in *Collier's* magazine and which he was subsequently very keen
to get implemented. The proposed solution was indeed unique and
bold because there was no example in the annals of water disputes
between nations where two countries ever partitioned their common
rivers whereas all examples point towards nations sharing them.[62] It
came about because the two countries were utterly incapable of jointly
managing their water resources as Black obliquely referred to it in the
following statement: 'for some time it has been evident to me that no
further progress could be made ... unless the Bank took the initiative ...'.
The question arises whether the Bank took the initiative that it did
on its own or at the behest of the parties. It had the mandate 'to
perform such other services as might be mutually determined to be
appropriate. [It] could thus assist in solving problems without being
in the position of an arbitrator.'[63] In other words, to play such an
active role, it needed the consent of the parties. According to Michel,
'[a]ctually, there are good reasons for believing that the parties asked
General Wheeler to make a proposal as a means of breaking the

deadlock.'[64] If Pakistan made a request, it is understandable because it was the weaker party and had no other option. Why India, which was *ab initio* opposed to the third party involvement in its disputes with Pakistan, agreed is a mystery.

In any case, India was delighted with the Bank's proposal as it conceded the Indian demand of 'proprietary rights' over the waters of the Eastern Rivers in perpetuity and divested Pakistan of any rights over them, including those based on historic uses. Gulhati divulges in his book that soon after India received it, Khosla informally told Wheeler that it would be acceptable to India, though as an after-thought he hastened to add that the latter did so in 'all good faith and sincerity but, ... somewhat naively'.[65] Less than two months after the Bank submitted its proposal, Nehru signalled his government's willingness to accept it in these words:[66] 'In the interest of a speedy and constructive settlement and in the spirit of good-will and friendship that has guided my government ever since the beginning of this controversy, we accept the principles of the Bank proposal as the basis of agreement.' At the same time, driven by the concern to not give the impression that what the Bank had proposed was quintessentially on the lines of the Indian demand, he pretended to accept it with a heavy heart, as if in doing so his country was making enormous sacrifices. Nehru's statement in the matter, as shown below, is quite revealing:[67]

> The Bank Proposal requires India to give up the use of a large part of the waters flowing through her own territory and thus to abandon, for all time, any hope of the development of a considerable portion of the extensive arid lands in India which has no possible source of water supply other than the Indus system of rivers and which will therefore remain a desert for ever. Its acceptance would also imply a very heavy financial burden for my Government; not only would it involve the payment of large sums of money to Pakistan, but would also make new developments in India much more expensive than if all the waters running through her territory and indispensable for her normal development could have been utilized therein.

As opposed to India, Pakistan opposed the Bank proposal tooth and nail. In fact, it was so shocked by it that its delegation in Washington showed no reaction for many days, forcing General Wheeler to ring up Hamid. The leader of the Pakistan delegation reportedly told him that 'as far as Pakistan was concerned, Working Party might be considered as having come to an end and most of the Pakistan delegation was leaving shortly.'[68] Lilienthal observes in his journal that he learned 'that the Bank's proposal hit the Karachi government like a bombshell.'[69] The Pakistan government decided to sack Hamid, the leader of the Pakistan delegation, on the grounds that 'he made too many concessions'.[70] It also fired Nasir Ahmed, the secretary of the Ministry of Water and Power, who had been participating in the negotiations.[71] It rejected the Bank proposal stating that it did not allocate any share in waters of the Eastern Rivers to it. Its representatives also tried to convince the Bank officials that, 'India has other systems from which water would be available. Pakistan has not. Money for water is no proposition—not a question of sentiment or principle.'[72] Its delegation started to make preparations to return to Pakistan. However, Black intervened and stopped it from doing so. In the middle of May 1954, the Pakistan Prime Minister, Muhammad Ali Bogra, addressed the following letter to Black in which he formally rejected the Bank proposal on the grounds that it was not fair:[73]

The Government of Pakistan are in accord with these principles [of the Bank Proposal]. They regret to note, however, that Bank's Proposal as regards the division of supplies does not in fact meet the test of fairness laid down by the Bank and unless necessary adjustments are made my government will not be able to accept it.

The Bank Proposal also states that, 'it is desirable, so far as practicable, to avoid control of supplies by India over waters on which Pakistan will be dependent'. I must say that I have been gravely perturbed by the application of this concept suggested by the Bank representative. I cannot understand how it is practicable under a plan that meets the test of fairness to cut-off the supplies which Pakistan had traditionally received from the Eastern Rivers .... It became clear ... and as I believe that Bank management must now appreciate, that it is neither practicable nor equitable to cut-off Pakistan's historic supplies from the Eastern Rivers. The Government of Pakistan cannot visualize with equanimity the possibility of implementing a plan which would affect its vital interests adversely for all time to come. Adjustments in this regard will also, therefore, be necessary if the proposed plan is to meet the test of fairness ....[74]

Was India really making the sacrifice that Nehru asserted in his letter referred to above? In our opinion, it was not. Consider the following, India made a claim to the entire flow of the Eastern Rivers and 7 per cent of the Western Rivers. What the Indian claim on the Western Rivers signifies is that India was actually staking a claim to almost 40 per cent the entire flow of the Chenab River as this is what the 7 per cent of the Western Rivers comes to. This claim was unjustified because the Jhelum and Indus Rivers are far removed from the Indian plains where their waters can be used for irrigation purposes and are thus useful only in mountainous Jammu and Kashmir where their utility for irrigation is very limited. As opposed to this, the Eastern Rivers are of real use to it, not only for irrigation purposes in Punjab but also Rajasthan (which incidentally lies outside the Indus Basin). Simply put, what India was demanding from Pakistan was to virtually not only renounce its claim to the entire flow of the Eastern Rivers on which it was entirely dependent for irrigation, but also a good part of the Chenab River. India knew that its claim was highly exaggerated and that Pakistan would never agree to it. It advanced it because as a good negotiator, it deemed it necessary to put forward its maximalist demand, though from day one it was aiming to appropriate the entire flow of the three Eastern Rivers. This is amply testified by the fact that India imposed the May 1948 Joint Statement on Pakistan and Nehru repeatedly asked the Bank to make Pakistan accept this as a binding agreement.

When we examine the Bank proposal, we realize that it was in essence nothing but a rehash of the Indian demand which explains why

India accepted it hastily and readily abandoned its claim to the waters of the Chenab River. This is testified by a number of contemporary observers of the scene. For example, the Indian Counsel General in the US at that time, Arthur Lall, openly acknowledged that the Bank proposal, 'was substantially the Indian plan'.[75] It is true that Lilienthal refuted his statement by calling it 'most unfortunate'. Nevertheless, he too could not resist observing that '*even if it were,* this is a way to stir up troubles with so sensitive and defensive a man as Hamid'.[76] Gulhati confesses that the Bank officials held similar views as it comes out in the following statement:[77] 'What I was more concerned with, however, was the feeling that the Bank proposal *was very favourable to India,* which appeared to have developed among some persons in the Bank.' He also admits that the Bank 'was proceeding on the premise that the Bank Proposal *was very favourable to India*'.[78] He however tried to negate this impression by contending that '[t]his feeling, no doubt, grown partly because of the relative readiness with which India had accepted the Bank Proposal and had even acquiesced, somewhat readily, in the Bank's offer to Pakistan made in its memorandum of 21 May 1954, "to bring about adjustments" to the Bank Proposal "acceptable to both sides".'[79] The following observation by Gulhati, which is quite revealing, nonetheless betrayed the falsity of his explanation:[80]

> Generally, I sleep well and long; I retire at about 10 p.m. and have to make no particular effort to go over. But the night of 5th February, 1954, stands out in my memory as one of the few nights when I lay tossing in bed and, in spite of repeated efforts, could not go to sleep until well after 2 a.m. The proposal made by the Bank earlier that afternoon was to me much more than an unusually important event with which I was officially concerned. It was, indeed, a matter of great personal significance and my mind was unusually exercised.
>
> There was much that I was thankful for, it seemed to me that the *Bank Proposal was a vindication, in essence, of the stand that India had maintained in regard to the water dispute for the preceding six years,* a stand for which I was responsible, in part. Indeed the Bank Proposal was more than a vindication of the Indian outline plan which (I could not help regret) Khosla had undermined to some extent in making his 'sporting offer'.

Nehru, too, was pleased with the Bank proposal as it is revealed from the following statement that he made to Vice-President Iliff, in which he stated that he was keen to travel to Pakistan and, along with Ayub (whom he utterly disliked), sign the Treaty: 'The only point from the conversation which needs to be recorded is that he [Nehru] is most anxious that the Indus Waters Treaty, when it has been agreed, should be signed in Rawalpindi by President Ayub Khan and himself. This, he said, would give him an opportunity to make a gesture which he wished to make, namely of visiting the President of Pakistan in the Pakistan capital.'[81] Last but not the least, writing almost fifty years after the Treaty was concluded, the American writer, Dennis Kux, grudgingly agreed that the Bank proposal was 'closer to the Indian stance'.[82]

E) ANALYSIS OF PAKISTAN'S CLAIM TO THE EASTERN RIVERS

We pause to examine whether or not Pakistan's reservations to the Bank proposal were justified. We begin by examining the soundness of Pakistan's claim to waters of the Eastern Rivers based on historic uses. To do so, we need to know the principles of international river law that existed in the 1940s and 1950s when the dispute between the two countries arose. The basic principle at that time was (and remains so today) that each nation,[83] being equal and sovereign, is entitled to put an international river to use that traverses its territory without causing substantial impairment to the right of other riparians to do the same. However, where it seriously affects the existing uses of other riparians, international law requires all riparians to desist from unilaterally undertaking new projects. In other words, every riparian must recognize the interests of other riparians in the full exploitation of a common river. This is because an international river creates a partnership which necessitates a cooperative approach by riparians so that each may derive its due share of benefits from it. The following survey of the thinking of the contemporary international jurists demonstrates that they lent their support to the principle of equitable apportionment in their writings.

We begin with the views of jurists from the English-speaking world. The celebrated writer, Oppenheim, in his classical study on International Law (vol. 1, *Peace*, 5th edition, 1937) had this to say on the subject:[84]

> But the flow of not-national, boundary, and international rivers is not within the arbitrary powers of one of the riparian States, for it is a rule of International Law that no State is allowed to alter the natural conditions of its own territory to the disadvantage of the natural conditions of the territory of a neighboring State. For this reason a State is not only forbidden to stop or to divert the flow of a river which runs from its own to a neighboring State, but likewise to make such use of the water of the river as either causes danger to the neighboring State or prevents it from making proper use of the flow of the river on its part.

Similarly, J. L. Brierly in his book (*Law of Nations*, 2nd edition, 1936) expressed himself on the issue in these words:[85]

> The practice of States, as evidenced in the controversies which have arisen about this matter, seems now to admit that each State concerned has a right to have a river system considered as whole, and to have its own interests weighed in the balance against those of other States; and that no one State may claim to use the waters in such a way as to cause material injury to the interests of another, or to oppose their use by another State unless this causes material injury to itself.

Again, Hall in his treatise (*International Law*, 8th edition, 1924) advanced the following view in the matter:[86]

> Obstruction or diversion of the flow of a river by an upper riparian State to the prejudice of a lower is alleged to be forbidden on the principle that 'no State is allowed to alter the natural conditions of its own territory to the disadvantage of the natural conditions of the territory of a neighboring State', and the same principle applies to the use of the river so as to cause danger to a lower riparian State.

As far as jurists from the non-English speaking world at that time are concerned, their views were no different. For example, the leading French jurist, Paul Fauchille, while supporting Max Huber, who held that the upper riparian was forbidden to do anything which diminished the natural state of the stream to the lower riparian, endorsed it in the following words:[87]

> It is neither to the upstream riparians, nor to the downstream riparians, that it is necessary to give preference exclusively. What it is necessary to do is to harmonize the rights which belong to both so that each may exercise his rights to an extent which does not cause the disappearance of those of the others ....
>
> The truth is that for States, as for individuals, neighborhood entails certain obligations and imposes certain restrictions on the exercise of rights. The riparian of a watercourse has the right to do all the acts which present to his coriparians only the inconveniences and troubles inseparable from being neighbors, but he ought to abstain from all those which exceed such inconveniences and difficulties. That is the principle ....
>
> If, on the subject of the difficulties which can arise from the industrial or agricultural exploitation of an international course, it is not possible to establish in advance exact rules for their solution, there are nevertheless certain rules which can be taken for granted *a priori* ....
>
> A riparian cannot accomplish in the part of the river upon which it borders acts which would have as their result the drying up and complete suppression of a watercourse at its entry into the territory of another State; there, again, there would be an actual interference with the territory itself of the other State.

Professor M. Kaufmann of the University of Berlin also lent his support in the following statement:[88]

According to international law, a riparian State, or its inhabitants

a) ought not to dissipate without purpose the water which another riparian State may have need;

b) ought not to use, even for lawful purposes, water without regard to the interests and to the needs of the other riparian States ...

d) Each riparian State ought not to use the waters without looking out for the opposing interests of the other riparian States in a just measure.

Last but not the least, the Italian jurist, Prospero Fedozzi, on the basis of a study of the various treaties on international streams, summarized the following trends:[89]

They constitute proof of the widespread sentiment that the States traversed by a watercourse do not have full autonomy in the treatment accorded it within their respective territories. To say that the upstream State cannot carry out works which alter sensibly the regime of the waters within the downstream State is certainly to affirm an important principle, but one of completely negative value, while what is essential is not to prevent the various uses of the water but to coordinate them in such a way that one use does not interfere with that by others.

The principle of equitable apportionment is to be found not only in the writings of the jurists of international law (as shown above) but also in international treaties in addition to judicial decisions of State courts. We propose to look at the examples drawn from the Nile, the Tigris, and the Euphrates, and the Rio Grande and Colorado Rivers which represent Africa, Asia, and Americas respectively. As far as the Nile River is concerned, Great Britain in 1925 and 1927, recognizing Egypt's 'natural and historic rights ... in the waters of the Nile', conceded to the latter, 'a prior right to the maintenance of her present supplies of water for the areas now under cultivation and to an equitable proportion of any additional supplies which engineering works may render available in the future'. The Nile Waters Agreement of 1929 was concluded in light of this rule. It forbade, *inter alia*, the construction of any irrigation or power project on the Nile River or its branches in Sudan or Uganda which would, 'entail any prejudice to the interests of Egypt, either reduce the quantity of water arriving in Egypt, or modify the date of its arrival, or lower its level'.[90] On 17 May 1951, Sayed Abdel Rahman

Abdoun, Sudan's undersecretary for irrigation, issued a statement in which the Sudanese government reaffirmed Egypt's above-mentioned rights.

As far as the Americas are concerned, in the early twentieth century, the US diverted waters of the Rio Grande River which led Mexico to protest against it. In 1906, the US concluded a treaty with Mexico whereby it undertook to construct, at its own expense, a storage dam on American soil and guarantee delivery of a fixed quantity of water matching the amount of water that the Mexican farmers had lost as result of the diversion.[91] Then, on 3 February 1944, the US and Mexico signed a treaty which provided for an equitable apportionment of waters of the Rio Grande and the Colorado Rivers to both of them. It also protected the existing irrigation uses in the US and Mexico, and guaranteed their further expansion in both the countries. During the Senate hearings for ratification of the treaty, the US Department of State justified the treaty on the grounds that Mexico had entitlement under it. When the detractors of the treaty argued against it on the grounds that without it the US would not be able to put the waters to its own use, Dean Acheson, the assistant secretary of state, rejected it on the following grounds:[92] 'The logical conclusion of the legal argument of the opponents of the treaty appears to be that an upstream nation by unilateral act in its own territory can impinge upon the rights of a downstream nation; this is hardly the kind of legal doctrine that can be seriously urged in these times.' Towards the end of the nineteenth century, the US had propounded the Harmon doctrine[93] which entitled a state with the right to do whatever it liked with the international rivers that flowed through its territory (incidentally it was on the basis of this doctrine that India, on 1 April 1948, stopped the flow of waters from the canal headworks located in India to Pakistan). In spite of the US proclamation, the doctrine in question always had a doubtful status in international law though some countries occasionally invoked it to justify their claims. However, with the ratification of the treaty, the US formally jettisoned the Harmon doctrine.

As far as Asia is concerned, in 1920, France and Great Britain, in their capacity as the mandatory powers of Syria and Iraq respectively, agreed that the execution of any irrigation plan in Syria—which contemplated the diminution of the waters of the Tigris and the Euphrates in any considerable degree at the point where they enter Iraq—would be jointly investigated. Subsequently, in 1946 Turkey and Iraq entered into a protocol 'for the purpose of assuring and maintaining a steady supply of water and the control of the flow of the two rivers [Tigris and Euphrates] in order to avoid the dangers of floods …'.[94] The protocol allowed investigation and construction of storage works on Turkish soil for the benefit of Iraq. There are many other international treaties which enshrine the principle of equitable apportionment but these examples should be sufficient. There is not a single treaty governing international rivers concluded during the nineteenth century which ignored the principle of the protection of established uses of the downstream riparian states.

The principle of equitable apportionment has also informed judicial decisions, particularly of the US courts. Under the American constitution, each state exercises sovereign rights over inter-state rivers except in matters pertaining to navigation and commerce, which the federal government regulates. While dealing with disputes between states concerning the use of waters of rivers not related to navigation and commerce, the US Supreme Court has recognized and applied the principle of equitable apportionment.[95] Thus, whenever confronted with the argument by an upper riparian state that it enjoyed an absolute right to do with waters originating through its territory whatever it liked, the Court always rejected it in favour of the principle of equitable apportionment to interstate rivers, besides consistently ruling that existing beneficial uses of waters must be accorded priority over new projects.[96] Courts of countries like Germany, Italy, and Switzerland have also uniformly applied the same principle in their decisions.

Moving to the subcontinent, the Rau Commission in 1942 confirmed the principle of the primacy of established uses as the

governing principle for the apportionment of waters of the Indus
Rivers. The Rau Commission was established in 1941 to consider
a complaint by Sindh against Punjab in which it alleged that the
latter was making unauthorized withdrawals from the Indus Rivers
that would further increase after the completion of the Bhakra Dam
(which it was planning to build on the Sutlej River). The Commission
examined the international practice in the matter and in its report
approvingly cited the following verdict which was given in the river
dispute between Wyoming and Colorado:[97] 'In the general interest
of the entire community inhibiting dry, arid territories, priority
may usually have to be given to an earlier irrigation project over
a later one; 'priority of appropriation gives superiority of right.'
Pursuing the matter further, the Commission went on to add:
'the rights of the general units concerned in this dispute must
be determined by applying neither the doctrine of sovereignty,
nor the doctrine of riparian rights, but the rule of 'equitable
apportionment', each unit being entitled to a fair share of waters and
its tributaries.'

These incipient principles of international law regarding sharing
of waters of rivers common between upper and lower riparians and
the jurisprudence of courts regarding river disputes between states,
in a federal setup like the US, demonstrate that Pakistan's claim to
the historic supplies of the Eastern Rivers was well-founded. Hence,
Pakistan's reservations to the Bank proposal, which rejected its claim,
were justified. Notwithstanding this fact, the Bank put its weight
behind the proposition that 'historic withdrawals of water must be
continued, but not necessarily from existing sources' which had an
eerie similarity to the May 1948 Joint Statement which, enshrining
the Indian point of view, asked the West Punjab government 'to tap
alternative sources.' To win over Pakistan, the Bank tried to dangle
before it the carrot of increased availability of water and promotion
of economic development, if it accepted its proposal.[98] Not content
with that, it also enlisted the services of the British government
to try to convince the Pakistan government of the benefits that it

would accrue:[99] a) India would not do anything underhand to the Chenab River, some of whose water it was planning to divert to the Indian side; b) it would pay for the costs of the construction of the replacement works; and c) Pakistan's existing uses would be guaranteed for the duration of the transition period. Apprehending that the Pakistan government may not go for these incentives, the British government at the same time warned the Pakistan government of the consequences of its refusal in these words:[100] 'True, Pakistan, under the Bank proposal, will have less water than she would have if it could be assumed that the existing situation would continue. But this is unfortunately an unrealistic assumption, and a more likely and imminent possibility is that India will cut Pakistan's existing supplies far below what the Bank proposal would assure to her'; and therefore, it, 'would be disastrous from every point of view if discussions now broke down and responsibility could be attributed, however, unfairly, to Pakistan.' When the Pakistan government refused to take the bait, the Bank resorted to blackmail and pressure. For example, it warned Pakistan's foreign minister, Sir Zafrulla Khan, who was in Washington at that time that if Pakistan did not resume negotiations, the reasons for the failure of talks would be made public; and that it would have to bear the blame.[101]

Simultaneously, the Bank, in concert with India, moved to mount pressure on Pakistan to acquiesce to its proposal. Thus, it sent a memorandum, dated 21 May 1954, to Pakistan through which it made clear that progress with the co-operative work was possible only on the basis of its proposal taking the division of waters as the starting point. It asked the latter to indicate within one week its acceptance of the proposal, if it desired further progress in the matter. As far as India was concerned, there existed an understanding of 13 March 1952, according to which it had agreed to continue the existing supplies of water to Pakistani canals. India was keen to replace it with an ad hoc agreement in order to operate the Bhakra Canal which it completed in 1954. Pakistan and India tried to negotiate an agreement for the purpose but failed to do so. Making this and Pakistan's non-acceptance of the Bank proposal an excuse, India issued an ultimatum to Pakistan to give a positive reply to the foregoing Bank memorandum by 21 June, failing which, it threatened to declare the 13 March 1952 understanding null and void. Pakistan was unable to do the Indian bidding by the indicated date.

Consequently, in a letter addressed to Black, issued the same day the ultimatum expired, India denounced the understanding of 13 March 1952. It also indicated in that letter that if Pakistan agreed to proceed on the basis of the Bank proposal, it would be prepared to consider, among others, arrangements for renewed co-operative work. Pakistan resisted the pressure mounted by the duo but not for long and ultimately succumbed to it when Pakistan's foreign minister, Sir Zafrulla Khan, in a letter addressed to Black and dated 28 July 1954, indicated his government's acceptance of the Bank proposal in these words:[102] 'My Government has given careful consideration to the Bank's proposals. It is clear that the Bank's proposals impose great sacrifices on Pakistan. Nevertheless, in the interests of a peaceful settlement my Government accepts the Bank's proposals *in principle* as the basis for agreement on the assumption that a workable plan can be prepared on that basis ....'

F) BANK DICTATES: SMOKING GUN EVIDENCE

There is a piece of hard evidence which shows that it was the Bank that forced Pakistan to accept its proposal. In the course of our digging into the Bank's archives on the Indus waters negotiations, we stumbled upon two typed, fullscape pages, seven paragraphs long, of an unsigned draft of Pakistan's reply to the Bank proposal.[103] On the left hand side of the top of the page the words 'Shoaib Draft of 12 July' is written in red ink which signifies that Mohammad Shoaib, who was Pakistan's executive director in the Bank at that time, gave the document to the Bank. On the right hand side, on top of the page the following comment is written: 'The alterations in red ink restore the text of the Sommers [the Bank's legal counsel] draft of 2 July, and shows the differences between the two versions.' It ends with the initials 'WI [which stands for William Iliff], 13 July.' Below these observations appear the text of the draft of Pakistan's reply to the Bank's proposal with alterations to different paragraphs.

One alteration which interests us relates to paragraph 4 which reads as follows: 'The Government has given careful consideration to the Bank's proposals. It is clear that the Bank's proposals impose great sacrifices on Pakistan. Nevertheless in the interest of a peaceful settlement, my Government *is willing to participate in arrangements for cooperative work in preparation of a plan on the basis of the Bank's proposals* on the assumption that a workable plan can be prepared on that basis which will provide, from flow of the Western rivers, all the uses ....'[104] One can clearly see that the italicized part of the paragraph is struck down with red ink and substituted with the following, 'accepts the Bank's proposals in principle on the basis of an agreement'. With this change, the altered sentence reads as follows: 'Nevertheless in the interest of a peaceful settlement, my Government accepts the Bank's proposals in principle on the basis of an agreement.' This marks a sea change in the purport of the Shoaib's draft because whereas the original text is unclear as to the acceptance or rejection of the Bank proposal, the altered text firmly commits Pakistan to its acceptance, though in principle. As shown above, two weeks after Iliff changed Shoaib's draft, Sir Zafrulla Khan sent Pakistan's letter accepting the Bank proposal to Black, which adopts *mutatis mutandis* the Bank's language as shown below: 'Nevertheless, in the interests of a peaceful settlement, my Government accepts the Bank's proposals in principle as the basis for agreement ...'. This unequivocally shows that the Bank forced Pakistan to accept its proposal.

C
O *of Shoaib's Draft*
P *of 12/vii*
Y

The alterations in red ink
restores the text of the
Sommers Draft of 2 July, and
Show the difference between
the two versions.
WI
13/vii

(16A)

Dear Mr. Black:

I refer to the Prime Minister's letter of 14th May, 1954,
and to the discussions and exchange of views that have taken place
between us with respect to the Bank's proposals dated February 5,
1954, regarding the Indus Basin Waters.

2.   It was explained by the Bank that with the dependable
flow of the supplies of (the) Western Rivers only, and by means of
suitable links, Pakistan would, in the Bank's view, be enabled to
maintain (the) uses historically supplied from (the) Eastern Rivers and
bring most of the Sutlej Valley canals up to an amount equivalent to
allocations, without invading (the) supplies required to maintain the
historical withdrawals on (the) Western Rivers or (the) supplies required
for projects in progress on those rivers, including Thal and Kotri.
It was further explained that there would be substantial additional
supplies which would be available for planned uses at Gudu and Sukkur.
India would bear the cost of the necessary works to the extent of the
benefit to be derived by her (therefrom) *later on* under arrangements which would
assure Pakistan of payment as (the) work progresses.

3.   My Government has given careful consideration to the Bank's
proposals.   It is clear that the Bank's proposals impose great
sacrifices on Pakistan.   Nevertheless, in the interest of a peaceful
settlement, my Government [is willing to participate in arrangements *accepts the Bank's proposals in principle as (the)*
basis for agreement for cooperative work in preparation of a plan on the basis of the
Bank's proposals] on the assumption that a workable plan can be prepared
on that basis which will provide, from flow of (the) Western rivers, all

Figure 3.1: Shoaib's Draft and Iliff's Alterations (I)

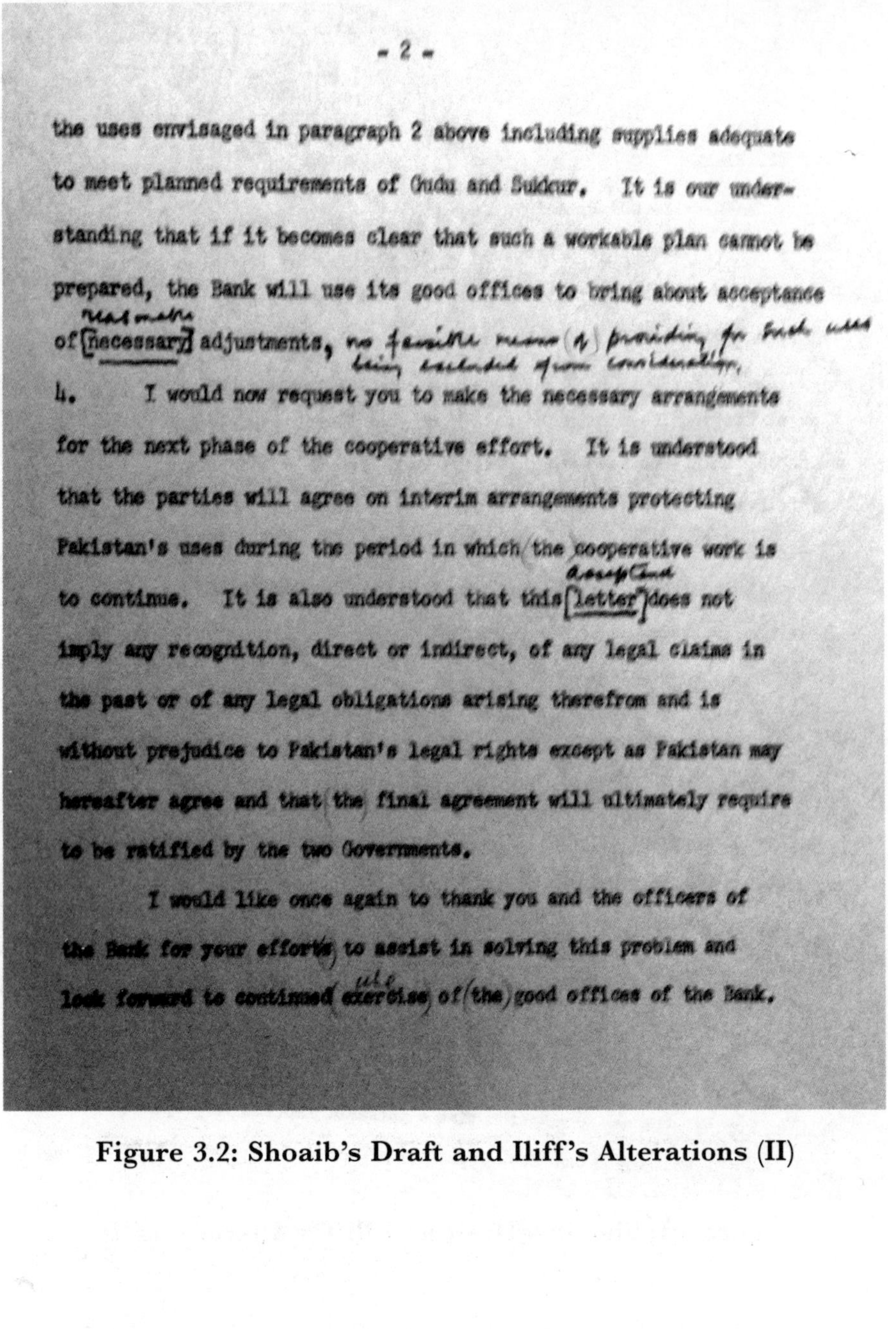

- 2 -

the uses envisaged in paragraph 2 above including supplies adequate

to meet planned requirements of Gudu and Sukkur.   It is our under-

standing that if it becomes clear that such a workable plan cannot be

prepared, the Bank will use its good offices to bring about acceptance

of [necessary] adjustments, *no feasible means (s) providing for such uses being excluded from consideration.*

4.   I would now request you to make the necessary arrangements

for the next phase of the cooperative effort.   It is understood

that the parties will agree on interim arrangements protecting

Pakistan's uses during the period in which the cooperative work is

to continue.   It is also understood that this [letter] does not

imply any recognition, direct or indirect, of any legal claims in

the past or of any legal obligations arising therefrom and is

without prejudice to Pakistan's legal rights except as Pakistan may

hereafter agree and that the final agreement will ultimately require

to be ratified by the two Governments.

I would like once again to thank you and the officers of

the Bank for your efforts to assist in solving this problem and

look forward to continued exercise of the good offices of the Bank.

Figure 3.2: Shoaib's Draft and Iliff's Alterations (II)

G) CROSSING THE RUBICON

Pakistan soon discovered a big flaw in the Bank proposal. It found that the link canals which were to carry water from the Western Rivers for replacement purposes would not be able to meet the shortages of water during critical periods. In its opinion, what was needed was the provision of storage on the Western Rivers in order to meet these shortages during the lean periods. This was indeed a costly affair which the Bank proposal had not provided for. To prove the soundness of its case, the Pakistan government engaged an American consulting firm of engineers called R. J. Tipton to make an independent appraisal of its case. Following the commissioning, the firm undertook the assignment and submitted a report which, proceeding on the basis of the actual flow of supplies recorded at ten-day intervals for ten years, confirmed Pakistan's contention. In its view, the Bank proposal had simply compared the aggregate supplies of waters of one year which was better than normal, namely during 1936–37, with the aggregate seasonal withdrawals by Pakistan in that year. In other words, it failed to take account of both seasonal and year-to-year fluctuations. In light of this finding, it concluded that the Bank proposal did not meet the standard of fairness under international law; that it did not equitably apportion waters of the Indus Rivers between Pakistan and India; that it did not promote the principle of use of water resources for the most effective development; that it would permanently deprive certain areas of Pakistan of water supplies; that it would fail to maintain historic withdrawals; that it negated pre-Partition planned uses; and that it would affect the future of Pakistan's development potential. The relevant part of the report is worth quoting:[105]

> [The] studies disclose that the Bank's proposal does not meet the standard of fairness laid down by the Bank Representative; that the proposal would not have the effect of apportioning the uses of the waters of Basin in accordance with the principle of equitable apportionment; that the proposal would be contrary to the principle

which allows water resources to be used as most effectively to promote development.

The studies disclose that the proposal would have the effect of invading existing uses of Pakistan, some of which date back many years; it would interfere with the development in Pakistan planned before Partition; and it would limit significantly the amount of additional water that could be made usable in Pakistan by the construction of new engineering works. At the same time, the proposal would fully protect India's existing uses on the Eastern rivers; would allot to India the total amount of water planned before Partition for the area which is now in India; and would allot to India, at the expense of Pakistan's existing and future development, substantial additional water supplies which can be made available by new engineering works.

In light of the Tipton report, the prime minister of Pakistan, in a letter addressed to Black, rejected the Bank proposal on the grounds that it did not meet the test of fairness as laid down by the Bank. He made it abundantly clear that, 'as the Tipton report independently demonstrates and as I believe the Bank management must now appreciate, that it is neither practicable nor equitable to cut-off Pakistan's historic supplies from the Eastern rivers. The Government of Pakistan cannot visualize with equanimity the possibility of implementing a plan which would affect its vital interest adversely for all time to come.'[106] On the basis of these observations, he demanded adjustments before Pakistan could accept the Bank's proposal. India was most unhappy with this development because it was of the opinion that Pakistan's demand for the provision of storage went beyond replacement work and was in the nature of development work. It had no objection to Pakistan building as many storage dams on the Western Rivers as it liked, as long as it was not asked to pay for them. Consequently, it made it clear to the Bank that it was not prepared to contribute more than the Bank proposal had envisaged. As far as the Bank was concerned, its view approximated to that of India on the question of storage. In other words, it did not regard storage as replacement work though it conceded that it was necessary for the full development of the Indus Basin. Hence, it too did not think that India should contribute to the building of storage dams if Pakistan undertook to do so; that its contribution should be restricted to the cost of replacement projects only.

Facing a sharp difference of views in the matter between Pakistan and the 'Bank-India' combination, Pakistan asked the Bank to assign independent consultants to evaluate whether or not storage was required to meet its needs. Initially, the Bank baulked at the suggestion and in tandem with the Pakistani and Indian teams, got ready to prepare a comprehensive plan on the basis of its proposal. Pakistan, however, persisted, forcing the latter to accede to its demand by modifying its proposal by incorporating a new 'terms of reference' for the resumption of negotiations whose relevant part was as follows:[107]

> In the event that the flow supplies of the Western Rivers are found to be inadequate (taking into consideration improved operational methods possible under a system of interlinked canals) to meet the uses envisaged ... above, the plan will outline the feasible means that might be adopted to meet any deficiencies.
>
> The planning will include consideration of, and recommendations with regard to, the engineering works required, the costs involved and the sharing thereof, the arrangements for the period of transition and all other pertinent matters.

The new 'terms of reference' were obviously a concession to Pakistan as it opened the possibility of the Bank considering the construction and financing storage dams. Subsequently, in order to press its view on the absolute need for storage dams on the Western Rivers, the Pakistani government succeeded in convincing the Bank to commission independent consultants to undertake a study of the quantity and timings of the availability of water without storage. This was a big success for Pakistan as the Bank decided to entrust the New York-based consulting firm, TAMS (Tippetts-Abbett-McCarthy-Stratton), to prepare such a study. The latter did so and found that the Bank's assumption in 1951, that there was enough water in the Indus Rivers for the future needs of the two countries, was not well-founded and that there was a need for a storage facility on the Western Rivers. In light of these findings, it concluded that Pakistan's assessment was justified.

Following the submission of the TAMS report, the Bank on 21 May 1956 issued an aide-mémoire in which it acknowledged that, '[t]here would be consistent shortages in *rabi*, occasionally beginning in late September and extending into early April of a degree, duration and frequency which the Bank Group could not regard as "tolerable".'[108] It was consequently of the view that without storage facilities, the surplus of water in the Western Rivers would not even be sufficient to meet the replacement needs in the early and late *kharif* period. It therefore felt a need for an adjustment in its proposal in order to assure Pakistan of timely supplies of water to eliminate the possibility of shortage. It envisaged the adjustment to be either through timely deliveries from the Eastern Rivers or construction of storage facilities on the Western Rivers to be paid for by India. It preferred the latter course and for this purpose suggested that, 'the adjustment should be in the form of storage on the Western Rivers ... [and] flow of the Western Rivers should be exploited to the maximum possible extent, and that minimum inroads should be made on Pakistan's limited storage capacity.'[109]

Following the issuance of the aide-mémoire, the focus of discussion shifted from the Bank proposal to the adjusted proposal (i.e. Bank proposal and the aide-mémoire). Iliff proposed that the two countries prepare separate plans on the basis of the adjusted proposal for which he spelled out the criteria and a time frame (which he fixed at 15 October 1956). Pakistan, however, was reluctant to give its consent to the adjusted plan on account of its fear that without adequate and firm guarantees to finance the storage facilities on the Western Rivers, it could be left with the surrender of the Eastern Rivers to India and at the same time be burdened with the obligation to build those facilities from its own resources. Commenting on Pakistan's dilemma, Sommers rightly observed that its focus had shifted from seeking water from the

Eastern Rivers to seeking high financial price from India, or what he termed as shift from 'cusecs to rupees'.[110] India, on its part, was keen to secure Pakistan's agreement so that it would have an absolutely freehand regarding the Eastern Rivers for which purpose it exerted a lot of pressure on Pakistan through the Bank. In fact, it was not prepared to enter any formal discussions about the implementation of the Bank proposal until Pakistan expressed its unconditional and unambiguous acceptance of the adjusted proposal. Faced with this situation, the Bank openly sided with India in the matter and put pressure on Pakistan to accept the adjusted proposal. Incidentally, this pressure tactic was evident, even in the aide-memoire whose paragraph 11 clearly stipulated: 'The Bank feels that if, by 31st March 1957, the Bank should see no reasonable prospects for a settlement on the basis of the Bank's proposal, with an appropriate adjustment, the Bank would have to consider whether the employment of its good offices could make any further contribution to a solution.'

Given Pakistan's reluctance to accept the adjusted proposal, Iliff decided to undertake a visit to Pakistan to bring its leadership round to the idea of accepting it. However, he failed to make the visit and instead sent Wheeler in his place. While in Pakistan, the latter had a meeting with, among others, Prime Minister Chaudhri Muhammed Ali. On his return to Washington, he submitted a report to Black, which clearly reveals Pakistan's serious concerns about financing the storage facilities. Given the fact that it provides insight into the thinking of Pakistan's leadership of that time, and the type of diplomacy that it conducted, we reproduce it here below *in extenso*:[111]

> He [Mohammed Ali] made all of the following points which, of course, we have heard ad nauseam already from the Pakis:
>
> a) He himself in the course of his direct negotiations with the Indians in 1951 could have got a better deal than the Bank Proposal.
>
> b) The Indian claim to the exclusive use of the waters of all the rivers running through Indian territory could never be sustained on the basis of international law. But the Bank Proposal, anyhow, so far as the three Eastern Rivers were concerned, would be a moral seal of

what was regarded as an impartial international organization on a very shaky Indian legal position.

c) The Bank Proposal had two most serious consequences for Pakistan. First, her development potential in the field of irrigation was sadly crippled and secondly, even to the extent that this crippled development could be carried forward the fact that it would have to rely on storage was going to make it infinitely more expensive.

d) He thought that the Bank overrated the seriousness of the Indian intentions, especially in regard to their proposal to tap the Chenab.

e) He felt that the Bank Proposal had made a complete departure from the Lilienthal concept and that the Bank had in effect made a determination basing itself not on engineering but on political considerations.

To all of the above points I returned the stock answers and the PM then went on to say that all that he had been talking about up to now was, of course, past history and that he felt that the only course open to Pakistan was to look to the future and to try and salvage what it could on the basis of the Bank Proposal and with continued Bank participation.

The PM then questioned me at some length about what the Bank had in mind as the probable amount of the Indian financial liability. On this I declined to be drawn. I said that this was precisely one of the points to which the continued co-operative work was to be devoted. The PM went on to say that while he understood that the Indian financial liability would obviously have to be related to the cost of a 'replacement' plan he hoped that in the course of working out a replacement plan the Bank would not lose sight of the Master Plan concept.

Even though Wheeler, as seen above, was reluctant to discuss the financial aspect during his trip to Pakistan, the Bank knew very well that it could make Pakistan accept the adjusted proposal only if it arranged finances for the works that the latter needed to undertake on the Western Rivers. Conscious of this fact, Iliff had, even before the issuance of the aide-memoire, seriously contemplated approaching the US and other Western countries to gauge if they were prepared to chip in with financial assistance to supplement the Indian contribution which, in his opinion, would be greatly inadequate to meet the expenses of said works.

Towards this end, in early 1956, he contacted the officials of these countries to apprise them of the benefits of success and the potential fallout due to failure of the Indus Basin negotiations. Following the encouraging response of these feelers, he decided to approach the British government, which he discovered was favourably disposed towards the idea, on condition that there was a sound engineering plan. As to the level of their interest in the successful outcome of the Indus Basin negotiations, the following statement by Gulhati is noteworthy:[112] 'Iliff also found that the Commonwealth Relations Office was closely following the situation in India and Pakistan and considered that, of all disputes between the two countries, it was most essential to find a solution for the water dispute. It advised the Bank to retain the matter under its umbrella so long as there was any chance of reaching the settlement.' Iliff was in touch with other possible donors, particularly the US. By early 1959, the latter had assured the Bank they would provide substantial financial support and by August 1959, it had committed a sum of $517 million towards the Indus Waters agreement which Eisenhower described as 'one of our more worthwhile projects'.[113]

In the meantime, despite all the pressure, Pakistan stood its ground and refused to accept the adjusted proposal unless it was satisfied with the financial package to implement it. With this backdrop, Iliff travelled to Pakistan where he met Prime Minister Hussain Shaheed Suhrawardy in Lahore. Following the meeting, he sent a report to Black which shows him trying to exploit Pakistan's lack of options rather than attempting to resolve the conflict over the water dispute. Reproduced below is the relevant part of the report, which is quite insightful:[114]

5. He [the Pakistan PM] then went on to criticize the Bank's conduct of its good officer functions. It was Pakistan on which the Bank had been putting pressure all along to concede, concede, concede, and a

point had now arrived where Pakistan's patience was exhausted. At this point, I interjected that of course if the Pak Government felt that the Bank had been in any way delinquent or unsatisfactory in the conduct of its assumed mission, he had only to say so and the Bank, of course, would not dream of imposing its intervention against the wishes of Pakistan. Meanwhile, however, I suggested that the manner in which the Bank carried out its good offices was a matter for the Bank and not for either of the parties. The PM then stated that in his view the proper course of action was to have this question referred 'ab initio' to the ICJ. Here again, an attack developed on Mr Nehru on the basis that this Mr Nehru, who had so often expressed devotion to the principles of law and justice in the conduct of international relations, might perhaps be given a dose of his own medicine. On this, I stated that the legal aspects of the matter were outside the scope of the Bank's interest but I reminded him that equally strong views on the legal merits were held by the Government of India.

The PM again reverted to his conviction that the Government of Pakistan was not prepared to rely on any assurances by India in the matter of the canal waters, even if these were embodied in the international treaty to which India was a signatory. I said that if the PM genuinely held this view, then there was nothing more for me to do but pack my bag and return to Delhi and inform India that Pakistan was unwilling to enter into any engagement with the Government of India, and then return to Washington and recommend to Mr Black that the Bank should fade out of the picture.

Pakistan wanted India to pay in proportion to the benefits that it was to derive, which in concrete terms meant not only for the replacement canals but also for the storage facilities. India, on its part, refused to go beyond the commitment that it had made—to pay according to terms of the original Bank proposal which meant payment for the link canals but not for the storage facilities. Consequently, a deadlock in the matter ensued between the two countries. In order to resolve it, the Bank decided to ask each one of them to prepare a separate, cost effective plan to deal with the issue. Since Pakistan wanted storage facilities both for the replacement and development purposes, the Bank tried to convince it to consecrate the storage facility on the Jhelum River for replacement purposes and dedicate the Indus River, with or without storage, for development purposes. It made this proposal as, in its opinion, it would reduce the cost of replacement works which in turn would make it acceptable to India.

Pakistan had different ideas. In the July 1958 meeting held in London, it proposed what came to be known as the 'London plan' which envisaged the construction of two dams, one at the Jhelum River for replacement purposes and the other at the Indus River for development purposes. Additionally, it proposed the construction of three smaller dams on the tributaries of the Jhelum and the Indus Rivers and a series of link canals. It put the estimated cost of these works at $1.12 billion.

India objected to Pakistan's plan on the grounds that the cost of the replacement works was excessive. Therefore, it submitted an alternative plan which was a rewrite of the plan that it had put forward back in 1953. It envisaged the transfer of water from the Chenab River, through a diversion tunnel, to the tributary of the Ravi River at Marhu and that of the Chandra (a tributary of the Chenab River) to the Beas River, and a storage dam at Dhiangarh. As a quid pro quo for this arrangement, it promised to deliver, out of Pakistan's total needs of 10 million acre feet of water, some 5 million acre feet of water at Ferozpur and Merala. As for the remaining 5 million acre feet, it was of the opinion that Pakistan could get it from the link canals that it was in the process of constructing. It justified its plan on the grounds that it was much cheaper than that of Pakistan.[115]

Pakistan, however, rejected the Indian plan on the grounds that the link canals it was building did not fall in the category of replacement works and were in fact a diversion of water that it had had even before Partition, from one irrigation system to another. It also rejected it on the grounds that the location of these works on Indian soil would make it dependent on India which was not acceptable to it beyond the transition period, irrespective of its cost and whether or not it was enshrined in a treaty. As far as the Bank was concerned, it favoured the Indian plan on the grounds of its cost effectiveness but did not press for it because it was contrary to the concept of independant development and operation by each country of its portion of the Indus Basin which its proposal enshrined. Consequently, it decided to jettison the Indian plan and work on the basis of a plan that Pakistan would approve.

In October 1958, a *coup d'état* took place in Pakistan as a result of which, Ayub Khan, the commander-in-chief of the armed forces, took over the government. This was an utterly unforeseen situation whose implications for the ongoing negotiations were not clear. Towards the end of 1958, Black travelled to Pakistan to gain a sense of where the new government stood on the ongoing Indus Basin negotiations. He met the new Pakistani leader over there and reportedly 'the two men had hit it off famously'.[116] Soon after, in a major development, Mueenuddin, Pakistan's pointman in the Indus Basin negotiations, in a letter addressed to Iliff, unconditionally accepted the adjusted proposal in these words:[117]

> As you know, it is the position of my Government that they have already accepted the Bank Proposal of February 5, 1954, and the Aide-Memoire of May 21, 1956, as the continuing basis for reaching a co-operative solution of the Indus Waters question with the assistance of the Bank.
>
> However, to remove any possible doubts, my Government do now accept without condition or reservation the Bank Proposal of February 5, 1954, and the Aide-Memoire of May 21, 1956, as the continuing basis for reaching a co-operative solution of the Indus Waters question with the assistance of the Bank.

It is not clear what made Ayub write this letter. Gulhati thinks that the latter did so either to abort the possibility of the Bank inclining towards the Indian plan based on the Marhu tunnel proposal or because of his reappraisal of the situation.[118] In our judgment, the explanation lies in the fact that, unlike the previous civilian governments, his government was strong and hence, decisive. Besides, Black must have offered an incentive in the form of a financial package for the implementation of the adjusted proposal which he had, by this time, almost shored up. The argument that Ayub's government might have been driven by the fear that the Bank would accept the Indian plan does not hold because it was an absolute non-starter. This is simply because, as discussed above, it would have made Pakistan dependent on India. It is indeed strange that Gulhati, who was a participant in the negotiations, was ignorant of this stark reality. In any case, Pakistan's acceptance of the adjusted proposal was a watershed event as it meant that for the first time since the start of the negotiations, there was absolutely no ambiguity where Pakistan stood on the question of dividing the Indus system of rivers. It also meant that the deck was now clear for movement towards a water treaty.

But before the parties got down to drafting a treaty, they had to sort out two important matters: India's financial contribution for replacement works in Pakistan and the transition period for the transfer of water supplies from the Western Rivers to replace historic withdrawals from the Eastern Rivers. As for the first issue, the Bank proposed to resolve it through adjudication or arbitration. Pakistan favoured a quasi-judicial approach but India rejected it. The matter was then taken up bilaterally between the two countries. Pakistan estimated the sum to be more than a billion dollars whereas Gulhati, who was India's pointman in the Indus Basin negotiations, put it at $80 million. Nehru thought that Pakistan's demand was 'quite unrealistic, even so far as India's capacity to pay was concerned: Pakistan could "look to the moon" if those were her ideas.'[119] He was of the view that the waters of the Eastern Rivers legally belonged to India and that any financial contribution that it intended to make was an *ex gratia* payment. He estimated that the Indian financial liability for a plan incorporating the Marhu tunnel proposal should not exceed Rs 70 to 75 crores. The Bank, on its part, put it at $250 million, spread over ten years. When the latter conveyed this figure to Nehru, he refused to accept it. He referred to India's claim of $600 million from Pakistan resulting from the Partition settlement out of which, according to him, $42 million were directly attributable to the award by the Arbitral Tribunal of the capital value of the canal system. The Bank and India then settled on a sum of $175 million as India's contribution in cash payable, annual, and equal instalments over a period of ten years. This arrangement took place between the Bank and India bilaterally without involving Pakistan.

As for the transition period, in its proposal the Bank had calculated five years for the necessary link canals to be built. India had fixed 1962 as the end of the transition period as, in its opinion, three out of five years had already elapsed. Pakistan regarded this date as arbitrary and wholly incompatible with the adjusted proposal. It contended that when the Bank put forward its proposal, it did not envisage any storage on the Western Rivers on the grounds that link canals would be enough to meet Pakistan's water needs. It maintained that subsequently, the Bank modified its plan through the aide-memoire which, in addition to link canals and other works, provided for storage. In its view, the transition period had to be calculated, keeping in view the time needed for constructing and stabilizing the system of works envisaged in the adjusted proposal which, it argued, would begin only after the finalization of the plan with Pakistan and after India started contributing towards the cost of works included in the plan and the schedule of its payment.[120] Pakistan wanted this period to be ten years whereas India thought that a shorter period would do. Gulhati, for example, thought that ten years was too long a period for the construction of the Mangla Dam which Pakistan had undertaken on the Jhelum River. In his opinion, it could be completed in not more than eight years. In the face of these differences between the two countries, the Bank volunteered to put forward its own proposal on the condition that if either government refused to consider it, the Bank would pull out by 'mak[ing] a public statement as to why it had done so'.[121] The two countries, however, were not amenable to this condition and ultimately struck a deal bilaterally, though with the help of the Bank.[122] They settled on a ten year transition period which was to run from April 1960, to March 1970, and another three years as grace period from April 1970, to March 1973, but with a penalty clause.[123]

Around the time (May 1959) the two countries struck a deal on India's financial contribution and the transition period, an important development took place when the Bank agreed to provide more finances for storage on the Western Rivers than it had earlier agreed. We know that it had envisaged two dams in Pakistan, the Mangla Dam on the Jhelum River and the Rohtas Dam on a tributary of the same river. Meanwhile, Pakistan had envisaged, in addition to the Mangla Dam, the much larger Tarbela Dam on the Indus River instead of the smaller Rohtas Dam. If it were to embrace Pakistan's plan rather than its own, it would have been obliged to carry an additional burden of $200 million over and above the considerable amount it had already pledged towards the Indus Basin settlement. Black baulked at the idea when he learned of it but eventually gave in to Ayub's demand.[124]

Given the fact that these additional funds were, like most of the remaining contribution, a mere grant, how do we explain this largesse? Was the Bank trying to be a 'good Samaritan' or was there something else at play? The answer lies in the fact that the Bank's principal backer, the US, which was the main provider of this largesse, was happy with Pakistan for conceding the three Eastern Rivers to the non-aligned India which it hoped would help rope the latter in to the capitalist fold (see Ch. 3 (J)).

H) TOWARDS A WATER TREATY

The first step in framing a water treaty was the drawing up of the Heads of Agreement. Towards this end, Iliff, in May 1959, submitted before the parties a draft which covered a number of areas such as the division of the rivers, a system of works on the Western Rivers, the Indus Basin Development Commission, the Indian contribution to Pakistan, etc.[125] He explained the significance of these heads in terms of the degree of agreement that representatives of the parties had reached on an *ad referendum* basis and that they were now ripe for consideration by the two governments. He further explained that they were to serve as the basis of the text of a treaty. He admitted that they were not couched in legal language and that the parties intended to transform them into a treaty. The parties worked out several Heads of Agreement and three drafts of the treaty. Counting from the date of submission of the Heads of Agreement, it took more than fifteen months to finalize the treaty. Pakistan proposed to add a paragraph to the Heads of Agreement which stipulated that if the parties failed to agree on a plan and its implementation, or on a quasi-judicial tribunal to resolve their differences, or the Bank decided to withdraw its good offices, each party would be at the liberty to submit the issues relating to the Indus system of rivers to the ICJ. This appeared to be a reasonable proposition. India, however, refused to accept it.[126] This was not surprising because, as noted before, it had always resisted resolution through adjudication.

A contentious issue between the two countries was India's use of the waters of the Western Rivers before they crossed into Pakistan. Of all the outstanding issues, this was perhaps the hardest to resolve because India enjoyed the status of an upper riparian and these rivers traversed Jammu and Kashmir which was under dispute. We know that the Bank had stipulated in its proposal that '[t]he entire flow of the Western Rivers would be available for the exclusive use and benefit of Pakistan, and for development by Pakistan …'. However, India had serious reservations to this formulation because, in its opinion, 'it could be made to imply some vague rights for Pakistan on the Western Rivers even while these rivers are in Indian territory— rights of control, rights of interference, rights of veto, rights of consent, etc.'[127] India wanted to deny to Pakistan the right to waters of the Western Rivers while they flowed on the Indian soil. It wanted to have the right to use these waters in all possible manners:[128] 'Such uses from these rivers, while they flow in Indian territory, as are necessary for the economic well-being of the people living in the catchment of these rivers had to be safeguarded; these could not be treated as "undue interference". Accordingly, we had not only to ensure that Pakistan should have no right whatever on the Western Rivers while in Indian territory but also to make certain that the treaty would specify clearly Indian rights to all possible uses …'.

Pakistan, on its part, was opposed to granting India the right to build works on the Western Rivers because it was apprehensive of the latter's designs. Its principal concern was to ensure a free flow of water without let or hindrance, which it regarded as a sacrosanct principle. It was of the view that any concession in favour of India to construct works on the Western Rivers could lead to interference by India in the free flow of these waters. It was so apprehensive of Indian intentions that it was not even amenable to the idea of letting India generate hydroelectric power because, in its thinking, this concession could provide the latter with a handle to control waters of the Western Rivers. As far as India was concerned, it was not prepared to accept this kind of arrangement under any circumstances whatsoever. Unfortunately, the Bank too was not favourably disposed towards Pakistan's viewpoint for the simple reason that this would have resulted in an important natural resource remaining unutilized. This fact tilted the balance on the issue in favour of India.

During the talks held in August and September 1959 (in London) to deal with the Heads of Agreement, Mueenuddin sent a flier to Karachi in which he disclosed that India had made a demand for an unlimited right to develop uses 'from Indus and Jhelum above lake and Chenab above RL 2000 covering Jammu and Kashmir as well as Indian territory. They also demand storages.'[129] Pakistan was highly upset with this development because it thought that it went far beyond the Bank proposal according to which Pakistan was to surrender its rights over waters of the Eastern Rivers in return for India relinquishing all of its rights over waters of the Western Rivers except for (a) existing uses in Jammu and Kashmir as in 1947; and (b) future irrigation through minor extensions from existing channels or use of insignificant amounts of water from smaller feeder streams, completely ruling out new channels and storages. Pakistan rejected the Indian demand on the grounds that it would interfere, or make it possible to interfere with the flow of rivers. Mueenuddin explained it in these words:[130]

The essence of the Bank Proposal and the basic justification for
the division of rivers was to make the two countries independent
of each other in the operation of their supplies. Being the lower
riparian, Pakistan alone is vulnerable to interference by India. By
introducing for the first time at this stage new uses on the Western
Rivers, e.g. those in Himachal Pardesh and Punjab (India) as well as
unrestricted right to develop hydroelectric power from these rivers,
India has, while trying effectively to secure to herself the exclusive
use and development of the Eastern Rivers, sought to deny the
reciprocal independence to Pakistan which the Bank Proposal and
the Aide-Memoire promised to afford each country. Pakistan cannot
obviously accept a position which, despite the sacrifice on her part in
relinquishing permanently her rights on the Eastern Rivers, would
take away from her the only consolation she could look forward to in
agreeing to the division of rivers. In Pakistan's view the acceptance
by India of the Bank Proposal and the Aide-Memoire, in letter
and spirit, to form a firm basis for an International Water Treaty
should mean an unequal surrender of India's claim, if any, on the
Western Rivers.

Pakistan was so upset with this development that Ayub sent a subtle
SOS to Black in which he complained of the Indian demand. He
also drew attention to the commitment that the vice-president of the
Bank, Robert Garner, had made to Pakistan in 1954 to not allow
India to build any storage on the Western Rivers. In desperation,
Pakistan also approached the US State Department in Washington
and the Commonwealth Relations Office in London to intervene on
its behalf with the Bank, but both of them refused to oblige. The
Bank, however, made India back off from its extreme position but at
the same time refused to support Pakistan in its opposition of all types
of run-of-river hydroelectric works on the grounds that, 'it would
mean freezing for all time the available hydel potential of these rivers
in their upper reaches.'[131] Consequently, India got the right to build
run-of-river hydroelectric plants on the condition that it would let the
water flow and not interfere with it. In return, Pakistan got the right
to inspect such works.[132]

While embarking upon the drafting of the water treaty, the dispute over Jammu and Kashmir was bypassed by writing it in such a manner that no party could gain support for its position or erode that of the other. Similarly, they decided not to mention any work that Pakistan was constructing in the disputed territory nor provide any indication that India had agreed to it. Despite this understanding, the issue of constructing dams and reservoirs in Azad Kashmir (by Pakistan) presented considerable drafting difficulties. Nehru feared that if he gave formal consent to the construction, for example, of the Mangla Dam, Pakistan might construe it as a waiver, by India, of its claim of sovereignty over Azad Kashmir or what India termed as Pakistan occupied Kashmir (PoK). In this vein, he referred to the protest that India had already lodged with the United Nations against the construction that Pakistan had already undertaken. Iliff tried to show a way out of the impasse[133] by suggesting a clause in the proposed treaty along the following lines: 'Nothing in this Treaty should be construed as prejudicing the rights or claims of either India or Pakistan in any territorial dispute.' Nehru disagreed with the proposed text on the grounds that India did not consider Kashmir a territorial dispute. It was, in his view, an Indian territory which Pakistan had illegally occupied. He felt, however, that it was possible to devise a formula to take care of this problem. Subsequently, Pakistan and India successfully negotiated the following text during the London talks:[134]

> Nothing contained in this Treaty and nothing arising out of the execution thereof shall be construed as constituting a recognition or waiver (whether tacit, by implication or otherwise) of any rights or claims of either of the Parties other than those rights or claims which are expressly recognized or waived in this Treaty; and each of the Parties agrees that it will not invoke anything contained in this Treaty, or anything arising out of the execution thereof, in support of any of its own rights or claims or in disputing the rights or claims of the other Party other than those rights or claims of the other Party which are expressly recognized or waived in this Treaty.

This formulation found its way into the Heads of Agreement. Later, the two countries decided to incorporate it in the draft Treaty rather than accept it through an exchange of notes. They also mutually decided to amplify the above text to read it as follows:[135]

> It is expressly understood that this Treaty relates only to the rights and obligations of each Party in relation to the other with respect to the use of the waters of the Rivers and matters incidental thereof; and nothing contained in this Treaty and nothing arising out of the execution thereof shall be construed as constituting a recognition or waiver (whether tacit, by implication or otherwise) of any rights or claims of either of the Parties other than those rights or claims which are expressly recognized or waived in this Treaty; and each of the Parties agrees that it will not invoke anything contained in this Treaty, or anything arising out of the execution thereof, in support of any of its rights or claims or in disputing the rights or claims of the other Party other than those rights or claims of the other Party which are expressly recognized or waived in this Treaty.

Mueenuddin subsequently expressed his government's dissatisfaction with the revised text as, in his view, it did not adequately protect Pakistan's position on Kashmir. He wanted the following provision to be added to it:[136]

> The rights and obligations of each of the Parties under this Treaty apply to all the territories which at the time are under its actual control; but neither the provisions of this Treaty nor any steps taken as permitted in this Treaty, or to promote compliance therewith, shall be construed as affecting in any way the position of the Parties as to the right to exercise such control.

Pakistan also proposed, as an alternative, a number of amendments to the above formulation whose basic thrust was to underline the fact that India was de facto but not *de jure* in control of Jammu and Kashmir. It did so in order to drive home the fact that whereas it was conceding to India the right to do certain things in the part of Jammu and Kashmir under its control (e.g. withdraw water from the Western

Rivers for irrigation purposes and build hydroelectric plants and storages), the latter was refusing to grant a similar right to Pakistan in the part under its control. In other words, it maintained that though it was recognizing India's de facto position in the Indian-administered Jammu and Kashmir, India was denying this status to it in the part administered by it. The Bank was not much amused with Pakistan's amendments as evidenced from the following observation by Black:[137]

> In my view we should tell Pakistan that with legal advice, she has following two choices: (a) maintaining intact her position on Kashmir, and losing a water Treaty or (b) of conceding some rather nebulous legal point on Kashmir (which may be adequately covered by the formula in para. 4 above), and getting a water treaty. Moreover if there is no water treaty, Pakistan will not only lose the prospects she has of substantial financial assistance from friendly governments, but her economic and her external financial credit will suffer a severe blow.

The Bank conveyed these views to Mueenuddin and Pakistan's ambassador to Washington. Given the fact that the room to manoeuvre for Pakistan in the matter was limited, the Bank succeeded in its attempt to make Pakistan drop its amendments and accept the text as drafted. The issue of non-recognition of Pakistan's de facto control of Azad Kashmir came up in the *Kishenganga* case when India argued that it was not under any legal obligation to avoid adverse impact of the KHEP on the territories that do not form part of Pakistan because the Treaty does not apply to regions which are under the de facto control of Pakistan.[138] It is true that Pakistan did not suffer, as a result of the above-mentioned legal lacuna, in its position regarding Azad Kashmir but this is because, fortunately for it, the Arbitration Court sided with Pakistan on this issue. However, it shows that Pakistan's apprehensions in the matter were not unwarranted and the Bank was not justified in rejecting Pakistan's objections. The Bank also made Pakistan drop several law propositions that its foreign minister wanted to insert which had bearings on certain articles of the draft Treaty as revealed below:[139] 'Black is glad to learn that you regard the issue of academic importance .... He is, however, deeply concerned that you have chosen nevertheless to link this issue with the Indus Treaty in that, at this late date, you propose to add to the Indus record an amplification of Pakistan's legal position, and are insisting that this should be done before the signature of the Indus Treaty.'

We have seen above that, before the consideration of the Heads of Agreement, Pakistan and India had agreed on the transition period which could go up to thirteen years. Pakistan now demanded the insertion of a provision for the transition period to be extended in case of *force majeure* affecting the implementation of the treaty, resulting from the failure of friendly governments to provide necessary funds for the completion of works. It also wanted the said provision to require India to desist, in such an eventuality, from withdrawing additional water over and above the replacement capacity developed at that time. India refused to oblige Pakistan on the grounds that India could not be made to pay for situations resulting from arrangements to which it was not a party; that such a provision would be a factor of instability in the treaty; and that it could not abandon its rights over the Western Rivers except under an arrangement in which the division of rivers was final and it enjoyed early freedom to fully develop waters of the Eastern Rivers.[140] However, at the insistence of Iliff, the parties decided to postpone consideration of the matter till further deliberations.

Subsequently, Pakistan jettisoned its proposal after Iliff assured it of the availability of funds for replacement works. Furthermore, he instead proposed the inclusion of a provision stipulating the extension of the transition period in case of an emergency arising out of the difficulty in acquiring the machinery or material needed for construction of works. India was not amenable to this demand either. Following Iliff's mediatory role, the parties then settled on a provision limiting the *force majeure* factor to the outbreak of large scale international hostilities arising out of causes beyond Pakistan's control. This relief was available to Pakistan only during the first five years of the transition period.[141]

The parties also envisaged the establishment of a Permanent Indus Commission—on the pattern of the Canadian-American International Joint Commission on St. Lawrence and Great Lakes and the Mexican-American International Boundary Commission on Rio Grande—to oversee the implementation of the Treaty. There were, however, sharp differences between them on a number of issues, particularly the powers of the Neutral Expert and the Court of Arbitration, so much so that on several occasions, the talks almost came to a breaking point. At one point, Pakistan's legal advisor proposed to invest the Neutral Expert with the power to issue an interim injunction in a situation where there was delay between the occurrence of a dispute and its submission to arbitration. He wanted both the parties to enjoy this right. The purpose of his proposal was to stop India from causing irrevocable damage to Pakistan through its dilatory tactics. This did not curry favour with India or the Bank. The former was furious with Pakistan's legal advisor's 'legalistic' approach and the latter told Mueenuddin that the legal advisor should, in the future, only speak through the Pakistan delegation.[142] In the face of their opposition, the proposal fell through. In hindsight, we can say that Pakistan's legal advisor's apprehensions were well founded because India, more often than not—taking advantage of the existing lacuna in the Treaty regarding this matter—continues with a project against Pakistan's legitimate objections in order to present Pakistan with a fait accompli. It is not clear from the record available in the Bank's archives why the Pakistan government did not support its legal advisor on this point. Pakistan also pleaded for a neutral chairman to head the Permanent Indus Commission but India would not agree to it. Eventually, the idea was abandoned.

After eight years of protracted and arduous negotiations,[143] the Indus Waters Treaty was finally concluded. It was signed on 19 September 1960, in Karachi.[144] Ayub signed it on behalf of Pakistan and Nehru on behalf of India. Vice-President Iliff signed it on behalf of the Bank because Black was not available due to an illness. Additionally, the representatives of Australia, Canada, Germany,

New Zealand, Pakistan, the UK, the US, and the World Bank signed an international financial agreement which created an Indus Basin Development Fund of about $900 million for the purpose of financing the construction of irrigation and other works in Pakistan.[145] The Treaty also brought the controversy regarding the May 1948 Joint Statement to an amicable end through an exchange of letters by Pakistan and India included in the Treaty as Annexure A. As expected, even this arrangement was not free from controversy. India insisted on calling the May 1948 Joint Statement 'an Agreement' whereas Pakistan termed it 'a document'. The two countries resolved the controversy by stating in their letters that the arrangement made in May 1948 'shall be without effect as from 1st April 1960'. India in its letter called it 'the Inter-Dominion Agreement on the Canal Water Dispute signed at New Delhi on 4th May 1948', and Pakistan termed it 'the document on the Canal Water Dispute signed at New Delhi on 4th May 1948'.

I) APPRAISAL OF AYUB KHAN'S ROLE

The signing of the Indus Waters Treaty was greeted with scathing criticism in both Pakistan and India.[146] India's opposition and local media criticized Nehru for adopting a policy of appeasement and surrender towards Pakistan in addition to abandoning India's national interest. In Pakistan, the opposition and the media alleged that Ayub 'sold' the three Eastern Rivers for the sake of money and that a democratically elected government would have never done it.[147] The controversy was so persistent that in the five years after its conclusion, the Treaty became an issue during the 1965 presidential election which pitted Ayub against the combined opposition candidate, Mohtarma Fatima Jinnah.[148] The wound of the loss of the three Eastern Rivers is so deep in the psyche of the people of Pakistan that even today passions run very high whenever the issue surfaces. The participants more often than not hold Ayub responsible for the loss. It is amazing that more than fifty years after the conclusion of the Treaty, Pakistanis have predominantly failed to make an objective assessment of Ayub's role in the matter. This is a reflection on the depth of the wound that the surrender of the three Eastern Rivers has inflicted on the Pakistani nation. This raises the following questions. Did Ayub really sell the three Eastern Rivers for the sake of money? Would a civilian government have behaved differently than the military government of Ayub Khan?

In addressing these questions, we observe that it was the civilian government of Firoz Khan Noon and not that of Ayub which committed the 'original sin' of accepting the Bank proposal to surrender the three Eastern Rivers to India. Therefore, Ayub should not be blamed for this, as many of his critics do, without realizing that he did not take the fateful decision but his civilian predecessor did. One may argue that he could have certainly gone back on the proposal, if he so desired, without being in breach of any obligation. Juridically speaking, the point is well taken because according to the terms agreed upon between the parties beforehand '... each side [was] free to withdraw at any time'[149] from the negotiations. However, morally and politically speaking, it was very hard to do so because two and half years had elapsed since the Bank had put forward its proposal and during this period, Pakistan had continued to negotiate the understanding that it would form the basis of the settlement. It's no wonder Black blackmailed the parties by threatening to release the details of the negotiations whenever he felt that they were creating hurdles in the way of the settlement. Besides, Ayub's predecessor government had signalled its express consent to the Bank proposal only three months before. With this in mind, it would have been extremely embarrassing, to say the least, for Ayub to jettison the Bank's proposal.

Another and perhaps more weighty reason why Ayub embraced the Bank's proposal was his realization that it represented the best available option for Pakistan under the circumstances. He must have reached this conclusion because Pakistan, despite its best efforts, had failed to bring India round to the idea of submitting the water dispute before the World Court and had no hope that its reference to the UN Security Council, which it threatened to do, would achieve the desired results. Black helped him adopt a pragmatic approach by reportedly making the following observation:[150] 'If you can fight [a] war with India and win it, then I would advise you not to sign the Treaty. But if you cannot, then under the circumstances, it is the best deal.' In other words, conscious of the fact that the odds were terribly stacked against Pakistan, he decided to act on the basis of realpolitik. It goes to his credit that in an emotionally charged atmosphere where virtually all his advisors, technical and administrative, were incapable of clear thinking and were tendering unrealistic advice on the issue, he kept his cool and accepted the Bank-brokered deal as we see from the following excerpt from his autobiography which we reproduce *in extenso*:[151]

> But before I write of the negotiations with Eugene Black, I should like to describe the confrontation that I had with our own technical experts and administrators. I sensed that they did not fully realize the gravity of the situation and were asking for [the] moon when we were in a position of weakness all along the line. They were also trying to dictate policy and were trying to take up extreme positions. Some thirty or forty of them were assembled in Government House, Lahore, where I addressed them. I said: 'Gentlemen, this is of far reaching consequences to us. Let me tell you that every factor is against Pakistan. I am not saying that we should surrender our rights but, at the same time, I will say this: that if we can get a solution which we can live with, we shall be very foolish not to accept it. Now when I say that, I am in fact saying to myself because I shall have to take the responsibility for the solution.
>
> 'The responsibility does not lie on any one of you, so let me tell you very plainly that the policy is going to be mine. I shall consult you whenever I am in doubt regarding technical details, but if any one of you interferes with policy, I shall deal with him myself. This problem, if not tackled properly, may mean the end of the country. I mean every word of it. So, don't let any one make any mistake about it ....'
>
> When one is dealing with a sensitive problem of this nature, one has to be realistic and judge the situation dispassionately in order to

formulate a rational approach. Very often the best is the enemy of the good. We abandoned the chase of the ideal and accepted what was good after a careful and realistic appreciation of the overall situation. Had we not done that, we might have drifted into a conflict at a time when many factors were against us. The basis of the agreement, therefore, as far as we were concerned, was realism and pragmatism. Emotions had no place in it, nor could they be allowed to have any place where the future and safety of millions of people depended on a solution.

As for the charge that Ayub sold the three Eastern Rivers to India in return for money, it is true that he got a handsome financial package, arranged by the Bank, to develop alternative sources of water supply for irrigation and other purposes.[152] However, instead of blaming him for this he should be appreciated because losing the three Eastern Rivers, which Pakistan would have done in any case because India was determined to appropriate them, and to be left without adequate funds to develop alternative sources, would have been absolutely unpardonable. Incidentally, he was not alone in committing this 'crime' because all of his civilian predecessors (without exception) had contemplated this approach. For example, his immediate predecessor, Firoz Khan Noon, while accepting the Bank proposal, had submitted a financial package along with it. Similarly, Chaudhri Muhammad Ali's thinking was no different in the matter as borne out by the following report that Iliff sent to Black after his meeting with the latter following his government's acceptance of the aide-memoire:[153] 'Prime Minister had revealed that his government was contemplating to take the matter to the Security Council if necessary. It is clear from the foregoing that the purpose of the Prime Minister's comment was to force the Bank to fund the construction of storage facilities rather than roll back the gains made from the negotiations. If this was not the purpose there was no point in raising these objections at such a late stage.' This line of thinking is visible from the day the Bank got involved in the negotiations. This is testified by Lilienthal who claims that all Pakistani governments were prepared to accept his plan but were reticent to do so openly out of the fear that they might surrender the Eastern Rivers without any financial compensation, with the result that they might be forced to develop the alternative sources from Pakistan's own resources. This is how he put it in his journal:[154]

> On Tuesday, May 6 [1952], I met with Eugene Black, head of the World Bank, for a short time prior to the 'ceremony' that opened the negotiations between the engineers of Pakistan and India regarding my Indus proposal. He said that while he was in India and Pakistan he 'learned' what was at the bottom of trouble over water. It was who was to pay for developing the alternative sources for Pakistan.

It may be fair to suggest that Pakistanis have utterly failed to properly assess Ayub's contribution to the making of the Treaty. As shown above, he is unfairly blamed for selling the three Eastern Rivers because Pakistan not only had no option in the matter, but also the policy he pursued was nothing but a continuation of the one all his predecessors had followed. We would, on the contrary, argue that it was a master stroke on his part to make the Bank arrange huge funds, which were almost entirely gratis, for the development of alternative sources. In our estimation, the Treaty is a crowning achievement of his rule because it adequately protected Pakistan's water rights in the Indus Basin. It is almost certain that in its absence, India would have not only gobbled up the three Eastern Rivers but also appropriated a good part of the Western Rivers as well by building a large number of dams and reservoirs under the so-called Harmon Doctrine—the declassified documents of the Bank reveal that India desired a free hand to do so. If it has not been able to do it so far, it is because Pakistan's water rights on the Western Rivers are protected under the Treaty. It must also be said that if Ayub had not signed the Treaty and Pakistan was to negotiate a water treaty with India today, given the power balance in the subcontinent, it would almost certainly not get the kind of generous terms that it got in the Treaty.

J) BANK'S MOTIVE TO SIDE WITH INDIA

We have seen that the Bank sided with India regarding its demand for the allocation of the three Eastern Rivers but would be prudent to ask why. Before addressing this question, there is a prior one (which we raised in the beginning of this chapter but deferred it for a later treatment) that needs to be answered. It relates to the reason why the Bank became involved in a complicated and tricky water dispute between Pakistan and India in the first place. Black, under whose stewardship the Bank had achieved the remarkable feat of concluding the Treaty successfully, has suggested that it did so because it was inspired by the ideal of peace (though it is another matter that Lilienthal says that the Bank blamed him for having gotten them into it[155]). He is not the only one to make this claim. There are others who share this viewpoint. Two prominent writers, in a joint article, have stated that '[t]he story of the Indus Treaty was the story of the Bank's commitment to development, its "investment in peace", and its success in "development diplomacy".' They go on to add that the Bank's role was 'proactive, neutral, pragmatic and fair'.[156] Another writer has suggested that the Bank's role was that of a 'benign go-between'.[157] If, on one occasion, Black explained the Bank's role in terms of peace promotion, on another, he offered a different explanation: 'not only did the Indus issue stand in the way of constructive action by the Bank on the subcontinent, it was a cause for concern in a wider sense. With two of the Bank's potentially largest customers so bitterly opposed, the world's investors might regard the Bank's own prospects with a skepticism that would cripple its ability to raise money through borrowing.'[158]

The idea that the Bank was nothing but a 'good Samaritan' and 'an honest broker', guided by an altruistic consideration to promote peace and to establish its own credentials as a sound banking institution is a bit dubious. This is particularly so, when we juxtapose these claims with the fact that it mobilized more than a billion dollars in grants and loans from Western nations (for Pakistan) for the loss of the Eastern Rivers; devoted about eight long years to expensive, stressful, and often frustrating negotiations; and went beyond the mandate that its creators had fixed for it at the Bretton Woods Conference in 1945. Here the question arises, if the above explanation does not make sense, then what does?

In our judgment, the answer is rooted in the international political context at the time the Bank got involved in the dispute. We need to remember that the dispute arose and intensified during the Cold War when the capitalist and communist blocs were fiercely jockeying with each other for domination and hegemony in the world as attested, for example, by the war in Korea. With this backdrop, the US must have realized that if the water issue between Pakistan and India was not resolved and the two countries went to war, it would provide a golden opportunity for the communists to fish in the troubled waters of South Asia. Lilienthal obliquely refers to this danger in his article where he says that a war between Pakistan and India would 'undoubtedly put the USA into another and bigger Korea'. But perhaps, more important than that was his fear that the US had already lost China to communism and was scared of losing India too, as he expressed it in the following words:[159]

> India presents the United States and democracy an opportunity, such as we had and missed in China 25 years ago .... It is probably too late in China now.
>
> But it is not too late in India. The same conditions of impoverishment and need; the same huge population; the same potential threat if this segment of humanity goes against us, *or, even worse, lines up with the Communists.* Let us not repeat the course in China. In our interest, and the long run as well ....
>
> We never had China to 'lose'. To demonstrate this, and the things that we can learn from that important proposition, point to India. It is now neutral, somewhat unfriendly to us, but far from hopeless.

What Lilienthal wrote in his journal is a continuation of what he wrote in his article, 'Are We Losing India?', that he published on his return from a visit to the subcontinent.[160] He started the article with the statement: 'We are witnessing today what may be the beginning of the end of friendly relations with India, just as only a few years ago ... we began the process of losing out in China. If we lose out in India, as we have in China, all hope for us on the continent of Asia is gone.' Continuing this line of thought, he observed: 'Our course, in general, seems to me plain: to encourage and aid Nehru and his people in the development of a modern nation.' He concluded it with the observation: 'All we want is a good working relation with a vast and struggling young nation. But we are not in the position of a suppliant, begging for India's favor. America can survive if it must without India, or without Asia, for that matter.' Before proceeding to India (and Pakistan), Lilienthal met with the State Department and other Administration officials who hoped that he would bridge the gap between India and the US.

He also met the Secretary of State, Dean Acheson, and President Harry S. Truman, both of whom encouraged him to undertake the trip. The former said to him: 'I'm very glad you are going; it can be very helpful, very useful, in a number of ways.' Though *en passant* he commented on the personality of Nehru whom he called 'slippery' and 'a monumental snob', in addition to observing that he '[l]ikes to be surrounded by beautiful and dumb women, the more beautiful and the less intelligent, the happier'.[161] In short, we can say that Lilienthal who went to the subcontinent as the '*Collier's* writer' actually landed there to further the American establishment's agenda which the US State Department couched in these words:[162] 'In all of Asia it is now the only nation that is large enough and has the power potential to resist a determined Communist military effort with any possibility of success. If India should fall into Communist power, a consolidation of that power throughout Asia would be inevitable. *If we are to have an effective policy in Asia, therefore, India must be the keystone of that policy.*' The National Security Council policy paper (no. 98/1) on South Asia, which President Truman approved on 22 January 1951, reiterated the same agenda in these words:[163] '*The loss of India to the Communist orbit would mean that for all practical purposes all of Asia will have been lost. This would constitute a serious threat to the security position of the United States.*'

Given this agenda, the US was worried about Nehru who had absolutely no mind to align India with the West against the communist bloc and was determined to stay neutral in the East-West power struggle. We know that India became a member of the Non-Aligned Movement in 1961. However, its proclivity towards non-alignment was visible in the early 1950s. If we keep this political context in mind, we have a better understanding of the motives behind the Bank's initiative on the water issue. The US wished to woo India and keep it on its side, therefore, Lilienthal vociferously recommended against the exercise of the right to self-determination that the UN Security Council had envisaged as a means to resolve the Kashmir issue. He also recommended against adjudication of the dispute by the World Court. He justified this recommendation on the grounds that 'it would antagonize India'[164] which, he argued, was not in the US interests.[165] It may be contended here that the US and the Bank are two different and entirely separate entities and that they should not be confused with each other. This position only existed at the rhetorical level. As for the practice, it is common knowledge that the Bank, more often than not, acts like a subset of US foreign policy.

We have seen that the Bank chose to break the deadlock between Pakistan and India by allocating the three Eastern Rivers to India (and the three Western Rivers to Pakistan). The allocation of the three Eastern Rivers to India was *grosso modo*, a rehash of the plan that the latter had submitted earlier and which caused an impasse in the talks as India was not prepared to compromise. This signifies that the Bank was siding with India. This disposition is also evident from the fact that, as seen above, it dictated changes in Pakistan's draft reply to its proposal and which Pakistan had to accept. Did it do so because it was convinced that India's claim to the Eastern Rivers was justified or was it pursuing some other agenda? The answer to this question can be found in the motive that led to its involvement in the Pakistan-India water dispute in the first place. That motive was, as seen above, that it wanted India as an ally because it was engaged in a struggle for global hegemony during the Cold War.

This analysis, however, is problematic. It fails to take into account a strategically placed Pakistan. In other words, it implies that the Bank was operating in a vacuum and that Pakistan did not matter in the game. The Bank was not oblivious to the reality that was Pakistan. However, at that point in time Pakistan was in a precarious security and economic situation and was keen to be America's friend and ally. Consequently, its alignment with the US was Hobson's choice. In this context, even before it joined CENTO [Central Treaty Organization] and SEATO [Southeast Asia Treaty Organization], the US was taking it for granted as attested by the following statement of Abul Hasan Ispahani, Pakistan's ambassador to the US:[166] 'We are taken much for granted as good boys who would not play ball with communism or flirt with the left; boys who would starve and die rather than even talk to communists ... we were treated as a country that did not seriously matter. On the other hand, the US Government paid much attention to India.'

The Indians, unsurprisingly, do not agree with this explanation. They believe that Pakistan rather than India was the real beneficiary of its relationship with the US. The former Indian foreign secretary, Jagat Mehta, claims that '[t]he fact that Pakistan had become a US ally in the CENTO and SEATO gave it an intangible advantage and diplomatic leverage.'[167] This argument may have some substance as far as financing the works on the Western Rivers is concerned, as one writer puts it, the Bank was able to 'purchase' the Treaty because Pakistan was a US ally.[168]

It is true that Pakistan tried to use its clout with the American administration to affect Indian behaviour on the water issue but without success. When India denied any obligation under the 13 March 1952 understanding to refrain, while the cooperative work continued, from diminishing the water supplies available to Pakistan for existing uses, Pakistan tried to use its American connection to achieve its objective. Thus, its American legal counsel, John Laylin, prepared a memorandum through which it sought the American administration to make clear to India 'that any further economic assistance from the United States presupposes a prompt settlement of the water dispute or, as a minimum, that India confirms Pakistan's present uses'.[169] He also approached Black, urging him to contact Dean Acheson for the same purpose which he did not do. In any case, Pakistan did not succeed in its objective.[170]

There is no evidence to suggest that the Bank adhered to Pakistan's views on the water issue, particularly concerning the Bank's proposal. Incidentally, the Bank made its proposal in 1954 whereas Pakistan joined CENTO in 1955 and SEATO in 1956. One may argue that even if Pakistan joined the military pacts after the Bank made its proposal, it had decided to accept American military assistance in May 1954, and that had, for all practical purposes, made Pakistan a US ally. Even this argument is flawed because Pakistan accepted American aid after the Bank made its proposal which was in February 1954. Consequently, the argument that Pakistan used its military relationship with the US to influence the Bank is largely false. On the contrary, the close similarity between the Bank proposal and the Indian plan demonstrates that the Bank put forward its proposal in order to play Washington's game to rope India into the Western fold.

We have seen how Pakistan and India negotiated the Treaty. We know the points on which the two countries reached an agreement, most important of which was the surrender by the Pakistan government of all the proprietary rights that it had over waters of the Eastern Rivers to India in perpetuity, in exchange for almost unlimited rights over the waters of the Western Rivers. Apart from this, we practically know nothing of what is written in the Treaty. This is what we propose to do in the next chapter.

REFERENCES AND NOTES

1. Nehru became friends with Lilienthal during his visit to the US in 1949 where he had met him and invited him to visit India. He did so because in 1948, India had established the Damodar Valley Corporation on the lines of the TVA. Lilienthal had run the TVA for the development of a river system as a

unit and the Indian government was interested in utilizing his experience in developing water resources in India.

2. David E. Lilienthal, *The Journals of David E. Lilienthal: Venturesome Years, 1950–55*, vol. 3 (New York, Evanston and London: Harper and Row, 1966), 65.

3. Ibid. 75.

4. Ibid. 83.

5. Aloys Arthur Michel, *The Indus Rivers: A Study of the Effects of Partition* (New Haven and London: Yale University Press, 1967), 220.

6. Another 'Korea', *Collier's* (4 August 1951). The other article was entitled 'Are We Losing India?', 23 June 1951. The magazine in question started publication in 1888. Initially it came out under the title *Collier's Weekly*. However, after a few years the name was shortened to *Collier's*. It ceased publication on 4 January 1957 but was resumed in 2012.

7. Ibid. Emphasis original.

8. Ibid.

9. Ibid.

10. Ibid.

11. Commenting on this event after retirement from the Bank, Black made the following observation: 'In his article, Lilienthal said that maybe the World Bank could do something about this. I read this article. I bought the magazine. I don't usually buy magazines, but I bought this one. I picked up a telephone and called Lilienthal and told him I read this article, and I thought I would take steps about it. So he thought that was a fine idea.' Transcript of Interview with Eugene R. Black, *Oral History Program*, IFC Archives/World Bank, Brookings Institute, 6 August 1961, 48.

12. 'Covington and Burling' was originally known as 'Covington, Burling, Rublee, O'Brien and Shorb'. The lawyers who represented Pakistan included John Laylin, John Lord O'Brien, and Roger Fisher. Of these, Laylin continued to represent Pakistan throughout the Indus waters negotiations.

13. Lilienthal, *Venturesome Years*, supra note 2, 210. Liaquat Ali Khan, the Pakistan prime minister, told Lilienthal that '[u]nless the Kashmir issue is settled it is unreal to try to settle the issues about water ...' (Ibid. 83). If Pakistan agreed to separate the dispute over Kashmir from that of water, it did not mean that it decided to abandon or downgrade the former as understood from the following statement of President Ayub Khan: 'The very fact that Pakistan had to be content with waters of the three western rivers underlined the importance for us of having physical control over the upper reaches of these rivers .... In my mind, the solution of the Kashmir issue acquired a new sense of urgency on the conclusion of this Treaty.' Mohammad Ayub Khan, *Friends Not Masters: A Political Biography* (Lahore, Karachi, etc.: Oxford University Press, 1967), 113.

14. H. Graves Jr., 'The Bank as International Mediator: Three Episodes', in E. S. Mason and R. E. Asher (eds.), *The World Bank since Bretton Woods* (Washington DC: The Brookings Institution Press, 1973), 613.

15. India referred the Kashmir dispute to the UN Security Council on the advice of its first Governor General, Lord Mountbatten. The Indians refer to this act as 'Mountbatten's folly'. Similarly, India referred the Rann of Kutch dispute to an arbitration tribunal on the advice of the British prime minister, Harold Wilson, for which the Indian prime minister, Lal Bahadur Shastri, never forgave him.

16. The Bank and the Indian officials were also skeptical about Prime Minister Nehru's acceptance of the Bank's involvement. For example, the Bank's representative stated in 1951 that 'I have doubts about the likelihood of Nehru's being receptive of (sic) a proposal offering the Bank's good offices' (Undala Z. Alam, *Water Rationality: Mediating the Indus Waters Treaty*, PhD thesis submitted to the Department of Geography, University of Durham, 1998, 99). Similarly, when told of Pakistan's acceptance of the Bank's offer of good offices, Lilienthal observed that the prospect of Indian acceptance was 'probably remote' (Lilienthal, *Venturesome Years*, supra note 2, 224). Again, according to Sommers, the Bank's legal counsel, when he showed a draft of the letter in which the Bank was contemplating offering its good offices to the Pakistani and Indian governments to B. K. Nehru, the Indian prime minister's nephew and nominee of the Bank as executive director, he is reported to have commented: 'Well, it is a nice letter, the Prime Minister is expert at answering such letters and you'll get a very polite answer which won't say no but will mean no' (Transcript of Interview with Davidson Sommers, *Oral History Program*, IFC Archives/World Bank, Columbia University, 2 August 1961, 11). It needs to be mentioned here that B. K. Nehru has refuted this story in his memoirs. He has claimed that the Bank totally bypassed him and approached the prime minister without his knowledge because, 'they thought that, vigorous as I was in repelling any assault on our sovereignty or interference in our internal affairs, I might oppose the proposal' (Alam, *Water Rationality*, supra note 16, 100, footnote 64). The Sommers story may or may not be true but its thrust remains valid.

    Nehru, while accepting the good offices of the Bank, observed that he would have preferred direct bilateral negotiations with Pakistan as in other indo-Pakistan disputes. Letter of Prime Minister Nehru dated 6 April 1953, to Eugene Black. Ibid. 112.

17. The Indus waters dispute was not the only case in which the Bank played a mediatory role. It did so in some others as well, the most important of which were the Abadan Oil Refinery nationalization case and the Suez Canal compensation agreement. Interview with Sommers, supra note 16, 12–14.

18. Lilienthal, *Venturesome Years*, supra note 2, 200. Lilienthal also notes in his journal the reaction of another legal counsel of Pakistan, John Lord O'Brien,

to his proposal in these words: 'Mr O'Brian was very skeptical. He was full of "misgivings" lest this be another stall by the Indians, another means of gaining time, which is probably on their side.' Ibid. 234.

19. Ibid. 199.

20. Ibid. 235.

21. Ibid.

22. Ibid.

23. Ibid.

24. Ibid. Black's Letter to Nehru, 6 September 1951.

25. B. K. Nehru, *Nice Guys Finish Second* (New Delhi: Viking (Penguin Books), 1997), 254. Despite agreeing to the involvement of the Bank in the Indus waters dispute, Nehru tried to wriggle out of it or at least this is the impression that the Pakistan government got when in 1953 Nehru proposed to the Pakistan prime minister to put the dispute on the agenda of the meeting that he was scheduled to have with him. However, the latter opposed the proposal on the grounds that 'discussing the matter bilaterally while discussions were proceeding under the good offices of the Bank ...'. Official Memo by D. Sommers, dated 30 April 1953. File no. 1787278, Title: *Indus Basin Dispute-General Negotiations-Correspondence* 03, Start Date 2/1/53.

26. Salman M. A. Salman and Kishor Uprety, *Conflict and Cooperation on South Asia's International Rivers: A Legal Perspective* (Washington DC, World Bank, 2002), 45.

27. Lilienthal, *Venturesome Years*, supra note 2, 233.

28. Black's Letters to Khawaja Nazimuddin and Nehru, dated 8 November 1951.

29. Black's Letters to Khawaja Nazimuddin and Nehru, dated 13 March 1951.

30. Lilienthal, *Venturesome Years*, supra note 2, 262. See also 'Lilienthal's Letter to Khosla', dated 13 December 1951. File no. 1787277, *Indus Basin Dispute-General Negotiations-Correspondence* 02, Start Date 3/1/52.

31. Lilienthal, *Venturesome Years*, supra note 2, 269.

32. Ibid. Lilienthal expressed his astonishment in these words: 'Why the draftsman of Black's letter, the general counsel, Sommers, didn't use my language I will never know.' Ibid.

33. Michel, *Indus Rivers,* supra note 5, 227; Lilienthal, *Venturesome Years,* supra note 2, 290.

34. Black's Letters to Khawaja Nazimuddin and Nehru, dated 13 March 1952.

35. Gulhati claims that he subsequently learned that the Bank was conscious of the unilateral character of this provision but gave it a bilateral touch in order to give the impression of impartiality and for the reason that, 'conceivably, it might have restraining influence on Pakistan in some of the sensitive situations which, the Bank thought, then existed' (N. D. Gulhati, *Indus Waters Treaty: An Exercise in International Mediation* (Bombay, Calcutta, etc.: Allied Publishers, 1973), 101, footnote). He however does not explain what 'sensitive situations' the Bank had in mind.

36. Ibid. 100.

37. Lilienthal, 'Another 'Korea', supra note 6.

38. Gulhati, *Indus Waters Treaty*, supra note 35, 101.

39. It was thought at that time that the water issue will be resolved in a matter of six months or so. (Ibid.) Gulhati suggests that, '[i]t is not difficult to imagine that, at that stage, the Bank felt it was undertaking a straightforward engineering job and had little idea of the tremendous complexities it had to face later. I was told subsequently that no one in the Bank then knew what it was getting into' (Ibid. 98). The vice-president of the Bank, Iliff, also had a similar assessment as revealed in his Oral History interview: 'Iliff's mediation was not only between India and Pakistan, but with Black as well since he more and more wanted to abandon the whole exercise—having believed in the beginning that [an] agreement could be reached in a reasonably short time (Undala Z. Alam, 'Notes from a Conversation with Sir William Iliff', 10 June 1970, *Archives-Iliff, Sir William A. B.*). Commenting on the Bank's involvement in the Indus waters dispute, Black himself made the following observation in the Oral History interview: 'I was naïve about this because I thought maybe we'd get all these engineers together and sit around a table. You could work out a big master plan, because I felt that engineers were different from other people, that they were interested in combating nature, that they were above politics, they did not much care about politics. I thought if he'd get all of them together, we could do it. But I was naïve in that because I didn't realize the feeling between the two countries and the historical difficulties involved' (Transcript of Interview with Black, supra note 11, 48–9).

40. The leader of the Pakistan delegation was M. A. Hamid, the former chief engineer of the West Pakistan Irrigation Department from 1947–52. Other members were Pir Muhammad Ibrahim, irrigation advisor to the Government of Pakistan; Ahmad Hasan, chief engineer of Bahawalpur; A. Rahman, irrigation advisor to the Public Works Department of the NWFP; and M. S. Quraishy, superintendent engineer from Sindh. Given the differences between Pakistan's provinces on the water issue, the Pakistan government took care to give representation to the provincial stakeholders who could possibly be affected by the ongoing negotiations. However this practice was not emphasized from 1954 onwards, particularly after the merger of the four provinces into one unit called West Pakistan.

41. The leader of the Indian delegation was A. N. Khosla. Other members were N. D. Gulhati, chief of the Natural Resources Division of India's Planning Commission and deputy secretary to the Ministry of Natural Resources and Scientific Research; and J. K. Malhotra, officer on special duty in the same ministry.

42. The leader of the World Bank delegation was General R. A. Wheeler, the engineering advisor and former chief of the corps of engineers of the US

army. Other members were Dr Harry W. Bashore, an irrigation engineer and former US Commissioner of Reclamation; and Neil Bass of the World Bank staff who was formerly associated with the TVA.

43. Black's Letters to Khawaja Nazimuddin and Nehru, dated 8 November 1951.

44. Gulhati, *Indus Waters Treaty*, supra note 35, 105.

45. Ibid.

46. Ibid.

47. Manav Bhatnagar, 'Reconsidering the Indus Waters Treaty', *Tulane Environmental Law Journal*, vol. 22, issue 2, Summer, 2009, 295.

48. One former Indian bureaucrat is of the view that Lilienthal apparently had the blessings of the American government when he suggested the idea of a joint development and operation of the Indus Basin because before proceeding to Pakistan in February 1951, he met Secretary of State Dean Acheson, with whom he shared his idea. J. N. Dixit, *India-Pakistan: In War and Peace* (London: Routledge, 2002), 134–5.

49. Michel, *Indus Rivers*, supra note 5, 223.

50. Jagat S. Mehta, 'The Indus Waters Treaty: A Case Study in the Resolution of an International River Basin Conflict', *Natural Resources Forum*, vol. 12, no. 1 (1988), 75.

51. While the working party was utterly immobilized because of sharp differences between the Pakistani and Indian delegations, Lilienthal records in his journal that Khosla suggested 'that the parties did not need anyone from outside, they could work it out among themselves in a meeting of Indian and Pakistani engineers'. He further notes that this made Laylin furious who commented that 'this was part of the old run-round; the Indians could, in this way, present outrageous problems they wouldn't dare present before a third party who had experience in international water disputes.' Lilienthal, *Venturesome Years*, supra note 2, 313.

52. Michel, *Indus Rivers*, supra note 5, 228.

53. *Bank Proposal for a Plan for the Development and Use of the Indus Basin Waters*, dated 5 February 1954.

54. Ibid.

55. Michel, *Indus Rivers*, supra note 5, 233.

56. Ibid. 234.

57. *Bank Proposal for a Plan*, supra note 53. There is an interesting revelation by Gulhati who claims that when Garner, Wheeler, and Bass showed the Bank proposal to Lilienthal, he was happy to learn that his article in the *Collier's* was at its origin. According to him, however, he was disappointed to know that 'there was no opening in it for "an Indus Engineering Corporation—the supranational international agency" which, he had hoped, he would be asked to head.' Gulhati, *Indus Waters Treaty*, supra note 35, 139, note*.

58. The Pakistan government was unclear about the implications of the expression 'insignificant' in the Bank proposal. Referring to it, Pakistan's Minister for Interior Mushtaq Ahmed Gurmani expressed the fear that it could lead to 'opening [the] way for India establishing future claims on Jammu Kashmir waters'. Subsequently, the chief minister of Punjab, Firoz Khan Noon, who later on became the prime minister of Pakistan, asked the Bank Vice-President Robert Garner to explain the 'exact scope of the words 'relatively insignificant consumptive uses', employed in the Bank Memorandum of 7 October 1954. According to Noon, the latter clarified that the expression was 'not meant to convey more than really insignificant uses, such as minor extensions of irrigation from existing channels or the use of insignificant amounts of water from small feeder streams, etc. [And that] these words certainly did not cover the construction of new channels or major extensions of existing channels. Nor did they cover the construction of new major works, such as storages, dams, tunnels, etc.' (Letter of the Chief Minister of Punjab, Firoz Khan Noon, to the Minister of Interior, Mushtaq Ahmed Gurmani, dated 20 October 1954, File no. 1787280, Title: *Indus Basin Dispute-General Negotiations-Correspondence* 05, Start date 8/1/54).

Whereas Pakistan's interior minister and Punjab's chief minister were uncertain about the meaning and scope of the expression 'insignificant', Pakistan's foreign minister, Sir Zafrulla Khan, had absolutely no qualms in accepting it without seeking any clarification: 'I am to point out that the Bank's proposals of February 5 1954, state that "the entire flow of Western rivers would be available for the exclusive use and benefit of Pakistan, and for development by Pakistan, *except for the insignificant volume of Jhelum flow presently used in Kashmir.*" The last 12 words relating to the State of Jammu and Kashmir are clear and must be adhered to.' *Letter of the Foreign Minister of Pakistan to Black,* dated 24 August 1954. Emphasis added. 'Indus Basin Irrigation Water Problem: Working Draft of Introductory Memorandum on Preparation of Workable Plan under the Good Offices of the International Bank', December 1954, 135. File no. 1787280, Title: *Indus Basin Dispute-General Negotiations-Correspondence* 05, Start Date 8/1/54.

59. *Bank Proposal for a Plan,* supra note 53.
60. Ibid. The Bank used the expression 'fair' four times in its proposal to justify it. It gives the impression as if it did so with a feeling of guilt towards Pakistan.
61. Ibid.
62. According to one study, till the end of the twentieth century, roughly 150 countries, many of them enjoying less than cordial relations between them, had shared international water systems. Aaron T. Wolf et al., 'International River Basins of the World', *International Journal of Water Resources Development* (1999), vol. 15, 392.

63. Black's letter to Pakistan's Prime Minister, Khawaja Nazimuddin, dated 8 November 1951. Commenting on the Bank's expanded mandate one writer has observed that it were the differences between the two parties that 'the Bank was forced to play a more active part in working out a solution. The Bank pursued its own inquiries into the facts, and it was the Bank which, at various stages, suggested principles upon which the agreements might be based—a process which might be described as "continuing conciliation". The "General Heads of Agreement" of 1957 and the Bank's plan of 1959, which formed the basis for the Treaty, were two of the landmarks in the Bank's active participation in resolving the conflict.' A. H. Garretson, R. D. Hayton and C. J. Olmstead (eds.), *The Law of International Drainage Basins* (Dobbs Ferry, New York: Oceana Publications, 1967), 477.

64. Michel, *Indus Rivers*, supra note 5, 234.

65. Gulhati, *Indus Waters Treaty*, supra note 35, 145.

66. Nehru's Letter to Black, dated 25 March 1954.

67. Ibid.

68. Bashir A. Malik, *Indus Waters Treaty in Retrospect* (Lahore: Brite Books, 2005), 130.

69. Lilienthal, *Venturesome Years*, supra note 2, 534.

70. Ibid. Lilienthal thought that Hamid's sacking was unjustified because, in his opinion, 'the poor fellow's trouble was he was so afraid that he not only made no concessions, but he didn't face the facts of the Pakistan situation nor force his superiors in Karachi to do so.' He also explained India's take on Nasir's sacking in these words: '[Khosla] said the Pakistan delegation, headed by Hamid, would not counter the impression that everything that India suggested was completely wrong, would not inform their people that Pakistan simply could not have everything and still regard this as a negotiation, or that anything less than what they "insisted" upon would have to be settled for in the end. As a consequence the Bank proposal, being at such variance with this whole-hog picture, made them quite indignant and aroused, and they still are.' Ibid.

71. Ibid.

72. Alam, *Water Rationality*, supra note 16, 127.

73. Bogra's Letter to Black, 14 May 1954, cited by Malik, *IWT in Retrospect*, supra note 68, 130–1. Michel has raised the question of how Pakistan rejected the Bank proposal while it could not do the same to the draft that India presented to it in the summer of 1948 and which subsequently came to be known as the Delhi Agreement. He answers it by suggesting that it was, *inter alia*, because of its acceptance of the American military assistance in May 1954, which made it feel strong. He goes on to elaborate this point by suggesting: 'Had Pakistan felt as weak in May 1954 as she had in May 1948 or even in May 1952 when the Indus Waters Treaty negotiations began, she might well have been tempted to accept the Bank's plan as quickly as India had. But perhaps for the first time in her young experience Pakistan began to feel that time might be on her side.

Indeed the prompt Indian acceptance may have suggested to Pakistan that she should hold out for further concessions, either from India or from the Bank, on the storage issue.' Michel, *Indus Rivers,* supra note 5, 241.

74. Lilienthal notes in his journal dated 2 June 1954, that the Bank Vice-President Garner, in his meeting with him, attributed the failure of the Pakistan government to accept the Bank proposal to the fact that it 'was now so weak at home (they had just had a complete eruption in East Bengal, with threats of secession, charges of treason, etc.) that they can't agree to anything'. He then goes on to claim that he had a solution to Pakistan's political difficulties which consisted of joint development, with the Bank support, of the Ganges-Hooghly rivers in East Pakistan which could take care of the disaffected Bengalis. He, however, says that he did not suggest it to Garner because, in his opinion, this wasn't the best time to do so. Lilienthal, *Venturesome Years*, supra note 2, 511–12.

75. Ibid. 473. Emphasis added.

76. Ibid. Emphasis Added.

77. Gulhati, *Indus Waters Treaty*, supra note 35, 175. Emphasis added.

78. Ibid. 179. Emphasis added.

79. Ibid. 175–6.

80. Ibid. 142. Emphasis added.

81. Memo by Iliff of the Meeting between Black, Nehru and Himself, dated 3 May 1960. File no.1787284, Title: *Indus Basin Dispute-General Negotiations-Correspondence* 09, Start Date 1/1/60.

82. Dennis Kux, *India-Pakistan Negotiations: Is Past Still Prologue?* (Washington: US Institute of Peace Press, 2006).

83. As far as arid regions are concerned, the principle of equitable apportionment applies which has the following three rules: (1) supplies of water for existing uses be protected from interference by withdrawals of water supplies for new uses of any sort; (2) unused water supplies available for development be shared equitably among riparians; (3) costs of development undertaken for the common benefit of two or more riparians be shared in proportion to the new benefits derived by each one of them. See *UN Economic Commission for Europe: Legal Aspects of the Hydro-Electric Development of Rivers and Lakes of Common Interest,* Geneva, January 1952. See File no. 1787277, Title: *Indus Basin Dispute-General Negotiations-Correspondence* 02, Start Date 3/1/52.

84. Hearings on Treaty with Mexico relating to the Utilization of the Waters of the Certain Rivers, *Committee on Foreign Relations*, US Senate, 79th Cong., 1st Sess., Part 5, 1743. We want to acknowledge that we have drawn liberally on this hearing for the present survey.

85. Ibid. 1743–4.

86. Ibid. 1743.

87. Ibid. 1744.

88. Ibid.

89. Ibid.

90. Exchange of Notes on the Use of the Waters of the River Nile for Irrigation Purposes, *7 May 1929, Treaty Series* no. 17, Cmd. 3348 (London: House of Commons–Parliamentary Papers online: His Majesty's Stationary Office), 3.

91. Convention between the US and Mexico, dated 21 May 1906, <www.ibwc.gov/Files/1906Conv.pdf>.

92. Hearings on Treaty with Mexico, supra note 84, 1762. See also, the Letter of Pakistan's Counsel General in New York, L. Shaffi, entitled 'Fact and Fiction of the Indus Basin Irrigation Water Dispute', which he wrote to a prominent New York publisher, A 1995–165, 5–8. File no. 1787278, Title: *Indus Basin Dispute-General Negotiations-Correspondence* 03, Start Date 2/1/52.

93. The doctrine owes its nomenclature to the US Attorney General Harmon who, while dealing with the water dispute between the US and Mexico over the Rio Grande River, issued an opinion in 1895 in which he enunciated the principle that a state had an absolute right to do whatever it liked with the waters of a river while it flowed through its territory (Garretson et al., *Law of International Drainage Basins*, supra note 63, 20–8. For fuller treatment of the subject, see Stephen C. Mccaffrey, 'The Harmon Doctrine, One Hundred Years Later: Buried, Not Praised', *Natural Resources Journal*, vol. 36, no. 3 (Summer 1996) 549–90). The opposite of the Harmon Doctrine is the Absolute Territorial Integrity Doctrine which states that an upper riparian state cannot interfere with the natural flow of the watercourse into the lower riparian state. It was invoked in the *Trail Smelter* case between Canada and the US though the issue involved trans-frontier pollution and not an international watercourse.

94. Protocol to Treaty of Friendship between Turkey and Iraq, dated 29 March 1947.

95. See for example *Kansas v. Colorado,* 206 US, (1907), 46, 97; 185 US (1902), 125, 146–7; *North Dakota v. Minnesota* 263 US (1923), 365, 373. The Court described such a dispute as creating 'a situation which, if it arose between independent sovereignties, might lead to war'. *Missouri v Illinois,* 200 US (1906), 486, 518.

96. *Wyoming v. Colorado,* 259 US (1922), 419, 464–5.

97. See Letter of Pakistan's Counsel-General in New York, supra note 92.

98. The Bank dangled the carrot in these terms: 'The Bank proposal is no arbitrary compromise, arrived at by mathematically splitting the differences between the two sides. It is a plan based on concepts of its own, which produce a fair and economic result …. The bank proposal is simple, workable and fair. It will effectively promote the economic development of the Indus Basin and will benefit both countries by substantially increasing the amount of usable water available to each of them.' *Proposal for a Plan*, supra note 53.

99. Alam, *Water Rationality,* supra note 16, 127.

100. Ibid.

101. Graves, 'Bank as International Mediator', supra note 14, 616.

102. Gulhati, *Indus Waters Treaty,* supra note 35, 165. Emphasis added.

103. For the text of Shoaib's Draft of 12 July and Iliff's Correction, see File no. 1787280, Title: *Indus Basin Dispute-General Negotiations-Correspondence* 05. Start Date 8/1/54.

104. Emphasis added.

105. Letter dated 15 March 1954, covering an American Engineer's Appraisal of the Bank Proposal. Indus Basin Irrigation Water Problem: Working Draft of Introductory Memorandum on Preparation of Workable Plan under the Good Offices of the International Bank, December 1954, 104. File no. 1787280, Title: *Indus Basin Dispute-General Negotiations-Correspondence* 05, Start Date 8/1/54. For more details see file nos. 1788098–100, *Indus Basin Dispute-Tipton Study-Documents* 01–03.

106. An American Engineer's Appraisal, supra note 105,110–11.

107. World Bank Press Release no. 380, Michel, *Indus Rivers,* supra note 5, 243n.

108. *Aide-Memoire,* 21 May 1956, para. 6 (c) (iii).

109. Ibid. paras. 8–9.

110. Gulhati, *Indus Waters Treaty,* supra note 35, 199.

111. Wheeler to Black, dated 28 June 1956, File no. 1787281, Title: *Indus Basin Dispute-General Negotiations-Correspondence* 06, Start Date 1/1/56.

112. Gulhati, *Indus Waters Treaty,* supra note 35, 200.

113. Dennis Kux, *The United States and Pakistan: Disenchanted Allies* (Karachi: Oxford University Press, 2001), 104.

114. Iliff's Meeting with Pakistan's Prime Minister in Lahore on 11 June 1957, File no. 1787282, Title: *Indus Basin Dispute-General Negotiations-Correspondence* 07, Start Date 8/1/57.

115. According to Gulhati, India's alternative plan could achieve the results envisaged in the Bank proposal at about 27 per cent of the cost of works included in the Pakistan plan. It would, in his view, require a transition period of no more than eight years as opposed to the fifteen years contemplated by the Pakistan plan. Gulhati, *Indus Waters Treaty,* supra note 35, 250.

116. Graves, 'Bank as International Mediator', supra note 14, 621.

117. Letter of Mueenuddin to Iliff, dated 22 December 1958, File no. 1787282, Title: *Indus Basin Dispute-General Negotiations-Correspondence* 07, Start Date 1/1/57.

118. Gulhati, *Indus Waters Treaty,* supra note 35, 250–5.

119. Report of Iliff's Meeting with Nehru, dated 9 December 1956, File no. 1782281, Title: *Indus Basin Dispute-General Negotiations-Correspondence* 06, Start Date 1/1/56. The unrealistic nature of Pakistan's demand reminds one of its proposal of zero allocation of water to India from the Western Rivers and 70 per cent from the Eastern Rivers.

120. Mueenuddin's Letter to Iliff, dated 10 September 1957, File no. 1787282, Title: *Indus Basin Dispute-General Negotiations-Correspondence* 07, Start Date 8/1/57; see

also, Mueenuddin's Letter to Iliff, dated 20 May 1957, File no. 1782281, Title: *Indus Basin Dispute-General Negotiations-Correspondence* 06, Start Date 1/1/56.

121. Notes from Black's Conversation with Gulhati, also present in the meeting were Iliff and Malhotra, dated 30 June 1960. File no. 1787284, Title: *Indus Basin Dispute-General Negotiations-Correspondence* 09, Start Date 1/1/60.

122. Following for the PM from Black, dated 7 July 1960. Ibid. Black apparently played a critical role in convincing Nehru to make concessions to Pakistan. The following argument by Black seems to have done the trick:

> 5. [A]t the same time, you, I am confident, realise that there are far wider issues involved and that the consequences of a break-down of these negotiations, no matter where the fault might lie, when so much has already been accomplished, would be an international disaster of the first magnitude.
>
> 6. I see in the present situation a critical opportunity for another manifestation of generosity and magnanimity on the part of India, which, I am sure, would be acclaimed by international opinion, and regarded as another act of statesmanship on your part. Ibid.

123. The agreement envisaged an extension in the transition period at Pakistan's request. However, in case the latter did so, India's financial contribution was to be reduced at the following rates: 5 per cent for one year, 10.25 per cent for two years, and 15.76 per cent for three years.

124. How did Ayub achieve this remarkable feat? He explains it in his autobiography. According to him, his water experts were of the opinion that requirements of replacement and development demanded a bigger dam than Rohtas Dam which only the Tarbela Dam could fulfil. The difference of cost between the two was to the tune of $200 million which was quite a staggering amount at that time. He knew that Black would hit the roof when he would hear of it. And so he did. To convince him of the soundness of his case, this is what he told him:

> 'I have been around these areas which are going to be affected by the withdrawal of waters by India. People have told me very plainly that if they have to die through thirst and hunger they would prefer to die in battle and they expected me to give them that chance. Our *jawans* and the rest of the people feel the same way. So this country is on the point of blowing up if you don't lend a helping hand. This is a human problem of a grave nature and cannot be blinked away.
>
> 'What we are being called upon to do is to barter away naturally-flowing waters into our canals for storage water, and the history of storage is that it begins to silt the moment it is completed. Besides, we are going to be put back by about ten years or so by building these storages and link canals. All this effort could have been put to more constructive effort. So, we are making great sacrifices.

'I know certain countries have been very kind in offering us assistance, but unless we get our additional needs of water, apart from replacements, there is going to be chaos in this country. So a dam at Tarbela is a must.'

Ayub observes that when Black heard it he thought that Ayub had made his task very difficult because he did not know how he would persuade the donors to cough up another $200 million. He wanted some time to ponder over the matter to which Ayub replied: 'Must you have time to think over an obvious thing like this?' He says that 'we went over it again and again till Black ... agreed to support our demand and said that he would ask the donor countries for the additional sum—the difference between Rohtas and Tarbela. In the end we got a promise of over 740 million dollars.' Ayub Khan, *Friends Not Masters*, supra note 13, 110.

125. In May 1959, Iliff proposed a Heads of Agreement which provided as follows: Division of waters as in the Bank Proposal of 1954; a system of replacement-cum-development works to be constructed in Pakistan, their location and capacity, etc., to be determined by Pakistan, without Indian participation or agreement (but with advice from the Bank); the transition period and additional Indian withdrawals, during this period, to be worked out somewhat loosely: an Indus Basin Development Commission, consisting of two representatives each of India and Pakistan with a neutral chairman, to modify the transition period and, as and when necessary, the additional Indian withdrawals to be set out in the treaty; an ad hoc sum to be fixed as the financial contribution by India towards the cost of replacement; and the Indus Basin Development Fund to be established and administered by the Bank.

Gulhati, *Indus Waters Treaty*, supra note 35, 257–8.

126. Mueenuddin's Letter to Iliff, dated 25 July 1957. *Indus Basin Dispute-Working Paper submitted re Heads of Agreement*, Coor, 1.

127. Gulhati, *Indus Waters Treaty*, supra note 35, 263.

128. Ibid. 264.

129. Mueenuddin's Flier to CRO London, 21 August 1959, File no. 1787283, Title: *Indus Basin Dispute-General Negotiations-Correspondence* 08, Start Date 1/1/59.

130. Mueenuddin's Letter to Iliff, dated 10 September 1957, File no. 1787282, Title: *Indus Basin Dispute-General Negotiations-Correspondence* 07, Start Date 8/1/57.

131. Office Memo from Iliff to Wheeler, dated 20 July 1959, File no. 1787283, Title: *Indus basin Dispute-General Negotiations-Correspondence* 08, Start Date 1/1/59.

132. Despite reaching an agreement with India on the question of storage on the Western Rivers, Pakistan was so apprehensive of the Indian intentions that it sought guarantees from the Bank that the latter would not operate the storage works maliciously. The Bank plainly refused to oblige Pakistan according to Iliff: 'I pointed out to the Foreign Minister that if Pakistan continued to apply the criterion of the harm that might be done in the event of storage

works being operated maliciously by India, he placed me in an impossible situation. The Indians would not listen to any representation from me that it was their intention to break an international agreement in which they had entered.' Consequently, Pakistan dropped this demand (Memo by Iliff on a talk with Pakistan Foreign Minister Manzur Qadir, dated 3 May 1960. File no. 1787284, Title: *Indus Basin Dispute-General Negotiations-Correspondence* 09, Start Date 1/1/60). Indeed the Bank was justified to refuse to entertain Pakistan's request. However, Pakistan's apprehensions were not unfounded as the subsequent developments, particularly in the *Baglihar* and *Kishenganga* cases, showed. Gulhati has tried to explain Pakistan's behaviour as a typical case of 'down-stream neurosis' (Gulhati, *Indus Waters Treaty*, supra note 35, 311). In our judgment, this is an unfair comment as Pakistan's apprehensions were totally justified. Let us not forget that in addition to the *Baglihar* and *Kishenganga* cases India also stopped the supply of water to Pakistani canals on 1 April 1948.

133. Messrs. Black and Iliff's Meeting with Nehru in New Delhi, 13 May 1959, File no. 1787283, Title: *Indus Basin Dispute-General Negotiations-Correspondence* 08, Start Date 1/1/59.

134. Meeting of Pakistan Ambassador in Washington and Mueenuddin with Black and Iliff, dated 9 December 1959. Ibid.

135. Ibid.

136. Ibid.

137. Ibid.

138. For details, see chapter 'Terms of the Treaty', under 'Territorial Scope'.

139. Iliff's Message to Pakistan's Foreign Minister, 12 September 1960, File no. 1787284, Title: *Indus Basin Dispute-General Negotiations-Correspondence* 09, Start Date 1/1/60. Something similar happened on the eve of the signing ceremony when Pakistan asked the Bank to transmit a letter to India that related to articles IV (15) and XI (1 & 2) which had been agreed upon in the tripartite meeting and whose language Pakistan's foreign minister was satisfied with. Black was 'astonished and shocked' to learn about it and thought that it would amount to re-opening substantive points. The Bank warned Pakistan that, 'responsibility for delay, or for a break, would rest on the shoulders of Pakistan and on an issue which in the eyes of the world would be regarded as of insignificant importance'. Black urged Pakistan to withdraw the letter and let the matter drop and in case it refused to do so he made it clear that '[he] will *not*, repeat *not*, be willing that the Bank should act as a post box for a communication, which ... is likely to lead to a last minute rupture and on a point that has no substantive importance ...'. Message from Iliff to Shoaib, undated. Ibid. Emphasis original.

140. Gulhati, *Indus Waters Treaty*, supra note 35, 269–70.

141. For details of this provision, see chapter 'Terms of the Treaty'.

142. Washington Working Party: 3rd Meeting, 4 December 1959, File no. 1787283, Title: *Indus Basin Dispute-General Negotiations-Correspondence* 08, Start Date 1/1/59.

143. On the conclusion of the Treaty, J. K. Malhotra, member of the Indian delegation, penned the following poem (adapted from Wordsworth's valedictory sonnet to the river Duddon) for Iliff which we discovered in the Bank archives (It must have given considerable amusement to Iliff):

> For backward, Indus, as I cast my eyes
> I see what was, and is and will abide.
> Still glides the Stream and shall forever glide.
> The Form remains, the Function never dies,
> While they, the brave, the mighty and the wise,
> The men who, in their morn of youth, defied
> the elements, must vanish—Be it so;
> Enough if something from their hands have power,
> to live and act and serve the future hour
> and if, as towards their silent tomb they go,
> through Love, through Hope and Faith's transcendent dower,
> they feel that they are greater than they know.

See J. K. Malhotra's letter to Iliff, dated 28 April 1959. Ibid.

144. Memorandum by Iliff of a meeting dated 3 May 1960 between Black, Iliff and Nehru, File no. 1787284, Title: *Indus Basin Dispute-General Negotiations-Correspondence*·09, Start Date 1/1/60. Graves states that Pakistan agreed to sign the Treaty because it 'had a government sufficiently strong to risk the domestic political consequences of an agreement'. Graves, 'Indus Waters Treaty', supra note 14, 624.

145. Its breakdown was as follows: Participating governments' share was equivalent to about $640 million, India's about $175 million, and the Bank's $80 million in the shape of a loan. India got a sum of $56 million for the Beas Dam. The Indus Basin Development Fund Agreement did not mention the amount of India's contribution. It was simply announced that India 'would agree to contribute towards the cost of these works'. The Agreement stated that India's contribution would not change irrespective of any alteration in the par value of any currency. Gulhati notes that this provision proved useful for India because the British currency in which India was to make payment to Pakistan was devalued in November 1967, with the result that India saved about £6 million (Gulhati, *Indus Waters Treaty*, supra note 35, 284). In other words, it signifies that Pakistan suffered a loss of £6 million because of the provision.

146. Apart from the media, the Treaty also came under severe criticism in the Indian *Lok Sabha* where most of the speakers who took part in the debate blamed the Indian government for pursuing a policy of appeasement and surrender to Pakistan, in addition to abandoning the Indian interest. For example, Congress party MPs from Punjab and Rajasthan, Iqbal Singh and H. C. Mathur, termed

the Treaty disadvantageous to India and that their home states had been let down. Ashok Guha, another Congress MP, stated that the Indian interests were sacrificed to placate Pakistan. Ashok Mehta, leader of the PSP, observed that it was a peculiar treaty as Pakistan, which was a water surplus area, would let the water of the Indus River flow to the sea whereas India would be short of water supplies.

K. Warikoo, 'Indus Waters Treaty: View from Kashmir', <www.jammu-kashmir.com/insights/insight20060601a.html>; see also, *Pakistan's Water Resource Strategy*, (Washington DC: World Bank (PK-34081), 2005), 7; Shaista Tabassum, *River Water Sharing between India and Pakistan: Case Study of the Indus Waters*, Regional Centre for Strategic Studies (RCSS Policy Studies 24), 31–2.

147. Whereas the Pakistanis accused Ayub of selling the Eastern Rivers, the Indians made allegations against their government regarding the Western Rivers. Responding to the Indian legislators' criticism in the *Lok Sabha*, Nehru stated on 30 November 1960, that, 'we purchased a settlement, if you like; we purchased peace to that extent and it is good for both countries.' K. Warikoo, 'View from Kashmir', supra note 146.

148. This is what she said: 'In reply to my reference to the Indus Waters Treaty, Mr Ayub Khan said that he did get Rs. 900 crores for it. I ask him, is that adequate price for losing permanently the water for all times? Is it not frittering away our permanent water rights? Is it not a fact that this treaty represents the lowest ebb of his will because it signifies a lack of determination to fight for our water rights on which our prosperity, progress and future depends?'

149. Black's Letters to Khawaja Nazimuddin, dated 13 March 1952.

150. Mr Shams-ul-Mulk, the former chairman of WAPDA, disclosed this fact to this author in an interview that he gave to him in his office in Islamabad on 5 March 2012.

151. Ayub Khan, *Friends Not Masters*, supra note 13, 109–10, and 112.

152. The Indus Basin Development Fund Agreement that Pakistan signed with the Bank for financial assistance for works on the Western Rivers carried the following clause in its preamble: 'And whereas, in concluding the Treaty, Pakistan has been influenced by the consideration that financial assistance of the nature and amounts specified hereinafter will be made available to Pakistan ...'.

153. Iliff's meeting took place at the Governor's House in Lahore on 11 June 1957, see File no. 1787282, Title: *Indus Basin Dispute-General Negotiations-Correspondence 07*, Start Date 8/1/57. It is noteworthy that the Pakistan prime minister used the same tactics during the meeting that he had with Wheeler as borne out by the report that he sent to Black. See, Wheeler to Black, 28 June 1956, File no. 1787281, Title: *Indus Basin Dispute-General Negotiations-Correspondence 06*, Start Date 1/1/56.

154. Lilienthal, *Venturesome Years*, supra note 2, 311.

155. David E. Lilienthal, *The Journals of David E Lilienthal: Creativity and Conflict, 1964–67*, vol. 6. (New York, Evanston and London: Harper and Row, 1976), 236.

156. Syed Kirmani and Guy Le Moigne, *Fostering Riparian Cooperation in International River Basins* (Washington DC: World Bank, 1997), 5.

157. G. T. Keith Pitman, 'The Role of the World Bank in Enhancing Cooperation and Resolving Conflict on International Watercourses: The Case of the Indus Basin', in Salman M. A. Salman and Laurence Boisson de Chazournes (eds.), *International Watercourses: Enhancing Cooperation and Managing Conflict* (Washington, DC: World Bank, 1998), 155.

158. Eugene Black, 'The Indus: A Moral for Nations', *The New York Times*, 11 December 1960. The water dispute between Pakistan and India posed lending problems for the Bank. We have already mentioned in this study that when Pakistan requested the latter to finance the Kotri Barrage project on the Indus River, India objected to it. Similarly, when India asked it to finance the Bhakra-Nangal Dam project on the Sutlej River, Pakistan opposed it. Pitman, 'Role of the World Bank', supra note 157, 159.

159. Lilienthal, *Venturesome Years*, supra note 2, 51, 54. Emphasis added.

160. *Collier's*, 23 June 1951.

161. Lilienthal, *Venturesome Years*, supra note 2, 60–1.

162. Robert J. McMahon, *The Cold War on the Periphery: The United States, India and Pakistan* (New York: Columbia University Press, 1994), 19. Emphasis added.

163. Ibid. 104. Emphasis added.

164. Lilienthal, 'Another "Korea"', *Collier's* (4 August 1951), supra note 6. Besides the US, other Western countries were also extremely keen to woo India. For example, Mr Pearson, the Canadian foreign minister at that time warned Black that the Bank's policy of refusing loans to India on water projects until it had resolved its disputes with Pakistan was not good as it ran the risk of damaging relations with India. He advised him to 'avoid taking any position in relation to this very controversial issue' for the sake of maintaining good relations with the latter. Iliff's (Director Loan Department at that time) Letter to Joseph Rucinski, 2 March 1950, File no. 1787277, Title: *Indus Basin Dispute-General Negotiations-Correspondence* 02, Start Date 3/1/52.

165. Whereas the US sided with India on the water issue in its bid to win it over, it also tried to keep Pakistan in good humour by supporting it on the Kashmir issue as evidenced by the position that it took in favour of Pakistan during the UN Security Council debates on Kashmir. The explanation for this delicate balancing act on the part of the US must have been the desire to keep both of them on its side.

166. McMahon, *Cold War*, supra note 162, 71.

167. Mehta, 'The Indus Waters Treaty', supra note 50, 73.

168. According to Michel, the Bank would not have been able to 'purchase' the Treaty if Pakistan was not an American ally. Michel, *Indus Rivers*, supra note 5, 241.

169. Lilienthal, *Venturesome Years*, supra note 2, 323.

170. Similarly, John Foster Dulles, the US secretary of state, responding to Nehru's request for financial assistance stated that 'as a preliminary, India should settle outstanding disputes with her neighbours in order to be able to reduce armament expenditures.' According to Sommers, the latter specifically mentioned water dispute as one of such disputes (Office Memorandum from D. Sommers, dated 22 May 1953, File no. 1787278, Title: *Indus Basin Dispute-General Negotiations-Correspondence* 03, Start Date 2/1/53). It is obvious that Dulles must have done it to please Pakistan but nothing positive came out of it.

4

Terms of the Treaty

A) UNIQUENESS OF THE TREATY

THE TREATY IS A BRIEF DOCUMENT WITH A SHORT PREAMBLE AND twelve articles.[1] However, with eight elaborate Annexures and four Appendices attached to it, it turns out to be a very lengthy document. It is also a complex and prolix Treaty[2] principally because the Annexures are highly technical in nature.[3] This is because Pakistan had all manners of safeguards built into the Treaty to prevent India from abusing it, so much so that at times it specified in detail the amount and timing of water that must flow from the Western Rivers. For example, paragraph 18 of Annexure E of the Treaty specifies that, in case of disagreement between the two commissioners on the timing of the annual filling of conservation storage and the initial filling below the dead storage level, India may carry it out according to the following schedule: a) if the site is on the Indus, between 1 July and 20 August; b) if the site is on the Jhelum, between 21 June and 20 August; and c) if the site is on the Chenab, between 21 June and 31 August at such rate as not to reduce, on account of this filling, the flow in the Chenab Main above Merala to less than 55,000 cusecs. Commenting on this aspect of the Treaty, Ramaswamy Iyer made this observation:[4]

> The first reason is the density of the technical details in the treaty, which provide ample opportunities for differences among engineers. It is interesting to compare this treaty with the Mahakali treaty between India and Nepal or the Ganges treaty between India and Bangladesh. The latter two are non-technical documents that are easy to understand, even for non-engineers. On the other hand, while

the main part of the Indus treaty is fairly slim and not too dense, the devil is in the detail: the treaty is accompanied by Annexures and Appendices of a highly technical and opaque nature. It is these Annexures and Appendices that determine the overall character of the treaty. Facetiously speaking, one could say that this is not a treaty between two governments, but a treaty between two sets of engineers. The engineers on the two sides can have a field day disagreeing on the meaning of and precise application of the various technical features and criteria that the Annexures and Appendices contain. ... The treaty provides a happy hunting ground for technical disagreements.

Another notable feature of the Treaty is that apart from Pakistan and India, the World Bank is also a party to it. It is perhaps the only international water treaty to which it is a party. Recognizing the contribution that it made through its mediatory role, the Arbitration Court in the *Kishenganga* case paid tribute to it in these words:[5]

[T]he Court considers it appropriate to note the extraordinary contribution of the World Bank to the conception, mediation, negotiations, drafting and financing of the Indus Waters Treaty, an instrument vital to the life and well-being of hundreds of millions of people of India and Pakistan. The conclusion of the Indus Waters Treaty in 1960, in which the leaders and staff of the World Bank lent vital support to the Parties, was and remains a great achievement of international cooperation.

Being a party to the Treaty, however, does not signify that the World Bank enjoys the same juridical status as Pakistan and India because, unlike them, it is bound only by articles V and X which respectively relate to the financial and emergency provisions; and Annexures F, G, and H which respectively relate to the role of the Neutral Expert, the Court of Arbitration, and the transitional arrangements. Given the fact that it has fulfilled its obligations covered by the financial provisions and the transitional arrangements, article V and Annexure H have, for all practical purposes, become redundant. In other words, the Bank is today only bound by Annexures F and G which deal with the adjudicative provisions.

Still another striking feature of the Treaty is that it is a partitioning and not a water-sharing treaty. It divides the Indus Rivers and allocates three rivers called the Western Rivers to Pakistan and three called the Eastern Rivers to India. By doing so, it seems to have consummated Partition. This feature distinguishes it from all other treaties relating to global drainage basins because whereas they share the waters of common rivers between riparian states, the Treaty divides them between Pakistan and India. The Treaty allocates all the waters of the Eastern Rivers 'for unrestricted use' to India. Pakistan is under obligation not to interfere with waters of the Sutlej Main and Ravi Main or their tributaries when they flow through Pakistani territory except for domestic, non-consumptive, and certain limited agricultural uses.[6] Similarly, it allocates all the waters of the Western Rivers 'for unrestricted use' to Pakistan. India is under obligation not to interfere with them while they flow on Indian territory except for domestic, non-consumptive, and certain limited agricultural uses as well as generation of hydroelectric power.[7] The Treaty allows India to build a maximum of 3.6 MAF storage on the Western Rivers within specified parametres whose details are laid down in Annexures C, D, and E.[8] The partitioning of the Indus Rivers between Pakistan and India and an adherence to the terms of the Treaty perhaps explains the success of the Treaty and the absence of an armed conflict between the two countries on the water issue.

B) SAFEGUARDS FOR PAKISTAN AND INDIA

As explained in a previous chapter, the Treaty owes its existence to the fact that Pakistan was wary of India's intentions with regard to the common rivers that flow through their territories and wanted an arrangement which could ensure the uninterrupted supply of water to it. If Pakistan had apprehensions regarding India's intentions, India too had its concerns. It was keen to get not only assured supply of water from these rivers but also an increased quantity in order to bring more land under cultivation. Consequently, the Treaty carries provisions which propose to take care of both countries' concerns.

Thus, according to the Treaty, each party agrees that it shall not materially change (through the non-consumptive use) the flow of water in any channel to the detriment of the other party. Similarly, it stipulates that each party, while carrying out any scheme of flood protection or flood control, avoid causing material damage to the other party. It specifically forbids India from carrying out any scheme on the Western Rivers that violates article III, which forbids India from interfering with their waters and binds it to let them flow towards Pakistan.[9] It obligates each party to use their best endeavours to maintain the natural channels of the rivers in such a way as to avoid any obstruction likely to cause material damage to the other party.[10]

The Treaty binds each party to prevent, as far as practicable, undue pollution of the waters of the Indus Rivers and to take all reasonable measures to ensure that before any sewage or industrial waste flows into them, it is treated in such a manner that it does not materially affect its uses. It is silent on transboundary environmental assessment which was introduced by the 1972 Stockholm Declaration on Human Development[11] (later updated by the 1992 Rio Declaration on Environment and Development 1992)[12] and formally codified by the UNECE Convention on Environmental Impact Assessment in a Transboundary Context 1991 (Espoo Convention).[13] It is understandable because there was not much consciousness about it in 1960, let alone the existence of a norm. The criterion of reasonableness that the Treaty lays down is the customary practice in similar situations.[14]

This provision is utterly vague as the Treaty fails to define not only what it means by 'undue pollution' but also the criterion of reasonableness. It demonstrates that those who drafted the Treaty did not pay sufficient attention to the question of pollution. Nor did it take into account the enormous pollution that was a result of industrialization, urbanization, and population growth in the years to follow. Given the fact that Pakistan is the lower riparian, it is bound to be the net recipient of pollution from India which could create tension in the relations between the two countries. In the absence of any clear guidance in the matter, the parties would have to sort out this matter bilaterally or, in case of failure, through the adjudicative process.[15] We are of the view that if a pollution issue involving the Treaty ends up before an arbitration tribunal, it is more likely to decide the case on the basis of present-day environmental standards because a clean environment has emerged as a principle of *jus cogens* or a peremptory norm in the international community, as demonstrated by a number of court decisions.[16]

The Treaty prohibits India from constructing any storage on the Western Rivers except as provided in the Treaty.[17] Following the allocation of the Eastern Rivers to India, the Treaty provided for a transition period during which Pakistan was to build the replacement works on the Western Rivers. This period was to begin on 1 April 1960 and end on 31 April 1970, and was extendable by another three years ending on 31 March 1973, whether or not the replacement works had been accomplished. In case of an extension of the transition period, the Bank was to pay, out of the Indus Development Fund, an amount specified in the Treaty.[18] During the transition period, India was obliged to release waters of the Eastern Rivers in accordance with the provisions of Annexure H (Transitional Arrangements). Incidentally, Pakistan was obliged to complete the replacement works within the period stipulated in the Treaty. It built two multi-purpose dams, namely Mangla and Tarbela, a system of eight new canals, remodelled existing canals, five barrages, and a gated siphon.[19]

The Treaty deals with design, construction, and operation of new hydroelectric plants that India is entitled to build on the Western Rivers. As far as the criterion for any new run-of-river plant is concerned, it prescribes it in great detail. Given the importance that the criterion relating to the gated spillway assumed in the outcome of the *Baglihar* and *Kishenganga* cases, it deserves a special mention here. The Treaty carries the following provision in this regard: 'If the conditions at the site of a Plant make a gated spillway necessary, the bottom level of the gates in normal closed position shall be located at the highest level consistent with sound and economical design and with satisfactory operation of the works.'[20] It states that 'to satisfy Pakistan that the design of a plant conforms to the laid down criterion India shall communicate to Pakistan the necessary information six months in advance of the start of the construction of river works connected with the plant.' Pakistan, on its part, is under obligation to communicate in writing to India, within three months of the receipt of the relevant information, any objection that it may have with regard to the proposed design. In case the two parties disagree on the question of whether or not the design of a plant conforms to the criterion laid down in the Treaty, then either party may resort to article IX (1) and (2) which deal with dispute settlement.[21] It also spells out the details of the storage works that India is entitled to construct on the Western Rivers. It lays down the aggregate storage capacity of the single or multi-purpose reservoirs that India is entitled to construct on each of the river systems.[22] It also spells out the procedure to satisfy Pakistan regarding the design of any plant, which is a verbatim reproduction of paras. 9–11 of Annexure D.[23]

Notwithstanding the fact that these provisions are unambiguous and detailed, their implementation has not been without problem. Pakistan often accuses India that it starts a run-of-river project on the Western Rivers surreptitiously and without informing it; that Pakistan learns about it either from the media or from its agencies tasked with information gathering; and that India fails to supply information about it in accordance with the criteria laid down in the Treaty. In its view, the construction of the Baglihar and the Kishenganga Dams are cases in point where, it alleges, India used dilatory tactics in order to complete them so that it could present Pakistan with a fait accompli. To take care of this problem, it proposed, in the non-paper that it presented to India in 2010 for the purpose of including the water issue in the Composite Dialogue, that the latter should not proceed with the implementation of any project on the Western Rivers until all the objections have been resolved. India, on its part, rejected these accusations and accused Pakistan of raising unnecessary and frivolous objections in order to stop it from going ahead with projects in hand.

Ramaswamy Iyer, the leading Indian writer on the water issue and whose views are quite close to those of the Indian government, states:[24]

> Pakistan's response to every bit of information given to it about proposed projects is completely negative; that its aim is not the fair and objective examination of the information supplied but the *blocking* of every project that India proposes; that what Pakistan wants to do is to prevent India from using the waters of the western rivers to the very limited extent allowed by the treaty; and that this is, in fact, a violation of the spirit of the treaty. However, such arguments, valid as they may be, are not easy to establish, and they might not command the immediate assent of an independent observer, whereas a complaint by a lower riparian that is being denied water by the upper riparian tends to find a sympathetic response.

As for Pakistan's demand contained in the non-paper, that India should not start a run-of-river project until its objections and queries have been taken care of, Iyer termed it reasonable on the grounds that the Treaty obligates India to supply, in advance, detailed technical information on the planned projects but makes its fulfilment contingent on Pakistan being genuinely interested in getting answers to its questions. He then passes the judgment that, '[i]t is India's contention that this is not the case, that "satisfaction" would never be reached, and that what Pakistan wants to do is to stall the project. As each set of questions is answered, more questions will be asked, or the answers given will be rejected as unsatisfactory. If this goes on endlessly, work on the project can never start, and the permissive provisions of the treaty will stand nullified.'[25]

He supports his contention by citing the Wullar Barrage case which, in his opinion, Pakistan has stalled for the last twenty years. What is the truth in this controversy? It lies somewhere in the middle. The problem with Pakistan is that it interprets various provisions of the Treaty in an absolutely literal sense, leaving no room at all for any departure from the text, with the result that it appears that it is averse to India building any structure on the Western Rivers. The problem was that India was inclined to denying or delaying information, as much as possible, on a run-of-river project in order to present Pakistan with a fait accompli. It is noteworthy that Pakistan's new Indus commissioner has decided to be more liberal in the interpretation of the Treaty provisions regarding India's run-of-river projects. For example, he decided not to raise any objection against the Miyar Dam that India is planning to build in the near future despite the fact that it has certain features which Pakistan deems detrimental to its interests. The reason for this change in Pakistan's attitude is the fact that the dam is small. Would India reciprocate in the matter by providing full and timely information to Pakistan regarding its future run-of-river projects?

C) PERMANENT INDUS COMMISSION

The Treaty provides for the establishment of a Permanent Indus Commission, for which the inspiration supposedly came from the International Joint Commission established by the US and Canada. It stipulates that each party would appoint a commissioner for Indus Waters who would be a high ranking engineer competent in the field of hydrology and water-use. He is to represent his government in all matters arising out of the Treaty and serve as the regular channel of communication on all matters relating to the implementation of the Treaty, particularly, for: a) furnishing or exchange of information or data provided for in the Treaty; and b) the giving of any notice or response to a notice provided for in the Treaty. Together, the two commissioners constitute the Permanent Indus Commission.[26] During the negotiations held under the Bank's auspices, Pakistan initially proposed a neutral chairman to head the Commission but India did not agree and Pakistan had to abandon the idea. The Treaty defines the purpose and functions of the Commission to establish and maintain co-operative arrangements for the implementation of the Treaty and to promote co-operation between the parties in the development of the rivers. It also provides that the Commission is 'to undertake, once in every five years, a general tour of inspection of the Rivers for ascertaining the facts connected with various development works on the Rivers'. Similarly, the Treaty stipulates that the Commission is 'to undertake promptly, at the request of either commissioner, a tour of inspection of such works or sites on the Rivers as may be considered necessary by him for ascertaining the facts connected with those works or sites.'[27]

India was utterly opposed to the idea of inspection by Pakistan in Jammu and Kashmir or on its territory on the grounds that it did not want the latter to police its actions. However, Pakistan wanted to be absolutely sure that India was not misusing the Treaty by developing additional acres of irrigation in the Indian-administered Jammu and Kashmir. Pakistan persisted with its demand until India finally agreed to the inspection clause, though it must be stated that Pakistan was not fully satisfied with the arrangement.[28]

The Commission is to meet regularly, at least once in a year, alternately in India and Pakistan in November or some other month as agreed to by the two commissioners. It also meets when either commissioner makes a request. To enable the commissioners to perform their duties, the Treaty obligates each government to accord, to the commissioner of the other government, the same privileges and immunities which are accorded to representatives of member states to the principal and subsidiary organs of the UN under sections 11, 12, and 133 of article IV of the 1946 Convention on the Privileges and Immunities of the United Nations.[29]

D) CONFLICT RESOLUTION PROCEDURE

The Treaty has a dispute settlement procedure which is unique as it categorizes issues that may arise from the implementation of the Treaty as 'questions', 'differences', and 'disputes'.[30] It begins by prescribing that in case any 'question' arises between the parties concerning the interpretation or application of the Treaty or the existence of any fact which, if established, might constitute a breach of the Treaty, the Commission will examine and try to resolve it by agreement. If the Commission fails to do so, then a 'difference' is deemed to have arisen. If, in the opinion of either commissioner, the 'difference' relates to part 1 of Annexure F, he may request the appointment of a Neutral Expert to deal with it in accordance with part 2 of the said Annexure. If para. 2 (a) does not cover it or if the Neutral Expert informs the Commission, in accordance with para. 7 of Annexure F, that in his opinion the 'difference' should be treated as a 'dispute', a 'dispute' is then deemed to have arisen which is to be decided in accordance with the provisions of paras. 3, 4, and 5 of article IX. In such an eventuality, the Commission shall, at the request of either commissioner, report the fact, as early as practicable, to the two governments stating the points on which the Commission is in agreement and the issues in dispute, the views of each commissioner on these issues and his reasons thereof.

The Treaty entitles the two parties to appoint the Neutral Expert by agreement. In case they fail to do so or fail to agree on a third party to do so, the Bank is then entrusted with this responsibility. The latter is also authorized to appoint a Neutral Expert during the transition period. The Treaty carries detailed provisions on the qualifications of the Neutral Expert to be appointed.[31] When the Neutral Expert deals with a 'question' he does not do so on appeal from the Commission but only because the latter has failed to resolve it. The determination made by the Neutral Expert is final and binding.

Article IX and Annexure F of the Treaty detail the role of the Court of Arbitration whereas Annexure D lays down the details about it. Annexure F identifies a list of twenty-three questions which fall within the purview of the Neutral Expert. Anything falling outside it is to be settled by the Court of Arbitration. If the Neutral Expert determines that the 'difference' under consideration or part of it does not fall under his purview, then the 'difference' or part of it becomes a 'dispute' and the Court of Arbitration is entitled to take cognizance of it. In case the Commission itself rules that a 'difference' is a 'dispute', the Court of Arbitration would take care of it. If any 'question' arises for determination which does not fall within the purview of the Neutral Expert, it is to be settled according to procedures which could involve the Court of Arbitration. The Court of Arbitration shall be established in accordance with Annexure G, a) by agreement between the parties, b) at the request of either party, if, after the start of negotiations, in its opinion, the 'dispute' is not likely to be resolved by negotiations or mediation; or c) at the request of either party, if, after the expiry of one month following receipt by the other government of the invitation, the party concludes that the other government is unduly delaying the negotiations. The Court is to consist of seven arbitrators of which two are to be appointed by each party while the remaining three are to be appointed by a complex procedure involving the World Bank, the United Nations, the president of the Massachusetts Institute of Technology, the rector of the Imperial College of Science and Technology in London, the chief justice of the United States,[32] and the Lord Chief Justice of England.[33]

The question as to which category a particular issue falls under can be a contentious matter as illustrated by the *Kishenganga* case. India challenged the admissibility of the second dispute (namely whether India can, under the Treaty, deplete or bring the reservoir level of a run-of-river plant below the dead storage level in any circumstances except in an unforeseen emergency) on the grounds that it did not constitute a 'dispute' because Pakistan did not follow the procedure laid down in the Treaty.

India contended that in view of the fact that there was a disagreement in the Commission on the question of disposition of the second dispute, Pakistan should have asked for the appointment of a Neutral Expert for his ruling whether it constituted a 'difference' or a 'dispute', and that it could bring the matter before a Court of Arbitration only if a Neutral Expert were to pronounce that it was not a technical matter within part 1 of Annexure F. It was of the opinion that every issue should be *ab initio*, considered a 'question' and hence, within the purview of the Commission. In case the Commission fails to consider it, either commissioner can, under article IX para. 2 (a), unilaterally refer the resulting difference to a Neutral Expert. If the commissioners disagree as to whether the difference is a technical matter and within part 1 of Annexure F, then the latter, under para. 7, Annexure F, can decide on the procedure to be followed in the matter. India was also of the view that as opposed to para. 2 (a), para. 2 (b), which does not entitle a single commissioner to deem an issue a dispute, applies if neither commissioner considers the difference a technical matter for a Neutral Expert, or if the latter determines that the matter falls outside his jurisdiction. In case of disagreement as to how to proceed, neither party can initiate arbitration and is obliged to request a Neutral Expert to determine if the difference is a dispute for submission before a Court of Arbitration.[34]

As far as the *Kishenganga* case is concerned, India argued that the two commissioners never agreed that the difference was not a technical matter for a Neutral Expert to consider nor did they ever request the latter to pass upon the proper disposition of the difference. It held that the Pakistan commissioner should have submitted the second dispute to a Neutral Expert for his determination. India further opined that his decision not to do so and instead unilaterally qualify it as a dispute usurped the role of the Commission and the Neutral Expert and caused the premature submission of the second dispute to a Court of Arbitration.[35] Taking issue with India, Pakistan justified the seizing of the Court on the grounds that it made extensive efforts to resolve the matter through negotiations and according to the procedural requirements of article IX. Invoking para. 2 (a), which entitled a Neutral Expert to decide his own jurisdiction, it maintained that either party could ask for the appointment of a Neutral Expert and in case it failed to do so, a Court of Arbitration was competent to decide the second question itself. It pointed out that prior to the present proceedings, India never argued that the second dispute was fit for a Neutral Expert to decide or that it constituted a 'difference' and that it consistently denied the existence even of a question in terms of para. 1 of article IX. It also pointed out that India repeatedly characterized matters relating to the second dispute as 'issues' rather than 'questions' and objected to any reference to the terminology of article IX. It therefore held that India was now estoppeled from arguing that there was a 'difference' to be resolved by a Neutral Expert.

Pakistan argued that India not only rejected the applicability of article IX but also never sought the appointment of a Neutral Expert. In its view, had the latter made such a request, it would have accepted placing the question (whether or not the difference fell within part 1, of Annexure F) before a Neutral Expert to decide. Since neither party made such a request, and the Indian commissioner expressly stated that there was no difference, para. 2 (a) did not apply. It asserted that having failed to request the appointment of a Neutral Expert at an appropriate time, India could not insist on it now. It concluded that acceptance of the plea against admissibility would amount to letting India frustrate the working of conflict resolution provisions in the Treaty.[36]

The Court examined the arguments of the parties and found Pakistan's viewpoint well-founded. It began by stating that para. 2(a) allowed either commissioner to refer a difference to a Neutral Expert if he deems it related to one of the identified technical matters and prefers the latter to resolve it. In its view, it is not enough for a commissioner to merely conclude that a difference will, at some point, be fit for a Neutral Expert to examine but at the same time make an actual request in order to avoid the procedural impasse that can arise. It invoked the December 1959 draft of the Treaty in support of its interpretation.[37] Next, applying para. 2(a) to the present case, it observed that neither commissioner called for the appointment of a Neutral Expert for the second dispute. Therefore, it decided to dispense with India's first objection to admissibility. It also brought out that at no point before the commencement of the present proceedings had the Indian commissioner deemed the second dispute a difference falling within the competence of a Neutral Expert. Citing the minutes of the relevant meetings of the Commission, it also brought out that India was of the opinion that the 'issues' raised by Pakistan could be the subject of further discussion within the Commission; that there were no differences as the design of the KHEP (Kishenganga

Hydroelectric Project) was in conformity with the Treaty; and that since depletion below the dead storage level is a general issue and not specifically related to the KHEP, there was no reason to consider that any difference had arisen. It also observed that para. 2 (a) ensured the appointment of a Neutral Expert where a party actually seeks it and that it did not serve to impose an additional, procedural hurdle for access to a Court of Arbitration. Nor was it prepared to accept that India's position (that the second dispute was fit for a Neutral Expert) would be relevant here even if India were now to call for the appointment of a Neutral Expert. In light of these observations, it concluded that having consistently argued in the Commission that no difference between the parties existed, India could not now assert that the second dispute is, in reality, a difference after all.[38]

Having disposed of the first objection to admissibility, the Court took up India's second objection to the second dispute whereby it pleaded that it involved highly technical issues fit for a Neutral Expert only. India understood it to mean that a technical question listed in part 1 of Annexure F must be decided by a Neutral Expert rather than by a Court of Arbitration. The Court disagreed with this interpretation except where a party makes such a request and a Neutral Expert deems himself competent in the matter. It also observed that with the exception of para. 2 (a), recourse to a Neutral Expert is permissive and not mandatory which, in its opinion, signified that the Treaty did not rule out mechanisms other than that of a Neutral Expert to deal with technical matters. Nor did it find any provision in the Treaty which prevented it from considering a technical question except if a party had actually asked for the appointment of a Neutral Expert. It was therefore of the view that it was not under any obligation to investigate its own jurisdiction or inform the Commission that a dispute involving technical matters be referred to a Neutral Expert. It also pointed out that the very composition of a Court of Arbitration attested to its competence in technical matters as one of its members is required to be a 'highly qualified engineer', and the Treaty did not stop the parties from appointing engineers as their arbitrators or as the chairman of the Court. Taking into account these observations, it adjudged that a Court of Arbitration could not declare a case submitted before it inadmissible, merely on the grounds that it involved a technical question.[39]

Before concluding the discussion on the categories of issues envisaged in the Treaty for conflict resolution, it should be mentioned that the parties have so far been involved in three cases, and in each case they have used different methods to resolve their issues. Thus, in the Salal Dam case, the parties determined the matter was a 'question' and decided to deal with it through bilateral negotiations and settled it at that level. In the *Baglihar* case, they characterized the issue as a 'difference' and took care of it through the Neutral Expert. Finally, in the *Kishenganga* case, the Court of Arbitration deemed both the issues involved as 'disputes' and resolved them through arbitration. The utilization of all the three methods, namely bilateral talks, the Neutral Expert, and the Arbitration Court, and the progression from 'question' to 'difference' to 'dispute' for settlement of the issue only shows that the matters are heating up between the two countries on the water front. Another negative aspect of the conflict resolution saga is that the two countries have squandered enormous amounts of money on litigation whereas they could have resolved their water issues bilaterally without much cost (provided they had conducted themselves in good faith). However, its positive side is that they have resorted to an agreed dispute settlement mechanism rather than violent means to take care of these contentious matters. The conclusion that we can draw from this discussion is that the conflict resolution mechanism in the Treaty, though prohibitively costly, stands out as a silver lining in an otherwise sad state of affairs between the two nations insofar as they are incapable of settling their disputes amicably through bilateral negotiations as demonstrated by the dismal progress of the Composite Dialogue so far.

Next, we propose to deal with the law, applied by a Court of Arbitration in the interpretation of the Treaty. In this respect, the Treaty specifies it as follows:[40]

Except as the Parties may otherwise agree, the law to be applied by the court shall be this Treaty and, whenever necessary for its interpretation or application, but only to the extent necessary for that purpose, the following in the order in which they are listed;
  a) International conventions establishing rules which are expressly recognized by the Parties.
  b) Customary international law.

The provision lays down that a Court of Arbitration is to apply international law but it is somewhat different from the one specified for the International Court of Justice. Whereas the judicial method brooks no departure from international law, the arbitral method does, if the parties so desire (however, in case they fail to do so, it is international law as specified in the statute of the ICJ which applies). Thus, the law to be applied by an arbitral court is incorporated in a compromise between the parties as is the case in the present situation.[41] That also explains why the provision does not refer to article 38 of the Statute of the ICJ which spells out the sources of international law that are bound to apply. On the contrary, unlike article 38, which mentions four sources of international law (international conventions, international customs, general principles of law recognized by civilized nations, and judicial decisions and the teachings of the most highly qualified publicists), it mentions only the first two and excludes the other two. Nor, once again, contrary to article 38, does it authorize the Court to decide a case *ex aequo et bono*. It obligates the Court of Arbitration to resort to the two sources mentioned above to the extent necessary for the Treaty. Incidentally, this issue came up for discussion in the *Kishenganga* case where the Court noted that 'the place of customary international law in the interpretation or application of the Treaty is subject to Paragraph 29.'[42] It is also noteworthy that the above provision begins with the expression '[e]xcept as the Parties may otherwise agree.' What does this signify? It signifies that even though the Treaty has specified the law that a Court of Arbitration is to apply, the parties have the liberty to set it aside in favour of some other arrangement. This provision is in-line with the tradition of arbitral procedure which allows the parties flexibility to apply any law that they desire.

The Treaty also stipulates that the parties shall not construe anything in the Treaty as establishing a general principle of law or any precedent.[43] The need for this provision is not clear because bilateral treaties or treaty contracts create rights and obligations only for the contracting parties and as such, do not lay down any general rules except when they operate through principles governing the development of customary rules.[44] At the same time, it must be underlined that, as one commentator of the Treaty has rightly observed, 'a provision of this nature cannot keep others from looking to the settlement as precedent or from deriving what general principles they choose from the terms agreed upon.'[45] Consider the following in this regard: the Treaty extinguished any claims that Pakistan and India had to the Indus Rivers based on prior appropriation. Does this fact constitute any contribution to customary international law on international rivers? Commenting on this aspect, the above-mentioned writer, rightly observes that the answer to this question would depend on how much weight one accords to bilateral treaties as a source of international law.[46]

E) TERRITORIAL SCOPE OF THE TREATY

The Treaty spells out its territorial scope by stating that it 'governs the rights and obligations of each party in relation to the other with respect only to the use of the waters of the Rivers'. Elaborating this idea, it provides that neither its text nor its execution 'shall be construed as constituting a recognition or waiver ... of any rights or claims whatsoever of either of the Parties other than those rights or claims which are expressly recognized or waived in this Treaty'.[47] To understand the meaning and purpose of this provision, reference to the background history of this article is necessary. When the negotiations had almost been completed in the World Bank-sponsored parleys and the parties were ready for a draft Treaty, it was feared that the unresolved dispute over Jammu and Kashmir could stand in the way of the settlement. The parties, therefore, tried to take care of this matter by finding a formula which, while avoiding reference to any territorial dispute between them, would protect their respective rights or claims to the disputed territory. In other words, the parties tried to bypass the dispute over Jammu and Kashmir through this article in the settlement of the water dispute. With this perspective, the provision became handy and was inserted in the draft Treaty. William B. Iliff, the chief negotiator on behalf of the Bank, acknowledged this fact in a letter that he addressed to N. D. Gulhati in these words:[48]

> My recollection of the understanding reached in the course of our conversations with the Indian authorities in Delhi is that ... India was concerned that the actual construction of a reservoir at Mangla should not carry an implication that India's sovereign rights in Jammu and Kashmir were in any way or to any degree eroded. I therefore wished to find a formula that would therefore protect her in this respect ... The general principle underlying the Bank approach was that neither party should, on the one hand, seek to gain, in or from the Water Treaty, any support for its own general position on the Kashmir issue, or, on the other hand, should seek to erode the general position of the other party.

Gulhati also confirms this background history of the provision when he states that for the sake of the smooth progression of negotiations, Pakistan and India agreed to some guiding principles at an early stage of the discussion relating to the draft of the Treaty, one of which he described as follows:[49] 'In the light of the disagreement between India and Pakistan on the status of Jammu and Kashmir, it was agreed that effort be made to write the treaty in such a manner as to bypass the problem of Jammu and Kashmir. There was no other way to reach agreement that would be accepted by the two parties.'

Pakistan and India agree on the background of this article. However they disagree with its scope. This question cropped up in the *Kishenganga* case where Pakistan was of the view that article XI was meant to regulate the rights and obligations of the parties with respect to the use of waters of the entire relevant area of the Indus Rivers, including those parts which are located in Jammu and Kashmir whether administered by Pakistan or India. It claimed that the World Bank confirmed this interpretation when it sought legal advice from the English barrister, Sir John Foster, on the following question that it had posed to the former:[50]

It is the intent that —

    a) The rights and obligations of India under the Treaty shall extend to acts and omissions in, or affecting, that portion of Jammu and Kashmir that is under the control of India.

    b) The rights and obligations of Pakistan under the Treaty shall extend to acts and omissions in, or affecting, the remainder of Jammu and Kashmir.

    c) The Treaty shall not affect the respective positions taken by the parties in the dispute over Jammu and Kashmir.

Question: Does the present draft accomplish the foregoing?

Pakistan further claimed that the Bank, in response to the question, assured it that its apprehensions were 'ill-founded' (Pakistan based this claim on a telegram dated 15 April 1960, that G. Mueenuddin sent to the Pakistan government in which he observed: 'In general, Foster's opinion was that our fears were ill-founded and the draft of the Treaty a) accomplished the common intent and b) excluded all other matters'.[51] On the basis of these claims, it submitted before the Court that it was entitled to argue that the KHEP would have an adverse effect in the areas located in Jammu and Kashmir administered by Pakistan.

India agreed with Pakistan that article XI, para. 1 skirts the issue of sovereignty over Jammu and Kashmir but it disputed its scope. It accepted that all the areas that the KHEP could adversely affect, whether in terms of hydroelectric, agricultural, or environmental, were located in the Pakistan-administered Jammu and Kashmir. It however, took the view that since it was not legally part of Pakistan, the latter was invoking the Treaty to support its claim and to dispute that of India to this territory. It contended that Pakistan was in violation of article XI, para. (1b) of the Treaty.[52] It also contended that it was not under any obligation, under the Treaty, to avoid adverse impact on the territories that do not form part of Pakistan because the Treaty does not apply to regions which are under the de facto control of Pakistan. To support its contention, it recalled that during the negotiations of the Treaty, Pakistan made a proposal which aimed to extend the application of the Treaty to 'all of the territories which at the time are under [a Party's actual control]' but it rejected this proposal and did not include it in the final text. As for Pakistan's assertion that the World Bank confirmed that the Treaty would apply to Pakistan-administered Jammu and Kashmir, India rejected it on the grounds that Pakistan mostly relied on internal correspondence to support this viewpoint which was not part of the *travaux préparatoires*, and that it relied on an interpretation which was made by Sir John Foster but was not the opinion of the Bank.[53]

Adjudicating the case, the Court observed that the sparse negotiating record in the matter under consideration was inconclusive. However, it held that since Sir John Foster did not represent the parties, his opinion was not determinative of the meaning of article XI. It further ruled that the parties' agreement to the text of the article, in full knowledge of Foster's interpretation, could be construed as acceptance of his view. It was of the view that the parties possibly rejected the proposed rider by Pakistan to extend the Treaty to 'all the territories which at the time were under [a Party's actual control]' because they wished to avoid overt reference to the division between them. However, it did not doubt that they shared the Bank's view that the Treaty should not and did not affect the question of sovereignty over Jammu and Kashmir.[54] The court further pointed out that the preamble of the Treaty refers to the parties' desire to attain 'the most complete and satisfactory utilization of the waters of the Indus system of rivers', in addition to fixing the rights and obligations of the parties regarding the use of 'these waters'. It concluded that the intent of the Treaty is 'to apply to the aggregate of the Indus river system and not only to those waters flowing through the uncontested territory'. It also observed that neither of the parties pointed out, nor did the court find, any provision that would exclude from the scope of the Treaty any portion of the Indus Rivers that flow through Pakistan and India. Continuing this line of thought, it brought out that if the Treaty were to exclude four rivers in the Indus system which flow through Jammu and Kashmir (namely Indus, Jhelum, Chenab, and Ravi), 'it would fall significantly short of providing the comprehensive solution sought by the Parties for the development and allocation of the waters of the Indus system.'[55] In light of these observations, it adjudged that the parties could use waters of the Indus system that flow, *inter alia*, through Jammu and Kashmir whether administered by Pakistan or India. It therefore held that Pakistan was entitled to object to the construction of the KHEP on the grounds that it would affect the flow of the river and uses of waters in the part of Jammu and Kashmir that it administers.

The significance of the Court's decision lies in the fact that as far as the construction of dams in the Pakistan-administered Jammu and Kashmir is concerned, India would not object by raising territorial claims which it had reportedly started to do, contrary to the Treaty's provisions. This came out recently when Pakistan made efforts to seek World Bank funding for the Diamer-Basha Dam that it proposes to build in its Northern Areas and the Bank reportedly asked the Asian Development Bank (ADB), which had tried to bring it on board for co-financing, to first seek a no objection certificate (NoC) from India. The WAPDA chairman, Syed Raghib Shah, officially confirmed this fact when he stated, before Pakistan Senate's Committee on Water and Power in 2013, that the international financial institutions had refused to fund the Diamer-Basha Dam because of Indian influence.[56] Incidentally, the previous WAPDA chairman, Shakeel Durrani, made a similar statement when he revealed that the World Bank, the ADB, and other multilateral donors retracted their commitment to finance the project and linked it to an NoC from India.[57] The media report did not explain why India made such a demand but perhaps it did so because it contends that the dam in question is to be located in Jammu and Kashmir which it claims is an integral part.[58] Media reports suggest that India has only informally communicated its objections to the Bank. According to a recent report, the Pakistan government, in the light of the *Kishenganga* award, took up the matter with the Bank which showed willingness to extend funding for the Diamer-Basha Dam.[59]

The Bank's position in the matter is untenable because it is contrary to its 'Operational Policy on disputed territories' which does not require an NoC from the other party:[60] 'The Bank may support a project in a disputed area if the government concerned agrees that, pending the settlement of the dispute, the project proposed for country A should go forward without prejudice to the claims of country B.' It is also untenable because it is in violation of article XI, para. 1 of the Treaty in whose drafting it played a critical role and which, as seen in the *Kishenganga* case, substantially suggests that the parties would not raise any territorial claims in the implementation of the Treaty. It signifies that if the Bank withholds financing on account of India's objections to the construction of the Diamer-Basha Dam it will be guilty of violating not only its Operational Manual but also of the Treaty. It needs to be added here that Pakistan's finance minister, Ishaq Dar, recently claimed that he was able to convince the Bank of Pakistan's case and that it had agreed to fund the dam.[61]

F) MISCELLANEOUS PROVISIONS

Despite dividing the Indus Rivers between Pakistan and India, the Treaty recognizes the need for the two countries to fully cooperate with each other for the optimum utilization of the water resources of these rivers. In this connection, it provides that a party will install, at the request of the other party, hydrological and meteorological observation stations within the Indus Basin and supply data obtained against the payment of costs by the beneficiary party. The Treaty also provides that a party will establish, at the request of the other party, new drainage works on its territory that may be required against the payment of costs by the beneficiary. This cooperation may extend to engineering works. In case a party undertakes them and said works interfere with the waters of the Indus Rivers, which affects the other party materially, the party planning it is obliged to notify the other party of its plans. Furthermore, it must supply data which will enable the other party to assess the nature, magnitude, and effect of the work. In case a work interferes with the waters of the Indus Rivers but the party planning it does not deem such work affects the other party materially, it is still bound to supply such data regarding the nature, magnitude, and effect of the work to the other party, on request.[62]

The Treaty entered into force on 1 April 1960, even though it was signed on 19 September 1960, and was ratified in December 1960.[63] In other words, it entered into force retroactively or more than five months before it was signed and more than nine months before the parties exchanged the instruments of ratification. This is an unusual feature because normally, a simple bilateral treaty comes into force the moment the two parties validly express their consent (to be bound by it). However, nothing stops them from agreeing upon some other date, if they so desire.[64] The date of the coming into force of the Treaty shows that Pakistan and India departed from the normal practice. The explanation for this anomaly lies in the fact that starting from 1948 until 31 March 1960, Pakistan and India had signed several agreements by virtue of which the latter was to supply water to the former from the Central Bari Doab Canal (CBDC). The Treaty also provides for its modification by the two governments through a duly ratified treaty concluded for that purpose.[65] It signifies that the parties cannot effect unilateral changes in the Treaty.

Given the fact that it was beyond Pakistan's financial capacity to build the replacement works required by the Treaty, the World Bank, in addition to providing help in the settlement of the water dispute also arranged the necessary funds for the works that Pakistan was required to undertake on the Western Rivers. Towards this end, India agreed to make a fixed contribution of £62,060,000 towards the costs of replacement works which were to be deposited in ten equal instalments to the Indus Basin Development Fund, established and administered by the Bank.[66] The Treaty stated that if the transition period was extended at the request of Pakistan, the Bank was under obligation to pay a specified amount to India from the Indus Basin Development Fund.[67] It laid down that if prior to 31 March 1965, large scale international hostilities arose out of causes beyond Pakistan's control, which adversely affected the execution of replacement works, and the Bank was satisfied with Pakistan's démarche, the parties, through the auspices of the Bank, were to consult each other in order to mutually agree whether or not the prevailing circumstances warranted any modifications in the Treaty and, if so, their nature and scope.[68] Similarly, if during the replacement works the inundation canals were damaged as a result of floods, the two Indus commissioners, through the good offices of the Bank, were to introduce provisional modifications to the replacement works. Following the exchange of ratification instruments by Pakistan and India, the two countries started implementing the Treaty.

The Treaty has been in force for more than half a century now. Its implementation has revealed many deficiencies from which it suffers. In the recent past, a group of former diplomats, academics, retired bureaucrats, water experts, civil society activists, etc., from both sides of the great divide met at an unofficial level and started a dialogue process termed 'Track II diplomacy' to identify those lacunae with the hopes of improving the Treaty. The process is highly useful because, since individuals participating in it speak in their own capacity and not as representatives of their countries, they can make bold recommendations. In the present instance, they have identified many areas where work needs to be done, some of which are as follows: watershed management; environmental issues and the importance of environmentally and ecologically necessary flows in the Eastern Rivers; the need for a dispute settlement mechanism that functions in a swift and timely manner; establishment of a working group on the impact of climate change on glaciers; and the establishment of a working group on data sharing and management.[69] The governments of Pakistan and India will indeed find these recommendations of great value whenever they decide to close the gaps in the Treaty.

In the meantime, to move forward, our next task is to examine how the Treaty has fared during its fifty odd years' legacy. This is what we propose to deal with in the next chapter.

REFERENCES AND NOTES

1. For the text of the Treaty, see Appendix 2.
2. Iliff relates a very interesting incident regarding the obscurity of the Treaty. He disclosed that just before the signing ceremony, Nehru told him that the previous night, while reading the Treaty, he came across one particular section which he did not understand. He asked Iliff to explain it to him to which the latter replied that it reminded him of a story concerning the famous poet Robert Browning. According to that story, one night Browning was at a dinner party where he found himself sitting next to a very attractive young lady who was a great admirer of his works. She asked him to explain to her the meaning of a particular passage in his poem entitled 'Sordello' which is a very obscure poem. Browning replied: 'My dear, when Sordello was written, only God and Robert Browning knew what it meant, and now only God knows.' He went on to state that he told Nehru: 'I'm afraid I find myself almost in the same situation with regard to this passage in the treaty.' Iliff, Transcript of Interview, *Oral History Program*, The World Bank/IFC Archives, Oral History Research Office, Columbia University (1961), 55–6.
3. Commenting on the technical nature of the Treaty, Raymond Lafitte, the Neutral Expert in the *Baglihar* case, made the following statement: 'the Treaty was negotiated and concluded during a period of tension between India and Pakistan ... [t]hose who drafted the Treaty aimed for predictability and legal certainty in the drafting of the Treaty so as to ensure its sound implementation. The wish for predictability and legal certainty is well illustrated by the technicalities of the Treaty and particularly of its annexures. The Treaty contains clear language and wording on how and to which extent India and Pakistan may be allowed to utilize the waters of the Indus system of rivers .... The Treaty also gives a clear indication of the rights and obligations of both Pakistan and India.' Baglihar Hydroelectric Plant, Expert Determination, *PCA*, 12 February 2007, 14.
4. Ramaswamy Iyer, 'Indus Treaty: A Different View', <http://www.upscportal.com/civilservices/blog/Indus-Treaty-Ramaswamy-R-Iyer>.
5. Kishenganga Arbitration, Partial Award, *PCA*, 18 February 2013, 134, para. 358.

6. Article II.
7. Article III.
8. Annexure E, para. 7. This is in addition to the storage facility that existed on these rivers before the coming into force of the Treaty. The details of 3.6 MAF that the Treaty allows to India are as follows:

Conservation Storage Capacity

(1)	(2)	(3)	(4)
River System	General Storage Capacity	Power Storage Capacity	Flood Storage Capacity
(a) The Indus	0.25	0.15	Nil
(b) The Jhelum (excluding the Jhelum Main)	0.50	0.25	0.75
(c) The Jhelum Main	Nil	Nil	As provided in para. 9
(d) The Chenab (excluding the Chenab Main)	0.50	0.60	Nil
(e) The Chenab Main	Nil	0.60	Nil

India claims that it has a storage entitlement of 3.6 MAF but has not built any storage facility and out of the 1.34 million acres that the Treaty permits it to use for irrigation purposes on the Western Rivers, it is currently (2010) using only 0.792 million acres which is well below the permissible limit. Sabharwal Sharat, 'The Indus Water Treaty', *The Speech of the Indian High Commissioner to the Karachi Council on Foreign Relations and the Citizen Friendship Forum* (3 April 2010), <www.india.org.pk/urdu/docs/SpeechbyMr. SharatSabharwalHighIndusWatersTreaty3April2010.pdf>.

9. Article IV, para. 2.
10. Article IV, para. 6.
11. See Principle 21, <www.unep.org>.
12. See Principle 19, <www.un.org>.
13. For the text, see <www.unece.org>.
14. Article IV, para. 10.
15. Brian E. Concannon, 'The Indus Waters Treaty: The Decades of Success, yet, will it Endure?', *The Georgetown International Environmental Law Review*, vol. 2 (1989), 73.
16. See chapter 6, 'Treaty in Action–II', under 'India Claims Carbon Credits'.
17. Annexures D (Generation of Hydroelectric Power by India on the Western Rivers) and E (Storage of Waters by India on the Western Rivers).

18. Article V lays down the exact amount that the Bank was to pay to India in case of extension of the transition period which was as follows: one year: £3,125,000; two years: £6,406,250; three years: £9,850,000.

19. *Water Security for India: The External Dynamics*, Report of the Institute for Defence Studies and Analysis (IDSA), <www.idsa.in/book/WaterSecurityforIndia>, 31.

20. Annexure D, para. 8 (e).

21. Annexure D, paras. 9–11.

22. Annexure E, para. 7.

23. Annexure E, paras. 12–14.

24. Ramaswamy Iyer, 'Pakistan's Questionable Move on Water', *Economic and Political Weekly*, 27 March 2010, 11.

25. Ibid.

26. Article VIII, paras. 1–3.

27. Ibid. para. 4.

28. 'Iliff, Transcript of Interview', supra note 2, 53.

29. Article VIII, paras. 5–6.

30. Article IX.

31. See Annexure F.

32. The US Chief Justice, while accepting the offer to appoint an arbitrator, clarified in his reply that he would appoint an American on the Arbitration Court for the reason that 'I would have no competence because I do not have sufficient acquaintance with the bench and bar of other countries to enable me to make a considered choice.', US Chief Justice's Letter, dated 8 September 1960, see File no. 1787797, Title: *Indus Basin Treaty-Court of Arbitration, Correspondence* 01.

33. Annexure G and accompanying Appendix carry the details.

34. Kishenganga Arbitration, supra note 5, 100–1, paras. 273–6.

35. Ibid. 101–2, paras. 277–8.

36. Ibid. 102–4, paras. 279–83.

37. Ibid. 180–2, paras. 478–80.

38. Ibid. 181–2, paras. 480–2.

39. Ibid. 182–4, paras. 483–7.

40. Annexure G, para. 29.

41. B. S. Murty, 'Settlement of Disputes', in Max Sorenson (ed.), *Manual of Public International Law* (London, Melbourne, etc.: Macmillan, St. Martin's Press, 1968), 690–2.

42. Kishenganga Arbitration, Final Award, *PCA*, 20 December 2013, 39, para. 111. The Court reaffirmed the point in *India's Request for Clarification or Interpretation*, dated 20 May 2013, where the parties referred to the World Court's case law on the question of admissibility. It noted that the World Court's practice in the matter is based specifically on its Statute and Rules which include substantive preconditions for the exercise of its interpretative power, whereas neither

the Treaty nor the Supplemental Rules lay down any condition for the party requesting clarification or interpretation except the filing deadline. It went on to declare that once a party makes a timely request, it, in accordance with para. 27 of Annexure D, 'shall, reassemble to clarify or interpret its Award'. *India's Request for Clarification or Interpretation*, dated 20 May 2013, Kishenganga Arbitration, *PCA*, 20 December 2013, para. 22.

43. Article XI, para. 2.

44. J. G. Starke, *Introduction to International Law* (London: Butterworths, 1989), tenth edition, 44–5.

45. R. R. Baxter, 'The Indus Basin' in A. H. Garretson, R. D. Hayton and C. J. Olmstead (eds.), *The Law of International Drainage Basins* (Dobbs Ferry: New York University School of Law, Oceana Publications, Inc., 1967), 476.

46. Ibid.

47. Article XI, para. 1 which covers it provides as follows:
    1) It is expressly understood that
       a) this Treaty governs the rights and obligations of each Party in relation to the other with respect to the use of the waters of the rivers and matters incidental thereto: and
       b) nothing contained in this Treaty, and nothing arising out of the execution thereof, shall be construed as constituting a recognition or waiver (whether tacit, by implication or otherwise) of any rights or claims whatsoever of either of the Parties other than those rights or claims which are expressly recognized or waived in this Treaty.

    Each of the Parties agrees that it will not invoke this Treaty, anything contained therein, or anything arising out of the execution thereof, in support of any of its own rights or claims whatsoever or in disputing any of the rights or claims whatsoever of the other Party, other than those rights or claims which are expressly recognized or waived in this Treaty.

48. Kishenganga Arbitration, supra note 5, 135, note 567.

49. N. D. Gulhati, *Indus Waters Treaty: An Exercise in International Mediation* (Bombay, Calcutta, etc: Allied Publishers, 1973), 265. As an example, he states that as Pakistan planned to construct Mangla Dam which was located in Jammu and Kashmir, the parties decided that 'there should be no mention in the treaty of any work to be constructed by Pakistan and no indication that India had agreed to, or had any responsibility in regard to, any such work.' He goes on to narrate that when on 9 August 1959, India lodged its third protest to the Security Council regarding the construction of the Mangla Dam by Pakistan, Iliff was very much upset with it but the Indian government assured him that 'this renewal of protest did not, in any way, alter the understanding reached between Black and the prime minister in New Delhi.' Ibid. footnote.

50. Ibid. 131.

51. Ibid. 136, note 569.

52. Ibid. 132, para. 356.

53. Ibid. 132–3, paras. 357.

54. Ibid. 136, para. 364.

55. Ibid. 137, para. 365.

56. 'IFIs Refuse to Fund Bhasha Dam under Indian Pressure, says WAPDA', *The News* (4 January 2013).

57. Syed Fazl-e-Haider, 'Pakistan Dam Dealt Funding Blow by India', *Asia Times (online)* (18 August 2012).

58. In 2006, the Indian minister of state for external affairs, responding to a question in the *Rajya Sabha*, stated that the 'Government conveyed, through diplomatic channels to the Government of Pakistan, its protest against the proposed construction of Basha Dam in territory that is part of the State of Jammu and Kashmir, which is an integral part of India.' John Briscoe, 'Troubled Waters: Can a Bridge be Built over the Indus', *Economic and Political Weekly* (11 December 2010), vol. XLV, no. 50, 31.

59. See report of the seminar entitled 'Analyzing Kishanganga Award' held by the Islamabad-based Institute of Policy Studies, <http://www.ips.org.pk/whats-new/92-seeminar/1748-ips-seminar-reviews-kishanganga-award>.

60. See OP 7.60 of March 2012, *World Bank Operational Manual.*

61. 'World Bank has Agreed to Finance Diamer-Basha Dam-Dar', *Pakistan Defence* (21 August 2013).

62. Article VII.

63. Article XII, para. 2.

64. Robert Y. Jennings, 'Treaties', in M. Bedjaoui (ed.), *International Law: Achievements and Prospects* (Paris, Dordrecht, etc.: UNESCO, Martinus Nijhoff, 1991), 139.

65. Article XII, para. 3.

66. Article V, para. 1.

67. Article V, para. 5.

68. Article X. For the history of this provision, see chapter 3 'World Bank: the Honest Broker?'

69. Shafqat Kakakhel, 'The Indus Waters Treaty: Negotiation, Implementation, New Challenges, and Future Prospects', *Criterion Quarterly* (Islamabad, April/June 2014), vol. 9. no. 2, 57–9.

5

Treaty in Action-I

SOON AFTER THE RATIFICATION OF THE TREATY INSTRUMENTS IN December 1960, Pakistan began implementing it on a fast track basis because it did not have much time to undertake replacement works after surrendering the Eastern Rivers to India. It started the Indus Basin Project which involved building dams, barrages, and link canals on the Western Rivers as envisaged in the Treaty. It completed the Mangla Dam and other replacement works by the cut-off date, 1 April 1970. It could not, however, undertake the construction of the other major dam, the Tarbela Dam, until 1968 (it was completed in 1976). In the first decade after the Treaty came into force, there were no water issues between the two countries, primarily because India did not build any structures on the Western Rivers. The problems started in the early 1970s when India decided to build dams on the Western Rivers electricity generation. There have been five disputes so far in which the two States have been involved and many more are expected to crop up with the passage of time as India has planned a series of such projects which are at various stages of their implementation. We propose to examine them below one by one.

A) SALAL DAM

a) History of the Dispute

The first dispute arose between the two countries when India proposed to construct a 690 MW hydroelectric power plant called the Salal Dam on the Chenab River located at Reasi in district Udhampur in Jammu. On 30 April 1970, the Indian Indus Water commissioner notified his Pakistani counterpart of the details of the

Figure 5.1: Salal Dam

proposed project along with its design. He did so under para. 9 of
Annexure D of the Treaty which lays down that '[t]o enable Pakistan
to satisfy itself that the design of a plant conforms to the criteria
mentioned in para. 8, India shall, at least six months in advance of
the beginning of construction of river works connected with the plant,
communicate to Pakistan, in writing, the information specified ...'.
The dispute coincided with the completion of the Mangla Dam's
construction and other replacement works by Pakistan. In other
words, the dispute over the interpretation of the Treaty started the
moment Pakistan had turned to the Western Rivers for its water
needs rather than the Eastern Rivers, which was the case in the past.
In a communication dated 17 July 1970, addressed to the Indian
commissioner, the Pakistan commissioner not only raised Pakistan's
objections to the proposed design but also expressed his dissatisfaction
with the adequacy of information about the design and requested
that he supply complete information. The two commissioners then
corresponded with each other for quite some time and also discussed
the matter in a number of meetings of the Commission held in
Islamabad and Delhi.

An examination of the correspondence between the two commissioners reveals the Pakistan commissioner's frustration over
his Indian counterpart's reluctance to part with information
regarding the design and his failure to fulfil promises that he made
in this regard. Exasperated with the Indian commissioner's dilly-
dallying tactics, the Pakistan commissioner, at the thirty-ninth
meeting of the Commission held in July 1974, conveyed to his
Indian counterpart the Pakistan government's desire to resolve the
matter under article 9(1) of the Treaty. The Indian commissioner
consented to take up Pakistan's proposal at the Commission's
next meeting.

In September 1974, the Pakistan commissioner undertook an
inspection tour of the site of the proposed dam. At the conclusion,
he handed over certain questions to his counterpart for consideration
by the Commission. The latter considered them at its fortieth meeting
held in December 1974, but the Indian commissioner refused to
discuss their substance on the grounds that his government wanted the
two governments, rather than the two commissioners, to take them up
directly. Subsequently, in a communication dated 19 February 1975,
the Indian commissioner conveyed certain changes in the design of
the plant to the Pakistan commissioner. In the Pakistan commissioner's
opinion, the proposed changes, instead of allaying Pakistan's
concerns, further heightened them. Following this development, he
conveyed Pakistan's objections to the modified design to his Indian
counterpart which related to paras. (a), (d), (e), and (f) of Annexure
D of the Treaty.[1]

b) Pakistan's Objections and India's Rebuttal

The first objection related to para. (a) which provides that, '[t]he
works themselves shall not be capable of raising artificially the water
level in the Operating Pool above the full Pondage Level specified in
the design.' Initially, Pakistan objected to the installation of iron gates
as, in its opinion, they could create hurdles in the free flow of water of
the Chenab River. Subsequently, realizing that their installation made

little difference to the capacity of the dam, it decided to withdraw this objection. However, it raised concerns regarding the design of these gates and the spillway. India wanted to construct twelve gates measuring 50 feet wide and 40 feet high, each for the purpose of controlling floods. In Pakistan's opinion, India was exaggerating the threat of floods. The Pakistani experts put the maximum discharge of water (at the point where India planned to construct the dam) at 4.5 lakh cusecs whereas Indian experts put it at 8 lakh cusecs. The former believed that the latter was exaggerating the figures in order to justify the installation of iron gates. India also justified the design of gates on the possible threat of quakes, failure of the dam to function, and failure of power-generating turbines due to mud clogging. It claimed that the Treaty permitted the use of engineering to achieve maximum benefits. The two commissioners debated and discussed the issue in the Commission's meetings held in Delhi and Islamabad but could not come to an agreement.

The second objection related to para. (d) which stipulates as follows: 'There shall be no outlets below the Dead Storage Level, unless necessary for sediment control or any other technical purpose; any such outlet shall be of minimum size and located at the highest level, consistent with sound and economical design and with satisfactory operation of the works.' The original Indian design provided for the construction of two sluices measuring 15x11 feet at a depth of 150 feet below the full reservoir level with a flow of 24,000 cusecs of water. Pakistan opposed the proposal at once on the grounds that the sluices were unnecessary and in violation of the foregoing para. India, on the other hand, justified them based on the grounds that a potential earthquake posed an emergency threat to the dam. The Pakistani experts disagreed with this approach contending that such a threat could be taken care of in the design of the plant. In their view, after the completion of the dam, there was no point in keeping these sluices open and doing so would be a violation of the Treaty. They felt that an emergency was not a technical issue and that such sluices could not serve the said purpose as they would not be effective in an emergency. Instead of meeting Pakistan's objections by taking remedial measures, India modified its design to increase the number of sluices from two to six and raise their height from 150 feet to 235 feet. Similarly, it decided to increase the quantum of the flow of water from 24,000 cusecs to 96,000 cusecs.

Pakistan protested it stating that it constituted an outrageous violation of the Treaty to which India responded that, in addition to ensuring the flow of water during the construction phase, these sluices were required for four to five years to observe the functioning of the dam. It was of the view that for these reasons it was not in a position to plug these sluices immediately but could plug two to four of them after five years. The Indian commissioner also justified the proposal on the grounds of sediment control. Since the original design did not accommodate sluices for the purpose of controlling sediment, the Pakistan commissioner felt that this purpose was later added to strengthen India's case and that they were not really required for the stated purpose. In his opinion, international experience also showed that the low level sluices were not good at controlling sediment except in a small area in their vicinity. He felt that since the Salal Dam did not have a live storage capacity it did not require sediment control, which was meant for prolonging the life of storage. He was of the opinion that in the absence of justification for the proposed outlets, the plant could be operated without any difficulty even without sluices.

The third objection related to para. (e) which provides as follows: 'If the conditions at the site of the plant make a gated spillway necessary, the bottom level of the gates in normal closed position shall be located at the highest level consistent with sound and economical design and satisfactory construction and operation of the works.' According to the Pakistan commissioner, the Salal Dam site was such that it did not require a gated spillway. He was of the view that the proposed design was in contravention of para. (e) of Annexure D. The Indian commissioner, on the other hand, defended the Indian position on the grounds that an un-gated spillway would increase the cost of the project by Rs. 14.8 crore. The Pakistan commissioner countered by contending that that argument was not relevant in the present situation as the work had to conform to the design criteria prescribed by the Treaty. He also rejected it on the grounds that wherever economical design was a relevant factor, the Treaty specifically spelled it out in the design criteria.

The fourth objection related to para. (f) which lays down: 'The intakes for the turbines shall be located at the highest level consistent with satisfactory and economical construction and operation of the plant as a run-of-river plant and with customary and accepted practice of design for the designated range of the plant's operation.' The design of the proposed dam placed the top of the intakes at a depth of 13.5 feet below the full reservoir level which is also the dead storage level. According to the Pakistani commissioner, since in normal practice a water cushion equal to half the diameter of the pen-stock is considered adequate, raising the level of the intake by 10.5 was enough. The Indian commissioner had justified the raising of the intake on the basis of the US Bureau of Reclamation's design standards. The Pakistan commissioner challenged the Indian placement by arguing that even by those standards, the proposed intake should not be raised above 5.1 feet.

c) The Settlement

The foreign secretaries held talks followed by four meetings of the two commissioners in which they tried to sort out the differences between their two countries. India refused to compromise at all, resulting in the failure of negotiations. Pakistan then offered to refer the matter to the Neutral Expert but India rejected it out of hand. Then the two commissioners made another effort (in April 1976) to settle the matter but failed to do so. This was followed by a meeting of the foreign secretaries in 1978 and they finally succeeded in clinching a deal. According to the terms of the agreement, the height of the dam was fixed at RL 1600 feet to store maximum of 303,300 acre feet of water; the iron gates were to be 50 feet wide and 30 feet high; and six sluices were to be installed at the height of RL 1365 feet which were to be closed with concrete plugs one year after filling the reservoir with water or in three years after the water had reached the spillway level, whichever was earlier.

The most contentious issue between the two sides related to the sluices because the Pakistan government feared that the Indian side could, in violation of the Treaty, use them to increase the dam's capacity to store more water, stop the flow of the water by emptying the dam, or trigger a flood in the Chenab River. Given these apprehensions, the Pakistan government had a clause inserted in the agreement by virtue of which India, after closing the sluices in case of an emergency emanating from a threat to the dam, was to inform the Pakistan government of the situation and then open the sluices. Following the agreement, India constructed the dam in two stages. The first was completed in 1987 and the second in 1995, with an installed capacity of 345 MW each.

B) WULLAR BARRAGE/TULBUL NAVIGATIONAL PROJECT

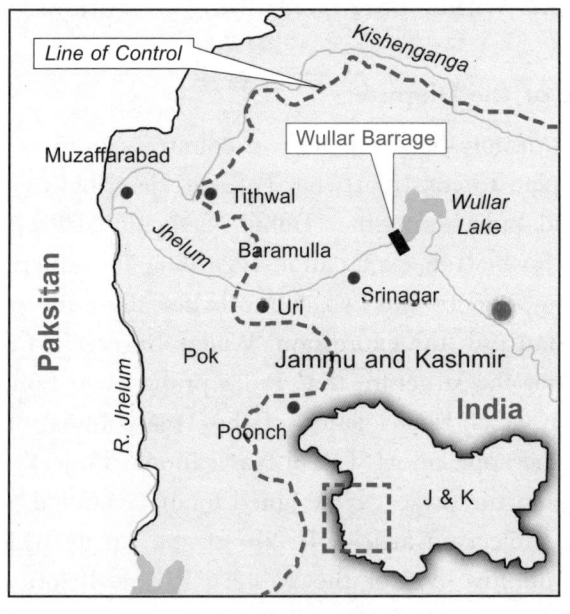

Map 5.1: Wullar Barrage

Figure 5.2: Wullar Barrage/Tulbul Navigational Project

a) History of the Dispute

The second dispute between the two countries arose as a result of the Indian plan to construct what Pakistan describes as the 'Wullar Barrage' and India terms the 'Tulbul Navigational Project'. Before proceeding further, an explanation regarding the difference in the nomenclature used by the two countries for the same project is in order. Pakistan uses the expression 'Wullar Barrage' to drive home the point that the structure that India proposes to build is meant for irrigation which is a violation of the Treaty. India, on the other hand, uses the appellation 'Tulbul Navigational Project' in order to emphasize that the project is designed for navigational purposes. It justifies the project on article III (2b) of the Treaty which permits it 'non-consumptive use' of the Western Rivers. Before proceeding further we propose to look at the facts of the case.[2] The Wullar Lake where the project is located is situated 25 kilometres north of Srinagar in the Bandipore district on the Jhelum River which flows into the lake from the south and out of it from the west. Its size varies from twelve to one hundred square miles. According to one view, the name 'Wullar' finds its origin in the Sanskrit word 'woll' which signifies an obstacle, such as the lake is an impediment in the way of the river.[3] According to another view, it is traced to a Kashmiri word *wul* which means a gap or a fissure.[4]

The Wullar Lake is not suited for a dam of any size because it would inundate the city and valley of Srinagar. For this reason, during the Indus Basin negotiations, Pakistan initially resisted the Bank proposal for India to have restricted use of the Western Rivers in Jammu and Kashmir but, as seen above, it ultimately gave in under pressure from the Bank. The strategic significance of the project lies in the fact that once it materializes, it could provide India with the means to intimidate Pakistan by releasing water that could ruin the entire Triple Canals Project, namely the Upper Jhelum Canal, the Upper Chenab Canal, and the Lower Bari Doab Canal.[5] In simple terms, storage on the Jhelum River at the site of the Wullar Lake, irrespective of the purpose, could endanger vast tracts of agricultural land in the Pakistani Punjab. Additionally, it could cause a serious shortage of electricity in Pakistan in case India decided to withhold water over an extended period of time, particularly during the dry season.

The proposed project comprises a 439 feet long barrage located at the mouth of the Wullar Lake with a gated weir, under-sluice and a 12-meter wide navigational lock. When completed, it would store roughly 300,000 acre feet of water.[6] According to the Indian government, the purpose of the project is to construct a control structure for the purpose of improving navigation in the Jhelum River during the winter season, for which purpose a minimum depth of 4.5 feet of water is needed. India justifies it on the grounds that it would solve the problem of navigation over a distance of 22 kilometres between the Wullar Lake and Baramulla in order to connect Srinagar with Baramulla.[7] Pakistan alleges that the purpose behind the project is to create a man-made lake to run the lower Jhelum Hydroelectric Plant and Uri Hydroelectric Plant on the Indian side of the Line of Control (LoC).[8] India conceived the project in the early 1980s and the work began in 1984. The Indian commissioner conveyed the information related to the construction of the barrage to his Pakistani counterpart in 1985. The latter raised objections to the proposed project while simultaneously seeking more details which India communicated in April 1986. The commissioners of the two countries then held negotiations to resolve the issue but failed to do so. Pakistan threatened to take the issue to the Neutral Expert under article IX (2b) of the Treaty. However, the two commissioners agreed to refer the matter to their respective governments for resolution. Subsequently, the representatives of the two countries met in November 1987, and the Indian representative provided his counterpart with a map and other related details. India also agreed to suspend the work. The two countries kept talking to each other on the subject but without success.

The next noteworthy development in the matter took place when the two countries decided to include the issue as one of the eight items in the Composite Dialogue that the two countries agreed to undertake following the success of Prime Minister Vajpayee's bus journey to Pakistan. It is noteworthy that when the peace process was derailed, as result of the Kargil episode, India neither denounced the peace process nor resumed work on the Wullar Barrage project. Subsequently, following the signing of the Islamabad Declaration in January 2004, the two countries decided to restart the peace process which included the Wullar Barrage issue. However, the peace process was interrupted again in the wake of the Mumbai terrorist attack in November 2009. When bilateral peace talks resumed again, they held several meetings on the Wullar Barrage.

From the start of the project to March 2012, there have been fourteen rounds of secretary-level talks between the two countries, which included five under the Composite Dialogue. Another round of talks was scheduled for 28–29 January 2013, but it could apparently not be held because the Indian secretary of water resources was retiring. India proposed March 2013 as the next date of the meeting but nothing came out of it. It was later scheduled for August 2013 but again did not materialize. Now, the two countries propose to take up the issue under comprehensive bilateral dialogue.

b) Arguments of the Parties and their Critique

Were Pakistan's objections against the Wullar Barrage well-founded? To answer this question, we need to examine the relevant provisions of the Treaty. We know that according to article III, India is obligated to let waters of the Western Rivers flow freely to Pakistan and is not permitted to interfere with them. As far as the Jhelum River is concerned, the article in question permits India to make use of its waters for domestic, non-consumptive, and agricultural uses in addition to generation of electricity. However, it forbids India to store water or construct any storage work except as laid down in Annexures D and E of the Treaty. Annexure E, sub-para. 8(h) which is relevant here makes allowance for '[s]torage incidental to a barrage on the Jhelum Main not exceeding 10,000 acre-feet.' Its sub-para. 9 permits India to construct such works on the Jhelum Main as it deems necessary for flood control and complete any works which were under construction on the effective date (read 1 April 1960)[9] subject to the condition that it does not construct any storage on the Jhelum Main. In case it does, it is confined to off-channel storage in side valleys, depressions, or lakes. It also stipulates that India shall release stored waters as quickly as possible after the flood recedes and return them to the Jhelum Main except those waters held in lakes, borrows-pits, or natural depressions. Sub-para. 10 stipulates that India shall only design and operate a storage work on a tributary of the Jhelum River, on which Pakistan has any agricultural or hydroelectric use, in such a manner as to not adversely affect the existing agricultural or hydroelectric use on that tributary.

The foregoing analysis demonstrates that the Treaty forbids India from constructing storage work on the Jhelum River in excess of 10,000 acre-feet of water. India wants to build a barrage with a capacity of 300,000 acre-feet of water which is thirty times more than the allowed capacity. No wonder Pakistan considers the project in violation of the Treaty. India, however, contests Pakistan's objection that it is engaged in constructing a barrage. It maintains that it is not building a storage work but merely a control structure meant to use natural storage[10] which, in its opinion, sub-par. 9(i) of Annexure E of the Treaty allows by proscribing a man-made storage but permitting one consisting of side valleys, depressions, or lakes. When we examine India's argument, it does not seem to be well-founded. The meagre information which is available to us, namely that the structure is 439 feet in length with a gated weir, under sluice, and a 12-metre wide navigation lock, hardly makes it a natural reservoir of water. As the title of the project—control structure—suggests, the project seems to be more of an attempt to control the flow of water of the Jhelum River than anything else.

India has also justified the project on the grounds that it is meant for navigational purposes during the winter months. It claims to draw support for this from article III(2) of the Treaty which, as seen above, permits it for 'non-consumptive use'[11] of the waters of the Western Rivers. It further argues that the structure in question would actually help Pakistan as water released during the lean period would augment supplies to it and reduce spills at the Mangla Dam by withholding it at other times.[12] When we examine the terms of the Treaty, we realize that article I(11) explicitly permits India to use waters of the Western Rivers for navigational purposes. However, it does not signify that India, under the garb of using waters of the Jhelum River for navigational purposes, is entitled to construct a barrage which is at cross purposes with the obligation that the Treaty imposes on it to let the waters flow unhindered and without interference. It goes without saying that with the barrage or control structure in place (as India calls it), India would be in a position to release or withhold water at its discretion.

Regarding the Indian argument that the Wullar Barrage would actually benefit Pakistan, through withholding of water supplies during the rainy season and releasing them during the dry season, it cannot be used as a justification for its construction without the prior consent of the Pakistan government, howsoever well-meaning the intentions of the Indian government may be in the matter. On the contrary, Pakistan had a serious problem with this argument because it felt that the barrage would give India control over the water supplies which it found utterly unacceptable. It also felt that, as stated above, the barrage had the potential to ruin the Triple Canals Project. These apprehensions were not unwarranted. Let us not forget that India, on 1 April 1948, shut-off water supplies from the Ferozpur headworks to canals in Pakistan at the critical *kharif*-sowing period with the result that not only did Pakistan's agriculture suffer but the city of Lahore, which was principally dependent on them for its daily use, was also severely affected. Additionally, power distribution to West Pakistan from the Mandi Hydroelectric Scheme was stopped. It is worth recalling that it was this incident, more than anything else, which made Pakistan opt for the division of the Indus Rivers with separate control by Pakistan and India over the Western and Eastern Rivers respectively. Given this background and the atmosphere of unabated suspicion prevailing in the relations of the two countries, Pakistan cannot afford to let India control its water supplies which are so essential for its agriculture, in addition to hydroelectric generation and defence.

The Indian argument in favour of a barrage for navigational purposes that would provide India with a handle to manipulate these waters cannot be justified on any legal or political grounds. It may not be out of place to mention here that Pakistan's apprehensions about India's intentions in the matter are further reinforced when we realize that the Wullar Barrage project was conceived before Partition but was shelved after the realization that a dam of any size would inundate a vast tract of land, including the valley and the city of Srinagar. As for the Indian designs behind the project, this is what a knowledgeable Kashmiri has to say:[13]

The idea of water transportation on the Jhelum river will create a basis for New Delhi's geopolitical strategists to manufacture arguments for improving navigation on the river and thus necessitate creation of more Wullar barrages which would neither help the people of Kashmir nor help the cause of confidence building and durable peace between New Delhi and Islamabad.

The bogey of the Wullar Barrage Project and its linking with the question of water transportation in the valley needs very careful handling by Kashmir's ruling elite. Not only the Kashmiris, but also those who have planned the Wullar Barrage Project know the argument that the project is meant to improve navigation on the Jhelum river during the lean season is nothing but a joke. *It is now well-known among the water resources management experts that the basic idea behind the Wullar Barrage Project is not navigation but its use as a geostrategic tool during negotiations with Pakistan.*

In 1991, the two countries reportedly reached an understanding on the issue which resulted in a draft agreement according to which India was to keep 6.2 metres of the barrage ungated with a crest level of 1,574.90m (5,167 feet) and was to forgo the proposed storage capacity of 300,000 acre-feet. In return, it was to be allowed to attain the full operational level of 5,177.90 feet.[14] However, the two countries failed to sign it. India held Pakistan responsible for this, accusing it of linking the signature with the outcome of the Kishenganga Hydro-electric Project dispute. In August 2007, the two countries agreed to refer the matter to a Pakistan-India body of technocrats. However, they also failed to resolve it. In February 2009, the Indian government, hoping to revive the project, initiated extensive consultations among the relevant Indian ministries. In May 2011, it offered to modify the design of the barrage but the Pakistani negotiators refused to go along with the offer and asked it to abandon the project altogether. This led to the Indian delegation staging a walkout from the meeting for about two and a half hours—a performance staged to force Pakistan to make concessions to India. As for the Indian offer's details, it consisted of leaving ungated one of the many bays that it proposes to build. The purpose of the offer was to allay Pakistan's fears regarding India's manipulation of the waters of the Jhelum River. Pakistan was, however, not prepared to accept it because, as explained by a former secretary of the Ministry of Water and Power, '[b]locking an un-gated bay is not a difficult thing to do once the main barrage structure is complete. There is no point in building a barrage when India's navigational requirements can be met through the existing natural Wullar lake.'[15] There has not been any headway in the matter. However, in the meantime, according to the information provided by the parliamentary secretary for foreign affairs, Palwasha Khan, to Pakistan's National Assembly in March 2012, India unilaterally resumed construction work on the project.[16] The Pakistan commissioner visited the project site in May 2013. There are indications that India is ready to make adjustments to the design of the barrage which may facilitate an agreement.[17]

C) BAGLIHAR DAM

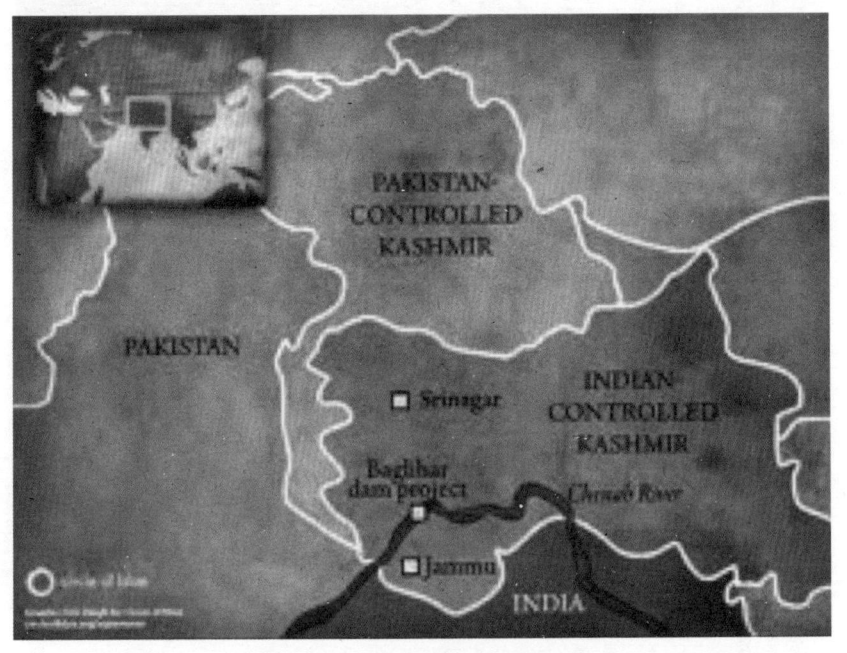

Map 5.2: Baglihar Dam

Figure 5.3: Baglihar Dam

a) History of the Dispute

The third dispute that has arisen between the two countries relates to the Baglihar Dam. It is a 450-megawatt power plant that India has built on the Chenab River, in the Doda district, in Jammu and

Kashmir. Conceived in 1992, its construction began in 1999. Right at the outset, Pakistan objected to the design of the plant on the grounds that it provided for gated spillways which, in its opinion, were in violation of the Treaty because they would increase the storage capacity of the plant. According to Pakistani experts[18], the Baglihar Dam would deprive Pakistan of about 7,000 cusecs feet of water per day and completely stop supplies for an uninterrupted period of twenty-six days during the critical months of December through February each year. They contended that it would adversely affect both agriculture and defence requirements of the country. They also held that in case of synchronized releases of water from Dul Hasti, Salal, and Baglihar Dams, the Bajwat area above the Merala headworks would be inundated. As far as the gated spillway structure which India was constructing was concerned, the Pakistan commissioner maintained that it did not figure in the earlier design which Pakistan gave its consent to and that India only introduced it subsequently and without Pakistan's consent. Disagreeing with his counterpart, the Indian commissioner was of the opinion that the modifications relating to the gated spillway structure were in accordance with the Treaty and that they would not affect the free flow of water to Pakistan.

To deal with the contentious issue of modifications and the question of an on-site inspection of the plant, Pakistan requested a special meeting of the Commission to which the Indian commissioner acceded during the Commission's meeting held in May–June 2001, but only after enormous pressure exerted by Pakistan. Despite the promise that the Indian commissioner made, India was reluctant to honour its commitments and for this purpose it used dilatory tactics. Incidentally, if India acceded to Pakistan's request for a meeting, it was in no way a concession to Pakistan but only a fulfilment of a legal requirement arising out of a number of provisions of the Treaty. Three are particularly relevant here. The first one is para. 4(c) of article VIII which enjoins the Commission to promote co-operation between the parties in the development of the Indus Rivers by undertaking (once in every five years) a general tour of inspection for the purpose of ascertaining the facts connected with various developments and works on these rivers. The second one is para. 4(d) of the same article which obliges the Commission to promptly undertake, at the request of the either commissioner, a tour of inspection of such works or sites on the Indus Rivers as it may deem necessary for ascertaining the facts connected with those works or sites. The third one is para. 5 of article VIII which requires the Commission to meet regularly, at least once a year, alternately in India and Pakistan in November or some other month mutually agreed upon by the two commissioners. It also obligates the Commission to meet at the request of either commissioner.

After much resistance by India, the Commission's special meeting was finally held in February 2003, a meeting Pakistan had asked for back in 2001. India reiterated its commitment to allow the latter to make an on-site inspection of the plant but made it contingent on the political and security situation in Jammu and Kashmir. Given India's past record in the matter and Pakistan's apprehensions that India wanted to buy time to complete the project by the date that it had fixed, which was 2004, Pakistan was wary of the latter's promise. To get out of the uncertain situation, during the February meeting of the Commission, Pakistan proposed[19] that the matter be submitted to a Neutral Expert for settlement. India rejected the plea for third party involvement, showing unambiguous preference for a negotiated bilateral settlement. It justified this approach on the grounds[20] that the practice had developed in the two countries. In other words, what it contended was that this practice had superseded the dispute settlement procedure laid down in the Treaty. Instead of rejecting India's counter proposal, Pakistan went along with it but with the caveat[21] that there should be a time frame for such a negotiated settlement and that if there was no settlement within the stipulated time period, the two countries would submit the matter to a Neutral Expert. The Indian commissioner baulked at Pakistan's proposal and rejected it out of hand.

This Indian line of reasoning was not surprising for the Pakistan commissioner as he knew that the mere idea of a third party's involvement in India's disputes with Pakistan was no less than a bugbear to India. His understanding of the Indian psyche was very accurate as borne out by India's negotiating behaviour on Kashmir and a host of other contentious matters with Pakistan where it has consistently spurned offers of mediation by friendly countries. A similar pattern of behaviour by India vis-à-vis its other South Asian neighbours is noticeable. We have already expounded the reason why India takes this position but, given the significance of the matter in the present context, it is necessary to reiterate. Its rejection of the third party's involvement is understandable because it feels that, given its preponderant position in one-to-one talks with Pakistan and other subcontinental neighbours, it can dictate terms to them which may not be possible in the presence of the third party. It possibly fears that the third party may impose a settlement unacceptable to it. And in case it dithers, it may lose face in the international community.

India adopted this policy right after independence though it temporarily departed from it when it took the Kashmir dispute with Pakistan to the UN Security Council where the latter turned the tables on it by getting the world body to adopt resolutions declaring Jammu and Kashmir a disputed territory and favouring the Kashmiris' right to self-determination through a UN-sponsored plebiscite to determine their future. Let us not forget that Indians have blamed Mountbatten for the 'folly' of making India involve the UN in the Kashmir dispute. India succeeded in getting the Kashmir dispute out of the UN's orbit through the following clause of the Simla Agreement which it imposed on Pakistan in 1972 after the Bangladesh war: 'The two countries are resolved to settle their differences by peaceful means mutually agreed upon between them.'[22] The only other occasion when India departed from this policy was when, following the border skirmishes with Pakistan in the Rann of Kutch, it accepted Harold Wilson's proposal for arbitration, though according to the veteran Indian journalist, Kuldip Nayar, Lal Bahadur Shastri never forgave the British prime minister for tricking him into it.[23] Barring these two exceptions, India has almost religiously pursued the policy of excluding third party involvement in its disputes with Pakistan.[24] India's viewpoint on the exclusion of a third party's role in Pakistan-India disputes may or may not be valid in light of the Simla Agreement but has certainly no validity with regard to the Indus Waters Treaty because it has its own dispute settlement procedure in article IX which is not superseded by the Simla Agreement.[25] Besides, the fact that the dispute settlement procedure laid down in the Treaty was never resorted to before the Baglihar Dam dispute, as India contended, does not signify that it had become obsolete and could not be invoked.

Thus, the Commission's February meeting failed as the Indian commissioner did not satisfy the Pakistan commissioner except in terms of oral promises. In the wake of this development, the Pakistan government served a notice on the Indian government (in May) to meet the following three conditions, namely[26] a) stop work on the project; b) allow an on-site inspection by the Pakistan commissioner; and c) resolve the dispute by 30 September 2003, failing which it reserved the right to approach the Bank for the appointment of a Neutral Expert. Following this notice, the Commission held its annual meeting where, on Pakistan's insistence, it discussed the Baglihar Dam issue despite the fact that it did not figure on the agenda. Pakistan made some progress as the Indian commissioner finally allowed Pakistani inspectors to make an on-site inspection in October 2003. Consequently, the latter undertook the visit and found the design of the plant to be in violation of the Treaty.

Following this finding, the Pakistan government in November 2003 served another notice[27] to the Indian government to suspend work and to negotiate a settlement by 31 December 2003, failing which it reserved the right to approach the Bank for the appointment of a Neutral Expert. This notice was the continuation of an earlier one, the only difference between the two being was that the earlier one also envisaged an on-site inspection which had already materialized. When India disregarded the second notice, Pakistan decided to approach the Bank but India requested that Pakistan give bilateral negotiations another chance to which Pakistan relented.[28] The two commissioners then met in January 2004, but the talks were abortive. They held another round of parleys in May 2004, but they too remained inconclusive. At this juncture, the Pakistan commissioner recommended that his government approach the Bank for the appointment of a Neutral Expert. He had already made such a recommendation in February 2003, in pursuance of which, the Pakistan government had served two notices to the Indian government in May and November of the same year.

Faced with the threat of referral of the issue to the Bank, the Indian government proposed another meeting of the two commissioners to which the Pakistan government agreed to the condition that the matter should be discussed under article IX(2) of the Treaty which provides for the appointment of a Neutral Expert. However, one day before the meeting, the Indian commissioner informed his counterpart[29] that the discussion in the meeting would take place on the basis of article VIII(5) which, as seen above, stipulates a routine meeting of the Commission. This upset the Pakistan commissioner because it meant that India would not discuss the issue for which Pakistan had called the meeting. Notwithstanding the futility of such a meeting, the Pakistan government gave the green signal to it perhaps because it did not want to be blamed for vitiating the recently-improved political atmosphere resulting from the Islamabad Declaration that Musharraf and Vajpayee had agreed upon to kick-start the stalled peace process. Another possible explanation for Pakistan's consent to the meeting could be that despite past experiences, it was hoping for a change in the Indian attitude in the negotiations. Pakistan, however, faced complete disappointment as India not only refused to stop work on the project but also reiterated its old offer to continue the open-ended bilateral talks without a time frame.

In June 2004, the two countries reached an understanding, according to which India agreed to provide additional data concerning the Baglihar Dam by July and settlement of the issue by November

2004. In the wake of this development, it appeared that the issue would soon be out of the way of the two countries' relations. India, however, failed to keep its word; Pakistan sent several reminders to Delhi to provide the promised data but to no avail. Pakistan, which had earlier approached the Bank for the appointment of a Neutral Expert, decided to make a final effort for a negotiated settlement before taking the Bank route. Pakistan's prime minister raised the issue with his Indian counterpart during the meeting held in New Delhi where they resolved to settle the issue bilaterally. Subsequently, each side proposed dates for a meeting but they could not come to an agreement. Then, in December 2004, India provided Pakistan with additional data, in its words, as a 'gesture of goodwill'. It gave the impression that it was doing so out of charity and not out of a legal obligation imposed on it by the Treaty. It asked Pakistan to finalize its objections, if any, to be followed by bilateral negotiations. This led to a difference of opinion in the Pakistan camp.[30] The Attorney General advocated a negotiated settlement on the grounds that arbitration would be expensive and that by the time the Neutral Expert would hand down his award, India would have completed the construction of the dam. He was also of the view that in case of an adverse award against Pakistan, India could ask for renegotiations of the Treaty. The Attorney General was opposed by the Pakistan commissioner who was in favour of referring the case to a Neutral Expert on the grounds that Pakistan had a very good case and that India could not be relied upon to demonstrate goodwill in reaching a satisfactory settlement.

In this controversy, the view advanced by the Attorney General did not appear to be sound as it banked on Indian goodwill which, in the circumstances that persisted at that time, hardly existed. This was attested by the fact that the Composite Dialogue that had resumed following the Islamabad Declaration in early 2004 had run aground before the end of the year. India had proposed 72 confidence-building measures out of which none was directed at the Baglihar Dam, a hot issue at that time. The Baglihar Dam issue did not figure in the Composite Dialogue. However, had India shown willingness to suspend work on the project temporarily, it would have radically transformed the ambience of the peace process. Instead, India not only refused to make such a gesture but also reportedly expedited the work on the project. The Attorney General's argument that, in case the award went against Pakistan, India could ask for renegotiations of the Treaty, showed that he was not certain about the soundness of Pakistan's case. However this, incidentally, was contrary to the opinion of the Pakistan commissioner and the foreign experts whom the Pakistan government had consulted in the matter.

According to press reports, foreign experts were of the opinion that Pakistan had a very good case to win. In any case, the Attorney General was wrong on this count because there is no principle of international law which requires a party to seek renegotiations of a treaty if it loses a case arising out of it. The Attorney General finally lost out to the Pakistan commissioner and the Pakistan government decided to approach the Bank for the appointment of a Neutral Expert under article IX (2a) of the Treaty. The foreign office spokesperson, Masood Khan, justified the decision on the grounds that, 'Pakistan was left with no other choice because the talks at the experts and water secretaries' levels between the two countries ended inconclusively.'[31] It took the Pakistan government almost five frustrating years of bilateral diplomacy before it moved the Bank. The Pakistan media criticized Islamabad for the delay. Mr Khan, however, rejected the criticism[32] on the grounds that the matter needed 'patience and tolerance'. He defended the Pakistan government by arguing that it wanted to exhaust all bilateral channels before approaching the Bank.

b) Did Pakistan Seize the World Bank Belatedly?

Should the Pakistan government have seized the Bank earlier? If so, why did it not do so? These are important questions which need to be answered. As for the first question, we propose a brief recapitulation of the history of negotiations on the Baglihar Dam to see if the charge is justified. Pakistan raised objections against the design of the dam at the outset. However, India not only did not address Pakistan's concerns but also did not allow the Pakistan commissioner to visit the site for a long time, though it was under legal obligation to do so.[33] Instead of dealing with the issue with a firm hand, the Pakistan government kept hurling threats at the Indian government to approach the Bank for the appointment of a Neutral Expert. India responded to these threats by simply ignoring them. The Pakistan government served a notice to India in May 2003. In the said notice, it threatened to seize the Bank in case its three demands were not met. Faced with this threat, India took the wind out of Pakistan's sails by letting the Pakistani experts make an on-site inspection. This took place after a lapse of six months. During the inspection, Pakistani experts found the design in violation of the Treaty. Instead of following up the threat, Pakistan served another notice in November in which it asked India to fulfil the remaining conditions by 31 December 2003, which India disregarded once again.

According to the secretary of Water and Power, Ashfaq Mehmood, when India ignored the second notice, Pakistan was about to move the Bank but did not do so at India's request as it wanted the matter to be resolved bilaterally. A meeting between the officials of the two countries took place in January 2004, but it was inconclusive. Another round of talks, at a technical level, took place in May 2004, which were also inconclusive. According to Ashfaq Mehmood, Pakistan at this point was about to move the Bank but did not do so because India again made a request for a bilateral settlement. India led Pakistan up the garden path by indicating the prospect of upgrading the talks to the secretaries' level. Pakistan was duly snared and fell in the Indian trap. In the secretaries' meeting, the two sides drew up a timetable according to which India was to provide additional data by July and the matter was to be resolved by November 2004. At that time, the Pakistani media was elated as the news of impending settlement made shrieking headlines. The facts on the ground were utterly different. India not only did not resolve the dispute, despite repeated reminders by Pakistan, but also failed to furnish the technical data by the due date.

Instead of moving the Bank, the Pakistan government pinned

hopes of a breakthrough on the meeting that Prime Minister Shaukat Aziz was scheduled to have with his counterpart in New Delhi on the sidelines of the SAARC (South Asian Association for Regional Cooperation) in November. The two leaders agreed to give a final try to resolve the dispute through the secretaries' meeting which took place in December 2004. The talks failed. The Pakistan foreign office spokesperson, Masood Khan, attributed the failure to India's unwillingness to 'effectively address and resolve' the issue as well as stop work on the project. The Pakistan government now took the stand that it would agree to a negotiated settlement only if India, among other things, suspended the work on the dam. Incidentally, this is what the Pakistan government had demanded from India through the two notices that it served in 2003. It goes without saying that the Pakistan government moved the Bank for the appointment of a Neutral Expert belatedly and that this procrastination was responsible for letting India complete the project.

As for the reason for the irrational delay on the part of the Pakistan government, we can only postulate. Perhaps it did so in the belief that it would be able to bring India round to the idea of an amicable bilateral settlement as it did in the case of the Salal Dam project. Perhaps it was deterred by the thought of prohibitive cost that the Neutral Expert's involvement would incur as testified by the controversy in the matter between the Attorney General of Pakistan and the Indus commissioner. Perhaps it was the fear of the unknown that stopped it from moving the Bank because the Neutral Expert would have taken the matter out of its hands and decided the issue as he deemed fit, which may or may not be to its liking.[34] Or perhaps, it was a combination of all these factors that resulted in the delay. Whatever the reason for the delay, the way the Pakistan government handled the matter it seemed as if it was dealing with the Indians for the first time, and was not aware that India was in the habit of playing games.

The lesson is that Pakistan should never, in a water dispute of this nature with India, dither because it not only provides the latter with an opportunity to complete the project but also makes it very difficult for the Arbitration Court or the Neutral Expert to order the dismantlement of an existing structure. It should not bank on the goodwill of the Indian government for an amicable settlement and after giving negotiations a reasonable chance to succeed, should not hesitate to avail itself of article IX of the Treaty.

c) Appointment of the Neutral Expert

Following the failure of bilateral talks, the Pakistan government was left with no choice but to move the Bank to appoint a Neutral Expert. It is interesting to note that, as seen above, after almost five years of failed bilateral diplomacy, India still did not want the matter to be referred to the Neutral Expert. Towards this end, it argued that the issue was under consideration of the Indus Commission and that it was still a 'question' and had not matured into a 'difference', a necessary condition for a referral to the Neutral Expert. The Bank on its part asked Pakistan to justify its stance which it did. India disputed it but the Bank came to the conclusion that the matter had matured into a 'difference'. It took the Bank three months' correspondence with the parties to do so. Finally, on 25 April 2005, it informed Pakistan and India of its decision to appoint a Neutral Expert and proceeded to do so in consultation with them. Since this was unchartered territory for the Bank to navigate as it had no precedent to follow in the matter, it decided to look to the procedures that the International Centre for Settlement of Investment Disputes (ICSID)[35] followed in the selection of arbitrators for the settlement of investment disputes.

It may be mentioned here that even after the Bank's decision to appoint a Neutral Expert, India tried its best to torpedo it. For example, the Indian prime minister stated in the Indian *Lok Sabha* that 'it was premature to appoint an expert since Indian and Pakistani technical experts could still resolve their differences.'[36] When in early 2005 Pakistan approached the Bank, the Indian water resources minister asked it to leave the two parties alone to resolve the remaining issues relating to the Baglihar Dam through dialogue.[37] At one point, India proposed to the Bank that it convene a meeting of the Indian and Pakistani water experts to resolve the matter rather than appoint a Neutral Expert.[38] At another point, it tried to stop it from seizing the matter by claiming that Pakistan's objections were political rather than technical in nature.[39] The explanation for India's desperate attempts to stop the Bank from involvement was its dread of not only the third party's role in a Pakistan-India dispute, but also a desire to see the adjudicatory clause of the Treaty become a dead letter. Pakistan, on the other hand, was keen to involve the Bank because it regarded it as the only hope to get out of the cul-de-sac in which it found itself after the failure of the bilateral talks. Besides, the Pakistan Commission was convinced that it had a winnable case. It had apparently based its view on the opinion of four international water experts whom it had consulted on the Baglihar Dam and who (by three to one) had supported Pakistan's objections to its design while the fourth one had only a minor objection to it.[40]

To select the Neutral Expert, the Bank, in accordance with the terms of the Treaty and the ICSID procedures, compiled a list of global, highly qualified and eminent engineers. While doing so, it was also to ensure that those who figured on the list did not have any conflict of interest in the matter. It was then to send names and curriculum vitae of three engineers from the list to the parties and ask them to identify, within fifteen days, those who were not acceptable to them. In case no agreement was forthcoming on one name, the procedure was to be repeated by sending out three more names. If the two parties selected more than one name, then it was to make a choice according to the alphabetical order. On the chance there was no agreement on a single name, even after the second round, the Bank itself was to appoint a Neutral Expert without any reference to the parties and the person selected did not need to be from among the names proposed in the first and second rounds.

In the present case, the Bank did not have to resort to the lengthy procedure adumbrated above to select a Neutral Expert because the parties agreed on one name in the first round. It is pertinent to mention here that it proposed names of three persons who belonged to Australia, Brazil, and Switzerland. It took the Bank five months to finalize a name after Pakistan approached it. The individual who got a nod from the parties was Professor Raymond Lafitte, a Swiss engineer who teaches at the Swiss Federal Institute of Technology in Lausanne. The proceedings in the case lasted for more than a year and a half and the Neutral Expert rendered his decision, termed 'expert determination', on 12 February 2007.[41] Before delivering the final determination, he submitted a draft determination, a practice unusual in the annals of international arbitration but not in other fora such the WTO, in order to provide an opportunity to the parties to make comments on it, both written and oral.[42] We propose to critically examine below how the two parties argued their case.

d) Pakistan's Objections

Pakistan raised objections to the Baglihar plant on the grounds that it violated clauses (a), (c), (e), and (f) of para. 8 of Annexure D, which stipulate the criteria for a run-of-river plant. Clause (a) spells out that the plant shall not be capable of artificially raising the water level in the operating pool above the full pondage specified in the design. Clause (c) prescribes that the maximum pondage in the operating pool shall not exceed twice the pondage required for firm power. Clause (e) specifies that if the conditions at the site of a plant require a gated spillway, the bottom level of the gates in normal closed position shall be located at the highest level, consistent with sound and economical design, and satisfactory construction and operation of the work. Clause (f) states that the intakes for the turbines shall be located at the highest level consistent with satisfactory and economical construction and operation of the plant as a run-of-river plant and with customary and accepted practice of design for the designated range of the plant's operation. Pakistan claimed that the Baglihar plant did not meet any of the criteria mentioned in these clauses.

As far as Pakistan's objections were concerned, the first one related to the maximum pondage or live storage which India had fixed at

37.5 M.m^3· Arguing that it exceeded twice the pondage required for firm power, Pakistan calculated it at 6.22 M.m^3. Pakistan also raised objection to the Indian calculation according to which 16,500 m^3/s was the maximum amount of water that could arrive at the dam. It calculated it at 14,900 m^3/s. The second objection related to the free board (or the vertical difference in elevation provided between the maximum reservoir level during a routing of the design flood and the dam crest) which Pakistan believed was excessive and provided India with the means to artificially raise the water level in the operating pool above the full pondage level. Therefore, it wanted the free board to be reduced in conformity with the Treaty. The third objection pertained to the level of the dam's power intake which it contended was not located at the highest level as stipulated in the Treaty. Pakistan's fourth and perhaps the most important objection related to the spillway of the dam. India had proposed a gated spillway in its design which Pakistan opposed on the grounds that it would endow India with the capability to control the flow of water. However, Pakistan's objection did not hold because the Treaty did not forbid a gated spillway. Going beyond the Treaty, India had proposed an orifice spillway or drawdown flushing which it justified on the grounds that the design of the chute, sluice, and auxiliary spillways were imperative for the safe passage of the design flood. Pakistan rejected it on the grounds that it was not located at the highest level as required by the Treaty.[43]

Though Pakistan's objections were technical in character and India replied to them in the same vein, the battle before the Neutral Expert was fought on legal grounds. Concretely put, the parties advanced their respective positions by invoking the principles of interpretation enshrined in articles 31 and 32 of the Vienna Convention on the Law of Treaties 1969, to elucidate various provisions of the Treaty. Thus, it is imperative to look at these articles:[44]

Article 31
General rule of interpretation

1. A treaty shall be interpreted in good faith in accordance with the ordinary meaning to be given to the terms of the treaty in their context and in the light of its object and purpose.
2. The context for the purpose of the interpretation of a treaty shall comprise, in addition to the text, including its preamble and annexes:
   (a) Any agreement relating to the treaty which was made between all the parties in connexion with the conclusion of the treaty;
   (b) Any instrument which was made by one or more parties in connexion with the conclusion of the treaty and accepted by the other parties as an instrument related to the treaty.
3. There shall be taken into account together with the context:
   (a) Any subsequent agreement between the parties regarding the interpretation of the treaty or the application of its provisions;
   (b) Any subsequent practice in the application of the treaty which establishes the agreement of the parties regarding its interpretation;
   (c) Any relevant rules of international law applicable in the relations between the parties.
4. A special meaning shall be given to a term if it is established that the parties so intended.

Article 32
Supplementary means of interpretation

Recourse may be had to supplementary means of interpretation, including the preparatory work of the treaty and the circumstances of its conclusion, in order to confirm the meaning resulting from the application of article 31, or to determine the meaning when the interpretation according to article 31:
   (a) Leaves the meaning ambiguous or obscure; or
   (b) Leads to a result which is manifestly absurd or unreasonable.

We can paraphrase these articles as follows. Article 31, which the Vienna Convention terms as the 'general rule of interpretation', stipulates that the interpreter of a treaty is under obligation to assign ordinary meaning to the terms of the treaty in their context and in light of its object and purpose. In other words, it emphasizes the textual approach of interpretation. In case this approach leaves the meaning ambiguous or obscure, or leads to a result which is manifestly absurd or unreasonable; or in case there is a need to confirm the meaning resulting from the application of article 31, the interpreter is allowed to have recourse to article 32, which the Vienna Convention terms as 'supplementary means of interpretation' and which includes the preparatory work *(travaux préparatoires)* of the treaty and the circumstances of its conclusion. Before we spell out how the Neutral Expert interpreted these articles, we will explain how Pakistan and India construed them.

e) **Point Counter Point**

i) India's Arguments

Dealing with the case, India relied on the 'general rule of interpretation' which, as seen above, emphasizes a textual approach whereas Pakistan invoked 'supplementary means of interpretation' with emphasis on preparatory work and circumstances of the conclusion of a treaty. To draw support in favour of its approach, India referred to the World Court's advisory opinion in the *Competence of the General Assembly for the Admission of State to the United Nations* where it had ruled:[45] 'If the relevant words in their natural and ordinary meaning make sense in their context, that is the end of the matter.' It also cited the following observation of the International Law Commission, which drafted the Vienna Convention on the Law of Treaties 1969, in support of its viewpoint:[46] 'The ordinary meaning is not to be determined in the abstract but in the context of the treaty and in the light of its object and purpose.' As for the object and purpose of a treaty, it argued that the preamble of a treaty was the place to look for it, and cited the following statement of Sir Gerald Fitzmaurice, the International Law Commission's special rapporteur on the Law of Treaties, in this regard:[47] 'Although the objects of a treaty may be gathered from its operative clauses taken as a whole, the preamble is the normal place in which to embody, and the natural place in which to look for, an express or explicit general statement of the treaty's objects and purposes. Where these are stated in the preamble, the latter will, to that extent, govern the whole treaty.' It also cited in support of its contention the International Law Commission's following statement based on the World Court's ruling in the *United States Nationals in Morocco*:[48] '[T]he Court has more than once had recourse to the statement of the object and purpose of the treaty in the preamble in order to interpret a particular provision.'

Dealing with the definition of the expression 'context' in article 31(2), India was of the opinion that neither of the two instruments covered in its sub-paras. (a) and (b) were present in the case. It further stated that by virtue of article XII (1) of the Treaty, the instruments contained in Annexure A, like other Annexures, form part of the Treaty. It pointed out that 'context' in article 31(2) provides that the '[it] shall comprise, in addition to the text [of a treaty], ... its preamble and annexes ...' It was of the view that Annexes or 'Annexures' are considered part of a treaty and not separate instruments such as those referred to in article 31(2). It held that the instruments contained in Annexure A to the Treaty do not fall in the category of documents referred to in article 31(2 a & b) of the Vienna Convention, though they constitute part of the context under article 31(2). It concluded that we had before us, 'the text of the treaty, including its preamble and annexes ...', as the relevant context for the interpretation of the Treaty.[49]

Next, dwelling on the Indus Waters Treaty, India accepted that the allocation of the Western Rivers to Pakistan and the Eastern Rivers to India is the fundamental principle on which the Treaty is based. However, it argued that according to article III (2) and Annexure D of the Treaty, this principle is dependent, among other things, on India's right to produce hydroelectric power through run-of-river dams on the Western Rivers. It did not consider this right as some kind of ornament but an integral part of the Treaty. It recognized that Pakistan apparently did not dispute this fact, at least directly, but was critical of its stand that 'the 'control/let flow' principle is indeed the fundamental principle underlying the treaty', and 'interference with the flow is explicitly permitted as an exception to the basic principle.' It also criticized Pakistan's contention that India's right to generate hydroelectric power on the Western Rivers 'should not be given a wide interpretation.' Giving its own interpretation of the Treaty, it put forward the thesis that no 'principle' such as 'control/ let flow' as contended by Pakistan exists. On the contrary, it argued that the true basic principle is contained in the preamble to the Treaty which aims at 'attaining the most complete and satisfactory utilization of the waters of the Indus system of rivers.' It also emphasized the fact that without incorporation of the right to produce hydroelectric power through run-of-river dams on the Western Rivers, India would not have accepted the Treaty. It pointed out that this right was not a 'mere exception, to be construed narrowly, as Pakistan contends.'[50] It rejected Pakistan's contention that the Treaty should be read in light of the events of 1948 (when India cut-off water supplies to Pakistan) and that it was drawn up to protect Pakistan from India's control of the 'tap', suggesting instead that its fears were unfounded as demonstrated by the fact that nothing of the sort had happened during the entire life of the Treaty, despite periods of great tension between the two countries.[51]

ii) Pakistan's Arguments

Pakistan totally disagreed with the Indian approach. Stating that it was conscious of the language of the preamble to the Treaty and the importance that India attached to the expression 'the complete and satisfactory utilization' of the Indus Rivers used in the preamble, it contended that it is a settled law that the preamble is not a substantive part of the Treaty. Therefore, in its opinion, it cannot control or supersede the substantive provisions of the Treaty. It furthermore argued that 'complete and satisfactory utilization' is strictly subservient to the framework of the Treaty and the substantive

limitations contained in Annexure D.[52] Arguing that the Treaty is
sacrosanct, it pleaded that the intent of the parties as expressed by
the clear language of the Treaty, rather than conceptions unrelated to
it or resource utilization perspective, should be the guiding principle
in the interpretation of the Treaty. It wanted the mutual wishes of
the parties as reflected in the Treaty to be paramount and other
considerations to be subordinated to them.

Taking into account these observations, it urged the Neutral Expert
to respect the Treaty norms because they, in its opinion, reflect what
the parties decided to be in their best interests. It emphasized that
by doing so, he would assure peace in the region.[53] Following the
enunciation of its approach to the Vienna Convention, it outlined its
position on the Treaty in which it emphasized, among other things,
the circumstances which led to its conclusion: 'It is important also to
recall the circumstances of the Treaty's conclusions and in particular
the events of 1948 when India cut-off the water supplies into West
Punjab .... The 1960 Treaty was drafted and concluded with the
view to preventing the repetition of such events.'[54] In light of these
observations, it summarized its position as follows:[55]

4. [T]he Treaty seeks to avoid conflict by trying to limit as much
   as possible, the *capability* of each country to interfere with waters
   allocated to the other. This was an objective of the Treaty precisely
   because India had already used its position as the upper riparian to
   interfere in the flow of water to Pakistan and Pakistan was therefore
   apprehensive that India would abuse its position as upper riparian
   in the future as well.

5. The fact that the Treaty seeks to minimize India's capability to
   interfere with the flow of the Western Rivers to Pakistan is not
   a matter of historical conjecture. This point is self-evident from
   a simple reading of the text of the Treaty which explicitly places
   the Treaty in the context of past water disputes between India and
   Pakistan and which provides detailed criteria for future projects so
   as to avoid subsequent conflicts.

6. More specifically, Paragraph 15 of Annexure D states that to the extent India chooses to operate a run-of-river hydropower project on any of the Western Rivers, the volume of water flowing into hydropower plant over a given seven-day period must equal the outflow of water over the same seven-day period from that plant. Similarly, the amount of water flowing through the project on a day to day basis must also remain within certain stipulated boundaries.

7. If the intent of the Treaty was only to ensure that India would not interfere with the flow of the Western Rivers to Pakistan there would have been no need for the Treaty to have provided any limitations with respect to the design of run-of-river plants. However, such design limitations are explicitly provided and elaborated in Paragraph 8 of Annexure D to the Treaty. The sole purpose of these limitations—for example, the requirement of Paragraph 8(e) that the spillways of a run-of-river plant must be ungated if possible—is to ensure that India has the *minimum possible capability* to interfere with the flow of the Western Rivers.

8. The intent of the Treaty to regulate not just the utilization of the waters of the Indus Basin but also the *capability* of the respective Parties to utilize those waters is made explicit not only by the provisions of Paragraph 8 of Annexure D to the Treaty but also by Art. I (11) of the Treaty. The said provision states as follows:

> The term 'Non Consumptive Use' means *any control or use of water* for navigation, floating of timber or other property, flood protection or flood control, fishing or fish culture, wild life or other like beneficial purposes, provided that, exclusive of seepage and evaporation of water incidental to the control or use, the water (undiminished in volume within the practical range of measurement) remains in, or is returned to, the same river or its Tributaries: but the term does not include Agricultural Use or use for the generation of hydro-electric power.

9. Again the point to note here is that the Treaty does not discriminate between 'control' or 'use' of water resources: instead, it defines both of them as equally dangerous from Pakistan's perspective and seeks to place equivalent limits on both. Admittedly, the term 'Non-Consumptive Use' as defined by the Treaty excludes the

use of water for the generation of hydro-electric power. However, what is evident from the text of this Article is that hydro-electric power generation has been excluded from the definition of the term 'Non-Consumptive Use' not in order to avoid the limitations placed upon non-consumptive usage, but in order to define those limitations more precisely, as indeed has been done in Paragraph 8 of Annexure D of the Treaty.

10. In conclusion, it bears repeating that the Indus Waters Treaty was not an 'Entente Cordiale' signed between fast friends with the common goal of exploiting a shared resource. Instead, it was a treaty signed between two hostile countries to demarcate and divide a very important resource into zones of exclusive jurisdiction. It was a treaty signed between two countries who did not trust each other. Its strict implementation between the Parties is one of the prerequisites for the maintenance of peace in vital region.

f) Neutral Expert's Determination and its Critique

As stated above, the Neutral Expert submitted a draft determination to the parties before making his final determination. The parties took advantage of this opportunity and submitted their comments on the draft determination. We know that of the four objections that Pakistan raised, the most serious was related to the low level orifice spillway that India had incorporated in its design effecting drawdown flushing. Pakistan rejected this technique on the grounds that it was in conflict with Annexure D, part 3, para. 8 (e & d) of the Treaty and in its place proposed an alternative design. The Neutral Expert disagreed the design proposed by Pakistan on the grounds that it would not be sufficiently safe and that with the passage of time (within two decades or even less) it would become dysfunctional. However, he agreed that Annexure D, part 3, para. 8 (e) of the Treaty did not permit drawdown flushing. Notwithstanding this finding, he justified the outlets incorporated in the Indian design for flushing on the grounds that they formed part of international practice and were state-of-the-art. He invoked Bulletin no. 115 of the ICOLD (International Commission on Large Dams) which supported the drawdown flushing in favour of his finding.[56] However, being uncertain of his stance, he advised Pakistan and India to 'develop jointly, without in any way questioning the Treaty, guidelines concerning sedimentation control in the waters of the Indus river system. This would also be extremely valuable for all the engineers in the world concerned by dam technology.'[57] He made this suggestion even though he confessed that it was beyond the scope of his mandate.[58]

Pakistan was terribly upset with this part of the draft determination. It challenged the findings of the Neutral Expert both in writing and orally. On the issue of drawdown flushing, it felt that, on the one hand, the latter considered it contrary to the Treaty while, on the other hand, he justified it by approving the low level orifice spillway for flushing of the reservoir. It therefore found a major contradiction and a fundamental error in his determination. As for the ICOLD Bulletin no. 115, it observed that while recommending drawdown flushing it also asked the concerned engineers to respect applicable institutional and legal constraints. Additionally, it argued that the parties made their submissions on the understanding that the Baglihar plant would be operated without drawdown flushing as this technique is not in conformity with the criteria laid down in the Treaty. It claimed that 'at least at one stage ... India had positively accepted that it was not entitled to drawdown below the dead storage level',[59] and that the draft determination proceeded on the basis that the Treaty prohibited drawdown flushing. It asserted that India subsequently changed its earlier stance by arguing that the drawdown flushing was permitted by the Treaty. Besides, it held that during the hearings 'India only outlined the arguments "it would have made" were the question of drawdown flushing before the Neutral Expert.'[60] It complained that the Neutral Expert did not communicate with it in writing or orally the comments that India made on the draft determination due to which it could not immediately respond to India's new arguments and hence, reserved its position in the matter. It, therefore, wanted the latter to make the determination by ignoring the new justification. It also criticized him for suggesting the two countries jointly find a solution to the issue of sedimentation control which was outside his mandate.

India was delighted with the draft determination. Its senior delegate Shankardass, at the far end of the proceedings, surprised Pakistan by challenging its contention that there was agreement between the parties that drawdown flushing was unlawful in terms of the Treaty. He asserted that the assumption that drawdown flushing is not allowed under the Treaty was not a legal determination and that such a conclusion could not be drawn from the Treaty. He particularly contradicted the Neutral Expert's statement that the Treaty proscribed the practice of drawdown flushing. Dwelling on this point, he argued:[61] '[T]he definition of Dead Storage in 2 (a) of Annexure D merely provides that the Dead Storage is the portion of the storage which is not used for operational purposes. In other words, there is no express or implied bar to use Dead Storage for other purposes such as occasional use for rehabilitation or maintenance purposes. Furthermore, India cannot use any water below Dead Storage Level for operational purposes as drawing down the reservoir below Dead Storage Level during the operation of power plant would damage the civil and electro-mechanical components.' He also argued that para. 8 (d) provides against outlets below the dead storage level unless it is necessary for sediment control or other technical purposes. He concluded that the use of the dead storage and outlets below the dead storage level are permitted for sediment control or other technical purposes such as drawdown flushing of the reservoir.

An analysis of the arguments of the parties shows that Pakistan was justified to complain that the Neutral Expert failed to furnish the rationale behind the conclusion that he made on the drawdown flushing and that his determination was contradictory. It was also justified to hold that the proceedings were conducted on the understanding that the drawdown flushing was in contravention of the Treaty. This is attested by the fact that the Neutral Expert admitted that para. 8 (e) did not allow use of such a technique. It is true that the Indian delegate, Shankardass, did not agree with this assumption and argued that the Treaty permitted drawdown flushing for rehabilitation and maintenance purposes. However, he was guilty of what Churchill termed as 'terminological inexactitude' when he discarded the said assumption because he then went on to admit that he made the foregoing submission to underline 'what [his delegation] would argue if and when this issue is discussed.'[62] This was a new position that India adopted but unfortunately, Pakistan did not rebut it. We can conclude from the foregoing that Pakistan's grouse against the Neutral Expert was totally justified.

Next, examining the final determination we realize that, caught in the dilemma to choose between two competing interpretations of the Treaty, the Neutral Expert decided to side with the one that India had advanced. He started with the premise that provisions in article 31 are not hierarchical and that interpretation of a treaty is an integrated approach. Proceeding thus, he held that the starting point of any interpretation is to assign ordinary meaning in good faith to the terms of the treaty and that the role of the context and object(s) and purpose(s) of the treaty is to help confirm or refine, and develop the ordinary meaning. He pointed out that the ordinary meaning is to be found in the text of the treaty and that the intention of the parties, which is in principle to be found outside the text, is relevant only to the extent expressed by the text. He also acknowledged the relevance of the circumstances which led to the conclusion of the treaty but only if the text is ambiguous, or obscure, or if it leads to a meaning which is obviously absurd or unreasonable. He claimed on the basis of decisions of various international courts and tribunals that a preamble defines in general terms the purposes and considerations that led the parties to conclude the treaty and that it also normally carries important indications about the object(s) and purpose(s) of the treaty.[63]

Following the exegesis of his outlook on the principles of interpretation, the Neutral Expert set out to apply these principles to the present case. He was of the view that the preamble of the Treaty explicitly defines its object(s) and purpose(s), which is to attain the most complete and satisfactory utilization of the waters of the Indus Rivers;

to fix and delimit the rights and obligations of each party in relation
to the other concerning the use of these waters; and to provide for
settlement of the questions arising from application or interpretation
of the Treaty. In his opinion, given the principle of integration, these
objectives are not to be read in isolation as they are complementary to
each other. He held that the preamble which enshrines the 'common
intention' of the parties should be instrumental in the interpretation
of the rights and obligations of the parties under Annexure D. He
reckoned that the principle of effectiveness (expressed in the Latin
expression *ut res magis valeat quam pereat*),[64] with emphasis on words
to give full effect and weight to object(s) and purpose(s) of a treaty,
should be used in the interpretation of the Treaty. In light of these
observations, he set aside Pakistan's plea based on the circumstances
surrounding the conclusion of the Treaty (which had set the stage for
Indus water negotiations under the auspices of the World Bank and
which had eventually led to the conclusion of the Treaty) because,
in his estimation, it would deprive the Treaty of its object(s) and
purpose(s) in addition to altering the scope and recourse to article 32,
rejecting it on the grounds that the wording of article 31 is very clear,
leaving no room for ambiguity or uncertainty. He therefore did not
feel the need to base his decision on the circumstances surrounding
the conclusion of the Treaty because, in his opinion, they come into
play only to confirm the meaning of the rights and obligations under
the Treaty, which were not needed in the present case.[65]

Taking into account these observations, the Neutral Expert made
his final determination in which, on three points, he pronounced
in favour of Pakistan and on one point in favour of India. This is
how he dealt with each objection: as to the first one which related to
maximum pondage or live storage of the dam, the Neutral Expert
rejected the viewpoints of both the countries as being contrary to
the criteria laid down in the Treaty. He thought that it should be
fixed at 32.56 M.m^3 and the corresponding dead storage level at el.
836 m asl, or one meter higher than the level of the Indian design.
Regarding the peak discharge of the design flood, he decided to

retain the figure of $16,500\ \text{m}^3/\text{s}$. He also justified it on the grounds of climate change. As for the second objection which pertained to the free board which Pakistan regarded as excessive and amenable to manipulation by India, to the detriment of Pakistan, the Neutral Expert agreed with Pakistan's argument that the crest level should be kept to the lowest and directed India to reduce the free board from 4.5 metres to 3 metres.

As for the third objection which related to the level of the power intake, the Neutral Expert accepted Pakistan's plea and made the determination that the location of the power intake was lower than that laid down by the Treaty. Hence, he decided in favour of raising it from the existing elevation of 818 metres to 821 metres. Regarding the fourth objection which related to the installation of a low level orifice spillway, the Neutral Expert rejected Pakistan's contention and upheld that of India on the grounds of hydrology, sediment control, topography, geology, and seismography. He also justified it stating that 89 per cent of all modern dam structures with a designated discharge of 14,000 cubic metres per second have them. In other words, he justified India's contention that contemporary international practice dictated their design. He also justified it by reasoning that its deletion could create the risk of flooding the upstream shores and that the elevated dam crest alternative could be expensive.[66]

Following the announcement of the decision by the Neutral Expert, both the countries claimed victory. Pakistan's minister for water and power, Liaquat Jatoi, describing the verdict as a 'great victory', stated that India was under a 'moral, legal and political obligation ... to accept the World Bank's decision.'[67] Similarly, India's minister for water resources, Saifuddin Soz, observed that, 'we are happy overall. The dam structure is intact, the changes are only minimal.'[68]

Did Pakistan really win the case as claimed by Mr Jatoi? Judged in terms of numbers, Pakistan was a clear winner as out of the four objections that it had raised, the Neutral Expert decided three in its favour. However, this cannot be a litmus test of success because Pakistan's principal objection related to the low level orifice spillway and possibility of flushing below dead storage level which it deemed contrary to the terms of the Treaty. Consequently, it is this objection whose rejection or acceptance is the real measure of Pakistan's failure or success regarding the case. Judged by this criterion, it is obvious that Pakistan lost the case because the Neutral Expert decided in favour of India on this point.[69] If Mr Jatoi still claimed victory, it is obvious that he did so for political reasons. Notwithstanding this claim, the Musharraf government criticized the decision on the grounds that it was not in consonance with the Treaty. More specifically, it stated that it reserved 'the right to take up the spillway issue any time at an appropriate forum.'[70] As expected, the Pakistan government never raised the issue at any other forum.

Was the Neutral Expert justified in siding with India on the dam design with low level orifice spillway or drawdown flushing technique? To answer this, we attempted to find out how he applied articles 31 and 32 of the Vienna Convention. We have seen above that his starting point for the interpretation of the Treaty was the preamble of the Treaty because, in his opinion, it incorporates the 'common intentions' of the parties. Next, he resorted to the principle of effectiveness with an emphasis on words to give effect to the object and purpose of the Treaty. As for the circumstances of the conclusion of the Treaty, he did not have to resort to it because, in his opinion, it would deprive the Treaty of its object and purpose. In our judgment, he was perfectly justified to start with the preamble to find out the object and purpose of the Treaty but unjustified to have recourse to the principle of effectiveness because it cannot be used for rewriting the Treaty which is what he did in the present case. He also faltered when he stopped with the preamble and did not explore what the annexes of the Treaty contain which he was under obligation to do. It was wrong on his part to use the preamble to override the substantive provisions of the Treaty.

Pakistan's legal counsel, James Crawford, rightly pointed out that if the Neutral Expert was to assign preponderant weight to 'the complete and satisfactory utilization of the waters' by the upper riparian at the expense of the lower riparian then the principal objective of the Treaty, which is to let the waters of the Western Rivers flow except as expressly provided by the Treaty, would be distorted. He was also unjustified to ignore the circumstances of the conclusion of the Treaty as para. 8 (d & e) of Annexure D did not suffer from any ambiguity. Article 32 clearly provides for such recourse not only in case of ambiguity but also in order to confirm the meaning resulting from application of article 31. It is clear that his determination was flawed and amounted to a miscarriage of justice because it was not well-founded in law. No wonder that the Arbitration Court in the *Kishenganga* case overturned the Neutral Expert's finding on the issue of dam design with drawdown flushing technique (for details see section A in chapter 6).

g) Why Did Pakistan Lose the Case?

This leads us to the most important question: why did Pakistan lose an air-tight case? In our judgment, there were three reasons for it. First, it did so because the Neutral Expert, being an engineer and having been involved in dam safety issues for a considerable period of time, carried a strong bias in favour of dams with drawdown flushing features. Secondly, Pakistan lost because its delegation, including the legal counsel, mishandled the case. Thirdly, it did so because its delegation behaved impolitely towards the Neutral Expert who reacted by deciding against it. As for the first explanation, it is attested by a number of statements that the Neutral Expert made in the draft determination and during the proceedings of the case.[71] For example, he justified a dam design with drawdown flushing technique in the draft determination on the grounds that it was '[i]n conformity with international practice and the state-of-the-art ...',[72] even though he agreed at the same time that the definition of the dead storage in the Treaty did not permit it. Similarly, he expressed his approval of this technique during the proceedings through the following statement: 'Just to point clearly that the basis of my decision will be the rules according to the engineering state-of-the-art ...'[73] This explanation is seconded by water experts who hold that Pakistan made the mistake of approaching the issue as a legal one whereas India looked at it as a purely engineering matter concerning the construction of a hydropower plant. For example, Asif Kazi, the honorary Vice-President of the ICOLD holds this viewpoint:[74]

> Prof. Raymond ... has acted as a pure professional engineer since he is trained to look at projects in the strictest sense of their operational efficacy and economic performance. Taking it for granted that the upper riparian would not resort to immoral or unethical practices, he failed to take into account the psyches and mindsets of the litigants in the context of their historic rivalry. Had he kept these factors in view, he might have concluded that, in the absence of spirit of cooperation, the only checks on an upper riparian to keep it from doing harm to the downstream country were constraints, as were proposed by Pakistan, in the shape of 'minimum needed sizes of water outlets to be located at the highest levels' to prevent emptying and refilling of reservoirs at will.

Similarly, Salman M. A. Salman shares this viewpoint:[75] 'First, as appeared from the composition of the two delegations, Pakistan seemed to have viewed the difference as largely a legal one, involving the interpretation of the Treaty, while India seemed to have viewed it as an engineering one, regarding hydropower plants.' Incidentally, he made a similar pronouncement to the present writer in the course of an interview during which he stated that Pakistan lost the case because its delegation was led by lawyers as opposed to that of India which was led by engineers.[76]

It is clear that the Neutral Expert carried a pro-dam bias. However, he had difficulty in justifying a dam design with drawdown flushing technique. Though he tried to do it on the basis of contemporary international practice and state-of-the-art engineering trends, he knew that this justification was far from convincing because it was not backed by the Treaty. Consequently, he was hard pressed to justify it. It was the senior Indian delegate, Shankardass, who provided him with a way out by arguing in the course of oral proceedings that, though the Treaty prohibited the use of the dead storage for operational purposes, 'there is no bar to use Dead Storage for other purposes such as occasional use for *rehabilitation or maintenance purposes.*'[77] This argument came of use to the Neutral Expert who seems to have used it in the final determination to justify a dam design with a drawdown flushing technique:[78] 'The reservoir drawdown below the Dead Storage Level will be done for *maintenance purposes.* It is commonly agreed in practice that maintenance is an absolute necessity, with its ultimate objective of ensuring the *sustainability of the scheme.*'

Dealing with the second reason why Pakistan lost the case, two incidents prove that they mishandled the case. The first took place when, as seen above, the senior Indian delegate, Shankardass, towards the end of the oral presentations, challenged Pakistan's contention that there was an agreement between the parties that the drawdown flushing was contrary to the Treaty. He asserted that the claim that the Treaty does not countenance drawdown flushing was not a legal determination and that the Treaty does not permit such a conclusion. Ashfaq Mehmood, secretary of the Ministry of Water and Power, and the leader of the Pakistan delegation, did not challenge and rebut Shankardass' arguments, which he justified as follows:[79]

I just want to make an observation, with your kind indulgence .... The rule of the game, which you specified, was that our mutual dialogue and answers and questions has come to an end (sic), and we will comment strictly on your determination. Since much has been said, we don't want you to give us any opportunity now to respond to it, but we would like you to kindly note that since we did not have any opportunity and we do not at this moment want to have because we know that this is the last meeting, you would not take any effect of what observations and comments and contentions have been made with respect to our submissions, and you will be kind enough to distill out what is relevant in respect to your determination on which we were supposed to make comment because there were a number of comments on which otherwise we would like to give a position, but I think it's not the time, but we only request you to distill it out in such a way that you only take effect of what relates to the determination.

In other words, instead of seeking to refute Shankardass' new argument, Ashfaq Mehmood let it pass on the grounds that the time for questions and answers had lapsed and that the parties were required to make comments only on the draft determination. He appealed to the Neutral Expert to ignore the Indian observations and comments on his delegations submissions and to take them into account only insofar as they related to the draft determination. It is not clear from the records whether or not he adopted this position after consulting the legal counsel. In any case, we know that at the end of the day, the Neutral Expert embraced the argument that Shankardass had advanced during the oral presentations to justify the drawdown flushing, namely for occasional uses such as rehabilitation or maintenance purposes. We believe that it was a huge mistake on the part of the Pakistan delegation to let India's new argument go unchallenged at the time of both written and oral comments and instead entreat the Neutral Expert to disregard it as irrelevant. The Pakistan delegation was entitled to respond to India's new point but its leader did not deem it appropriate to do so.

Another incident that proves that the Pakistan delegation mis-handled the case took place when the legal counsel, James Crawford, in his closing statement made on 28 May 2006, in London observed:[80] 'Had we known what we now know about this project, some of our fears might have been less because we were not aware it [Baglihar Dam] would silt up so quickly when we raised concerns.' During the oral presentations the Indian delegation used this statement to justify the drawdown flushing as demonstrated by the following statement of F. S. Nariman, the senior Indian legal counsel:[81] 'This is Mr Crawford saying this in London. Fears might have been less. This whole thing is on fear. We are not aware that it would silt up. So, something to be afraid of. So I ask myself what are you afraid of? It's going to silt up. You have five years, 10 years, 15 years, 20 years? That's my job. I'm building it. You're not financing it. What's the problem?' Continuing this line of thought, he further commented:[82] 'It's in my favor, not against me on this question of intent because a treaty has to be interpreted with a purpose. There must be a purpose in everything ...'. We do not know why Crawford made this statement. However, one thing is certain that it was contrary to Pakistan's position in the matter and highly damaging to its cause. It is not outside the realm of possibilities that this gaffe on the part of Pakistan's legal counsel influenced the Neutral Expert in his decision on drawdown flushing against Pakistan.

Dealing with the third reason which related to Pakistan delegation's rude attitude during the proceedings, we begin by recapitulating the incident at the origin of the Neutral Expert's supposed ire. During the oral hearings on the draft determination, Professor George Annandale, who was called up as a witness by the Pakistan delegation, made a presentation in which he argued that the Indian design for sedimentation control was not workable because it was not in conformity with the Treaty. Proceeding from this proposition he rhetorically stated:[83] 'Now that is so clear, there is no way around it. I cannot see, and I'm really—I'm going to ask you to give me an answer. I don't understand how you can get to any other answer but that. And that's the question. Do you have a reason, why did you do that? Why did you make that decision?' The Neutral Expert was so flabbergasted with this tirade that he inquired whether it was directed towards him to which Annandale replied in the affirmative. The Neutral Expert asked if it was useful to ask this in such a meeting, to which Annandale replied that he was trying to understand and not embarrass him though he gave him the right to duck the question, if he so desired. The Neutral Expert, feigning that he was not embarrassed, sheepishly observed:[84] 'I have never been in a meeting like this, so I would have to ask my lawyers.' At this point, F. S. Nariman, senior member of the Indian delegation, intervened:[85] 'This is not a cross-examination of a judge.' The Neutral Expert appreciated the intervention of the Indian delegate, exclaiming:[86] 'It is a very good question. Thank you very much.'

Then Ashfaq Mehmood entered the scene. He first clarified that the purpose of Annandale's statement was not to cross-examine the Neutral Expert but to understand the logic behind his draft determination. He explained that he too had not been able to understand the rationale behind the Neutral Expert's determination even though he had raised this point in the previous meeting and in his written statement on the draft determination. He pointed out that the Neutral Expert had promised to answer the question in his final report but failed to do so. He implored him to answer the question because, in his opinion, once his delegation knew the rationale it would be able to identify the fundamental error behind it and thus address it, failing which his country would get a final determination without going through this process. He persisted with his entreaty to the point that the Neutral Expert was forced to say that he might give an answer at the end of the presentation. The Neutral Expert then explained his method of proceeding in the matter:[87] 'I try to entertain your views, but it's clear at the beginning of this process I have said to you, I give you my draft determination, give me your remarks, and if I don't understand your remarks, I will ask you questions. It was absolutely not in the process that you are arriving with your arguments and asking me any questions. If it's something quite important, if I don't understand your point of view, I will ask you questions, not on the contrary.' Despite this explanation, Mehmood remained persistent, asking the Neutral Expert to explain his rationale. At this point, Nariman intervened to point out that the minutes of the previous meeting held in Paris provided for seeking clarification of specific points but not any discussion. He told Mehmood that if he wanted to subject the Neutral Expert to questioning he should have raised this point at the Paris meeting. He then went on to make the following statement which was the most damning censure of the Pakistan delegation:[88]

It has not been raised, and I would earnestly request you, that if this meeting is to have any meaning at all, all of us have to come and give you what we wish to present to you and not keep asking you questions and keep on which in our system of jurisprudence we call cross-examining the judge. This is not done, and it was an agreed procedure. The parties agreed to this procedure. It should not lead to any discussion. What is this except leading to a discussion ....

I have never heard of this sort of proceeding before any Neutral Expert, Arbitrator, call him what you like, or a judge. Either we follow this procedure that we agreed to in Paris or we don't follow it, and I respectfully submit that my learned friend ought to follow the procedure. Should not lead to any discussion. Now we are starting a discussion. The experts ask you for an answer, the leader of the delegation asked you for an answer. I can understand we ask for some answer. I'm entitled to ask him, but how can we ask you?'

Unfortunately, this was not the end of the Pakistan delegation's misery. Towards the end of the presentations, Nariman took another dig at the Pakistan delegation when he recounted the following anecdote:[89]

I have been practicing law for more than 55 years in courts, and when I was very young, only 12 years in profession, and much more impulsive than I am now, I appeared before a judge in the high court of Bombay. When I was arguing, the judge put to me a proposition which I knew to be wrong, and I impertinently told him I never heard of such a proposition in law, and where did you get it from?

The judge was a solemn fellow. He curled up his lip and he told me, Mr Nariman, you are young. There are many things you haven't heard of. When you get older, you may, if you are wise, get to hear and learn of the things that you haven't heard. Since then, sir, I have never dared to ask a presiding judge an explanation for his observations.

So, you will pardon me if I don't demand that you will tell me here and now reasons for the conclusions which we on this side may not immediately appreciate or understand. I will do only what practicing lawyers are trained to do; namely, make submissions for your consideration.

The foregoing incident which was quite unprecedented in the annals of international adjudication created an unpleasant ambience in the oral proceedings. Who was responsible for it? Mehmood felt justified to ask the Neutral Expert to explain the rationale behind the justification for the drawdown flushing because he had failed to do so even though he was asked this question during the previous meeting (no.4) and in Pakistan's written comments. The Neutral Expert, on his part, did not feel bound to explain it because, in his view, the parties could make observations and ask questions if they felt that he did not understand them and not the other way round. He pointed out that there was an agreement on this procedure between the parties.[90] Technically speaking, he had a point as Mehmood was not entitled to ask the question that he did at that stage of the proceedings. However, we cannot look at this aspect in isolation. We have to keep in mind the overall perspective. Mehmood had asked the question during the Paris meeting (no. 4) and through the written comments, but the Neutral Expert had failed to address it. With this perspective, he was not at fault in seeking an answer to the question.

There is another perspective to keep in mind while sitting in judgment on this issue. As seen above, there was no way the Neutral expert could have decided in favour of India on the issue of drawdown flushing. He was therefore utterly inequitable and biased in the matter, thus, clearly at fault. In fact, it would not be wrong to suggest that he was guilty of the 'original sin' regarding this issue. However, we find Mehmood guilty of wrongdoing on a different count. He and Annandale, who both represented Pakistan, were rude towards the Neutral Expert. According to Jamaat Ali Shah, the Pakistan commissioner at that time and member of the Pakistan delegation during the Baglihar proceedings, he (Mehmood) adopted the attitude that he did because he felt that since the Neutral Expert had decided to favour India, he needed to be taught a lesson.[91] We are of the view that Mehmood's attitude towards the Neutral Expert was not appropriate. He should have tried to influence him through politeness and good manners. This might have borne fruit and changed his point of view. Furthermore, even if it did not work, Pakistan would have certainly lost the case but not its good name. It must be said here that the Indian delegation played its cards very well by posing as honest brokers and law abiders during the oral proceedings. The Pakistan delegation came out of this episode on the losing side.

Did this incident influence the Neutral Expert in his final determination? As we try to answer this question, we propose to share with our readers an important, relevant piece of information. In the course of our research, we were introduced to a gentleman by the name of Bashir Ahmad, who is advisor to the Pakistan Indus Commission, vice-president of the water and agriculture division in NESPAK, and a former member of the Pakistan delegation in the *Baglihar* case. During the course of our conversation, he revealed to us that he met Professor Raymond Lafitte, the Neutral Expert, in Russia in September 2007, where he had gone to attend the annual ICOLD conference. According to him, Professor Lafitte, while discussing the Baglihar issue on the sidelines of the conference, criticized the conduct of the Pakistan delegation during the oral proceedings of the case on the grounds that it violated the basic code of conduct applicable to a delegation participating in legal proceedings.[92] Learning of this stricture, we decided to get in touch with Professor Lafitte to seek his version of the story and also ask him whether the incident in question influenced his decision. We also decided to seek his response to the charge that he decided the question of drawdown flushing as an engineer and did not base it on the provisions of the Treaty. Towards this end we drew up a questionnaire and sent it to him. He was kind enough to respond to it. As to the first charge, he categorically denied that there was a tiff between him and the Pakistan delegation. As to the second charge, he stated that he decided the issue as an 'engineer being governed by the provisions of the Treaty'.[93]

After receiving his answers, we wrote to tell him that they were at variance with the record of the proceedings. He disagreed with us, stating that there was no contradiction between his statements and the record and challenged us to disprove it. We did that with chapter and verse in our next communication but this time he did not respond to our exposé. We sent him several reminders to which he finally replied on 3 June 2014, in which he made the following observation: 'I have not forgotten you. I would reply to you but it would take time. In the meantime, with lot of modesty I would refer to the conviction of Galileo *l'heliocentrisme est contraire aux Saintes Ecritures* (the sun as centre of the universe is in conflict with the Holy Scriptures)'. The reference was to a famous statement by the Italian scientist of the late sixteenth and early seventeenth centuries, Galileo Galilei, who offended the authorities of the Roman Catholic Church by declaring that sun and not Earth was the centre of the universe. The Church authorities tried him for this statement because, in their view, it was contrary to the teachings of the Holy Bible. The trial court convicted him and as a consequence imprisoned him. I do not understand the relevance of Professor Lafitte's statement to my questions. I wrote to him seeking explanation for this but he did not respond. I can only make a conjecture as to what he wanted to convey through it. Perhaps, he wanted to tell me that I considered myself infallible and was not prepared to give any space to the opposite viewpoint.

We know that Professor Lafitte's answer to the second charge was patently wrong because the Arbitration Court in the *Kishenganga* case reversed his decision on the drawdown flushing on the simple grounds that the Treaty does not permit construction of hydroelectric plants incorporating this feature on the Western Rivers. However, how do we reconcile the denial that our distinguished professor made to us that there was a tiff between him and the Pakistan delegation and the confession that he made to Ahmad that the Pakistan delegation was rude towards him? The reconciliation in our opinion is quite simple. It is that whereas Professor Lafitte could speak his mind freely to Ahmad as the conversation was informal, he could not take the same liberty with us as his comments were formal. Had he repeated to us what he told Ahmad, it could have led his critics to believe that the incident in question (involving the Pakistan delegation) might have influenced the final determination. That could have landed him in a troubling situation. He therefore decided to play it safe by altogether denying the incident. Despite the denial, his complaint to Ahmad and the record of the oral proceedings show that he felt offended by the Pakistan delegation's behaviour. Did this affect his judgment? It cannot be said with certainty but given that human beings tend to be influenced by irrational impulses, it is more than likely that it did. Besides, his silence in the face of the irrefutable evidence that we provided him concerning the existence of a tiff between him and the Pakistan delegation perhaps also points towards this conclusion. Irrespective of whether or not the Pakistan delegation's misbehaviour influenced his decision, his wrong decision on drawdown flushing could have easily put him in the category of Radcliffe had it become a precedent for the future hydropower plants on the Western Rivers.

Given the fact that the Neutral Expert's finding on drawdown flushing wasn't *strictu sensu*, based on the terms of the Treaty as shown above, could it be justified on the grounds of an *ex aequo et bono* decision (A settlement that is just and equitable in disregard of existing law)?[94] The Neutral Expert did not make such a claim. However, had he done so, it could not have been accepted because according to the jurisprudence of international courts a judge is not entitled to base his decision on this principle unless the parties to the dispute entrust him with that kind of mandate. In the present situation, the Neutral Expert, in the absence of an express or implied consent of the parties to base his findings on this principle, clearly went beyond his mandate. Resultantly, instead of calming the situation between two antagonistic neighbours, his finding has further exacerbated the situation between them as underlined by Professor Robert Wirsing:[95] 'Lafitte's call for conspicuously modest alterations to the dam's design, and his insistence on assigning more importance to the dam's efficient and cost-effective operation (the heart of the Indian argument) than to its strict adherence to the [Indus Waters Treaty's] detailed, albeit ambiguous provisions aimed at restricting New Delhi's ability to control the river's flow (of utmost concern to Pakistan) seemed more likely to fuel existing tensions over the future of the Indus's waters.'

Before concluding this section, we would like to raise some important questions, both with regard to Pakistan's debacle and the Neutral Expert's determination. How do we explain Pakistan's debacle, particularly when the Pakistan Indus Commission claimed from the start that it had a winnable case? It is noteworthy that Sheraz Jamil Memon, the additional commissioner, had reportedly stated that three out of the four reputed international water experts that the Commission had consulted, to assess the soundness of its objections to the dam design, had categorically supported Pakistan whereas the fourth one had expressed a minor reservation.[96] With this backdrop, troubling questions arise on the competence of those consultants and that of the Canadian hydropower expert, Peter Joseph Rae, who was entrusted with the task to prepare the case. The question also arises whether, while making an appraisal, they took into consideration the question of the relevance of the contemporary international practice in dam building. And if so, did Pakistan's legal team plead the case keeping this factor in mind?

This is not the end of the matter. Serious questions also arise with regard to the role of the Neutral Expert. On what basis did he make the determination? It is undeniable that he must have been hard pressed to choose between two contending viewpoints. There was India which was keen on building an orifice spillway on the grounds of flushing out silt to prolong the life of the dam. Then there was Pakistan which opposed it on the grounds that it would provide India with a handle to control the flow of water to Pakistan at critical moments causing enormous damage to its agriculture, defence, and hydroelectric generation. Given the Neutral Expert's preference for the Indian point of view, how did the Neutral Expert base his determination on the contemporary international practice rather than the Treaty provisions *strictu sensu*?

To begin with, it must be understood that the Neutral Expert was not in a position to adjudge the objections raised by Pakistan alone because, being an engineer, he did not possess the requisite expertise to address the legal questions relating to them. He needed the assistance of a legal expert in the matter. For this purpose, he approached a former senior counsel to the World Bank and professor at the Graduate School of Higher Studies, Geneva, Madame Laurence Boisson de Chazournes. According to the former legal vice-president of the Bank, Salman M. A. Salman,[97] India cleared her name straightaway whereas Pakistan took time to do so. He also revealed that the Bank had originally suggested to Raymond Lafitte that he submit the names of three legal experts for selection by both Pakistan and India, but he refused to do so. The Bank then suggested that he choose one name and inform the parties accordingly which he did. He justified the selection of Madame de Chazournes on the grounds that she was, in his opinion, a lady of high integrity and neutrality. But why did he prefer her over other specialists of international law with impeccable credentials in terms of integrity and neutrality? Is it because she had been associated with the Bank? If so, it is hardly a sound reason to do so on its own. There must have been some other reason for her selection. Unfortunately, he did not explain the selection process. Given the importance of the case at hand, in our judgment, his decision to consult only one legal authority, howsoever well-intentioned, was flawed. This was all the more so when the decision-making authority happened to be a single individual as was the case in the present situation.

Before concluding this section we would like to make certain submissions about the legal expert. First, we do not know what part Madame de Chazournes played in the Neutral Expert's determination of the question of drawdown flushing. However, we know that the Treaty does not permit it in hydropower plants on the Western Rivers as testified by the award of the Arbitration Court in the *Kishenganga* case which unanimously ruled against it not only in the case under consideration but also in all future projects of this kind.[98] In light of this, it wouldn't be unreasonable to presume that Madame de Chazournes must have given an opinion against it. If so, it signifies that the Neutral Expert simply ignored this and issued a decision based on considerations unrelated to law. This means that consultation with a legal expert by a Neutral Expert is a mere formality. This is indeed a worrying development as far as the future cases of similar kind are concerned.

Secondly, irrespective of Madame de Chazournes' actual role in the present case, we are of the view that it is wrong on the part of a Neutral Expert to consult only one legal expert. In our judgment, he or she should preferably consult three in order to take the element of arbitrariness or bias out of the equation. It is true that, theoretically speaking, it would still be up to him or her to decide which opinion to accept and which to reject. However, practically speaking, he would have no option but to accept the majority opinion. In such an eventuality, the losing party would hardly have a cause to complain regarding having been given a raw deal, as is the feeling in the *Baglihar* case. Thirdly, it would also be preferable to take away the right to appoint legal experts from the Neutral Expert and entrust it to another entity, say the president of the International Court of Justice. We would suggest that Pakistan and India should jointly work on these submissions in order to improve the performance of the Neutral Expert. India should not stand in the way because it may have been a beneficiary in the *Baglihar* case, but it may not be so in the future.

h) Allegation of Stealing Pakistan's Water

Following the decision by the Neutral Expert, India filled the Baglihar Dam reservoir. This created a heated controversy between the two countries because, according to the Pakistan commissioner, India did so after the monsoons between 19 August and 5 September when the flow of the Chenab River was low. It was, in Pakistan's opinion, not only a clear breach of the Treaty which mandates the filling to take place between 21 June and 31 August but also of the actual understanding that he had reached with his counterpart according to which India was to undertake the filling during the rainy season. He observed that according to the Treaty, India was under obligation to release 55,000 cusecs feet of water downstream at the Merala Head but it released between 30,000 and 35,000 cusecs feet which, in his opinion, created a drought-like situation in Pakistan.[99] He claimed that as a result of India's filling of the dam at the wrong time, Pakistan suffered a loss of 0.2 million acre feet of water. He did not want India to compensate Pakistan monetarily but with 'water for water', as he put it, during the next *rabi* season. He cited the precedent of the Salal Dam in which case, according to him, India compensated Pakistan with water because it filled the reservoir out of season.

The Indian officials initially disregarded the Pakistani claim and denied that India had violated the Treaty. They maintained that India filled the reservoir within the stipulated time frame and that there were fewer rains that year which could have impacted the overall discharge of the river. Ramaswamy Iyer, though not a spokesperson but a close associate of the Indian government, defended the Indian position as follows:[100]

There is in fact a conundrum in this regard in the Treaty. When the filling of the reservoir begins, there is bound to be a brief interruption of flows (i.e. zero flows) to the other side until the waters rising against the dam will reach the outlets, i.e. the spillway gates. (If, as desired by Pakistan, the gates had been placed higher, the water would have taken still longer to reach them.) However the Treaty requires a minimum flow to be maintained during the initial filling. How can this be done? Please note that the minimum flow stipulation is for flows above Merala in Pakistan and not for release from Baglihar. Apart from flows through Baglihar, there are other stream flows beyond the dam. Based on its own calculations of those flows, India has to satisfy itself that the flow above Merala is likely to be 55,000 cusec or more, even if the flow from the Baglihar is zero. It is clear that that there can be no certainty in such calculations. If during the initial filling, the flows from other streams beyond the Baglihar dam do not add up to 55,000 cusec, there is nothing that India can do—unless it is to stop the filling, reopen the closed diversion outlets and let the water flow through, in which case the reservoir cannot be filled at all. That is a reductio ad absurdum.

Iyer also questioned the extent and duration of the shortfall, in case there was indeed a shortfall. Based on the Indian calculations according to which the flow above Merala during the filling could not have been less than 40,000 cusecs, he thought that Pakistani reports of the shortfall were highly exaggerated. He also contended that in the absence of a monitoring mechanism, there was no way of ascertaining the precise shortfall, if it ever took place. Furthermore, he argued that the shortfall in question, which was one-time only, might have caused some difficulty but could not be described as a disaster or a major deviation from the Treaty. Finally, he disregarded the idea that India deliberately timed the filling to inflict maximum affliction on Pakistan. He firmly believed that India completed the filling within the prescribed time frame. He rejected the intended implication that India could have done the filling a bit earlier—say in July—on the grounds that in the presence of the time frame given by the Treaty there was no basis for a different one. He thought that Pakistan raised the issue, which he described as a huge controversy, as a reaction to the unfavourable decision of the Neutral Expert.

The Pakistan commissioner visited New Delhi in October 2008, where he held talks with his counterpart on the issue as a result of which India apparently accepted Pakistan's claim. However, the Indian Indus Commission wanted to first visit the river site for verification purposes before taking a final decision on the question of compensation. The Pakistan commissioner was at a loss to understand how the Indians would be able to verify the Chenab flow for August and September. In his opinion, the only thing that Pakistan could do in this regard was to make the data of the flow at that time available to India which he revealed Pakistan had already done. Though India officially kept denying the Pakistani charge, the Pakistan commissioner claimed that unofficially it accepted it. Anne W. Patterson, the US ambassador to Islamabad, in a cable dated 3 November 2008, revealed by Wikileaks, confirmed Pakistan's claim. Stating that Pakistan 'was facing a 34 per cent water shortage' that year due to water scarcity in the Chenab River resulting in 'lower crop yields for winter' and 'extended blackouts across the country due to reduced hydropower production', she asserted that '[o]fficially, India dispels Pakistani claims but, unofficially, the Indian side admits that, 'structural constraints of Baglihar Dam, and weather constraints' have resulted in a reduction of Pakistan's share of water (sic).'[101] In any case, President Zardari and Prime Minister Yousaf Raza Gilani separately raised the issue with Prime Minister Manmohan Singh on 24 September and 24 October 2008 respectively. And, the security advisors of the two countries discussed it in New Delhi on 13 October 2008. Subsequently, when Manmohan Singh failed to take any action on the matter, despite promising to do so, President Zardari contemplated writing to the Indian PM as well as to Muslim countries and Friends of Pakistan for the purpose. However, the matter is now a closed affair because at the 105 meeting of the Commission held in May/June 2010, Pakistan reportedly decided to drop the issue after India pledged to put in place a proper consultation procedure to obviate such controversies in the future.[102]

REFERENCES AND NOTES

1. Muhammad Aslam Jan, 'Salal Dam: Pus Manzar aur Muzmairat', *Nawa-e-Waqt* (6 February 1978); Arif Nizami, 'Salal Moahida aur us ke Mauharaqat', *Nawa-e-Waqt* (10 February 1978).

2. We would like to acknowledge here that in writing this section we have drawn liberally on the following article: Ijaz Hussain, 'Pakistan and the Wullar Barrage', *Regional Studies* (Spring 1988), vol. VI, no. 2, 47–62.

3. Mir Abdul Aziz, 'Wullar and the Proposed Barrage', *The Muslim* (Islamabad, 24 October 1986).

4. See Wikipedia under 'Wullar Barrage', <http://en.wikipidia.org/wiki/Wullar_Lake>.

5. Aloys Michel, *The Indus Rivers* (New Haven and London: Yale University Press, 1967), 201 and 239. Michel suggests that Pakistan dragged its feet on the Bank plan to allow India a restricted use of the Western Rivers in the Indian-administered Jammu and Kashmir in order to get concessions for the reconstruction and payment of dam sites in Pakistan or Azad Kashmir. Ibid. 240.

6. Aslam Sheikh, 'India Told to Stop Work on Jhelum Barrage', *The Muslim* (29 September 1986).

7. M. G. Srinath, 'India Denies Pak Charge on Jhelum Barrage', *The Hindustan Times* (27 September 1987).

8. Zafar Bhutta, 'Pakistan-India Water Disputes: No Headway in Wullar Barrage Negotiations', *The Express Tribune* (13 May 2011).

9. It is noteworthy that the Effective Date means the date on which the Treaty took effect in accordance with the provisions of article XII, i.e. 1 April 1960.

This concession was made to allow India, 'to continue to irrigate, from the Western Rivers, those areas which were so irrigated, as on 1st April 1960.' N. D. Gulhati, *Indus Waters Treaty: An Exercise in International Mediation* (Bombay, Calcutta, etc.: Allied Publishers, 1973), 287.

10. M. G. Srinath, 'India Denies Pak Charge', supra note 7; see also, 'India Defends Jhelum Project', *The Telegraph* (Calcutta, 27 October 1986).

11. Article I (11) of the Treaty defines the term 'non-consumptive use' as follows: 'Any control or use of water for navigation, floating of timber or other property, flood protection or flood control, fishing or fish culture, wild life or other like beneficial purposes, provided that, exclusive of seepage and evaporation of water incidental to the control or use, the water (undiminished in volume within the practical range of measurement) remains in, or is returned to, the same river or its Tributaries; but the term does not include Agricultural Use or use for the generation of hydroelectric power.'

12. 'India Allays Pak Fears on Tulbul Project', *The Times of India* (27 September 1986).

13. Arjimand Hussain Talib, 'New Fantasy on Jhelum River', *Bharat Rakshak*, <forums.bharat-rakshak.com/viewtopic.php?p=403472>, posted on 30 September 2007. Emphasis original.

14. Shaheen Akhtar, *Emerging Challenges to Indus Waters Treaty: Issues of Compliance and Transboundary Impacts of Indian Hydroprojects on the Western Rivers*, Focus on Regional Studies, Institute of Regional Studies (2010), vol. XXIII, no. 3, 32–3.

15. 'India Offers to Amend Design of Wullar Barrage', *Dawn.com*, <beta.dawn.com/news/628533/india-offers…>.

16. 'India Resumes Construction Work on Wullar Barrage; International Arbitration…', *The Lahore Times* (14 December 2012).

17. Shafqat Kakakhel, 'The Indus Waters Treaty: Negotiation, Implementation, New Challenges, and Future Prospects', *Criterion Quarterly* (Islamabad, April/June 2014), vol. 9, no. 2, 53.

18. 'What does Baglihar Portend?', in Ijaz Hussain, *Dimensions of Pakistan-India Relations* (Lahore: Heritage Publications, 2006), 233.

19. Ibid. 234.

20. Ibid.

21. Ibid.

22. See para. 2, under (ii).

23. 'Wolfensohn Statement and the Baglihar Dam', in *Dimensions of Pakistan-India Relations*, supra note 18, 267.

24. India abstained in the voting on the UN Convention on Non-Navigational Uses of International Watercourses of 1997 because of inclusion, among others, of article 33 which provided for the third party involvement in the settlement of disputes if bilateral negotiations failed to bear fruit. It was of the view that the Convention contained an element of compulsion. In its opinion, it should

have left it to the parties to decide how to proceed in the matter and that any mandatory third-party procedure was inappropriate and should not have been included in a framework convention. See Press Release, <http://waterwiki. net/index.php/GeneralAssembly_adopts_Convention_on_the_Law_of_Non-Navigational_Uses_of_International_Watercourses>, cited in *Water Security for India: The External Dynamics*, Report of the Institute for Defence and Security Analysis (IDSA), 2010, 28. <http://www.indiaenvironmentalportal.org.in/files/book-WaterSecurity.pdf>.

25. For a detailed discussion on this point, see Ijaz Hussain, *Kashmir Dispute: An International Law Perspective* (Islamabad: Quaid-i-Azam Chair, National Institute of Pakistan Studies, QAU, 1998), 185–96.

26. 'Baglihar: Beware of Indian Designs', in Ijaz Hussain, *Dimensions of Pakistan-India Relations*, supra note 18, 250.

27. 'Baglihar under Spotlight Again', in ibid. 243.

28. 'Who is Responsible for the Baglihar Debacle?', in ibid. 261.

29. 'Baglihar: Beware of Indian Designs', in ibid. 250.

30. 'Pakistan's Baglihar Dilemma', in ibid. 254.

31. 'Who is Responsible for the Baglihar Debacle?', in ibid. 259.

32. Ibid. 259–60.

33. This fact seems to belie the oft-repeated claim made by certain writers that the Treaty has worked smoothly notwithstanding the ups and downs in the relations between the two countries.

34. The Pakistan government's thinking was a variant of the states' claim to decide whether or not to submit their disputes with other states to a judicial forum by dividing them into two categories, legal and political or justiciable and non-justiciable. In their view, only legal or justiciable disputes are disposed to settlement by courts. They also claim a right to decide which dispute falls in which category. They do so because they fear that a tribunal may render a decision contrary to their interests. See Ijaz Hussain, *Kashmir Dispute*, supra note 25, 235–9.

35. The ICSID was established in 1966. Its procedures are based on article 6(3) of UNCITRAL (United Nations Commission on International Trade Law) arbitration rules. When the Bank was seized for the *Baglihar* case there were 143 parties to the ICSID Convention (as of 25 July 2012, there were 147 Sates parties). Pakistan is a party to it since 1966 while India has not accepted it till today. The Bank designated the ICSID to coordinate the process of arbitration in the *Baglihar* case which both Pakistan and India accepted. The Neutral Expert appointed an engineer and a legal advisor to assist it to which both Pakistan and India gave their consent.

36. 'Significance of Appointment of Neutral Expert' in Ijaz Hussain, *Dimensions of Pakistan-India Relations*, supra note 18, 270.

37. Ibid.

38. Ibid. Pakistan presumably refused to go along with the proposal fearing it to be India's delaying tactic. However, given the fact that it had already lost so much time because of India's earlier dilatory tactics, it could have given this proposal a try. Had it done so, it might have been spared the enormous amount of money that it spent on the case as well as the adverse verdict of the Neutral Expert.

39. Ibid.

40. Ibid. 273.

41. The Neutral Expert provided the following explanation as to why the case took so long to decide: 'The points of difference referred by Pakistan were not trivial and their complexity required from the claimant Party as well as the respondent a major work of analysis and of synthesis to present their theses. The exchanges between the Parties were documented with great care; the oral presentations during three meetings and the visit to the site of the Baglihar and to the hydraulic laboratory of Roorkee were found to be of a high technical, scientific and legal interest. The process lasted one year.' *Baglihar Hydroelectric Plant, Expert Determination,* Lausanne, Switzerland, 12 February 2007, 109.

42. It was agreed in the introductory meeting which took place in Paris in 2005 that the Neutral Expert would provide the parties with a draft of his final report for their comments (*Baglihar Hydroelectric Plant, Expert Determination,* Protocol of Meeting no.1, June 9–10 2005, Paris, Annex 1 (no. 3.1), 2–3, paragraph 2.4). The Neutral Expert furnished the following explanation for the procedure that he adopted in this case: 'As is usual in the relationship between engineers, he provided the Parties with his final report in a draft form on the 2 and 3 of October 2006. The objective was to inform the Parties, as a courtesy, of his decisions, and to invite them for possible comments. The NE was conscious that however much care would be taken to strengthen his opinion, he would not totally preclude the possibility of omitting an important fact, if this should happen, he would review his opinion so as to give a sound and non-contestable determination based on application of the Treaty and the state of the art in the field of technology.' *Baglihar Hydroelectric Plant,* supra note 41, 4.

43. Ibid.

44. Although neither Pakistan nor India is a party to the Vienna Convention on the Law of Treaties, 1969 (though Pakistan signed it on 29 April 1970, but has not ratified it so far, while India did not even sign it (*Multilateral Treaties Deposited with the Secretary General* (Status as of 1 April 2009) (New York: UN, 2009), 525–26), both are nonetheless bound by its articles 31 and 32 because of their customary law status. There is considerable jurisprudence on the binding character of these articles. See Case Concerning the Territorial Dispute (Libya v. Chad), Judgment, *ICJ Reports* 1994, 21–2, para. 41; Oil Platforms (Iran v. US), Preliminary Objections, Judgment, *ICJ Reports* 1996 (II), 812, para. 23; Kasikili/Sedudu Island (Botswana/Namibia), Judgment,

ICJ Reports 1999, 1059. para. 18; Sovereignty over Pulau Ligitan and Pulau Sipadan (Indonesia/Malaysia), Judgment, *ICJ Reports* 2002, 645–6, paras. 37–8; Arbitration regarding the Iron Rhine (IJzeren Rijn) Railway (Belgium v. Netherlands), Award of the Arbitral Tribunal, *PCA,* 24 May 2005, 23, para. 45.

45. *ICJ Reports* 1950, 8.

46. 'Draft Articles on the Law of Treaties', *Yearbook of International Law Commission* (1966), vol. II, 221, para. 12.

47. Sir Gerald Fitzmaurice, 'The Law and Procedure of the International Court of Justice: Treaty Interpretation and Other Treaty Points', *British Yearbook of International Law* (1957), vol. 33, 228.

48. *Yearbook of International Law Commission*, supra note 46, 221.

49. *Kishenganga Arbitration, Counter-Memorial of India*, vol. 1, 23 November 2011, 104. We have invoked here an argument which Pakistan used in the *Kishenganga* case because the parties advanced the same arguments in both the cases.

50. Ibid. 104–5.

51. Ibid. 102.

52. *The Baglihar Hydroelectric Plant, Memorial of the Government of Pakistan,* 14 August 2005, 9, para.15.

53. Ibid. 9–10, paras. 18.

54. *Kishenganga Arbitration, Pakistan's Memorial*, vol. 1, 27 May 2011, 96.

55. *Baglihar Hydroelectric Plant, Pakistan's Memorial,* supra note 52, 10–1. Emphasis original.

56. ICOLD Bulletin no. 115 provides as follows: 'Bottom outlets may be used for under sluicing of floods, emptying of reservoirs, sluicing of sediments and preventing sediment from entering intakes, etc.' ICOLD Bulletin no. 115, 'Dealing with Reservoir Sedimentation', Guidelines and Case Studies, 1999, cited in *Baglihar Hydroelectric Plant, Expert Determination (Final Draft)*, Lausanne, Switzerland, 30 October 2006, 85.

57. *Baglihar Hydroelectric Plant (Final Draft),* ibid. 89.

58. Ibid.

59. *Kishenganga Arbitration, Pakistan's Memorial*, supra note 54, 104–5, para. 6.26. In support of its contention, Pakistan referred to the Indian delegate Shankardass' following statement: 'I want to place it on record that that issue [of drawdown flushing] has not arisen in this case. Our design, we have repeatedly said, is intended to be working with sluicing. The issue of whether or not flushing is permitted by the Treaty is not something which we have either argued, and that is not something—in other words that is something to be debated and argued and decided at some other point in time as to whether it is or it is not [allowed]; it does not arise in this case and does not need any comment, in my respectful submission.' Ibid. 104, footnote 229.

60. Kishenganga Arbitration, Partial Award, *PCA,* 18 February 2013, 128, para. 344.

61. *Baglihar Hydroelectric Power Plant*, Transcript of Meeting no. 5 with the Neutral Expert and the Delegations of India and Pakistan, vol. 2, 264–5.

62. Ibid. 266.

63. *Baglihar Hydroelectric Plant*, supra note 41, 16–17, paras. 12–16.

64. According to the International Law Commission, the concept signifies as follows: 'When a treaty is open to two interpretations one of which does and the other does not enable a treaty to have appropriate effects, good faith and objects and purposes of the treaty demand that the former interpretation be adopted'. See *Yearbook of International Law Commission*, supra note 46, 219, para. 6.

65. Ibid. 17, paras. 17–18.

66. Commenting on the verdict, the former secretary of Indian National Commission on Irrigation and Drainage, M. S. Menon, described it 'a treaty.' He termed the additional cost resulting from modifications in design as unacceptable and contended that India should reject Pakistan's repeated objections to its hydropower projects. He also argued that it should seek a review of the Treaty and in case of Pakistan's refusal, it should abrogate it. M. S. Menon, 'A Redundant Treaty', *The Hindu* (8 April 2007).

67. Ijaz Hussain, 'Not Treated According to the Treaty', *The Daily Times* (28 February 2007). The title of the article is weird. The explanation for this is as follows. We had entitled the piece 'Why did Pakistan lose the Baglihar case?'. The editor of the newspaper changed it to the present title. In our estimation, there was absolutely no reason to do so. Did he do so out of 'patriotism' or fear of the Pakistan government's possible adverse reaction which was claiming victory in the case? It is anybody's guess.

68. Ibid.

69. Commenting on the verdict, an Indian journalist observed that contrary to Pakistan's victory celebrations, 'Islamabad has lost on the most crucial point it wanted to erase from the Jammu and Kashmir flagship power project's design—the gated spillway.' Masood Hussain, 'Baglihar Dam: Spillways a Blow to Islamabad', *The Economic Times* (14 February 2007).

70. Ijaz Hussain, 'Not Treated', supra note 67. Musharraf's statement meant that Pakistan had the option to challenge the Baglihar award. Though the question is an academic one now, we are of the view that Pakistan could have done so on the spillway issue provided it could convince the Bank to open the case which would have been a highly doubtful proposition. See Ijaz Hussain, 'Can Pakistan Challenge Baglihar Verdict?', *The Daily Times* (14 March 2007).

71. At the outset of the proceedings of the case there were tell-tale signs that the Neutral Expert would render the decision on grounds other than those enshrined in the Treaty and in favour of India. Thus, according to Bashir Ahmad who was a member of the Pakistan delegation in the case, the Neutral Expert, during the introductory meeting held in Paris, made the observation that apart from storage, the dam has uses in controlling floods which raised

the eyebrows of the Pakistan delegation (He disclosed this information to the present writer in the course of an interview that he gave him at his residence in Lahore on 9 March 2014). Similarly, after a visit to the dam he remarked that it was a beautiful dam which Pakistanis interpreted as a tilt towards India.

72. *Baglihar Hydroelectric Plant (Final Draft)*, supra note 57, 88.

73. Ibid. 33.

74. Asif H. Kazi, 'Misusing the Indus Treaty', *The News* (1 July 2011).

75. Salman M. A. Salman, 'The Baglihar Difference and its Resolution Process—a Triumph for the Indus Waters Treaty', *Water Policy*, No. 10 (2008), 115.

76. The interview took place at Salman M. A. Salman's residence in Washington DC, on 22 August 2010.

77. *Baglihar Hydroelectric Power Plant*, Transcript of Meeting no. 5, supra note 61, 264–5. Emphasis added.

78. *Baglihar Hydroelectric Plant*, supra note 41, 100. Emphasis original.

79. *Baglihar Hydroelectric Power Plant*, Transcript of Meeting no. 5, supra note 61, 311–12.

80. Ibid. 296.

81. Ibid. 294–5.

82. Ibid. 297.

83. Ibid. vol. 1, 72.

84. Ibid. 73.

85. Ibid.

86. Ibid.

87. Ibid. 77.

88. Ibid. 80–1.

89. Ibid. vol. 2, 279–80.

90. It is noteworthy that during the introductory meeting in Paris it was agreed that the Neutral Expert would provide the parties with a draft of his final report for their comments which 'will be carefully examined and taken or not taken into account.' *Baglihar Hydroelectric Plant*, Protocol of Meeting no. 1, supra note 42, 2–3, para. 2.4.

91. This information was shared with the present writer by Jamaat Ali Shah in the course of the interview that he gave him at his residence in Lahore on 10 July 2014.

92. Bashir Ahmad narrated this incident to the present writer in the course of an interview. For details of the interview, see supra note 71.

93. We contacted the Neutral Expert on 15 March 2014 through an email and posed the following two questions: 1) Do you think that your decision on the drawdown flushing was justified in the light of the *Kishenganga* award which reversed it? Is the charge justified that you made the determination as an engineer and not on the basis of the IWT?; 2) What are your comments on the incident that took place during the oral presentations on the draft

determination involving the Pakistan delegation and you in which Ashfaq Mehmood seemed to pester you with certain questions? Did that influence your determination?

He replied to the first question partly in English, partly in French in these terms: 'My decision was that of engineer being governed by the provisions of the Treaty. *Aucune clause du Traité interdict cette utilization de la tranche morte*' (No clause of the Treaty bars the use of the dead storage). '*Je ne connais pas les raisons pour lesquelles la Cour d'Arbitrage n'a pas retenu mon démonstration*' (I don't know why the Arbitration Court did not retain my determination). He replied to the second question entirely in English in these terms: 'I don't think that any incident happened during the 5 meetings with the Parties under my chairmanship. It was not discussions between the participants but hearings. The oral presentations on both technical and legal aspects were of very high standard, demonstrating competence and honesty, in a spirit of goodwill and courtesy, but also including clarity and firmness.'

94. The concept is based on article 38(2) of the Statute of the World Court which provides: 'This provision shall not prejudice the power of the Court to decide a case *ex aequo et bono*, if the parties agree thereto.' For details, see Monique Chemillier-Gendreau, 'Equity', in Mohammad Bedjaoui (Gen. ed.), *International Law: Achievements and Prospects* (Paris, Dordrecht, etc.: UNESCO, Martinus Nijhoff Publishers, 1991), 273.

95. 'Resource Dispute in South Asia: Water Security and the Potential for Interstate Conflict', *American CIA Report* (1 June 2009), 5.

96. 'Significance of Appointment of Neutral Expert', in Ijaz Hussain, *Dimensions of Pakistan-India Relations*, supra note 18, 273.

97. Salman M. A. Salman imparted this information to the present writer in an interview. For details, see supra note 76.

98. The Neutral Expert did not think that his determination was flawed. In fact, he thought that he had done a great job. Consider his following statement: 'The NE considers that his decision has not been rendered against one or the other Party. His opinion is that, in fact, specific Parties emerge successfully from the treatment of this difference: the Authors of the Treaty. The Treaty is the successful document.' *Baglihar Hydroelectric Plant*, supra note 41, 109.

99. See, Annexure E, para. 18(c).

100. Ramaswamy Iyer, 'Troubled Waters', *The Friday Times* (22–28 June 2012).

101. 'Water Issues could Sweep Away Indo-Pak Peace Process', <beta.dawn.com/news/638350/secret-us-cables-accessed-by-dawn-through-wikileaks-water-issues-could-sweep-away-indo-pak-peace-pr>. Observation Original.

102. Ramaswamy Iyer, 'Troubled Waters', supra note 100.

6

Treaty in Action-II

A) KISHENGANGA DAM

a) History of the Dispute

THE FOURTH DISPUTE THAT HAS ARISEN BETWEEN THE TWO COUNTRIES over water issue relates to the Kishenganga Hydroelectric Project (KHEP) which India is building on a major tributary of the Jhelum River in the part of Jammu and Kashmir it administers, called Kishenganga. After it enters the territory administered by Pakistan, it is known as the Neelum River. The dam is located 160 kilometres upstream of Muzaffarabad (the capital of Azad Kashmir) and involves the diversion of water from a dam site on the Kishenganga/Neelum River through a 22-kilometre long tunnel to the Bonar Nallah which is another tributary of the Jhelum River. The diversion will change the course of the river by about 100 kilometres and will then join it through the Wullar Lake near the town of Bandipore in the Baramulla district. As a result of the diversion, the Neelum and Jhelum Rivers will meet in the Indian-administered Jammu and Kashmir rather than at Domail near Muzaffarabad where they currently meet. Pakistan regards the 330 megawatt dam project as a violation of the Treaty. In 2004, it raised a number of objections as a result of which India revised the design of the dam. However, Pakistan was not satisfied with the revised design and consequently raised six objections which related to gate structure, height and size, level, diversion plan, storage capacity, power intake, and free board. The Pakistan government made efforts to get the Kishenganga Dam issue resolved through bilateral negotiations but failed. On 22 October 2009, following an inter-ministerial meeting, it decided to go for international arbitration. However, despite this decision, the ministry of law sat on the file for six months and it was only in April 2010, that it was able to intimate its decision to India.[1]

The two countries' representatives then met in July 2010 in New Delhi and agreed to refer the case to the International Court of Arbitration. Subsequently, the two countries requested the UN Secretary General to constitute, in accordance with the Treaty, an international court to take up the Kishenganga issue which he did towards the end of 2010.[2]

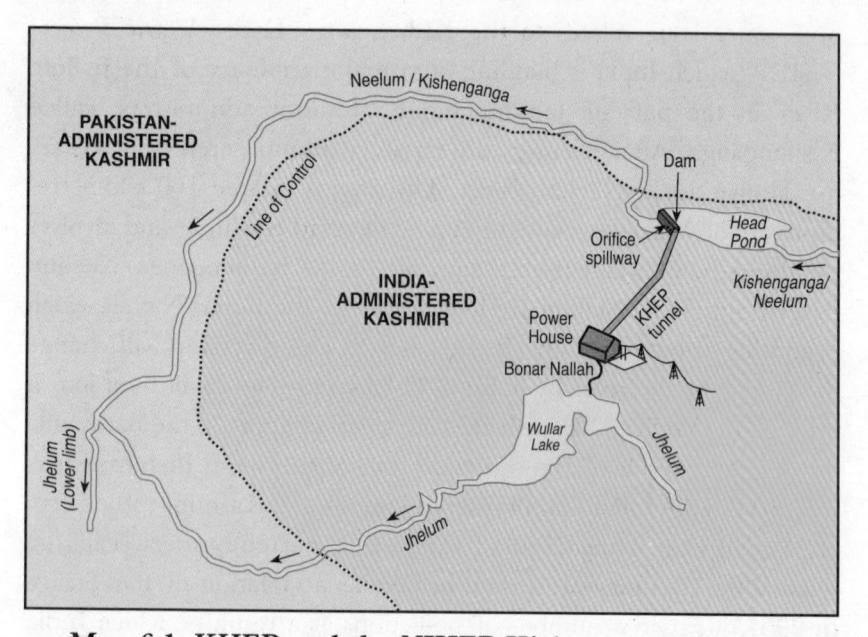

Map 6.1: KHEP and the NJHEP Kishenganga Dam (I)

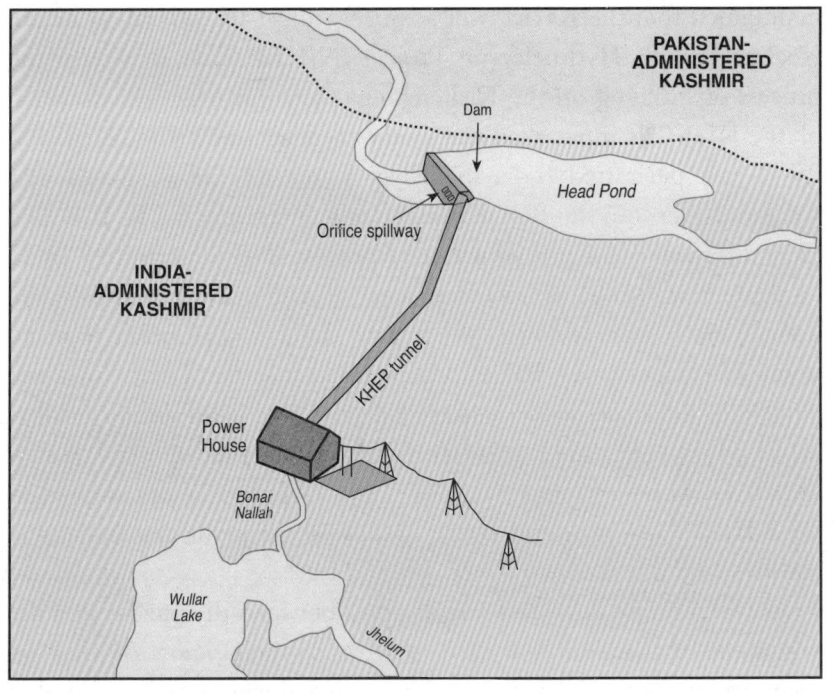

Source: Pakistan's Memorial, volume 2, Figure 9.

Map 6.2: KHEP and the NJHEP Kishenganga Dam (II)

b) Point Counter Point

i) Pakistan's Arguments

The Court was seized of two issues or what it termed to be 'disputes'. The first one related to whether India was entitled to divert the waters of the Kishenganga/Neelum River into the Bonar Nallah through the KHEP. Pakistan raised the objection that the inter-tributary transfer of water breached India's legal obligations which it owes to Pakistan under the Treaty, as interpreted and applied in accordance with international law, particularly India's obligation under article III (2) which requires it to let flow all the waters of the Western Rivers and forbids any interference with those waters and article IV (6) which obliges it to maintain natural channels of the rivers. Simply put, it maintained that the KHEP will adversely affect the operation of the Neelum-Jhelum Hydroelectric Project (NJHEP) that it was in the process of building on the Kishenganga/Neelum River downstream of the KHEP. It thus asked the Court to pronounce whether or not the Treaty permitted India to deliver waters of the Kishenganga/Neelum River into the Bonar Nallah while operating the KHEP. The second dispute related to whether or not the Treaty permitted India to deplete or bring the reservoir level of run-of-river hydroelectric plants below dead storage level in any circumstance other than an unforeseen emergency. Pakistan contended that the Treaty does not permit India to do so because it will give the latter impermissibly broad control over the flow of the river waters allocated to Pakistan under the Treaty. In its view, the second dispute was not limited to the KHEP alone; it extended to any run-of-river plant that India may construct on the Western Rivers in the future.

As we examine the nature of disputes between the parties and the arguments that they marshalled to advance their points of view on them, we notice that they repeated in essence the same philosophy that they had adumbrated in the *Baglihar* case. Since we have dealt with this aspect at considerable length in the last chapter, we desist from repeating the exercise and restrict ourselves to providing a summary. Pakistan began by recalling the history of the lengthy Bank-sponsored negotiations which finally led to the conclusion of the Treaty. It then pointed out that the Treaty proceeds on the philosophy that parties would develop and harness the water resources of the Indus Basin independently of each other and for this purpose it divides the six rivers between them and fixes and delimits each party's rights and obligations in this regard. It accepted that the Treaty should be interpreted first of all by assigning ordinary meaning to the text, and in case of doubt the precise delimitation of the parties' respective rights should be the guiding principle.

Regarding whether or not the Treaty entitles India to divert waters of the Kishenganga/Neelum River into the Bonar Nallah, Pakistan accused the latter of violating three principles: a) general obligation to 'let flow' waters of the Western Rivers; b) requirements pertaining to the permissible use of waters for the generation of hydroelectric power; and c) obligation to use the best endeavours to maintain natural channels of the Western Rivers. It was of the view that article III imposes an obligation on India to 'let flow all the waters of the Western Rivers' and 'not to interfere with these waters' before they flow into Pakistan which, in its view, constitutes 'fundamental principle underlying the Treaty' and carries 'existential importance'. It accepted certain exceptions to this obligation, like the construction of run-of-river plants for the generation of electricity, but pointed out that Annexure D imposes strict restrictions on their design and operation. It relied on the *travaux préparatoires* of the Treaty to justify its position, particularly Black's letter of 6 February 1960, addressed to Pakistan's finance minister in which he stated: '[T]here is no doubt and no reservation in the mind of anyone, either in the Indian delegation or the Bank that [the Treaty imposes obligation] on India to allow to flow down *all* waters of Western Rivers (emphasis original).' It claimed that restrictions on hydroelectric plants are important because without them India would have controlled the flow of waters at will. It held that the preamble's emphasis on the most 'complete and satisfactory utilization' of the Indus waters did not signify unilateral rights of use or development. Hence, India could not impose on Pakistan what it considers to be the optimal development. It also asserted that the Treaty does not entitle India to deliver water from one tributary into another for power generation because article III obliges it to 'let flow' all the waters, and para. 15 of Annexure D requires it to deliver into the river below the hydroelectric plant the same volume of water that is received in the river above the plant within a given 24 hour period.[3]

Next, Pakistan contended that the diversion of the Kishenganga/ Neelum River is incompatible with para. 15 (iii) of Annexure D because while it permits delivery of water released below a hydro-electric plant into another tributary, it does not permit permanent diversion of the entirety of waters in order to create a potential for the generation of hydroelectric power that does not naturally arise from the flow of the river within its course. Besides, the Treaty allows only the diversion of waters of a tributary of the Jhelum River when 'necessary', a term that it conceived as an 'emergency exit.' It did not deem the KHEP diversion to be 'necessary' since it failed this test and that established by the *Iron Rhine* and *Gabčikovo-Nagymaros* judgments in terms of obligations related to international environmental law.[4] It also contended that even if the Treaty permits the construction of the KHEP, its article III (2) restricts the use of electricity to the drainage basin of the Jhelum River. It cited the 15 September 1959, Heads of Agreement which became article III in the Treaty in support of its contention. In its view, if interpreted otherwise, India's potential uses would be limitless, depleting a resource which is critical for Pakistan. It also rejected India's premise that para. 15 (iii) was introduced in the Treaty with the KHEP in mind. It insisted that the Indian CWPC's (Central Water and Power Commission) letter of 16 May 1960, on which India bases its claim, did not support it. It rejected the Indian assertion that in anticipation of the KHEP, Annexure E did not prohibit inter-tributary transfers on the grounds that if India had genuinely intended to realize such a project, it would have sought its express inclusion in the Treaty. It also argued that para. 15 (iii) does not cover the KHEP because it is not located on a tributary of the Jhelum River. In its opinion, only the KHEP's dam is located on the tributary because the power plant is located 23 kilometres away in a separate catchment area. It maintained that the power house and the tributary above the site needed to be located in the same catchment area.[5]

Pakistan also differed with India regarding the meaning of the expression the 'then existing use'. It argued its case on this point: in its view, the expression means the use existing at the time of the water's release into another tributary. In other words, while planning run-of-river projects, India must take into consideration the specific plans to which Pakistan is firmly committed. India knew since December 1988 that Pakistan was engaged in planning the NJHEP and since 1990, also had the details concerning it. The agricultural uses downstream of the KHEP depend on the flow of water from India and the KHEP would disrupt current agricultural projects. Though much activity is dependent on tributaries, irrigation also depends on the Kishenganga/Neelum River and it has developed plans to expand the area by pumping water from the main river. Since the Treaty accords Pakistan priority on the tributaries of the Jhelum River with regard to agricultural and hydroelectric uses, India needs to adjust its uses so that Pakistan's uses are not affected now or in the future. Selecting a cut-off date for their assessment by India would freeze Pakistan's development and subvert the bargain enshrined in the division of the Indus Rivers. The Treaty protects Pakistan's usage as it exists from time to time from the moment of delivery of the diverted waters.[6]

Pakistan also accused India of breaching article IV (6) of the Treaty which imposes the 'best endeavours' obligation on the parties to 'maintain the natural channels'. It pleaded its case on this point as follows: India is under obligation to avoid, as far as practicable, creating obstructions to the flow of waters that cause material damage. Since the KHEP will divert waters from their natural channels, thus causing their deterioration, it will materially damage the environmental ecology of the channel. The plea that a fixed minimum environmental flow will avoid significant harm to the environment is not convincing. India failed to use its 'best endeavours' as it neither carried out an adequate Environmental Impact Assessment (EIA) nor shared information on the anticipated impact of its project despite requests to do so. It is noteworthy that Pakistan relied on the *Pulp Mills* judgment[7] to support its foregoing contention. Pakistan also contended that the KHEP would adversely affect its agricultural and hydroelectric uses as it would divert the entire flow of the waters of the Kishenganga/Neelum River during the lean period and up to $58.4m^3/s$ during the high flow season, thus significantly reducing power generation and revenues from the NJHEP and other hydroelectric power plants that Pakistan may build on the Kishenganga/Neelum River in the future.[8]

As for the second dispute, which did not pertain to the KHEP in particular but was general in character, it focused on the question of whether the Treaty permits India to bring the reservoir level of a run-of-river plant below the dead storage level in circumstances other than unforeseen emergencies and especially drawdown flushing. Pakistan answered in the negative on the grounds that, if it does, India can then manipulate the water flowing into Pakistan. To prove its point, it referred to the specific provisions contained in Annexure D which impose restrictions on India which it asserted were part of the bargain enshrined in the Treaty even though the parties were aware of the need for sediment control and drawdown flushing. It drew the Court's attention to the fact that India acknowledged in the Indus Commission's meetings in 1995 that these restrictions '[are] a major handicap in efficient operation of sediment control.'[9] Dealing with the *Baglihar* case where the Neutral Expert had, as seen above, allowed India to use the drawdown flushing technique and which India had invoked now, Pakistan argued that it differed from the present case as it involved a different hydroelectric plant on a different river. It also asserted that the Neutral Expert exceeded his competence in deciding the question of the permissibility of drawdown flushing and that the parties did not refer this difference to him nor did he provide Pakistan with an opportunity to address the question. It concluded that since the Neutral Expert's determination was *ultra vires* of his competence, it was neither final nor binding and hence should not be given any weight.[10]

ii) India's Arguments

Contrary to Pakistan, India argued that the proposed diversion of the Kishenganga/Neelum River was in consonance with the Treaty. It began by contending that all the provisions of the Treaty must be interpreted in light of its object and purpose as in the preamble which it specified puts emphasis on the 'most complete and satisfactory utilization of the waters.' It pleaded that the Neutral Expert confirmed India's interpretation in the *Baglihar* case. It claimed that the planned diversion would be in conformity with this purpose because it would allow India to realize the full power generating potential of the upstream stretch of the Kishenganga/Neelum River. It also claimed that it would benefit Pakistan's hydroelectric uses, though not the NJHEP but those further downstream of it. Spelling out its perspective on the Treaty, it put forward the thesis that though the Treaty divides the Indus Rivers between the parties, at the same time it endows each party with significant rights in the rivers which are allocated to the other. Dealing with article III (2), it was of the opinion that it expressly entitles India to use the waters of the Western Rivers to generate hydroelectric power, though subject to the provisions of Annexure D. It acknowledged this as an exception to India's obligation to 'let flow' waters of these rivers but at the same time added that the KHEP fell within this exception.[11]

India argued that the KHEP conformed to specifications laid down in Annexure D and that para. 15 (iii) does not forbid it from permanently diverting the entirety of the waters of the Kishenganga/Neelum River into another tributary. It brought out that it realized the possibility of such diversion in the mid-1950s and in consequence it ensured during the negotiations that the Treaty contained a provision to that effect. It referred to a letter of the CWPC to prove that it was contemplating hydroelectric projects on the Kishenganga/Neelum River involving inter-tributary diversions of water. It further contended that not only were the framers of the Treaty aware that India had envisaged such a project in the Jhelum Basin, but Pakistan also recognized such a possibility as demonstrated by Pakistan Cabinet's decision of 15 February 1960 in which it recommended incorporation of a clause in Annexure D 'so that the uses in the Azad Kashmir [are] not affected adversely by inter-tributary diversions.' Similarly, it observed that the Pakistan commissioner stated in 2005 that Annexure E on storage works did not permit inter-tributary transfers but Annexure D on run-of-river plants did.[12] Regarding para. 15 (iii) which envisages transfer of water from one tributary of the Jhelum River to another on condition that it is 'necessary', India drew attention to the preamble of the Treaty which contemplates

the 'most complete and satisfactory utilization of the waters.' In its
view, the expression 'necessary' was intended to refer to that which
was 'optimal for power generation'. It argued that the 'most complete
and satisfactory utilization' cannot be attained without taking
advantage of the difference in elevation between the Kishenganga/
Neelum River and the Bonar Nallah. According to it, the definition
of the term 'necessary' given in the Order for Interim Measures of
23 September 2011 (i.e. 'required, needed or essential for a particular
purpose') was consistent with India's interpretation of the text that
diversion is justified for power generation. Commenting on the
importation of the principle of environmental harm that Pakistan had
tried to bring into the meaning of 'necessary', India rejected it on the
grounds that the issue concerned the generation of electric power
and not the protection of the environment.[13]

India also claimed that para. 15 (iii) only protects Pakistan's
'then existing uses' downstream in terms of agricultural and hydroelectricity which, it contended, did not exist on the Kishenganga/
Neelum River. Expanding on the issue, it revealed that it requested
Pakistan for information on the 'then existing uses' in 1994 when it
conceived the KHEP as storage work and again when it changed it to a
run-of-river plant. The latter provided a certain figure of the irrigated
land but failed to specify the area. Nor did the Indian inspection team
observe such works during the tour of the area. Besides, it argued that
agriculture in the Neelum valley is based on rainfall and channels
fed by side streams rather than the Kishenganga/Neelum River. It
further stated that Pakistan announced they would build the NJHEP
in 2008 whereas the CWPC wrote a letter for construction of the
KHEP in 1960 and produced a document entitled 'KHEP Project'
in 1971 which, it claimed, demonstrated that India had the project in
mind. On the other hand, it claimed that the latter did not intimate it
about the NJHEP until 1989. Despite the communication of technical
information on the KHEP in 1994, Pakistan only assured it in 2005
that it (Pakistan) would provide relevant information regarding the
impact of the KHEP on the NJHEP. It concluded that the NJHEP

fails Pakistan's own definition of the 'then existing use' as it neither made a firm commitment nor showed an active engagement in the project until 2007–08. It further added that during the inspection tour of 2008, there was no preparatory work in sight. Besides, it construed the phrase 'then existing' to mean that it needed to take Pakistan's downstream uses only up to a critical cut-off date for the purpose of finalizing its hydroelectric design. It rejected Pakistan's interpretation according to which India is to adjust its hydroelectric operations on the Kishenganga/Neelum River to the latter's downstream uses as they develop on the grounds that it would not only run contrary to India's express entitlement to use waters of the Western Rivers to generate electricity, but also lead to a waste of the vast amount of resources invested in the KHEP.[14]

Dwelling further on the expression the 'then existing uses' it reasoned that even if the NJHEP constituted such a use, it would not be adversely affected by the KHEP as the latter would divert less than 1 per cent of the total volume of waters of the Western Rivers. It maintained that notwithstanding the KHEP, during the high flow period the NJHEP would receive water in excess of its discharge capacity, and during the lean period it could be operated from waters of the numerous tributaries that flow into the Kishenganga/Neelum River between the KHEP and the NJHEP. As for the lean period, it observed that the NJHEP would actually receive more water than the KHEP. It also reasoned that in case of adverse effect in terms of hydroelectric power generation by the NJHEP, the storage work that Pakistan intends to build on the Kishenganga/Neelum River at Dudhnial between the KHEP and the NJHEP would mitigate the adverse effect to the hydroelectric power generation by the NJHEP from the water that it would release during the lean period. It added that the Kohala hydroelectric plant that Pakistan intends to construct would also take care of the adverse effect from the increased flow of water in the Jhelum River resulting from the diversion of the Kishenganga/Neelum River by the KHEP. It was of the view that the KHEP would not have a significant adverse effect on the NJHEP.

On the contrary, it would have a net positive effect in terms of generating hydroelectric power in the region as it would increase the flow of water into Pakistan's proposed Kohala hydroelectric power project, thereby increasing its capacity to generate electricity during the winter and substantially offsetting the loss of power generation capacity at the NJHEP. The relevant flow data from the area does not demonstrate any material adversity to Pakistan's hydroelectric use. As far as Pakistan's flow data is concerned, it is not dependable and its calculations suffer from inconsistencies, oversight, and other shortcomings. Even if there is an adverse effect on hydroelectric power at the NJHEP, it would be minor.[15]

Regarding Pakistan's plea for restricting the uses of waters to its drainage basin, India rejected it on the grounds that article III (2d) referred to the use of waters for the generation of hydroelectric power and not the use of electricity generated. In other words, it held that the Treaty did not stop the transportation of the power outside the drainage basin where it is generated. It also pointed out that prior to this Pakistan had not raised this objection to the four projects whose power was used outside their respective basins.[16] Dealing with international environmental obligations, India contended that neither the Treaty as a whole nor article IV (6) provide any basis for enshrining them in the Treaty. Consequently, it contested the Court's competence to determine the alleged breaches of these obligations. Invoking the *travaux préparatoires* of the Treaty and the surrounding provisions in article IV (6) it clarified that the foregoing article was meant to prevent India from placing temporary bunds or dikes across the Eastern Rivers and not to stop inter-tributary transfers. It rejected the application of international environmental law principles to the interpretation of article IV (6) or the present dispute as a whole. It intimated to the Court that notwithstanding this position, it had at once complied with the provisions of article IV (6), its domestic environmental regulations, environmental customary international law obligations binding on it, and the international standards applicable to engineers in the design and operation of hydroelectric projects. Besides, it indicated that it had commissioned a comprehensive Environmental Impact Assessment (EIA) in 2000 according to which, it contended, the KHEP would not have any significant adverse environmental impact on the Kishenganga/Neelum River as it would, at all times, release at least a minimum 'environmental flow' of 3.94 cumecs at the site.[17]

In regards to the second dispute, India began by contesting the Court's jurisdiction to determine this question on the grounds that it fell within the competence of the Neutral Expert. In its opinion, Pakistan should have referred it to that forum. Dilating on the merits of the case, it rejected Pakistan's fears that the drawdown flushing would lead to control of the Western Rivers by India as unfounded. It contended that Pakistan's argument resulted in ignoring the Treaty's concern for the generation of hydroelectric power on the Western Rivers by India on the basis of an evolving state-of-the-art technology. It advanced the view that, 'the framers of the Treaty were mindful of the rapid evolution of the technology and therefore enshrined the "state-of-the-art" concept in the Treaty'. It drew attention to the fact that drawdown flushing was not known or accepted as a sediment control practice in 1960 but had since become a state-of-the-art technique.[18] It urged the Court to reject the second dispute as it constituted an appeal against the determination made in the *Baglihar* case and declare it as inadmissible. It referred to the *Baglihar* determination concerning drawdown flushing as an 'authoritative precedent'. Commenting on Pakistan's statement that India did not argue the permissibility of drawdown flushing during the proceedings of the present case but changed its position during the written comments and oral hearings, it observed that the draft determination prompted it to re-examine its position and submit a new one and that Pakistan chose not to reply to the new Indian argument.[19]

c) **Partial Award and Court's Reasoning**

On 18 February 2013, the Court announced its award which was unanimous but partial. It made pronouncements on the two disputes with which it was seized. On the subject of the first dispute, it decided that the KHEP constitutes a run-of-river plant under the Treaty and that India is accordingly entitled to divert water from the Kishenganga/Neelum River for the purposes of power generation. At the same time, it directed India to maintain, while operating the KHEP, a minimum flow of water in the Kishenganga/Neelum River at a rate which the Court indicated that it would determine subsequently. With reference the second dispute, it decided that the Treaty does not permit India to reduce the water level in the reservoirs of run-of-river plants located on the Western Rivers below the dead storage level except in case of an unforeseen emergency. It clarified that the foregoing ruling did not apply to plants already in operation or under construction to which Pakistan had not objected. It also declared that it would render its final award by the end of the year.[20] It is pertinent to point out that the Court dealt with the question of the scope of the Treaty in terms of sovereignty over Jammu and Kashmir because the parties raised the question as to the status of the territory administered by each of them and whether and how far the Court's award would have bearing on it. The Court pronounced in favour of Pakistan and proceeded with the merits of the case. Since we have discussed this aspect elsewhere[21] we propose to skip its examination here. It is also equally pertinent to mention that as far as the second dispute is concerned, India also raised objections to its admissibility which the Court rejected. This matter has also been discussed elsewhere in the book.[22]

Dealing with the merits of the first dispute, the Court observed that the Treaty expressly permits India to transfer water from one tributary of the Jhelum River to another for the purpose of generating hydroelectric power. It reached this conclusion on the basis of para. 15 (iii) of Annexure D. As to the question whether India's right to generate hydroelectricity curtails its usage on the Western Rivers to the drainage basins of those rivers, it declared that article III (2) restricts what India may do with waters of the Western Rivers but not with the products that may be generated from their use. In its view, the Treaty does not impose a geographic restriction on the use of electricity or any other product resulting from the use of the waters. Examining article IV (6) of the Treaty, which obligates the parties to maintain the natural channels of the rivers and its effects on the inter-tributary transfers, it explained that this provision involves maintenance of the physical condition of the channels of the rivers but not the volume and timing of the flow of water in these channels. Nor did it consider the KHEP to be an 'obstruction' foreseen by article IV (6). It did not think that para. 15 (iii) allows the occasional diversion of water in the course of operation rather than diversion as an integral part of the design and operation of the plant. It found justification for this view in the CWPC's letter dated 16 May 1960, which showed that India was contemplating, at the time the Treaty was concluded, a diversion scheme on the Kishenganga/Neelum River.[23]

Next, the Court considered whether the KHEP met the conditions for inter-tributary transfer which it specified as follows: (a) It must be a run-of-river plant; (b) be located on a tributary of the Jhelum River; and (c) conform to para. 15 (iii). As to condition (a), it observed that a 'run-of-river plant' is a term of art which the Treaty has defined and found the KHEP falling within that definition. As to condition (b) which was controversial because the generation of hydroelectricity was to take place at a distance of 23 kilometres from the Kishenganga/Neelum River, the Court conceded that the 'ordinary meaning' does not cover the term 'run-of-river plant' and 'special meaning' under article 31 (4) of the Vienna Convention on the Law of Treaties is needed to that end. In light of this observation, it concluded that the KHEP should be regarded as located on the Kishenganga/Neelum River. Regarding condition (c), the Court explained that it comprises of two elements: the criteria of 'necessity' and the place where the water delivered from the Kishenganga/ Neelum River is released after passing through the power house. For the first element, it found, on the basis of *travaux préparatoires* of the Treaty, the KHEP's inter-tributary transfer as 'necessary' because the scale on which India wanted to generate the power could be done on this location only by using the 665 metre difference in elevation between the dam site and the place where water is released. As to the second element, it pointed out that by releasing water into the Bonar Nallah after it passes through the power house, the KHEP fulfils the requirement that water released below the plant be delivered into another tributary.[24]

Dealing with the central question of the first dispute, whether or not the inter-tributary transfer would adversely affect the 'then existing' agricultural or hydroelectric use by Pakistan on the downstream reaches of the Kishenganga/Neelum River, the Court decided to seek guidance from article 31(1) of the Vienna Convention on the Law of Treaties which, as seen above, lays down as follows: 'A treaty shall be interpreted in good faith in accordance with the ordinary meaning to be given to the terms of the treaty in their context and in the light of its object and purpose.' The Court first examined the text of para. 15 (iii) and held that given its focus on the operation of hydroelectric plants the 'then existing' uses should be determined on an ongoing basis throughout the operational life of a plant. It then examined the context of para.15 (iii) and came to the conclusion that the provision falls within a continuum of design, construction, and operation. It observed that various provisions of the Treaty should be interpreted in a mutually reinforcing fashion because, in its opinion, it would not make sense if the Treaty permitted a plant to be designed and operated in a certain manner and then prohibited its operation in the very manner for which it was designed. Finally, the Court delved into the object and purpose of the Treaty and found that it accords to Pakistan unrestricted right to use waters of the Western Rivers, and to India to generate hydroelectric power.[25]

Dealing with the application of the Treaty to the KHEP, the Court began by delving into implications of the approaches that the parties had advocated. It was of the opinion that Pakistan's 'ambulatory' approach would signify that even if Pakistan cleared a run-of-river project as consistent with the design specifications of the Treaty, India could still be prevented from operating it because of new uses by Pakistan. Deeming this approach as a source of uncertainty and potential for wastage, it was of the opinion that it could discourage India from undertaking a hydroelectric project on the Western Rivers. Turning to the approach that India had advocated, according to which Pakistan's uses would be determined at the time India communicates a 'firm intention' to proceed with a project, it held that identifying a critical date would be difficult but regarded it possible to identify (on the basis of design, tenders, financing, public consultations, environmental assessments, governmental approvals, and construction) a 'critical period' to indicate a firm intention to proceed with a project. It rejected an exclusive 'critical period' approach as, in its view, it could result in a race in which each party would try to create uses that would freeze out future uses by the other.[26]

Considering both the Pakistani and Indian approaches as unsatisfactory, the Court ruled that a proper interpretation of the Treaty should combine elements of both. Hence, it decided to fix the critical period when the parties not only planned the KHEP and the NJHEP but also undertook concrete measures to realize them. It concluded on the basis of the evidence adduced by the parties that the KHEP reached this period during 2004–06 and the NJHEP during 2007–08. In light of this finding, it accepted that the Treaty protected India's right to divert waters of the Kishenganga/Neelum River to construct the KHEP but did not deem this right to be absolute as it was subject to the constraints specified in the Treaty and the relevant principles of customary international law. It opined that whereas para. 15 (iii) accords India the right to construct and operate hydroelectric projects involving inter-tributary transfers, it equally obliges the latter to operate them in a way that would not adversely affect Pakistan's then existing agricultural and hydroelectric uses. It justified this ruling with the view of emphasizing the absolute need to make the entitlements of both the parties effective as far as possible.[27]

The Court also held on the basis of customary international environmental law that Pakistan was entitled to receive a minimum flow of water from India in the Kishenganga/Neelum riverbed at all times because, in its opinion, the Treaty is to be applied in light of the principles of contemporary international law. Though it decided that the Treaty required the preservation of a minimum flow of water downstream of the KHEP, it however failed to fix the precise volume of flow to be preserved because the parties did not provide it with sufficient data. Consequently, it decided to defer the issue to the final award and requested the parties to provide additional data relating to the impacts of a range of minimum flows at the KHEP on (for India) (a) power generation at the KHEP; (b) environmental concerns from the dam site to the LoC; and (for Pakistan) (a) power generation at the NJHEP; (b) agricultural uses of water downstream of the LoC to Nauseri; and (c) environmental concerns at and downstream of the LoC.[28]

As far as the second dispute, which related to the depletion of the reservoir below the dead storage level at the KHEP and generally at any future run-of-river plant concerned, the Court began by first of all clarifying the question of reservoir depletion. It observed that it was linked with the permissibility of controlling sediment through the drawdown flushing procedure. It briefly reviewed the process of sedimentation in the reservoirs of hydroelectric plants and the various approaches to sediment control, including drawdown flushing. It then went on to examine the permissibility of depletion below the dead storage level in light of Annexure D in the context of the Treaty as a whole. To begin with, it underlined that one of the primary objectives of the Treaty is to limit the storage of water by India but not the volume of the dead storage. Consequently, it ruled that India is entitled to include dead storage of any capacity in the design of any run-of-river plant or storage work because it is understood to be truly dead—an area to be filled once and not thereafter—so that it is not subjected to manipulation. Secondly, it pointed out that by virtue of para. 8 (d) of Annexure D, the Treaty imposes design restrictions on the low-level outlets which are required for reservoir depletion and that such restrictions only make sense if depletion is equally restricted. Lastly, it considered beyond debate that those who drafted the Treaty intended to accord India the right to generate hydroelectric power. In its opinion, this right could only be protected by allowing the latter to develop this resource on a sustainable basis.[29]

Next, taking up para.19 of the Annexure E which provides that '[t]he Dead Storage shall not be depleted except in an unforeseen emergency', the Court held that the Treaty prohibits not only depletion of the reservoirs of run-of-river plants below the dead storage level but also correspondingly, drawdown flushing. It also pointed out that the Treaty imposes restrictions on the permissible variation in the volume of flow in a river above and below a hydro-electric plant; and that these restrictions could be incompatible with drawdown flushing in certain reservoirs and in certain flow conditions. Concerning the distinction made by India between operation and maintenance to justify drawdown flushing in the face of Treaty restrictions, it rejected it on the grounds that when the Treaty does not specify a category of 'maintenance purposes' it cannot be invoked to free a party from restrictions that the Treaty explicitly imposes. It also examined whether it is possible to have sustainable hydroelectric power generation without drawdown flushing. On the basis of the technical data presented by the parties and the testimony of experts, it came to the conclusion that drawdown flushing is one of a number of techniques available for sediment control and that generating hydro-electricity is possible without flushing. While handing down the award, the Court made sure that it did not prejudice India's rights already covered by the Treaty. Consequently, it declared that its ruling on the second dispute should not be interpreted to cast doubt retrospectively on any run-of-river plant already in operation, or any plant already under construction on the Western Rivers whose design has not been objected to by Pakistan.[30]

Before proceeding further, it is necessary to point out that during the proceedings of the case, Pakistan approached the Court to issue an order to restrain India from continuing work on the project. As expected, India opposed it. However, the Court concurred with Pakistan's viewpoint and on 23 September 2011, issued an order in which it indicated interim measures. It was of the view that during the pendency of the case proceedings, the works on the dam were likely to advance to a point where it would be significantly more difficult and costly to the potential prejudice of any prescription that it might make to restore the flow of the Kishenganga/Neelum to its natural channel. Therefore, to 'avoid prejudice to the final solution ... of the dispute', it forbade India from proceeding with the construction of 'permanent works on or above the Kishenganga-Neelum riverbed that may inhibit the full flow of that river to its natural channel', until it rendered the award. However, it allowed India to utilize the temporary diversion tunnel that it claimed to have completed at the Gurez site and construct and complete temporary cofferdams to permit the operation of the temporary diversion tunnel for the reason that such tunnel provisionally constituted a temporary by-pass. It also allowed India to temporarily dry out the riverbed of the Kishenganga/Neelum at the Gurez valley and excavate the riverbed and proceed with construction of the sub-surface foundations of the dam. In the wake of the announcement of the partial award, the Court's order on interim measures was also vacated.

d) Final Award and Court's Reasoning

Following the announcement of the partial award, India, by virtue of para. 27 of Annexure G of the Treaty,[31] on 20 May 2013, made a request to the Court to clarify or interpret para. B.1 of its award relating to the second dispute which barred India from drawdown flushing below the dead storage level at the KHEP and all run-of-river projects in the future, at once. On 20 December 2013, the Court gave its decision in which it reaffirmed its earlier ruling that the prohibition on reduction below the dead storage level is of general application on the grounds that it 'is not dependent on the particulars of given site or project, that is, to use India's term, the prohibition is not "site specific" but general.'[32] Then, on 21 December 2013, the Court in accordance with its earlier declaration made on the occasion of the partial award, rendered its final award. Pronouncing that India should have access to at least half of the average flow at the KHEP site during the driest months of the year, it fixed the minimum flow at 9 cumecs.[33] It admitted this approach was somewhat severe in environmental terms but justified stating that it represented an appropriate balance between the needs of the environment and India's right to power generation. How did it reach these conclusions? We propose to examine this question on the basis of the arguments that the parties adduced to support their viewpoints.

Dealing with agriculture in the Neelum valley, Pakistan admitted that it was not dependent on water from Kishenganga/Neelum River but rather on rain and that increase in agricultural productivity would depend on a lift irrigation system. It further stated that the future development of agriculture was predicated on the uninterrupted flow of water for which it did not, however, provide any data. Invoking para. 15 (iii) of Annexure D which expressly protects its agricultural uses it urged the Court to make some allowance for the future development in the field. India, on its part, contended that for minimum flow to be taken into consideration, Pakistan needed to establish two facts: namely that agriculture depended on river water and that such use in the future will be adversely affected by the KHEP, which, in its opinion, is 'unidentified, unplanned and unsubstantiated' and that it has ample water for any such development.[34]

Pakistan and India presented extensive hydrological data to indicate the level and patterns of flow in the Kishenganga/Neelum River. The difference in the data of the two countries was small. Pakistan submitted plans that it had made to increase the use of lift irrigation in the Neelum valley for agricultural uses but failed to provide a quantitative estimate of the likely scale of such development. As for the environmental impact of the KHEP, there were sharp differences in the assessment reports presented by the parties. Pakistan approached the issue on holistic lines, on the basis of interaction between a range of environmental indicators and tried to capture the complexity of interactions within the river ecosystem. India, on the other hand, presented a simpler assessment in which it based its conclusions on limited data relating to the habitat for fish species.[35]

The Court began by defining its task which it described as the determination of a minimum flow that will mitigate adverse effects on Pakistan's agricultural and hydroelectric usage as a result of the operation of the KHEP and the preservation of India's right to operate the KHEP and maintenance of the priority it acquired for the KHEP. It underlined that in the performance of this duty it would give due regard to 'the customary international requirement of avoiding or mitigating trans-boundary harm and of reconciling economic development with the protection of the environment.'

Dealing with the parties' submissions on the hydrology of the Kishenganga/Neelum River which underpinned all other calculations, the Court remarked that there was little difference in the data of the parties, though it observed that Pakistan's estimates were slightly higher. It noted that it would have no effect on its decision regarding the minimum flow. It recommended to the parties that they practice undertaking quality assurance on hydrological data and later on sharing it with each other through the Permanent Indus Commission.[36] As for the agricultural uses, it observed that although Pakistan had provided plans to increase the use of lift irrigation in the Neelum Valley, it had failed to provide a quantitative estimate of the scale of its development. With this backdrop, the Court observed that it was not in a position to take account of agricultural uses, though, in its opinion, the minimum flow ultimately adopted would ensure adequate water for development.[37]

Dealing with Pakistan's hydroelectric uses and especially the NJHEP, the Court admitted that the diversion of water by the KHEP would somewhat reduce the downstream generation of hydroelectric power under almost any minimum flow regime.[38] Regarding the environmental impact of the KHEP, noting the differences in the assessment reports of the parties, it was of the view that there was no single correct approach in the matter. However, it held that Pakistan's report, though not perfect, was more appropriate for a project at the magnitude of the KHEP. At the same time, it pointed out that Pakistan's approach in these proceedings was not commensurate with its own historical practices. In consequence, it encouraged both the parties to focus more on the environmental aspects of their hydro-electric projects. Taking the hydrological tables and the anticipated effects of various flow scenarios presented by the parties into account, it provisionally suggested on the basis of environmental considerations a flow of 12 cumecs.[39]

Next, dealing with the question of fixing the minimum flow meant mitigating the effects of the KHEP on power generation at the NJHEP, it identified two considerations that it indicated it would take into account. First, it underlined that it was bound to recognize the priority that the KHEP enjoyed as decided in the partial award. On this point, it decided that India should have access to at least half of the average flow at the KHEP site during the driest months of the year. Secondly, it concluded that although customary international law required the provision of minimum flow, the Treaty limits its operation only for the interpretation and application of the Treaty. In consequence, it did not deem that the Treaty permitted it to apply customary international law to negate rights expressly granted in the Treaty. On the contrary, it considered itself under obligation to mitigate significant harm. Proceeding on the basis of the hydrological data, it determined a flow of 9 cumecs at the KHEP as sufficient to maintain the natural flows throughout the dry months. It admitted that this approach was somewhat severe in environmental terms but defended it as an appropriate balance between the needs of the environment and India's right to generate power.[40]

Dealing with the effects of a 9 cumec minimum flow, the Court pointed out that India would enjoy 51.9 per cent of the flow at the KHEP dam site during January and a higher proportion in other months. On the basis of these figures it admitted that the minimum flow would reduce electricity generation at the KHEP by 19.5 GWh per month from October to March but indicated that it would result in an annual reduction of only 5.7 per cent which it held would not render the KHEP economically unviable. At the same time, it accepted that due to lack of certainty, in any attempt to predict environmental responses to changing conditions and the potential for climate change, the determination of minimum flow was open to reconsideration. It specifically recognized the right of either party to do so through the Permanent Indus Commission and the mechanism laid down in the Treaty, if it deemed it necessary, seven years after the diversion of the Kishenganga/Neelum River.[41]

e) Critique of the Award

Pertaining to the partial award, the Indian government, public, and the media were jubilant as they deemed it a victory for India. For example, the ministry of external affairs described the award as a reaffirmation of the validity of India's position.[42] Similarly, a former bureaucrat, A. K. Bajaj, who was closely associated with the case from the beginning, lauded it as the correct interpretation of the relevant provisions of the Treaty.[43] Again, leading Indian newspapers such as the *Hindu* and the *Indian Express* portrayed the Court's verdict as favourable to India.[44] Later, when the Court delivered the final award, senior Indian legal counsels who represented India in the case termed the award as an absolute victory for India.[45] As opposed to India, Pakistan's opposition and the officials dealing with the NJHEP considered the partial award as a defeat for the country and subjected the government in Islamabad to criticism. The leader of the opposition in the Senate, Ishaq Dar, blamed the PPP government and the Musharraf regime for the debacle.[46] Similarly, a group of opposition leaders expressed their reservations over the decision and demanded that the facts be presented before the nation.[47] Again, the CEO of the NJHEP Company, General (R) Zubair Ahmad, stated that the award would affect the viability of the NJHEP and reduce electricity generation by 10 per cent resulting in a loss of $141.3 million annually.[48] However, when the Court rendered the final award, the PML(N)-led government, which had criticized the partial award as an opposition party, now called it a 'big victory' for Pakistan.[49] Did Pakistan lose and India win the case as was portrayed in the two countries?

To answer this question, we propose to examine the award. We know that there were two disputes under consideration by the Court. As for the first dispute, namely whether the KHEP is a run-of-river plant and India is entitled to divert water from the Kishenganga/ Neelum River for power generation by the KHEP, the Court decided in favour of India by allowing it to go ahead with the diversion of water as it had contemplated in its design. This obviously marked a defeat for Pakistan which it could have avoided had it taken serious steps in the construction of the NJHEP earlier than India and claimed priority for its project. This award is the result of various Pakistani governments failing to take timely and adequate measures over the years to construct reservoirs and dams. The fear is that such catastrophes will continue to strike Pakistan unless the governments in power mend their ways.[50] Pakistan also suffered a debacle regarding the determination of the minimum flow of water in the final award because it failed to furnish any data on the current or anticipated agricultural uses of water from the Kishenganga/Neelum River, forcing the Court to decide the matter on the basis of hydroelectric and environmental factors. The Pakistan commissioner, Asif Baig Mirza, justified this lapse on Pakistan's part on the grounds that there are no Kishenganga/Neelum River-based agricultural uses at present nor is there any chance of developing them in the near future because of topographical and economic reasons. He believed that submitting fake data would have, in addition to bringing a bad name to Pakistan, spoiled their case.[51]

Pakistani analysts severely criticized Pakistan's legal team for this failure.[52] They were of the opinion that Pakistan could have secured more flow of water if it could support its claim with the required data of an international standard, particularly relating to environment, hydroelectric, and agriculture uses.[53] The question is, if agricultural uses do not exist at present and are not likely to develop in the near future, why did the Pakistan delegation mention them in the first place? This failure led to embarrassment for Pakistan. Nevertheless, the award on the first dispute was not totally bad news for Pakistan. The Court had directed India to release a minimum water flow of 9 cumecs in the Kishenganga/Neelum River compared to India's concession of a mere 4.25 cumecs which Pakistan deemed woefully inadequate. The Pakistan commissioner thought that the award partially safeguards Pakistan's interests because, against Pakistan's demand for release of quantum of water that would generate 94 per cent of the installed capacity of electricity, the quantum fixed by the Court would generate 90 per cent of its installed capacity or an overall loss of 10 per cent.[54] He admitted that the release of 9 cumecs would not be enough to maintain the environment of the Neelum valley but justified it on the grounds that the award was a compromise.[55]

In terms of the second dispute, namely whether India could deplete or bring the reservoir of a run-of-river plant below the dead storage level in any circumstance except in an unforeseen emergency, the Court ruled in favour of Pakistan. This certainly constitutes a major victory. In our estimation, this ruling has the potential to avert the enormous catastrophe that could have resulted if the Court had endorsed the *Baglihar* decision on drawdown flushing for the future run-of-river projects on the Western Rivers. To understand the significance of this point, it may be pointed out that in the wake of the *Baglihar* decision India had started to claim that the Neutral Expert's ruling had assumed the status of a precedent and that India would be entitled to build dams on the Western Rivers on the pattern of the Baglihar Dam. It took this line of reasoning in the *Kishenganga* case as well, urging the Court to follow this precedent.[56] Pakistan, on the other hand, was most unhappy about this ruling and thought that the Neutral Expert had decided the *Baglihar* case outside the Treaty framework. It asked the Court to reject India's argument in the present case, which it did. By overruling the Neutral Expert's decision, the Court rectified the distortion that could have crept into the Treaty. The dean of Indian commentators on water issues, Ramaswamy Iyer, seems to agree with this assessment. He has maintained that Pakistan seized the Court because of the second dispute which was much more important to it than the first one as it wanted to appeal the Neutral Expert's decision, but could not do so for lack of a provision in the Treaty. In his opinion, Pakistan decided to use the *Kishenganga* dispute to make an indirect appeal against drawdown flushing, and it succeeded.[57] John Briscoe reaches the same conclusion by arguing that the Court's award has restored the Treaty to its pristine shape:[58]

> But, as the Christian Brothers told me when I was growing up in South Africa, the Lord works in mysterious ways. In this case there is no doubt that India has won the battle, but I think that it has, in fact, lost a far more important war.
>
> What is my reasoning? ... The Baglihar decision would appear to have provided India with a green light to build these projects with as much live storage as they chose (as long as they classified it as 'for sediment flushing'). What is enormously important is that the ICA has, according to early press accounts, addressed this issue head-on and, *de facto*, concluded that the Baglihar finding in this regard undercut the central compromise of the Indus Waters Treaty, was wrong and should not be applied to future projects. The ICA has apparently ruled that the design and operation of Indian hydropower projects on the Indus, Chenab and Jhelum cannot include more live storage than allowed under the IWT, even if the justification for such storage is silt management.
>
> This finding is of far greater significance than the one-off (and correct, in my view) finding relating to Kishanganga. It restores the central protection—put into question by the Baglihar finding—which Pakistan had acquired when Nehru and Ayub Khan signed the IWT in 1960.

Here we ask whether or not a Neutral Expert or a Court of Arbitration would follow the *Kishenganga* verdict on drawdown flushing in the future. Ramaswamy Iyer has asked the same question.[59] He has contended that each case is *sui generis* and the finding in one case has applicability to the difference or dispute in that case alone. However, he believes that, notwithstanding this fact, the finding assumes the status of a precedent because it would be absurd to adopt divergent principles in different cases. He then poses the question whether the *Kishenganga* award on drawdown flushing which is to be the guiding principle in the future cases retrospectively nullifies the Neutral Expert's finding in the *Baglihar* case. He tries to answer this question by saying that *strictu sensu* it cannot, but considers it odd to adopt a certain practice in the *Baglihar* case and reject it in future cases. To address the question that Iyer raised, we need to examine the theory and practice of the World Court in the matter. Article 59 of the ICJ's Statute states that '[t]he decision of the Court has no binding force except between the parties and *in respect of that particular case.*'[60] As for the practice of the Court, the latter adheres to the text of article 59 by denying the principle of *stare decisis* or 'precedent' but its jurisprudence demonstrates that this is no more than lip service. In reality, it recognizes the validity of the principle of *stare decisis* or 'precedent' by constantly referring to its previous decisions and declarations and has never admitted that any of its judgments or advisory opinions contradict each other. It is noteworthy that at least in one case, *Interpretation of Greco-Turkish Agreement of 1st December 1926 (Final Protocol, Article IV)*, it has expressly referred to its previous advisory opinion as a 'precedent'.[61] The reason for this practice is pragmatism, i.e. it is afraid that it would make its clients, which are essentially sovereign States, turn away from it in case of utter unpredictability in his decisions. Given this background of the World Court, the Court of Arbitration in the *Kishenganga* case should have adhered to the Neutral Expert's decision in the *Baglihar* case. However, it did not do so because it felt that the latter was not *strictu sensu* based on the Treaty. It therefore decided to jettison it to rectify the wrong it felt the Neutral Expert had done. We would conclude from this discussion that a Neutral Expert or a Court of Arbitration is most likely to follow the precedent set by the *Kishenganga* rather than the *Baglihar* case in the future.

Despite this conclusion, we would like to introduce a caveat. We understand that the *Kishenganga* verdict is not the last word on drawdown flushing because we have observed that the World Court has occasionally repudiated its previous decisions, at least insofar as their grounds are concerned. It has done this through a subtle process called 'distinguishing the law' which is prevalent in countries with common law traditions. For example, in the *Interpretation of Peace Treaties with Bulgaria, Hungary and Rumania* advisory opinion,[62] while dealing with the plea that the case was identical to the *Eastern Carelia* case[63], the Court rejected it on the grounds that the circumstances of the two cases were totally different. Another example is that of the 1971 *Namibia* advisory opinion[64] where South Africa had challenged the competence of the Court on the grounds that the relevant legal question related to an existing dispute between itself and other states. In support of its contention, it referred to the *Eastern Carelia* case where its predecessor, the Permanent Court of International Justice (PCIJ), had refused to pronounce on the foregoing question on the grounds that it was directly related to main point of the dispute actually pending before two states. It rejected South Africa's plea by distinguishing the two cases. It was of the view that one of the states concerned in that case was not a member of the League of Nations nor did it appear before the PCIJ whereas in the case before it, South Africa was not only a member of the UN but had also appeared before it and participated in its proceedings. These examples show that despite the Court's pronouncement in favour of Pakistan in the *Kishenganga* case on the second dispute, it may not have said the last word. Incidentally, the following observation by the Court in the *Kishenganga* case is a word of warning for Pakistan: 'If a prohibition on drawdown flushing would render any sustainable hydroelectric development impossible, the Court would consider this relevant in approaching any Treaty provision seeming to suggest such a prohibition.'[65]

Before concluding our discussion regarding this case, a final word is in order. With a 7/0 vote in favour of the award on both the disputes and with no dissenting opinion attached, the result looks like that of a tennis match rather than a tribunal verdict. The vote on the second dispute is particularly significant because the Court's decision on it is contrary to that of the Neutral Expert. This means that even the two nominees of India who sat in the case voted against India. This is something remarkable because in arbitral jurisprudence it is uncommon for an arbitrator to vote against the country that nominated it. Take the case of the *Rann of Kutch Arbitration Award* where Pakistan's nominee voted in favour of Pakistan and India's nominee in favour of India.[66] Similarly, the situation in the *Atlantique* case[67] was no different (though contested as a judicial suit before the ICJ, it smacked of arbitration insofar as Pakistan and India had nominated ad hoc judges to sit in the case)[68] because the nominees of Pakistan and India voted as if they were their masters' voice. How do we explain this unusual phenomenon of India's nominees voting against India and in favour of Pakistan in the present case? Does it signify that they really believed that the Neutral Expert's decision in the *Baglihar* case was erroneous? If so, they deserve to be applauded for their courage and independence of mind but this explanation looks highly implausible. Or, is it explained by the realization on the part of the invisible omnipotent forces in the international community that voting along the lines of the Neutral Expert's decision would imperil peace between the two nuclear-armed South Asian neighbours? We are not in a position to answer this question one way or the other with certainty but such an occurrence cannot be completely ruled out.[69]

Figure 6.1: Nimoo Bazgo Dam

B) NIMOO BAZGO AND CHUTAK DAMS

a) Basic Facts

In addition to the hydroelectric dams discussed above, India has also built two more dams in Jammu and Kashmir. The first one is the 57-metre high, 45-megawatt dam called Nimoo Bazgo on the Indus River near the village called Alchi which is some 70 kilometres from Leh. Located at more than 10,000 feet above sea level, it is one of the highest hydroelectric projects in the world. Started in 2005, India has already completed it and it is reportedly operational now. It is expected to meet the electricity needs of the Ladakh region. India also plans to extend it to the northern grid through a 220 KV transmission line from Leh to Srinagar and subsequently supply electricity to the Indian troops stationed at the Siachen Glacier. Pakistan raised five objections against the design and height of the dam which India outrightly rejected. Of these objections, perhaps the most important one related to the submerged gated spillway which, according to Pakistani water experts, will substantially reduce water flows in the Indus River. The second dam called Chutak is 42 metres high. Launched in 2005, it is located in the Kargil district on the river Suru which is a major tributary of the Indus River. Pakistan has cleared its design as in accordance with the parametres laid down in the Treaty because according to its water experts who visited the project site, 'New Delhi has made holes in the wall of the pondage as per the desire of Pakistan so that the water could not be stopped.'[70] The project was completed in August 2011. The two dams can reportedly store up to 120,000,000 cubic metres of water.

b) India Claims Carbon Credits

In March 2006, the Indian government applied to the Executive Board of the UNFCCC (United Nations Framework Convention on Climate Change) for carbon credits for the Nimoo Bazgo and Chutak Dams on the grounds that the Pakistan government had issued favourable transboundary EIA reports for them. Before proceeding further, a word about 'carbon credits' is in order. A carbon credit is a certificate that the UNFCCC Executive Board issues to countries or groups which have reduced greenhouse gases below their emission quota. It entitles the holder to emit one tonne of carbon dioxide which it can also trade in the international market at the prevalent market value.[71] The carbon credit system, a part of the Kyoto Protocol, aims at stopping the increase of carbon dioxide emissions into the atmosphere. While the Treaty is silent on the question of the environment, it does not signify that Pakistan and India can ignore the environmental aspect of any project, whether concerning irrigation or hydroelectricity or both, simply because the Treaty is silent regarding the matter. This is so because clean environment has now assumed the status of *jus cogens* or a peremptory norm in the international community.[72] For example, the World Court, in its judgment in the *Case Concerning Gabcikovo-Nagymaros Project* between Hungary and Slovakia, declared the norm of environmental law as a principle of *jus cogens* in the construction of dams which comprise hydroelectric projects as well.[73] The International Commission on Large Dams (ICOLD)[74] recommends states undertake an EIA report before the actual start of work on any large dam.[75] In light of these developments, the UNFCCC has made it mandatory for a state seeking carbon credits to submit a transboundary EIA report of the project in question.

c) Controversy about Jamaat Ali Shah

The executive board of the UNFCCC, which had approved the project design documents for the Nimoo Bazgo and Chutak Dams, accepted India's claim and agreed to award carbon credits to India. In August 2008, it awarded the latter a sum of $482,083 for seven years at the rate of $68,869 per year. It should be noted that the Pakistan commissioner never undertook an on-site inspection of the projects in question. He was reportedly planning to do so in May 2010 in order to assess whether or not the designs of the projects in question were in conformity with the parameters laid down in the Treaty. However, he could not make it. Later, he made a request to the Indian government to let him visit to the project sites but before he could do so, the Pakistan government removed him from his post. The Pakistan authorities were clueless as to who issued the EIA reports. The prime minister's office asked the ministries of water and power, foreign affairs, and environment to investigate the matter to find out the culprit. The Ministry of Water and Power[76] not only denied its responsibility in issuing the said reports but also claimed that no agency or ministry ever shared the relevant documents relating to the reports with it. Besides, it pointed out that it was the job of the Pakistan Environmental Protection Agency (EPA) to conduct an environmental impact assessment. The latter also washed its hands of the whole affair. It claimed that since the Indian projects were of strategic importance, it could not have intervened unless some agency sought professional advice from it, which it claimed none did.

In this context, the Ministry of Water and Power initiated a probe for which purpose it asked the secretary of WAPDA, Imtiaz Tajwar, to make a preliminary investigation into the matter which he did. According to his report, the commissioner, Jamaat Ali Shah, remained silent on the Nimoo Bazgo Dam issue between 2002 and 2009, and did not raise any objection either with India, or in the internal meetings, or in the brainstorming sessions with the concerned officials. It also found him guilty of conspiring with India by allowing it to complete the project and of not taking any steps to stop India from continuing the work.[77] Shah, on his part, claimed that he asked the Indian side on more than a dozen occasions to allow him to visit the Nimoo Bazgo and Chutak projects but was not allowed to do so. He also claimed that in July 2010, he asked the Ministry of Water and Power and the special assistant to the prime minister on water, Kamal Majidulla, to move the International Court of Arbitration (ICA) on Nimoo Bazgo but, according to him, the concerned authorities remained unmoved. According to one news item carried by *Dawn* in its issue of 16 April 2011, Pakistan's intelligence agencies seized the record of at least two federal ministries to investigate whether or not there was an institutional lapse in the matter.[78] According to another news item, Pakistan's Federal Investigation Agency (FIA) was probing the accounts of Shah who had allegedly fled to Canada. It was also contemplating contacting Interpol for his arrest.[79]

In the meantime, Shah returned to Pakistan on his own, rendering the need for the Interpol's help unnecessary. The Ministry of Water and Power then set up a three-member committee headed by WAPDA's chairman, Syed Raghib Abbas Shah, to investigate the top officials of the Pakistan Indus Commission, particularly its head, Shah. Following the investigation, the committee exonerated Shah of all the charges leveled against him because, according to it, he had no role in causing the delay in moving against India.[80] Following the exoneration by the committee, Shah reaffirmed his innocence in the matter and showed readiness to appear before any forum, including any court of law, if summoned in the case. On the basis of records from meetings with

India, he further claimed that whatever the Commission did was the result of teamwork and he alone could not be held responsible. He held officers appointed on recommendation during the Musharraf regime responsible for the construction of the dam by India. He warned that if anyone tried to blame him in the future he would take him to court on defamation charges.[81] In the meantime, the concerned authorities, in light of the scrutiny of the official records, reportedly claimed that no Pakistani agency ever issued an EIA report and that India could have provided fake information to the UNFCCC Executive Board in order to earn carbon credits.[82] Towards the end of 2011, in a knee-jerk reaction, the Pakistan government decided to file a lawsuit in the ICA against the UNFCCC for wrongfully awarding credit carbons to India. However, during the secretaries-level talks held on 5 July 2011, Pakistan informed India of its decision not to do so. The reason for the reversal of the earlier decision was apparently the fact that India had already completed the dam.[83]

Nonetheless, India's quest for carbon credits did not end there. Following the determination by the Neutral Expert in the *Baglihar* case, India again made a claim for carbon credits on the grounds of a favourable transboundary EIA report. However, before the executive board of the UNFCCC could approve it, the matter stalled because Arshad Abbasi, an advisor to the Sustainable Development Programme Institute (SDPI) reportedly thwarted the Indian attempt by filing a complaint with the executive board in which he argued that India was not entitled to carbon credits for the Baglihar Dam because it had started the project much before the Kyoto Protocol came into existence.[84]

In addition to building the foregoing dams, India has embarked on a very ambitious plan to construct a large number of run-of-river plants of various sizes on the Western Rivers. Pakistani experts have estimated that they are over a hundred in numbers.[85] Some are at a conceptual stage and would take time before they are realized, while others are at an advanced stage of planning and are ripe for construction. The Pakistan foreign office, in July 2013, informed the National Assembly of India's intention to build four hydropower projects on the Western Rivers in the near future: Ratlé (850 MW), Miyar (120 MW), Lower Kalnai (48 MW), and Pakul Dul (currently of 1000 MW capacity, India proposes to increase it to 1500 MW).[86] The commissioners of the two countries have held meetings on these proposed dams. Pakistan requested India to provide information on the designs of these projects at the planning stage but India did not do so till 2012. It reportedly has major objections to the Ratlé Dam and minor ones regarding the remaining three. The Ratlé Dam deserves special mention because of its size and the fact that the former Indian prime minister, Manmohan Singh, and the ruling Indian Congress president, Sonia Gandhi, jointly inaugurated it. Located on River Chenab near Drabshala village in the Kishtwar district, it is situated between Dul Hasti which is upstream and Baglihar which is downstream.

Pakistan's objections to the Ratlé Dam were based on free board, which it believes should be less than the Indian plan envisages; the spillway gates which it deems should be higher than presently planned; and the intake structure which it wants to be at the highest level. Though the Miyar Dam has low spillway gates, which Pakistan believes to be in breach of the Treaty, it has decided not to raise any objection to it because, given its size, it would not have much of an adverse impact on Pakistan. The foregoing decision signals a change in Pakistan's attitude because, in the past, it always insisted on a strict application of the Treaty. The explanation for this change, according to Pakistani water experts, lies in the fact that Pakistan's new commissioner, Asif Baig Mirza, has decided not to insist on a literal interpretation of the Treaty and to be accommodating towards the Indian projects if their impact on Pakistan is not adverse.

Pakistan's new attitude is not restricted to the Miyar Dam case alone. The new commissioner has also tried to impress upon Pakistan's media the importance of demonstrating responsibility when reporting on the water issue. For example, commenting on certain reports regarding the release of excess water by India during the 2014 floods without informing Pakistan, the commissioner defended

India by admitting that it not only informed Pakistan on time but also provided information about its quantum as well. He also declared that if, despite India's intimation, we fail to take effective measures in minimizing the estimated loss, India should not be blamed. He went on to observe that unwanted water coming from India damages them before damaging Pakistan. He concluded by requesting the Pakistan media to project factual information on the issue and to investigate thoroughly before disseminating the information.[87] This is a new and indeed welcome development, though it's unclear whether the new Pakistani approach would reduce the acrimony between the two countries on the water front and help in the smooth functioning of the Treaty. The new commissioner is quite Micawberish in the matter. However, we are sceptical because as shown elsewhere, the balance of power between the two countries and India's ambitions in the region, rather than the attitude of the new commissioner, would be the determining factor in the matter.[88]

Given India's rapid dam building on the Western Rivers, our scepticism appears to be well founded. According to media reports, Pakistan is of the view that the designs of the Kishenganga and Ratlé projects are faulty for being in conflict with the parameters of the Treaty. The Pakistan commissioner contends that if India executes the Ratlé Dam project with the present design, the water flow of the Chenab River would be reduced by 40 per cent at head Merala. Similarly, he holds that the present Kishenganga Dam design would hurt Pakistan's legitimate water interests. He raised Pakistan's objections in the Commission meetings but to no avail. The two countries have traditionally tried to sort out such differences at the political level in case the Commission fails to deliver. However, given the strained relations between the two countries at this point in time, the Pakistan government has decided not to avail itself of this remedy and has instead opted for arbitration. Towards this end, at the end of 2015 it formally wrote a letter to the Indian government in which it sought its consent for the appointment of a Neutral Expert. The latter has so far not responded to its proposal. In the meantime, it has hired two American legal firms to further pursue the matter.[89]

C) RESTRUCTURING GOVERNMENTAL WATER-RELATED MACHINERY

The preceding survey of the cases in which Pakistan has been involved with India ever since the Treaty came into force reveals that there are indications of corruption. There are indeed some serious problems concerning the way the Pakistan government's machinery functions on the water front. In this section, we propose to identify the nature of the malady and suggest ways and means to remedy it. The first problem that we have identified is the lack of an institutional mechanism to properly handle issues related to water. This came out during the *Baglihar* case when, following the failure of the bilateral approach to resolve the matter, the Pakistan Indus Commission was in a quandary as to how to proceed.[90] The Attorney General of Pakistan reportedly wanted the Commission to continue with the bilateral approach whereas the commissioner favoured the Bank's involvement. The confusion was further compounded when the Indian authorities offered to hold secretary-level bilateral negotiations. The ministry of foreign affairs admitted on one occasion (at least) that Pakistan deferred the decision to seize the Bank at the request of the Indian government for another shot at bilateral settlement. Pakistan took the Indian bait without seeking a commitment from India that it would suspend the work on the project as long as the bilateral negotiations continued. Later on, we learnt to our chagrin that the promise of a bilateral settlement was an Indian ploy to buy time so that it could present Pakistan with a fait accompli on the issue. Pakistan made the concessions to India without realizing that time was of the essence as it would be difficult for the Neutral Expert to order the dismantling of a structure already in place, which was what India hoped to achieve.

Another problem that we have identified is the lack of a sound mechanism to nominate the right person for appointment to a key position. This flaw came to the fore when the Bank forwarded three names to Pakistan and India and asked them to select one for appointment as Neutral Expert. Out of the three names, the two countries fortuitously agreed on Raymond Lafitte from Switzerland. Pakistan's reason for choosing him is not clear. Perhaps the fact that he belonged to a country which does not have a colonial past and is considered neutral in international politics was a factor. This criterion is flawed if it is not accompanied with other considerations which are essential for such appointments like the appropriate professional expertise of the person in question. According to Bashir Ahmad of NESPAK, who was hired by the Pakistan government as consultant on this case, Raymond Lafitte was the least suitable person because he is a dam safety expert and is therefore inclined to look at the case from the prism of security rather than the rights of the parties under the Treaty. He favoured David Blackmore from Australia (the other Bank nominee) who, in his opinion, was more suitable because he had worked on agreements between upper and lower riparians in the Mekong River Basin and Murray Darling Basin and would have consequently been more sensitive to Pakistan's concerns as a lower riparian.[91] Ahmad also revealed that Pakistan's prime minister at that time, Shaukat Aziz, did not provide him with an opportunity to convince him of his choice and approved Raymond Lafitte's name in a highly perfunctory manner.[92] Incidentally, this is a replay of what he did during the Lal Masjid operation when he reportedly refused to attend to the emergency situation at hand based on the frivolous plea that he was going out with his wife to have some ice cream.[93]

Still another shortcoming is the lack of coordination between different agencies charged with overseeing the implementation of the Treaty. The issue came to the fore when the executive board of the UNFCCC granted carbon credits to India on Nimoo Bazgo and the Chutak Dams. This created a lot of furore in Pakistan. Upon inquiry, it was revealed that some governmental agency issued an EIA report in favour of these dams. All relevant agencies and departments, namely the Ministry of Foreign Affairs, the Pakistan Environmental Protection Agency, and the Ministry of Water and Power denied ever having issued such a report.[94] It is still unclear as to who was

responsible for it. The above survey also revealed a lacuna to deal
with individuals who fail to perform their duty adequately. This came
to the fore in the Nimoo Bazgo case when the commissioner, Jamaat
Ali Shah, was charged with failure to visit the Nimoo Bazgo Dam site
and challenge India for undertaking its construction in violation of the
Treaty. Two commissions were constituted to investigate the matter.
The first one found the commissioner guilty as charged whereas the
second one exonerated him of all charges. Despite the clearance, this
episode showed a lack of effective mechanisms to take the appropriate
measures if an individual raises his voice and his superiors fail to
listen. Last but not the least, the survey demonstrated the negligence
of the Pakistan government in safeguarding Pakistan's water interests.
This was manifested in the *Kishenganga* case when the Court of
Arbitration declared that compared to India, we were late in seriously
undertaking the construction of the dam. This was a failure of various
Pakistani governments even though they knew or should have known
that India would build this dam as it had a specific provision inserted
in the Treaty to this effect.[95]

How do we rectify these oversights? The simple answer is that since
corruption often starts at the top, the remedy also lies in attending
to it at that level. We know that the Pakistan government indulges in
occasional rhetoric regarding Pakistan's water rights, but in practice
its actions do not match its words. Khawaja Asif, the federal minister
for water and power, stated that water was a matter of life and death
for Pakistan and vowed to aggressively protect it.[96] However, his
government has yet to concretely demonstrate that it is serious about
this claim. It certainly does not have a well-thought out water policy.
This is evidenced, for example, by the statement of Khawaja Asif
who declared that the Treaty is not favourable to Pakistan because it
was signed by a dictatorial regime and that Pakistan would seek its
revision. He made this statement without realizing that dictator Ayub
Khan did the country a great favour by timely signing the Treaty
and that without it, Pakistan would have been in a very precarious
situation vis-à-vis India. Similarly, Khawaja Asif's statement that his

government would seek its revision was made out of sheer ignorance and without realizing that he was playing into India's hands as the latter would gain from Pakistan doing so.[97] Hence, the first thing that the Pakistan government should do is to accord priority to the water issue and devise a clear cut long-terms policy on it.

The next step that the Pakistan government should take is to restructure different departments and agencies which are charged with the duty to get the Treaty implemented. The first and the foremost candidate in this regard is the Pakistan Indus Commission which was specifically created for this purpose. It is true that it was not given much importance in the past because the confict was not as predatory as it is now. This is evidenced by the fact that in the first two or three decades, India embarked on very few dams and reservoirs on the Western Rivers but now it has plans to put a large number of them in place. In this regard, the first thing that the Pakistan government should do is to appoint the most suitable person as the Indus commissioner. It must be said that most of those selected so far have left much to be desired because despite being qualified engineers, they lacked qualities that would enable them to act on the international stage. There is a need to appoint a person who, apart from being a competent water engineer, should have a good understanding of relations between Pakistan and India, and should be articulate. He should hold the post on a full-time rather than a part-time basis. Furthermore, he should be given a handsome salary package commensurate with his qualifications, failing which the commission will not be able to attract good candidates.

The commissioner should have direct access to the prime minister because it has been observed that at times when he urgently needs to consult the prime minister, he cannot do so because of bureaucratic hurdles. At the moment, he cannot access the prime minister without going through the secretary of the Ministry of Water and Power. Thus there is a need to devise a mechanism which takes care of such difficulties. Perhaps, the commissioner should not have administrative control of the secretary of the Ministry of Water and Power, and the secretary should be placed under the Cabinet Division which would allow him to bypass red tape and get matters under consideration promptly resolved. Next, there is a need to establish a permanent high-powered board composed of experts from all relevant fields to advise the commissioner in carrying out its functions. The Commission also needs to have state-of-the-art library facilities and a group of trained people to man it. Finally, he should also be entrusted with the responsibility to coordinate and oversee the water-related activities of different agencies and departments.

The foregoing analysis has demonstrated that the Treaty performed smoothly in the early decades of its existence, principally because India was building very few hydroelectric plants on the Western Rivers. Resultantly, the observers, at once Pakistani, Indian, and international, hailed it as a great success in an otherwise less than enviable state of relations between the two countries. However, India eventually decided to go on a dam building spree on the Western Rivers, apparently for hydropower generation. It has built Baglihar, Kishenganga, Nimoo Bazgo, and the Chutak Dams, and is planning to build more dams and reservoirs. As a result, the Treaty is expected to come under considerable strain in the future. In the meantime, another issue has cropped up relating to climate change, which has the potential to further bedevil relations between the two countries. This is what we propose to take up in the next chapter.

REFERENCES AND NOTES

1. Khalid Mustafa, 'India Set to Squeeze Pak Waters while Officials Sleep', *The News*, 12 May 2011. Sheraz Memon stated in a report that he submitted in his capacity as Pakistan's acting commissioner to the Senate Committee on Water in May 2011, that his predecessor, Jamaat Ali Shah, had cleared the 240-megawatt Uri-11 Dam which is part and parcel of the Kishenganga Dam project; and that the functioning of the Uri-11 is dependent on the Kishenganga Dam as the water diverted from the latter must reach the former. He contended that his predecessor's clearance had provided justification to India to build the Kishenganga Dam. In this backdrop, Pakistani experts were of the view that Pakistan's case was weak. They thought that it would be a replay of the *Baglihar* case where Pakistan seized the Bank when 80 per cent

of the work was already complete. Salman M. A. Salman, the Bank's former legal vice-president told the present writer that way back in 2002 he warned Pervez Hasan, Tariq Hasan, and Tariq Usman Haider, who were participating in a seminar on river law at the Hague Academy of International Law, that the *Kishenganga* award would be a replay of the *Baglihar* decision if Pakistan delayed availing itself of conflict resolution provisions of the Treaty. According to him, his prognosis was based on the fact that the Neutral Expert could not ignore the facts on the ground and that he could not order the dismantling of a structure already built. For the date and place of the interview, see chapter 'Treaty in Action-I', note 76.

2. Composed of seven members, the Court of Arbitration was headed by Judge Stephen Schwebel from the United States, formerly president of the International Court of Justice. The other members of the Court were as follows: Sir Franklin Berman KCMG QC (UK), Professor Howard S. Wheater FREng (UK), Professor Lucius Caflisch (Switzerland), Professor Jan Paulsson (Sweden), Judge Bruno Simma (Germany), and HE Judge Peter Tomka (Slovakia). Judges Jan Paulsson and Bruno Simma were nominees of Pakistan whereas Judges Lucius Caflisch and Peter Tomka were those of India.

3. *Kishenganga Arbitration*, Partial Award, *PCA*, 18 February 2013, 57–8, paras. 168–71, and 59–62, paras. 175–80. Pakistan rejected the relevance of the *Lake Lanoux* case, involving Spain and France over the dispute concerning the diversion of water from Lake Lanoux for the generation of hydroelectricity, that India had invoked in its favour, to the present case. Ibid. 63, paras. 181–2.

4. Pakistan invoked the following statement made in the *Iron Rhine* and the *Gabcikovo-Nagymaros* cases: '[T]he general obligation of States to ensure that activities within their jurisdiction and control respect the environment of other States or of areas beyond national control.' It also pointed out that the principle laid down by the World Court in the *Pulp Mills* case that 'States must exercise due diligence with respect to activities bearing an impact on the environment of other States' was also relevant to the interpretation of the expression 'necessary'. Ibid. 77, para. 222. See also *Kishenganga Arbitration, Pakistan Memorial*, vol. 1, 27 May 2011, para. 5.44.

5. *Kishenganga Arbitration*, Partial Award, supra note 3, 66–8, paras. 194–6; 73–4, paras. 211–15; and 76–8, paras. 220–4.

6. Ibid. 80–1, paras. 229–31; 83–4, paras. 240–1. See also Kishenganga Arbitration, *PCA Press Release*, 1 September 2012.

7. Pulp Mills on the River Uruguay (Argentina v. Uruguay), Judgment, *ICJ Reports* 2010.

8. *Kishenganga Arbitration*, Partial Award, supra note 3, 86–8, paras. 246–50; and 90–2, paras. 256–8; see also, *PCA Press Release*, supra note 6.

9. *Kishenganga Arbitration*, Partial Award, supra note 3, 108–10, paras. 292–6.

10. Ibid. 127–9, paras. 341–6.

11. Ibid. 58–9, paras. 172–74; and 62–5, paras. 183–89; see also *PCA Press Release*, supra note 6.

12. Ibid. 71–3, paras. 206–9.

13. Ibid. 78–80, paras. 225–7.

14. Ibid. 71–3, paras. 206–9; 78–80, paras. 225–7; and 81–3, paras. 232–8; see also, *PCA Press Release*, supra note 6.

15. Ibid. 84–5, paras. 242–4; and 88–90, paras. 251–3; see also the *PCA Press Release*, supra note 6.

16. Ibid. 68, paras. 197–8.

17. Ibid. 92, paras. 259–62; see also, *PCA Press Release*, supra note 6.

18. Ibid. 110–2, paras. 297–302.

19. Ibid. 129–30, paras. 347–9.

20. *Kishenganga Arbitration*, Partial Award, supra note 3, 201.

21. See chapter 'Terms of the Treaty.'

22. Ibid.

23. *Kishenganga Arbitration*, Partial Award, supra note 3, 137–42, paras. 368–80.

24. Ibid. 142–8, paras. 381–99.

25. Ibid. 148–54, paras. 400–13.

26. Ibid. 154–60, paras. 415–32.

27. Ibid. 161–7, paras. 433–44.

28. Ibid. 168–74, paras. 445–63.

29. Ibid. 188–93, paras. 495–503.

30. Ibid. 196–200, paras. 516–23; see also Kishenganga Arbitration, *PCA Press Release*, 19 February 2013.

31. Para. 27 of Annexure D of the Treaty provides as follows:
'At the request of either Party, made within three months of the date of the Award, the Court shall reassemble to clarify or interpret its Award. Pending such clarification or interpretation the Court may, at the request of either Party and if in the opinion of the Court circumstances so require, grant a stay of execution of its Award. After finishing this clarification or interpretation, or if no request for such clarification or interpretation is made within three months of the Award, the Court shall be deemed to have been dissolved.'

32. Kishenganga Arbitration, India's Request for Clarification or Interpretation dated 20 May 2013, *PCA*, 20 December 2013, para. 34.

33. The cumec is a unit of measurement for the flow of water, equal to one cubic metre per second (35.3147 cubic feet per second in the US customary units).

34. Kishenganga Arbitration, Final Award, *PCA*, 20 December 2013, 17–19, paras. 49–52.

35. Ibid. 19–27, paras. 54–70.

36. Ibid. 32–3, paras. 89–91.

37. Ibid. 34, paras. 93–4.

38. Ibid. 34, para. 95.

39. Ibid. 357, paras. 97–104.

40. Ibid. 38–41, paras. 105–13.

41. Ibid. 41–2, paras. 114–22.

42. Ashfak Bokhari, 'Kishanganga Verdict a Tilt in India's Favour', *Dawn (Economic and Business Review),* (25 February 2013).

43. Ibid.

44. Ibid.

45. *The Times of India* (25 December 2013).

46. Amir Wasim, 'Kishanganga Dispute with India: Government Criticised in Senate over Hague Court Verdict', *The News* (27 February 2013).

47. 'Kishanganga Project: Arbitration Court's Decision not a Legal Defeat for Pakistan says Farhatullah Babar', *The Lahore Times* (27 February 2012).

48. Khalid Mustafa, 'Kishanganga in The Hague: Flawed Strategy led to Pak Defeat, say Experts', *The News (Business News),* (20 February 2013). The Pakistan commissioner claimed that the diversion would reduce the quantum of water by 11 per cent in summer and 27 per cent in winter which would violate the Treaty as India is obliged to let flow water towards Pakistan without let or hindrance. He also claimed that the KHEP would reduce the power generation capacity of the NJHEP by about 11 per cent. Similarly, he contended that the diversion would lead to ecological disaster in the area. According to media reports, the diversion would reduce the river flow in Azad Kashmir for at least six months, causing irreparable damage to the environment, especially to Musk Deer Gurez Park which is a national park and hence deprive the Neelum valley where it is located of tourism revenues. It would affect about 200 kilometres of riverbed and about a 40-kilometre-long stretch of the Neelum River would become completely dry, which would violate the environmental laws according to which at least 70 per cent of river flow is to be protected in case any development project is undertaken. The Pakistan government some time back commissioned a reputed international firm to prepare an environmental damage assessment report which reportedly lends support to its contention. India, on its part, totally rejected Pakistan's objections to the design of the plant. It disagreed with the contention that the diversion would reduce the flow of water to Pakistan. It contended that the quantum of water would remain unchanged. The only difference that the diversion in its view would make would be that instead of meeting in Azad Kashmir as is the case at present, the Neelum and Jhelum Rivers would meet in the Indian-administered Jammu and Kashmir. India also rejected Pakistan's contention that the completion of the feasibility study of the Neelum-Jhelum project by Pakistan has created an acquired right in its favour. Ijaz Hussain, 'Kishanganga Dam Controversy', *The Daily Times* (18 June 2008).

49. See the statement of Khawaja Asif, the federal minister for water and power, *The Express Tribune* (21 December 2013).

50. According to Bashir Ahmad, Pakistan lost the case because even though we started the project earlier than India, we failed to mention it in the dossier that we presented before the Court. He explained it by saying that the Executive Committee of the National Economic Council (ECNEC), the highest body charged with approving or rejecting development activities reflective of long-term national priorities and objectives, approved the PC-1 of the project in March 2002, and the work began but this fact was not presented before the Court. He blamed Kamal Majidulla, the advisor to the PM on water, for this lapse as, in his opinion, he failed to engage the stakeholders including WAPDA and the Ministry of Water and Power for the purpose. This information was revealed to the present writer by Bashir Ahmad in the course of an interview that he gave him on 9 March 2014, at his residence in Lahore.

51. The commissioner shared this information with the present writer in an interview that he accorded him in his office in Lahore on 9 March 2014. See also Khalid Mustafa, 'Establishment Starts Looking into Kishenganga Fiasco', *The News* (29 December 2013).

52. Khalid Mustafa, 'India Emerges Winner in Kishenganga Case', *The News* (22 December 2013); Khalid Mustafa, 'Pakistan to Face $145 Million Yearly Loss', *The News* (23 December 2013); Ansar Abbasi, 'Marginal, not 'Big Victory' on Kishenganga, *The News* (24 December 2013); Ansar Abbasi, 'Charge Sheet against Pakistan', *The News* (26 December 2013). See also <http://www.ips.org.pk/whats-new/92-seeminar/1748-ips-seminar-reviews-kishanganga-award>.

53. Those who criticized Pakistan's performance were Dr Gregory L. Morris, Pakistan's technical expert in the Kishanganga case; Ashfaq Mehmood, former federal secretary of the Ministry of Water and Power; Shamila Mahmood, senior consultant, Pakistan Transboundary Water Organization. They were speaking at a seminar organized by the Institute of Policy Studies, Islamabad, in January 2014, on the *Kishenganga* award. See <www.ips.org.pk/whats-new/92-seeminar/1748-ips-seminar-reviews-kishanganga-award>.

54. Khalid Mustafa, 'India Emerges Winner', supra note 52.

55. Khalid Mustafa, 'Establishment Starts Looking', supra note 51.

56. During the proceedings before the Court, India invoked the decision in the *Baglihar* case, not as a binding precedent, but as 'a relevant and applicable precedent ... dealing with similar facts and law; and one that sheds authoritative light ... on the interpretation of the provisions in question.' It termed the *Baglihar* decision as an 'authoritative precedent'. It justified its reliance on precedents as a 'desirable and universally accepted practice'. *Kishenganga Arbitration*, Partial Award, supra note 3, 129, para. 348.

57. Ramaswamy Iyer, 'Jubilation at Indus Win is Premature', *The Hindu* (28 February 2013).

58. John Briscoe, 'Winning the Battle but Losing the War', *The Hindu* (22 February 2013).

59. Ramaswamy Iyer, 'Jubilation at Indus Win', supra note 57. Iyer has drawn readers' attention to another dilemma arising from the Treaty. He points out that while the Treaty permits India limited uses of waters of the Western Rivers including generation of run-of-river hydroelectric power, it imposes stringent engineering and operational conditions on such use in order to protect Pakistan from any harm. However, each condition is accompanied by the proviso 'consistent with sound and economical design and satisfactory construction and operation.' He contends that this kind of balancing act is easy to write in the Treaty but hard to practice because India insists in the Indus Commission meetings on the permissive provisions and the foregoing proviso whereas Pakistan emphasizes the restrictive provisions and de-emphasizes the proviso. He concludes from this that a too liberal interpretation of the proviso puts the protection afforded to Pakistan in jeopardy whereas an absolute insistence on stringent conditions to the neglect of the proviso virtually nullifies the permission granted to India to build hydroelectric plants. He further contends that if India is directed to build such plants without drawdown flushing resulting in their unsatisfactory operation, then it amounts to stopping it from building them, which is a departure from the Treaty (Ibid.). In our opinion, Iyer's presumption that the prohibition of drawdown flushing would result in an unsatisfactory operation of the plant is unfounded because the Court in the *Kishenganga* case rejected this argument on the grounds that alternative methods exist to adequately take care of sediment control. If, for the sake of argument, Iyer's plea that drawdown flushing is the only satisfactory and viable method available (which underlies his argument) is accepted, a Neutral Expert or an Arbitration Court would have a hard time reconciling between Pakistan's right to receive an uninterrupted flow of water and India's right to generate hydro-electric power.

60. Emphasis added.

61. *PCIJ*, Series B, No.16 (1928).

62. *ICJ Reports* 1950, 72.

63. *PCIJ*, Series B, No. 5 (1923).

64. *ICJ Reports* 1971, 23–4.

65. *Kishenganga Arbitration,* Partial Award, supra note 3,192, para. 509.

66. Rann of Kutch Arbitration Award (India and Pakistan), *PCA,* 19 February 1968. Nasrollah Entezam from Iran and Ales Bebler from Yugoslavia represented Pakistan and India respectively on the bench.

67. Aerial Incident of 10 August 1999 (Jurisdiction of the Court), Judgment, *ICJ Reports* 2000.

68. Article 30, para. 1 of the Statute of the *ICJ* provides that judges of the nationality of each of the parties are entitled to sit in the case before the Court. Availing themselves of this provision Pakistan and India respectively appointed Sharifuddin Pirzada and M. Reddy their as ad hoc judges in the case.

69. Sometimes one hears of invisible international forces influencing the outcome of a case at international legal fora though it must be said such talk remains in the domain of rumour rather a proven fact. One instance where such allegations were made was the *South-West Africa* case. To understand these allegations, we need to recapitulate the facts surrounding the case. After World War I, the League of Nations entrusted South West Africa, a former German colony, as a mandated territory to South Africa. Following the establishment of the United Nations, all the mandatory powers were required to place their mandated territories under the Trusteeship Council. All complied with the requirement except South Africa which refused to do so on the grounds that it was not legally bound to do so. The UN General Assembly tried to get South Africa to do the needful through advisory opinions of the World Court but did not succeed because of their non-binding character. In order to force the latter to comply with the UN obligation, Liberia and Ethiopia, former members of the League of Nations, in 1960 instituted a contentious case against it before the World Court. South Africa raised a number of preliminary objections but the Court in 1962 rejected them and affirmed its jurisdiction. Following this verdict the stage was set for the Court to pronounce on the merits of the case. However, instead of doing so, in 1966 it adjudged that there were two questions of antecedent character appertaining to the merits which needed to be adjudicated before pronouncing on the 'ultimate merits' of the case. This was nothing but a subterfuge to avoid giving a judgment on the merits. It did so because its composition (with eight judges for and seven judges against) was such that the verdict would have gone against South Africa. This would have forced the latter to leave South West Africa which was unacceptable to the West. Consequently, the president of the Court, Sir Percy Spender from Australia, apart from manipulating the proceedings by refusing to deal with the merits of the case as seen above, also barred the Pakistani judge, Sir Zafrulla Khan, who would have been most probably part of the majority of eight, from sitting in the case on the grounds that before his elevation to the Court, Liberia and Ethiopia had nominated him as an ad hoc judge. With the exit of Sir Zafrulla Khan from the bench the vote was divided with seven judges in favour and seven against. Then by virtue of his casting vote, the president rejected the applicants' claims by eight votes to seven. It was indeed a bizarre case. According to the grapevine at that time, Sir Zafrulla Khan did not sit in the case at the behest of the Western powers which were not prepared to allow South West Africa to become independent at that point in time for which the explanation is to be sought in Cold War politics. See Ijaz Hussain, 'Sir Zafrulla Khan–The Silent Judge', in Ijaz Hussain, *Issues in Pakistan's Foreign Policy: An International Law Perspective* (Lahore: Progressive Publishers, 1988), 129–32.

70. Khalid Mustafa, 'India Rejects Pakistan's Objections over Another Power Project', *The News* (9 October 2011).

71. 'What is carbon credit?', *CBS News*, <www.cbsnews.com/news/what-is-carbon-credit>.

72. A peremptory norm of general international law *(jus cogens)* is a norm accepted and recognized by the international community of states as a whole as a norm from which no derogation is permitted and which can be modified only by a subsequent norm of general international law having the same character. See article 53, *Vienna Convention on the Law of Treaties*, 1969.

73. The following observation by the World Court is relevant on this point: 'Neither of the Parties contended that the new peremptory norms of environmental law had emerged since the conclusion of the 1977 Treaty, and the Court will consequently not be required to examine the scope of Article 64 of the Vienna Convention on the Law of Treaties. On the other hand, the Court wishes to point out that newly developed norms of environmental law are relevant for the implementation of the Treaty .... These articles do not contain specific performance but require the parties, in carrying out their obligations, to ensure that the quality of water in the Danube is not impaired and that nature is protected, to take new environmental norms into consideration when agreeing upon the means to be specified in the Joint Contractual Plan.' *ICJ Reports* 1997, 67, para. 112.

74. The ICOLD has defined a 'large dam' as a dam which has a height of 15 metres or more (over the foundation); or is 5 to 15 metres high and has a storage volume of over 3 million cubic metres. *ICOLD Criterion/Dam*, <http://www.economypoint.org/i/icold-criterion-html>.

75. For example, the ICOLD, in one of its bulletins, makes the following recommendation: 'As a consequence of the potential transformations and modifications of the environment, and the possible negative effects mentioned, environmental changes must not be neglected in the planning and designing of large dams. Public awareness of these effects is now rightly substantial and in the majority of cases environmental impact studies should be carried out and should play a significant role in the decision-making process.' *ICOLD Bulletin 50* (Dams and the Environment: Notes on Regional Influences), 90.

76. Khaleeq Kiani, 'Agencies Seize Record of two Ministries', *Dawn* (16 April 2011).

77. Ashfak Bokhari, 'Failure to Defend Riparian Rights', *Dawn (Economic and Business Review)*, (3 December 2012); see also Khaleeq Kiani, 'Pakistan's Institutional Lapse Helps India Complete Water Projects', *Dawn* (30 July 2012). Shah revealed, in an interview on 26 October 2012 at his residence in Lahore, that Kamal Majidulla was at the heart of this controversy. He opposed the move to replace Pakistan's chief legal counsel, James Crawford of Oxford University, in the *Kishenganga* case by Kaiyan Kaikobad (now deceased), professor of international law at Brunel University. According to him, the motivation for replacement was monetary benefit that he hoped to gain. He claimed that when Shah thwarted him in his nefarious design, Majidulla used

some journalists to malign him by leaking false information to them. A Pakistani journalist too traced the genesis of the controversy to Kamal Majidulla who, according to him, created the Pakistan Trans-border Water Organization (PTWO) in September 2011, and placed the Pakistan Indus Commission under it with instructions that it was to report directly and exclusively to the PTWO. When Shah defied Majidulla and reported to the ministry of finance rather than the PTWO, the latter tried to limit the Commission's role in the legal battle against India by barring the Commission from reporting to the parent Ministry of Water and Power. He also reported that, following the *Kishenganga* award, the Pakistan government was likely to investigate the way the legal and technical team fought the case and more importantly the appointment of Kamal Majidulla because he had nothing to do with transboundary issues (Khalid Mustafa, 'Establishment Starts Looking', supra note 51). On 10 April 2015, we emailed these facts to Kamal Majidulla for his reaction but he did not reply. We repeated the request on 25 April but again he chose to remain silent. It may be noted here that, irrespective of Shah's role regarding the Nimoo Bazgo Dam case, the rightist lobby in Pakistan was generally hostile towards him for his alleged soft attitude towards India on the water issue. For example, *Jarrar,* the mouthpiece of Jamaat-ud-Dawa (JuD) in its issue of 5 March 2010, attacked Shah in which it alleged that he was Pakistani in body but his tongue spoke the language of Hindus. It justified its attack on him on the grounds that he had reportedly stated that India had not stolen Pakistan's water. Similarly, *Nawa-e-Waqt*, a rightist newspaper, castigated Shah for allegedly claiming that Pakistan was getting its share of water under the Treaty; that India was entitled to build dams on the Western Rivers; and that the reason for shortage of water in Pakistan's rivers was not that India was stealing water but because of less rain. Khaled Ahmed, 'Target: Jamaat Ali Shah', *The Express Tribune* (2 December 2012).

78. Khaled Ahmed, 'Target: Jamaat Ali Shah', supra note 77.
79. Khalid Mustafa, 'FIA to Contact Interpol for Jamaat Ali Shah's Arrest', *The News* (16 January 2012).
80. Khalid Mustafa, 'Pak Govt Helped India Build Dam on Indus River', *The News* (24 September 2011). It is indeed odd that one inquiry report held Shah guilty as charged whereas the other absolved him of wrongdoing. The present writer tried his best to get hold of the two reports but did not succeed.
81. 'Mush Weakened Baglihar Dam Case: Jamaat Shah', *The News* (19 March 2012). According to media reports the Senate Standing Committee on Cabinet Secretariat in 2014 sought the immediate arrest of Shah on the grounds that he did not protect Pakistan's water rights. It also stated that the federal government was in contact with Interpol for repatriation of Shah from Canada. *The Express Tribune* (1 October 2014).

82. Noor Aftab, 'Pakistan Authorities in a Fix as India Secures Carbon Credit', *The News* (14 January 2011).

83. Ashfak Bokhari, 'Failure to Defend', supra note 77.

84. Khalid Mustafa, 'Who "Cleared" India to Bag Multi-Million Dollar Carbon Credits?', *The News* (30 May 2011).

85. For details, see Appendix 3.

86. Ansar Abbasi, 'Pakistan to Move World Court if India Pursues Ratlé Dam Project', *The News* (20 July 2013); see also Khalid Mustafa, 'Another Blow to Pakistan's Water Interests', *The News* (10 July 2013); Khaleeq Kiani, 'Objections to Four More Indian Projects Raised', *The News* (28 September 2013).

87. 'India Informs Pakistan before Releasing Water, Baig', *Pakistan Defence* (2 July 2014), <http://defence.pk/threads/pak-to-ask-india-to-inform-before-releasing-dam-water.321938/>. See also, Khaled Ahmed, 'Getting Ready for a "Water War?"', *Friday Times* (8 March 2010).

88. See 'Conclusion' of the book.

89. Khalid Mustafa, 'Pakistan to Move WB for Neutral Expert after no Response from India', *The News* (9 December 2015). See also 'Pakistan Seeks Indian Consent for Neutral Expert', *The News* (18 November 2015).

90. The secretary of the Ministry of Water and Power, Saifullah Chattha, admitted during a press conference that Pakistan moved its case on Kishenganga in the International Court of Arbitration belatedly. 'Indus Waters Treaty not in Pak Interest: Asif', *The News* (26 October 2013).

91. The Indian NGO, South Asia Network on Dams, Rivers and People, rightly guessed before the verdict that Lafitte, being a supporter of large dams, would tilt towards India in deciding the case. Per Steineide Refseth, *The Indus River Basin, 1999–2008: An Intellectual History in Hydropolitics,* master's thesis submitted to the Institute of Archaeology, Conservation and History, University of Oslo, Sweden, Spring, 2013, 85, note 365.

92. According to Ahmad, Shaukat Aziz was leaving for an assignment when he gave the green signal for the appointment of Professor Lafitte. He disclosed this fact to the author in the interview that he gave to the author. For details of the date and place of the interview, see supra note 50.

93. Tahir Khalil and Asim Yasin, 'Shujaat Says Aziz was Fully Aware of Lal Masjid Operation', *The News* (20 March 2013).

94. Another example of the lack of co-ordination between different institutions was revealed during the media briefing on the *Kishenganga* case, when the top officials of the Indus Commission and the Ministry of Water and Power advanced conflicting viewpoints. Thus, whereas Additional Commissioner Sheraz Memon observed that India was not involved in blocking Pakistan's water, the minister and the secretary held the opposite view. Memon insisted that the Indian hydroelectric projects on the Western Rivers were not inflicting

loss on water interests of Pakistan. 'Indus Waters Treaty not in Pak Interest: Asif', supra note 90.

95. The federal minister, Khawaja Asif, admitted in a press conference that the previous Pakistani governments committed serious mistakes in protecting water rights as well as showed inability to develop water resources of Pakistan's rivers. Ibid.

96. Ibid.

97. Ibid.

7

Treaty and Climate Change

WHEN THE TREATY WAS SIGNED IN 1960, THE CONCEPT OF CLIMATE change was almost non-existent. At least, very few people talked about it. Consequently, it is hardly surprising that it does not figure in the Treaty at all. However, some fifty-odd years after the conclusion of the Treaty, things are looking entirely different. Climate change, which is held responsible for the recent devastating tsunamis, rapidly melting glaciers, threatening rising sea levels, mega floods, etc., is on top of the international agenda. It has impacted, among others, the South Asian region as well.[1] In this context, some Indian analysts are of the view that the Treaty deserves to be scrapped or rewritten because it was concluded without taking this phenomenon into consideration. To buttress their claim, they invoke article 62, para. 1 of the Vienna Convention on the Law of Treaties, 1969 which lays down that a treaty continues to operate as long as the circumstances, which existed at the time it was concluded and regarding which the parties did not foresee any change occurring, remain unchanged. Furthermore, in case a fundamental change occurs with regard to those circumstances, it could become grounds for terminating or withdrawing from the treaty provided those circumstances constituted an essential basis of the consent of the parties and that the change radically transforms the extent of the obligations of the parties under the treaty. Simply put, they claim that the Treaty was built on the assumption that a certain quantum of water would continue to flow through the Indus Rivers which is in the process of undergoing a shift as a result of climate change. Hence, they contend that the Treaty be scrapped or rewritten. Is this contention justified?

A) IS CLIMATE CHANGE REAL?

To address this issue we need to answer a number of questions. First, what is climate change and is it real? Secondly, if it is real, has it affected the subcontinent, especially the flow of water through the Indus Rivers? If the answer is in the positive, then is India entitled to terminate or withdraw from the Treaty on the basis of article 62, para.1 of the Vienna Convention on the Law of Treaties? To address this question, we begin by looking at the meaning of 'climate change'. It was Svante Arrhenius, a Swedish physical chemist (1859–1927), who was the first one to introduce this concept but he looked at it in terms of global warming. He advanced the view that doubling mixing ratios due to the burning of coal in the United Kingdom would increase the temperature by 5°C.[2]

Though the concept of 'global warming' is *strictu sensu* different from that of 'climate change' ('Global warming' generally refers to increases in surface temperature whereas 'climate change' encompasses not only global warming but also everything else which greenhouse gases will effect) yet the two concepts are, in common parlance, used interchangeably.[3] When we try to define 'climate change' we have two highly authoritative documents to work from. The first one is the United Nations Framework Convention on Climate Change (UNFCCC) 1994 which defines it as 'a change of climate which is attributed directly or indirectly to human activity that alters the composition of the global atmosphere and is in addition to natural climate variability observed over comparable time periods.'[4] The second one is the report of the Intergovernmental Panel on Climate Change (IPCC) which looks at it as 'a change in the state of the climate that can be identified (e.g. using statistical tests) by changes in the mean and/or the variability of its properties, and that persists for an extended period, typically decades or longer. It refers to any change in climate over time, whether due to natural variability or as result of human activity.'[5]

On the face of it, the two definitions differ considerably as the FCCC puts more emphasis on the role of human activity rather than on natural variability in effecting climate change in the atmosphere whereas the IPCC definition seems to apportion equal blame to both factors. However, on close scrutiny we discover that the IPCC's definition is no different from that of the FCCC as it also holds human activity principally responsible for climate change as we see from the 2007 Synthesis Report in the following paragraph:[6]

> Most of the observed increase in global average temperatures since the mid-20th century is very likely due to the observed increase in anthropogenic GHG [greenhouse gas] concentrations. This is an advance since the TAR's [Third Assessment Report] conclusion that 'most of the observed warming over the last 50 years is likely to have been due to increase in the GHG concentrations.' The observed widespread warming of the atmosphere and ocean, together with mass loss, support the conclusion that it is extremely unlikely that global climate change of the past 50 years can be explained without external forcing and very likely that it is not due to known natural causes alone. During this period, the sum of solar and volcanic forces would very likely have produced cooling, not warming. Warming of the climate system has been detected in changes in surface and atmospheric temperatures of the upper several hundred metres of the ocean. The observed pattern of tropospheric warming and stratospheric cooling is very likely due to the combined influences of GHG increases and stratospheric ozone depletion. It is likely that increases in GHG concentrations alone would have caused more warming than observed because volcanic and anthropogenic aerosols have offset some warming that would otherwise have taken place.

The 2007 Synthesis Report also signalled the emergence of a consensus in the global scientific community on human activity as the principal factor of climate change in the world today. The same year, certain members of the scientific community challenged it by putting forward a counter narrative. To understand the nature of this challenge and how it came about, we need to first trace the origins and nature of the IPCC. To begin with, the antecedents of the IPCC go back to the 1970 World Earth Day, the 1971–72 Stockholm Conference, and the 1980 and 1985 Villach Conferences which inspired its creation, though it was formally created by the World Meteorological Organization (WMO) and United Nations Environmental Programme (UNEP) in 1988 in pursuance of the United Nations General Assembly Resolution 43/53. In other words, the IPCC is an intergovernmental UN body which was mandated to prepare a comprehensive review and recommendations with respect to the state of knowledge of the science of climate change, social and economic impact of climate change, and possible response strategies and elements for inclusion in a future international convention on climate.

Today the IPCC's role has expanded as per the *Principles Governing IPCC Work* which states that it is to 'assess on a comprehensive, objective, open and transparent basis the scientific, technical, and socio-economic information relevant to understanding the scientific basis of risk of human-induced climate, its potential, impacts, and options for adaptation and mitigation. IPCC reports should be neutral with respect to policy, although they may need to deal objectively with scientific, technical and socio-economic factors relevant to the application of particular policies.' The IPCC has so far produced five assessment reports which came out in 1990, 1995, 2001, 2007, and 2014. Through its work on climate change it was responsible for creating the UNFCCC in 1994. It was awarded the Nobel Peace Prize in 2007 for its work on climate change.

Much before the emergence of the consensus, a group of European and American scientists got together in Milan in 2003 and founded a body called Non-Governmental International Panel on Climate Change (NIPCC). It became an ongoing affair after a workshop in Vienna held in April 2007, where the number of IPCC detractors swelled and more scientists joined the NIPCC, including some from the southern hemisphere. It was conceived and choreographed by Dr S. Fred Singer, Professor Emeritus of environmental sciences at the University of Virginia in the US. It claims to be an independent body with the avowed objective to break the IPCC monopoly on scientific debate on climate change.

The NIPCC has put out a counter narrative in the shape of a report entitled *Nature, not Human Activity, Rules the Climate*[7] in which it argues that when a nation faces an important decision which risks its economic future or its ecological future (as is the case in the present situation) it is a time-honoured tradition to set up a 'Team B' to examine the same original evidence in case that may reach a different conclusion. It alleges that the IPCC has a hidden agenda to support the FCCC thesis that global warming is due to anthropogenic reasons rather than natural causes and that the way to tackle it is to control the emission of greenhouse gases. Commenting on the various IPCC reports[8] it contends that the 1990 Summary utterly ignored satellite data which according to it, showed no warming; that the 1995 Report was doctored *after* its approval by the scientists in order to convey the impression of a human influence; that the claim made in the 2001 Report that the twentieth century showed 'unusual warming' was based on a hockey-stick graph; and that the 2007 Synthesis Report absolutely devalued the climate contributions from changes in solar activity which are likely to dominate any human influence. Dealing with the most important IPCC claim that 'most of the observed increase in global average temperatures since the mid-twentieth century is *very likely* (defined by the IPCC as between 90 to 95 per cent certain) due to observed increase in anthropogenic greenhouse gas concentrations' (emphasis original), it discards it by contending that a diametrically opposite conclusion is justified, i.e. that natural causes are very likely to be the dominant cause.[9]

The NIPCC is not the only body which challenged the IPCC consensus. There are others who have joined it. In 2007, the year the IPCC consensus emerged, 400 scientists signed a report entitled *Climate Depot Special Report* which was submitted to the US Senate in which they expressed scepticism concerning the role that human activity plays in effecting climate change. The report was updated in 2009 with the number of scientists increasing from 400 to 700. It was further updated in 2010 in which more than 1,000 scientists joined in. It is noteworthy that some of these scientists are high ranking current and former IPCC members. Besides, those who challenged the consensus are twenty times more in numbers than the IPCC scientists who were no more than 52 in 2007. Here are samples of their critique of the IPCC report.[10] For example, the well-known Princeton physicist, Dr Robert Austin, denounced it in these words: 'I view Climategate as science fraud, pure and simple.' Similarly, the IPCC scientist, Eduardo Zorita, publicly demanded that his colleagues Michel Mann and Phil Jones 'should be barred from the IPCC process ... [because] they are not credible', and called the IPCC process insular. Again, a leading UN author, Richard Tol, declared that the IPCC is 'captivated' and demanded the removal of the chairs of the IPCC and its working groups. When invited in 2010 to join the IPCC as the lead author for the next report, instead of accepting the invitation, he demanded its suspension.

How do we explain the onslaught by the NIPCC and others on the IPCC? Are they justified in challenging the consensus worked out by the IPCC? The NIPCC alleges that the main personnel and lead authors of the IPCC owe their appointments to their respective governments; that the IPCC Summaries for Policymakers are subject to approval by member states of the United Nations; that almost all the scientists associated with the IPCC are beholden to various governments which at once fund their research and pay for their IPCC-related activities; and that the IPCC drafting authors are paid for their travel and stay at exotic locations at government expense. The NIPCC sums up its position by claiming that the IPCC agenda is 'to find evidence of a human role in climate change ...; its organisation as a government entity beholden to political agenda ...; and the large and professional rewards that go to scientists and bureaucrats who are willing to bend scientific facts to match those agendas ...'.[11] It further claims that the IPCC had an activist agenda from the very beginning which consisted of justifying control of the emission of greenhouse gases, especially carbon dioxide. It refers to the following IPCC mandate to prove its point:[12] '[Its role] is to assess on a comprehensive, objective, open and transparent basis the latest scientific, technical and socio-economic literature produced worldwide relevant to the understanding of *the risk of human-induced climate change*, its observed and projected impacts and options for adaptation and mitigation.' In its view, this explains why the IPCC scientific reports are exclusively focused on evidence that points towards human-induced climate change.

Is the criticism against the IPCC justified? And, if not, what are the motives behind this vicious criticism? We do not deem it necessary to comment on the charges and counter charges that the two groups have made against each other. The critical factor in resolving this controversy is to ask whether or not the IPCC consensus is scientifically justified. The answer appears to be in the affirmative because the evidence in favour of the human activity influencing climate change is so overwhelming that there is no escape from the IPCC consensus. As for the motive behind the reaction to the IPCC, it is an attempt by fossil fuel companies to influence the debate in favour of their viewpoint. For example, there are credible reports that the NIPCC has numerous close links to Exxon-Mobil which is known for being an inveterate opponent of any legislation relating to climate change and is notorious for funding projects which deny global warming. It has been claimed that Exxon-Mobil has, since 1998, the year it launched a campaign to oppose the Kyoto Treaty, funded the Heartland Institute which finances the NIPCC activities. Besides, it is also claimed that Exxon-Mobil has generously funded S. Fred Singer and Craig D. Idso, the two key authors and principal leaders of the NIPCC.[13]

Irrespective of the controversy, what the nay sayers while challenging the consensus actually attacked was the question of what is causing climate change, i.e. whether it was human activity or nature. In their opinion, the change could be attributed at once to human activity or nature. They considered it 'voodoo science' to attribute it to human activity or nature because, in their opinion, it is very hard to establish whether it is due to one factor or the other. It signifies that the controversy centred on the cause of climate change rather than the question of whether or not climate change is real. In fact, there is a complete unanimity of views in the scientific community that climate change is for real. The fifth IPCC Assessment Synthesis Report issued in 2014 has confirmed this phenomenon in the following statement:[14]

Human influence on the climate system is clear, and recent anthropogenic emissions of greenhouse gases are the highest in history. Recent climate changes have had widespread impacts on human and natural systems ....

Warming of the climate system is unequivocal, and since the 1950s, many of the observed changes are unprecedented over decades to millennia. The atmosphere and ocean have warmed, the amounts of snow and ice have diminished, and sea level has risen ....

Each of the last three decades has been successively warmer at the Earth's surface than any preceding decade since 1850. The period from 1983 to 2012 was *very likely* the warmest 30-year period of the last 800 years in the Northern Hemisphere, where such assessment is possible *(high confidence)*[15] and *likely* the warmest 30-year period of the last 1400 years *(medium confidence)* ... (Figure 1.1).

Ocean warming dominates the increase in energy stored in the climate system, accounting for more than 90 per cent of the energy accumulated between 1971 and 2010 *(high confidence)* with only about 1 per cent stored in the atmosphere (Figure 1.2). On a global scale, the ocean warming is largest near the surface, and the upper 75 m warmed by 0.11 [0.09 to 0.13] °C per decade over the period 1971 to 2010. It is *virtually certain* that the upper ocean (0–700 m) warmed from 1971 to 2010, and it *likely* warmed between the 1870s and 1971. It is *likely* that the ocean warmed from 700 m to 2000 m from 1957 to 2009 and from 3000 m to the bottom for the period 1992 to 2005 ... (Figure 1.2).

Over the last two decades, the Greenland and Antarctic ice sheets have been losing mass *(high confidence)*. Glaciers have continued to shrink almost worldwide *(high confidence)*. Northern Hemisphere spring snow cover has continued to decrease in extent *(high confidence)*. There is *high confidence* that there are strong regional differences in the trend in Antarctic sea ice extent, with a *very likely* increase in total extent ... (Figure 1.3).

Over the period 1901–2010, global mean sea level rose by 0.19 [0.17 to 0.21] m (Figure 1.1). The rate of sea-level rise since the mid-19th century has been larger than the mean rate during the previous two millennia (*high confidence*) ....

The observed reduction in surface warming trend over the period 1998 to 2012 as compared to the period 1951 to 2012, is due in roughly equal measure to a reduced trend in radiative forcing and a cooling contribution from natural internal variability, which includes a possible redistribution of heat within the ocean (*medium confidence*) (Figure 1.4).

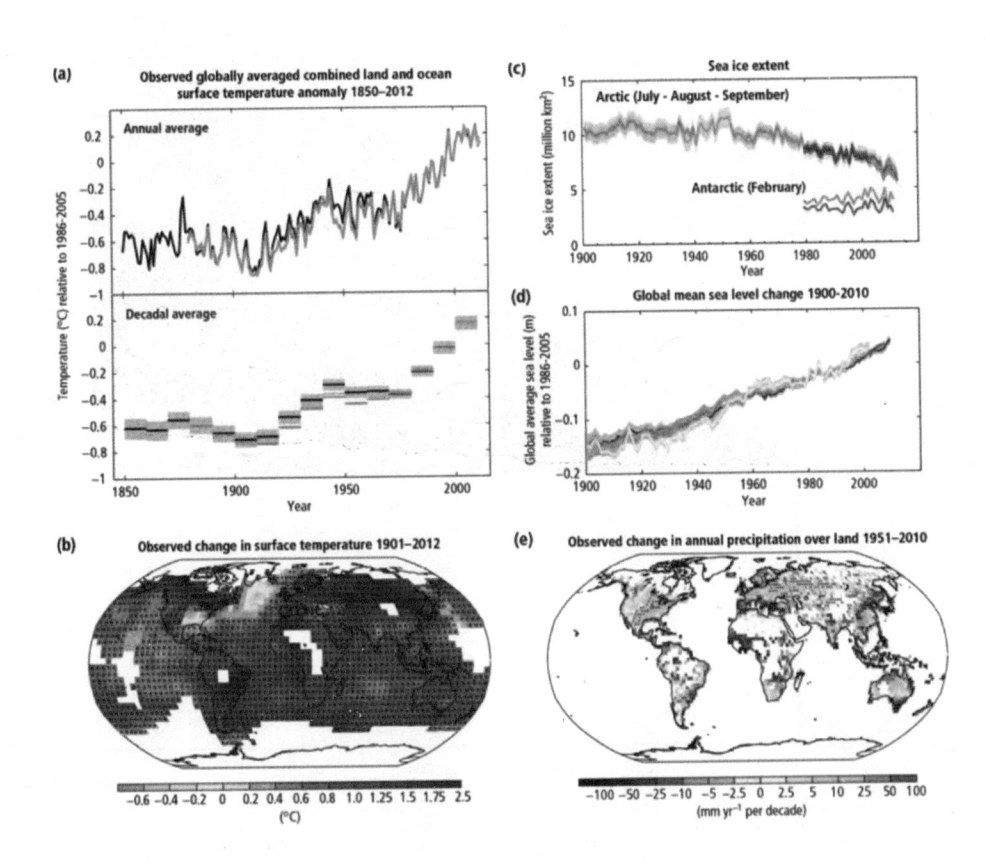

Graph 7.1: Rise in Global Mean Sea Level (1901–2010)

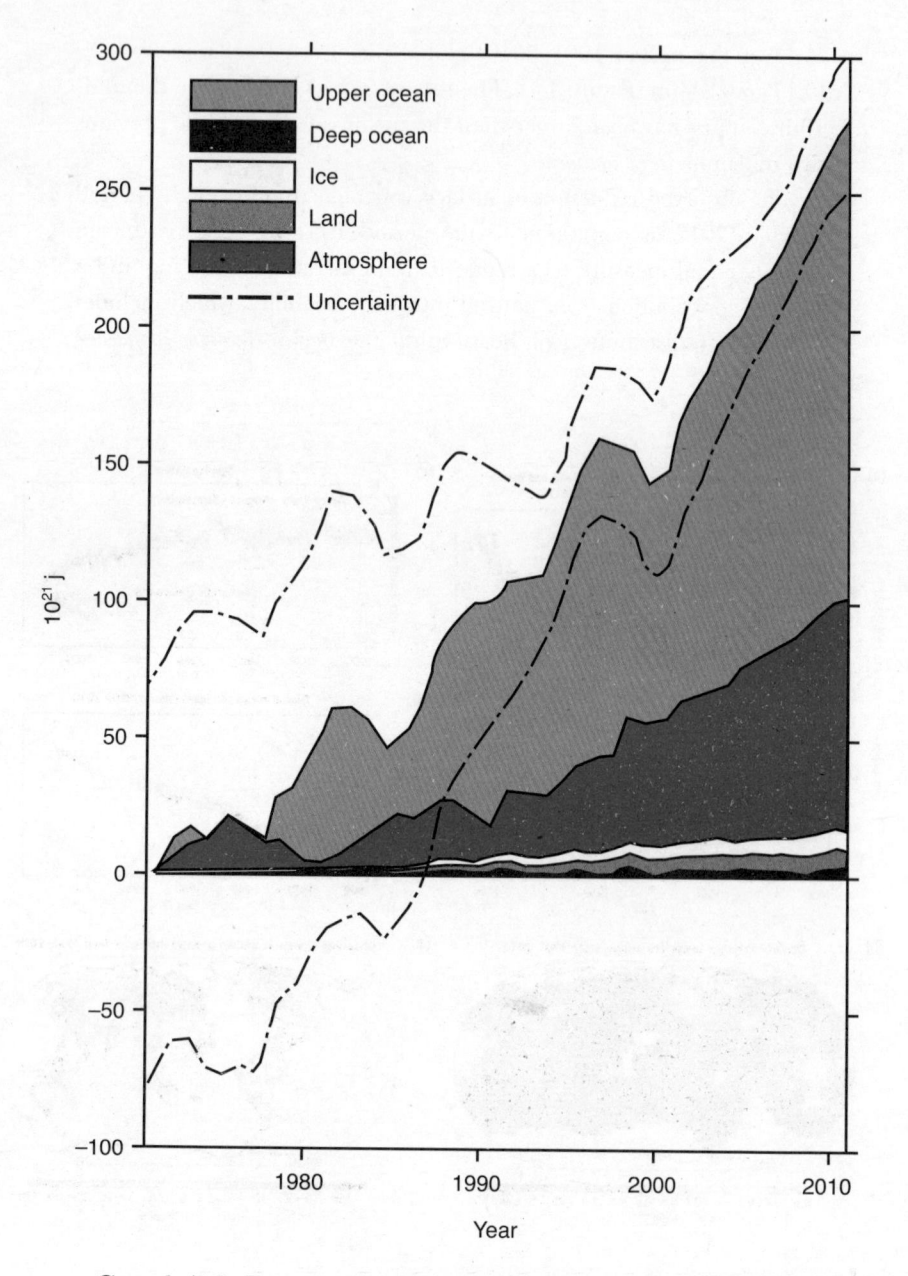

**Graph 7.2: Energy Accumulation within the Earth's
Climate System**

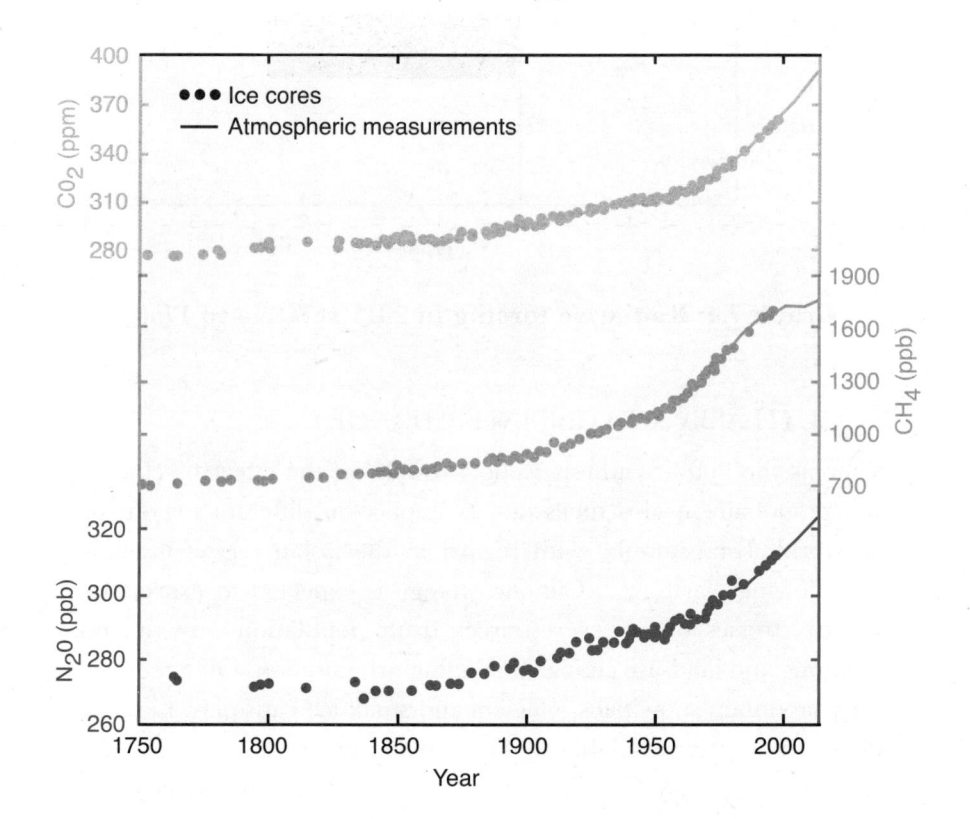

Graph 7.3: Globally averaged greenhouse gas concentrations

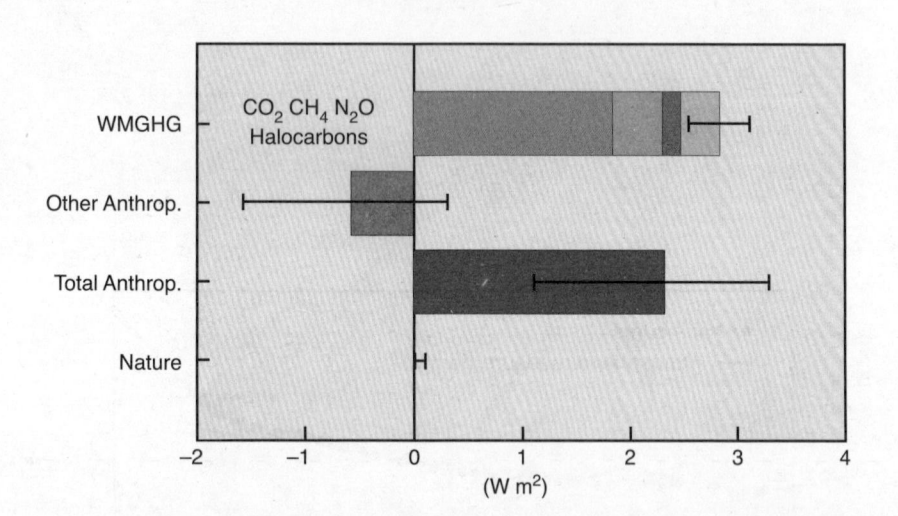

Graph 7.4: Radiative forcing in 2011 relative to 1750

B) SOUTH ASIA AND CLIMATE CHANGE

Whereas the 2007 Synthesis Report dwells on the effect of climate change globally, it also spells out its impact on different regions of the world. For example, with regard to the Asian region it made the following warning:[16] 'Climate change is expected to exacerbate current stresses on water resources from population growth and economic and land-use change, including urbanisation. On a regional scale, mountain snow pack, glaciers and small ice caps play a crucial role in fresh water availability. Widespread mass losses from glaciers and reductions in snow cover over recent decades are projected to accelerate throughout the 21st century, reducing water availability, hydropower potential, and charging seasonability of flows in regions supplied by melt water from major mountain ranges (e.g. Hindu Kush, Himalaya, Andes), where more than one-sixth of the world population currently lives.' As far as South Asia is concerned, the IPCC Working Group II made the following prediction:[17] 'Glaciers in the Himalaya are receding faster than in any other part of the world and, if the present rate continues, the likelihood of them disappearing by the year 2035 and perhaps sooner is very high if the Earth keeps warming at the current rate. [Their] total area will likely shrink from the present 500,000 to 100,000 km by the year 2035 .... The current trends of glacial melts suggest that Ganga, Indus, Brahmaputra and other rivers that criss-cross the northern Indian plain could likely become seasonal rivers in the near future as a consequence of climate change and could likely affect the economies in the region.' It backed its sensational claim with the following data:[18]

Glacier	Period	Retreat of Snow (Metre)	Average Retreat (Metre/Year)
Triloknath (H Pradesh)	1969 to 1995	400	15.4
Pindari (Uttaranchal)	1845 to 1966	2,840	135.2
Milam (Uttaranchal)	1909 to 1984	990	13.2
Ponting (Uttaranchal)	1906 to 1957	262	5.1
Chota Shigri (Uttaranchal)	1986 to 1995	60	6.7
Bara Shigri (H Pradesh)	1977 to 1995	650	36.1
Gangotri (Uttaranchal)	1977 to 1990	364	28.0
Gangotri (Uttaranchal)	1985 to 2001	368	23.0
Zemu (Sikkim)	1977 to 1984	194	27.7

The prediction about South Asia triggered a huge controversy among the global scientific community which challenged the IPCC's findings regarding the rate of recession and the date of disappearance of the Himalayan glaciers. For example, one study subjected the IPCC statement to severe criticism, firstly by rubbishing the idea that Himalayan glaciers were receding faster than those located in other parts of the world stating that the IPCC did not present any evidence in support of this contention. Next, it attacked the IPCC statement regarding the timetable of the disappearance of glaciers on the grounds that, 'a rough calculation would indicate that melt rates on the order of 20 times the current observed melt rate in the Himalayas would be required to remove all glaciers by 2035. The 2007 IPCC WG1 authors of 'Changes in Glaciers and Ice Caps' noted that 'the glaciers of High-Mountain Asia have generally shrunk at varying rates and that several Karakoram glaciers are reported to have advanced and/or thickened .... The cumulative glacier mass balance data from the high mountains of Asia show values that are, in fact, approximately mid-way between the global extremes.'[19]

Responding to the criticism, the IPCC issued a statement in which it accepted its mistake by observing that the paragraph in question was based on, 'poorly substantiated estimates of rate of recession and date for the disappearance of Himalayan glaciers'; and that, '[i]n drafting the paragraph in question, the clear and well-established standards of evidence required by IPCC procedures, were not applied properly.'[20] Jean-Pascal Van Ypersele, vice-chair of the IPCC added that the mistake did nothing to undermine the large body of evidence which demonstrated that climate was warming and that human activity was largely responsible for it. The World Wildlife Fund (WWF) on whose campaigning report the IPCC had based its findings[21] also issued a statement in which it clarified that, '[the original statement] was issued in good faith but it is now clear that this was erroneous and should be disregarded.'[22] However, it went on to add that it still supported the claim that widespread mass losses in the Himalayan glaciers are likely to continue in the twenty-first century. The IPCC also observed in the statement referred to above that the foregoing claim was, 'robust, appropriate and entirely consistent with the underlying science and the broader IPCC assessment.'[23]

A recent study based on research in the Himalayas by the Dutch scientist, Dr Walter Immerzeel,[24] shows that water levels in the rivers, instead of dropping, will increase during the twenty-first century. It concedes that the size of the glaciers in the Indus (and the Ganges) Basins will decrease but monsoon rains will increase resulting in greater discharge of water which will signal good news for water and food security for Pakistan, India, and Bangladesh. Further dwelling on the subject, the study specifies that the glacial melting will touch a peak around 2070 but subsequently the glacial discharge will begin to drop. However, there will simultaneously be an increase in precipitation which will compensate the drop in the glacier melt, leading to increase in the annual discharge from the river basins. The study emphasizes caution in treating the foregoing results.[25]

C) CLIMATE CHANGE AND INDUS RIVERS

This begs the question whether climate change has affected the flow of water in the Indus Rivers. As we try to address this question, we note that the source of water in the Western Rivers is the snow and ice melt from the Himalayas, the Karakoram, and the Hindu Kush (HKH) and the monsoon rains in the catchment areas of these rivers (we would like to clarify that unfortunately our study is restricted to the Western Rivers and does not cover the Eastern Rivers as India is a no-go area for Pakistani researchers. However, the data on the flow of water in the Eastern Rivers is not likely to be very different from the Western Rivers). The HKH (Himalayas, Karakoram, and Hindukush) is often called the 'water towers' of the Asian continent and its glaciers are the world's largest body of ice outside the polar ice caps. These glaciers release melt water which feeds ten large river systems across Asia including the Indus, Ganges, and Brahmaputra. In regards to the amount of water that the Himalayan glaciers contain the 2007 report of the Working Group II of the IPCC stated:[26] 'Himalayan glaciers cover about three million hectares or about 17 per cent of the mountain area compared to 2.2 per cent in the Swiss Alps. They form the largest body outside the polar caps and are the source of water for the innumerable rivers that flow across the Indo-Gangetic plains. Himalayan glacial snowfields store about 12,000 km of freshwater. About 15,000 Himalayan glaciers form a unique reservoir which supports perennial rivers such as the Indus, Ganga and Brahmaputra which, in turn, are the lifeline of millions of people in South Asian countries.'

According to a recent study by the ICIMOD (International Centre for Integrated Mountain Development), the Indus is the most glaciated of the region's major basins. It has 18,495 glaciers covering 21,193 km and contains about 2,696 km of ice which constitutes 44 per cent of the total ice reserves in the entire HKH region. Snow and glacial melt, for example, contribute more than 50 per cent of the total outflow of the Indus which constitutes a critical source of water during the summer when melt water comprises 70 per cent of the river's flow.[27] The melt water helps avert or alleviate a possible calamitous drought in case of feeble or failed monsoons. According to a study conducted by Dr Immerzeel et al., snow and glacier melt are extremely important in the Indus Basin as on the basis of total runoff generated below 2000m, the part of snow and ice melt is 151 per cent in the Indus as compared to 27 per cent in the Brahmaputra and 10 per cent in the Ganges. The study concludes that the glacier melt has much greater contribution to make to the Indus than other rivers of South Asia.[28] As for the monsoons, those from the southwest supply 70 to 90 per cent of water to the region, though there is considerable spatial and temporal variation in their activity. This is demonstrated by the fact that as opposed to most of Pakistan and North India, Northeast India and North Bangladesh receive plentiful rain.

Concerning whether or not the melting of glaciers in the HKH and the monsoon rains in the catchment areas of the Western Rivers have affected the flow of water in the Indus Rivers, the data collected by Pakistan's Water and Power Development Authority (WAPDA) over the years suggests a negative answer as shown below.[29]

Western Rivers Inflow at Rim Stations

Average (MAF)	Indus at Kalabagh	Jhelum at Mangla	Chenab at Merala	Total
Pre-Independence (1922–47)	89.25	22.55	23.47	135.27
Pre-Treaty (1947–61)	94.26	24.24	29.18	147.68
Pre-Mangla (1961–67)	87.41	21.54	24.92	133.88
Pre-Tarbela (1967–76)	83.57	21.31	23.72	128.60
Post-Tarbela (1976–2010)	89.69	22.59	26.09	138.37
Long-term (1922–2010)	89.51	22.64	25.51	137.66
Max	120.09	32.74	35.13	186.79
Min	63.19	11.89	17.85	97.16

(Source: Pakistan Indus Waters Commission)

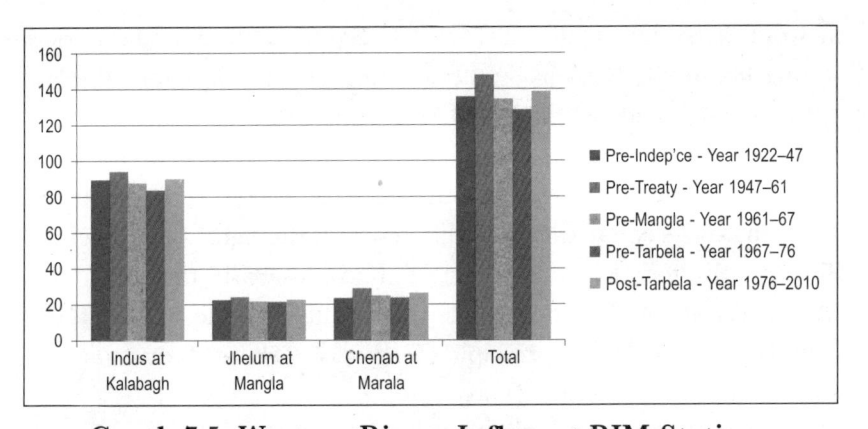

Graph 7.5: Western Rivers Inflow at RIM-Stations (Average MAF)

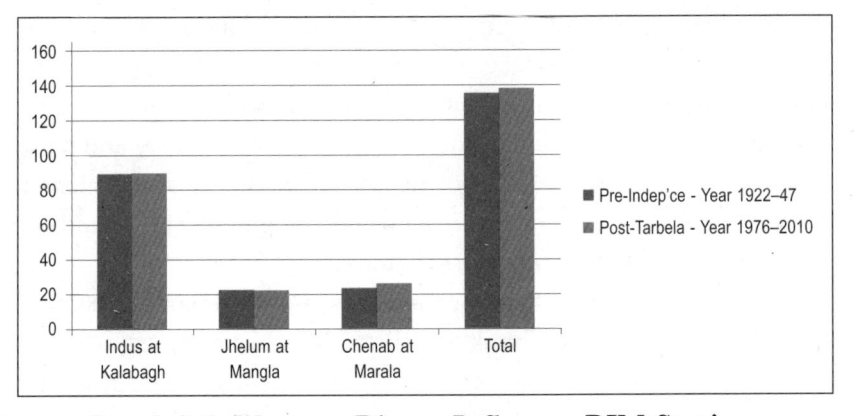

Graph 7.6: Western Rivers Inflow at RIM-Stations Pre-Independence vs. Post Tarbela (Average MAF)

The table above shows that the average flow of water in the Western Rivers was more or less constant during the period 1922–2010. This is attested by the fact that it stood at 135.27, 133.88, and 138.37 MAF during 1922–47, 1961–67, and 1976–2010 respectively. The only exception occurred during 1947–61 and 1967–76 when the fluctuation was quite noticeable as the average flow stood at 147.68 and 128.60 MAF respectively. The average flow stood at 138.37 MAF during 1976–2010 when climate change was supposedly at work (according to the 2014 IPCC Synthesis Report the period from 1982 to 2012 was likely the warmest 30-year period of the last 1400 years);[30] and at 135.27 MAF during 1922–61 when nobody talked of climate change (See figure 2). The closeness of the data shows that climate change has not affected the flow of water in the Western Rivers so far since the difference in the figures is negligible. Since the Eastern Rivers are part of the Indus Basin the conclusion *mutatis mutandis* applies to them as well. In this scenario, we feel there is no case to invoke the principle of *rebus sic stantibus* to question the continuing validity of the Treaty.

However, it does not signify that the situation with regard to flow of water in the Indus Rivers would remain frozen in time. In fact, in light of the 2014 IPCC Synthesis Report which foresees an acceleration of the phenomenon of climate change in the future, unless the international community takes remedial measures, the situation may undergo considerable transformation in the not too distant future. In such an eventuality, the Indians may try to scuttle the Treaty and have it replaced with a new one by invoking article 62, para. 1 of the Vienna Convention on the Law of Treaties. In our judgment, this would be an extremely unwise step because it took the two countries about a decade of serious and difficult negotiations and third party assistance in the shape of the World Bank to strike a deal. Next time, it may not be possible to achieve an agreement, especially since Pakistan would not be prepared, under any circumstances, to forego its rights over the Western Rivers. Therefore, the quest for a new treaty to replace the existing one would be a recipe for disaster as this could lead to political strife between the two countries more dangerous than the Kashmir dispute. The best course of action for them would be to continue with the Treaty and conclude an 'Indus II' through which they should take care of the lacunae in the Treaty that have come to the fore, particularly the issue of fluctuation in the flow of water resulting from climate change.[31]

However, climate change may not be the only issue between the two countries. There are now unfortunately voices in India, both at the official and unofficial levels, which favour a revision or termination of the Treaty. We propose to deal with this issue in the next chapter.

REFERENCES AND NOTES

1. According to the Global Climate Risk Index (2015), India and Pakistan in 2013 were respectively the third and the sixth most affected countries of the world in terms of climate change, <http://germanwatch.org/en/download/10333. pdf>.

2. M. H. Bukhari and Ejaz Ahmad Sayal, 'Emerging Climate Changes and Water Resource Situation in Pakistan', *Pakistan Vision*, vol. 12, no. 2, 240–1. There are different versions as to when the term 'climate change' was used for the first time. According to one version, it dates back to 1966 when an article appeared with the title 'Climate Change'. See 'NASA–What is in a Name? Global Warming v. Climate Change', <http://www.nasa.govt/ topics/earth.../climate_by_any_other_name.html>. According to another, its first use took place in a 1975 article by geochemist Wallace Broecker of the Columbia University's Lamont-Doherty Geological Observatory whose title read as follows: 'Climate Change: Are we on the Brink of a Pronounced Global Warming.' See 'Origin of the Phrase "Climate Change"? Blame it on Bush', *Gore Lied*, <http://algorelied.com/?p=3760>.

3. According to biologist Dave Miller, who worked in the US National Oceanic and Atmospheric Administration, it was George W. Bush who was responsible for the use of the expression 'climate change' rather than 'global warming' because the latter was politically incorrect nonsense. He explained this point in a letter that he addressed to the editor of *The Oregonian*, an online newspaper that comes out of the state of Oregon. According to him, it was the Republican pollster Frank Luntz who sold the phrase 'climate change' to President Bush on the grounds that it was less frightening than the expression 'global warming'. In his opinion, 'global warming' has catastrophic connotations whereas 'climate change' suggests a more controllable challenge. He maintains that Bush agreed with this suggestion. In consequence, the Republican political appointees at the National Oceanic and Atmospheric Administration forced scientists to always use 'climate change' rather than the accurate but alarming 'global warming.' 'NASA–What is in a Name?', supra note 2.

4. 'Climate Change' is generally defined as 'a change in the statistical properties of the climate system when considered over long periods of time, regardless of cause.' 'Glossary–Climate Change', Education Centre–Arctic Climatology and Meteorology, NSIDC National Snow and Ice Data Centre, Glossary, *IPCC: TAR WG1, 2001*.

5. *IPCC Synthesis Report 2007, Summary for Policymakers*, 30, <http://www.ch/pdf/assessment-report/ar4/syr/ar4_syr.pdf>.

6. *Climate Change 2007: Synthesis Report (adopted section by section at the IPCC Plenary XXII (Valencia*, Spain, 12–17 November 2007), 39.

7. S. Fred Singer (ed.), *Nature, not Human Activity, Rules the Climate: Summary for Policymakers of the Report of the Nongovernmental International Panel on Climate Change* (Chicago, IL: The Heartland Institute, 2008), iv.

8. Ibid. iii.

9. Ibid. iv.

10. See *Climate Depot Special Report: More than 1000 Scientists Dissent over Man-Made Global Warming Claims*. The report is available online at <http://www.climatedepot.com/a/9035/SPECIAL-REPORT-More-Than-1000-International-Scientists-Dissent-Over-ManMade-Global-Warming-Claims-Challenge-UN-IPCC-Gore>.

11. *Nature, not Human Activity, Rules the Climate*, supra note 7, vi.

12. Ibid. iv. Emphasis added.

13. Many news organizations like the Pulitzer-Prize-winning *Inside Climate News* and the *Los Angeles Times* recently unearthed a trove of documents which showed that Exxon-Mobil knew way back in the 1970s that carbon dioxide from the burning of oil and gas could have dire consequences for the earth. Instead of heeding the warning it funded the deniers of climate change. For example, from 1998 to 2005 it contributed about sixteen million dollars to various organizations to mislead the world in the matter. It came clean in a way in 2007 when it publicly acknowledged that CO2 was responsible for global warming and promised to desist from funding climate change deniers. It has now made an unqualified confession that climate change is real and caused by humans. Timothy Egan, 'Exxon Mobil and the GOP', *International New York Times* (9 November 2015).

14. *IPCC Fifth Assessment Synthesis Report 2014*, 5–8. For the figures see ibid. 83–4, and 86–7.

15. The Report indicates the various terms and their weight as follows: *virtually certain*: 99–100 per cent; *probability, extremely likely*: 95–100 per cent; *very likely*: 90–100 per cent; *likely:* 66–100 per cent. It also indicates that '[u]nless otherwise indicated, findings assigned a likelihood term are associated with high or *very high confidence*'. Ibid. 4.

16. *IPCC Synthesis Report 2007*, supra note 5, 49.

17. *Climate Change 2007: Working Group II: Impacts, Adaptation and Vulnerability*, chapter 10.6.2, 493. The Working Group II report carried the projection on the Himalayan glaciers but the IPCC final Summary for Policymakers dropped it.

18. Ibid. 494.

19. *Changing Glaciers and Hydrology in Asia: Addressing Vulnerabilities to Glacier Melt Impacts*, USAID Publication: 20 November 2010, 21.

20. *IPCC Statement on Melting of the Himalayan Glaciers,* 20 June 2010, Geneva, <www.ipcc.ch/pdf/presentations/himalaya-statement-20january2010.pdf>.

21. A USAID-funded study claimed that the original source material of the WWF pointed to the year 2350 rather than 2035 for the disappearance of the Himalayan Glaciers (*Changing Glaciers and Hydrology in Asia,* supra note 19, 17). It makes sense because scientists believe that it would take about four centuries before Himalayan glaciers completely disappear. For example, the Space Application Centre in its study on retreating Himalayan glaciers has this to say in the matter: 'As per the finding, average pace of retreat is 3.75 per cent a year so at this pace, it would take 400 years to melt all the glaciers in the Himalayan region,' 'Himalayan Glacier Controversy', *The Hindustan Times* (12 February 2013), <www.hindustimes.com/india-news/Ahmedabad/Himalayan-glacier-controversy/Article1>.

22. The WWF based its claim on the remarks made in 1999 by the Indian glaciologist Syed Hasnain, vice-chancellor of the Jawaharlal Nehru University, Delhi, and the chair of the International Commission on Snow and Ice Working Group on Himalayan glaciology to journalists of the magazines *New Scientist* and *Down to Earth.* The IPCC took its data from the non-peer reviewed material of the 2005 WWF report. Defending himself, Hasnain stated that it was not proper for the IPCC to include references from popular magazines or newspapers. He also claimed that he was misquoted in the *New Scientist* article and that he had said that only a subset of the Himalayan glaciers could vanish ('Claims Himalayan Glaciers could Melt by 2035 were False, says UN Scientist', *The Guardian* (20 January 2010); *Time (Science and Space),* <http://www.time.com/health/article/0,8599,1955405,00.html>). However, the *New Scientist* stood by its story. Murai Lal, who was the coordinating lead author of the IPCC Report's chapter on Asia, talking to David Rose of the *Sunday Mail,* acknowledged that there were no solid data to support the claim that Himalayan glaciers could dry up by 2035 but he thought that if he could highlight it, 'it will impact policy makers and politicians and encourage them to take some concrete action.' *Science News,* 02/23/13 issue, <https://www.sciencenews.org/blog/science-public/ipccs-himalayan-glacier-mistake-not-accident>.

23. *IPCC Statement,* supra note 20.

24. Dr Immerzeel is affiliated with the Utrecht University, Netherlands, and is a visiting scientist at the ICIMOD, Nepal.

25. 'Scientists Predict Warmer and Wetter Himalayas' (5 August 2013), <http://www.icimod.org/pq=11326>.

26. *Climate Change 2007, Working Group II,* 493, supra note 17.

27. *Connecting the Drops: An Indus Basin Roadmap for Cross-Border Water Research, Data Sharing, and Policy Coordination,* Stimson/SDPI, 2013, 21.

28. *Changing Glaciers and Hydrology in Asia,* supra note 19, 24–5.

29. For a detailed chart of the flow of water in the Western Rivers, see Appendix 4.

30. *IPCC Assessment Synthesis Report*, 2014, 5.

31. Realizing the absence of a provision on climate change in the Treaty, the participants in the Track II dialogue between Pakistan and India have recommended, among other things, a joint coordinated monitoring of the HKH glaciers and joint scientific studies of the patterns of their change. Shafqat Kakakhel, 'Indus Waters Treaty: Negotiation, Implementation, New Challenges, and Future Prospects', *Criterion Quarterly* (10 May 2014), vol. 9, no. 2, 44–60.

8

Treaty in Trouble

UNTIL QUITE RECENTLY THE INTERNATIONAL COMMUNITY, INCLUDING India,[1] was hailing the Treaty as a great success story on the grounds that no war had taken place between Pakistan and India over the water issue since the two countries signed it and that even when wars broke out (for reasons other than water), in 1965 and 1971, the two countries not only scrupulously observed the Treaty, but also did not pick holes in it.[2] However, a sea change in the situation appears to have taken place in recent years. Pakistan now occasionally accuses India of stealing its waters.[3] According to the Indian think tank, the Strategic Foresight Group, every proposal that Pakistanis have made since 1999 in track II diplomacy has 'either directly or indirectly refer[red] to water as a core issue'.[4] Pakistan's former chief of army staff, General Ashfaq Parvez Kayani, seemed to agree with this assessment as, during a visit to the US in 2010, he observed that water has replaced Kashmir as the core issue between Pakistan and India.[5] The changed atmosphere is perhaps best reflected in the fact that the two countries did not bother to celebrate the fiftieth anniversary of the signing of the Treaty which fell on 19 September 2010.[6] Nor did the Bank show much enthusiasm for the occasion if one were to believe one of its former legal advisors who claimed to have made a suggestion on these lines without success.[7] More disconcerting is the fact that voices have been raised in recent years in India, both at the official and unofficial levels, against the Treaty. This development is of particular concern to Pakistan because it essentially impacts the Western Rivers which, under the terms of the Treaty, almost exclusively belong to it. Why are the Indians opposed to it?

Before dealing with the grounds for antagonism, it is necessary to point out that various Indian critics approach the issue differently. They can be divided into two schools of thought. One school preaches that the Treaty is flawed but can be salvaged through changes in the existing text. The second school believes that the Treaty is beyond redemption and hence needs to be jettisoned and replaced with a new treaty. The late B. G. Verghese, former editor of the *Hindustan Times* and water expert at the Centre for Policy Research in Delhi, represents the first school.[8] Instead of discarding the Treaty, he favours[9] constructing Indus II on the foundations of Indus I. Starting with the contention that there is 'a possibly large untapped potential in the upper catchments of the three western rivers that are allocated to Pakistan but are under Indian control', he advocates that it should be fully surveyed and harnessed by the two countries through 'joint investment, construction, management and control'. He feels that this is absolutely necessary if Pakistan is interested in additional storage, flood control, and hydroelectric power. In his opinion, it will also render unnecessary what he terms as grotesque schemes like the Skardu Dam which he contends will spell doom for a proud civilization. Besides, he holds that India has a limited entitlement to the Western Rivers that Pakistan cannot deny nor can it freeze all further development. To achieve this objective, he deems it absolutely necessary to revise the Treaty which, he argues, article XII provides.[10] Moreover, the revision he conceives is based on building Indus II 'on the foundations of and without prejudice to the present Indus-I ...'.[11]

The second school is represented by Ramaswamy Iyer, the former water resources secretary of India who is associated with Delhi's Centre for Policy Research and a leading writer on water issues. He disagrees with the Verghese approach on the grounds that the Treaty is not a constructive, water-sharing treaty which could have been built upon and taken further by concluding Indus II, but, 'a negative, *partitioning* treaty, a coda to the partitioning of the land'[12] on the basis of which, he holds, neither cooperation can be built nor joint projects undertaken. He does not put much in store on article XII which Verghese alludes to because, in his opinion, its reference to modifications to the Treaty merely signifies that the Treaty can be modified by another treaty. He interprets it to mean that if a change is desirable, a new treaty is to be negotiated. This is no big deal for him because he thinks that even in the absence of article XII, a new treaty could have been negotiated by the parties. Nor does he accept Verghese's reference to article VII on 'future cooperation' on the grounds that though it refers to the 'common interest in the optimum development of the rivers'. The allusion, in his opinion, is unmistakably towards cooperation for the establishment of hydrological and metrological stations and sharing of costs of drainage works, and not to agriculture or electricity related joint projects. Therefore, he does not attach much importance to article VII, believing it encompasses a very narrow concept of cooperation. He concludes:[13] 'If we want a new relationship between the two countries on the Indus a totally new treaty will have to be negotiated; it cannot grow out of the existing treaty; and questions will immediately arise about the coexistence of two divergent treaties.'

Irrespective of the difference of approach by the two schools of thought, the fact remains that the Indians want to be rid of the provisions of the Treaty which are onerous for India. This debate is essentially confined to India (a sprinkling of foreign analysts have joined it too)[14] as Pakistanis have remained almost utterly aloof from it. In India the issue is mostly confined to non-governmental or unofficial circles. The government in Delhi has rarely made a statement against the Treaty (the one occasion it did was the terrorist attack on the Indian parliament in 2001).

Consequently, one may argue that we are perhaps according undue importance to the issue. Proceeding further, one may even argue that we should not bother with this one-sided debate and instead wait for the day when the government in Delhi takes up the matter with its counterpart in Islamabad. There may be some merit to this argument. However, we find that we should debate and discuss this issue for two reasons. First, because some of those involved in it are important individuals who shape public opinion and have clout with the government in Delhi.[15] Secondly, it is important to remember that many a time it is the unofficial or non-governmental circles who initiate a debate on a given matter which is subsequently taken over by their government. It is a kind of East India Co. syndrome (the Company became involved in India through commerce and trade and after it had conquered territory the government in London took over) which a country can only ignore at its own peril. We therefore propose to take up the issue in the following pages.

A) TERRORISM AS GROUNDS FOR TERMINATION

As for the grounds of termination or revision of the Treaty, the Indian critics advance several of them. Of these, the first one relates to terrorism which the Indian government invoked when the Lashkar-e-Taiba (LeT) attacked the Indian parliament in 2001. Recalling its high commissioner, Vijay Nambiar, from Islamabad, it stated that this was the first in a series of punitive measures that it had contemplated to take against Pakistan and that these could, among other things, include termination of the Treaty. Later on, the Indian foreign secretary and former high commissioner to Pakistan, S. K. Singh, while maintaining that Prime Minister Vajpayee was opposed to the idea of hot pursuit of militants inside Pakistan, added: 'He has other options in mind. For instance, ending the Indus Valley Water Treaty and starving Sindh and Punjab.'[16] Subsequently, Vajpayee himself repeated the same threat during a visit to Jammu and Kashmir when, addressing the state functionaries and political leaders, he told them that his government was seriously considering scrapping the Treaty.[17]

Vajpayee was not alone in threatening Pakistan, others followed him. A union minister issued a statement in which he said: 'When Pakistan cannot honour the Simla agreement and the Lahore Declaration, then why should we honour the Indus Waters Treaty.'[18]

Similarly, a former Indian high commissioner to Pakistan warned: 'Should we not consider measures to deprive Pakistanis of the water they need to quench their thirst and grow their crops? Should we not seriously consider whether it is necessary for us to adhere to the provisions of the Indus Waters Treaty?'[19] Subsequently, though much later, the Indian Institute for Defence Studies and Analysis (IDSA), a Delhi-based think tank, in its 2010 report entitled 'Water Security for India: The External Dynamics' advocated rescinding the Treaty in these words:[20]

> Use water as a political leverage and even consider unilateral abrogation of the treaty by applying the principle of 'state responsibility' as Pakistan continues to support terrorism .... In the draft provisions 'Responsibility of States for Internationally Wrongful Acts ... Art 22 of Chapter V states 'Countermeasures in respect of an internationally wrongful act'. Thus, in case a party to a treaty fails to abide by its international obligation, the other party can through the interpretation of Article 22 take possible recourse to several actions. India can consider the abrogation of the treaty so long as it is proportionate to infringement by the other side. It is well established that Pakistan aids and abets terrorist actions from its soil. India should quantify the damage it has sustained over the decades because of Pakistani support to terrorism and seek as a first step suitable compensation from Pakistan. Given that Pakistan will not comply with this, India can cite this as a reason and threaten to walk out of various bilateral agreements including the IWT.

Before dealing with the question whether or not India can terminate the Treaty on the grounds of what it terms as 'cross-border terrorism', committed by a group like the LeT or in Jammu and Kashmir, a word about the Indian threat to starve the people of Pakistan is in order. International law prohibits the use of water as a weapon. This is attested by article 54 (1 and 2) of the 1977 Additional Protocol I and article 14 of the Additional Protocol II to the Geneva Conventions of 1949. They forbid starvation of civilians as a method of warfare and prohibit attack, destruction, removal, or rendering useless objects indispensable to the survival of the civilian population such as agricultural areas for production of foodstuff, drinking water installations, supplies, and irrigation works in order to starve out civilians. International law also regards starving civilians as a war crime which is attested by article 8(2) (b) (XXV) of the 1998 Statute of the International Criminal Court (ICC). In light of these observations, the threat by Indian officials to deny water to the people of Pakistan is not only a violation of international law but also highly irresponsible.[21]

Taking up the question of terrorism, militant activity has more or less died down in Jammu and Kashmir following the signature of the Islamabad Declaration by Pakistan and India in 2004. As for India proper, after the Mumbai incident of 2008, no violent incident had taken place since. Things appeared to be peaceful until 27 July 2015, when terrorists attacked targets in Gurdaspur and on 2 January 2016, the Pathankot airbase. With this backdrop of recurrence of violence, we propose to address the issue here. We begin by examining what international law prescribes in the matter. We note that international law forbids aggression by one state against another. This comes out clearly from article 2, para. 4 of the UN Charter which unambiguously directs all states to, 'refrain in their international relations from the threat or use of force against the territorial integrity or political independence of any state, or in any other manner inconsistent with the Purposes of the United Nations.' In case a state commits aggression against another state, the latter, under article 51 of the UN Charter, enjoys the right self-defence.

It is clear that a state enjoys the right of self-defence because the territory which comes under aggression is part of it.[22] Here the question arises whether a state enjoys the right of self-defence over a territory which is under its control but whose status is internationally disputed. The jurists of international law do not think so. For example, the great jurist Derek Bowett in his classical study on self-defence has this to say on the matter: 'Once title to the territory itself is in dispute it can no longer afford a basis on which to rest the right of self-defence ...'.[23] Similarly, A. O. Cukwurah, another authority on the subject contends that a state is entitled to exercise the right of self-defence in case of aggression by a neighbour but not in a disputed territory.[24] Jammu and Kashmir, where India claims that Pakistan was involved in acts of 'cross-border terrorism' is not part of India, the latter's vociferous and unremitting assertion to the contrary notwithstanding. The fact of the matter is that the territory in question is a disputed one—a fact which not only a multitude of UN resolutions on Jammu and Kashmir[25] maintain but the international community also generally recognizes. In light of the foregoing, the Indian claim of 'terrorism' perpetrated by Pakistan in Jammu and Kashmir is not tenable. Nor is the claim of exercizing the right of self-defence.

Next, dealing with the acts of terrorism in India proper (and Jammu and Kashmir as well), we need to make a distinction between acts committed by the state and non-state actors. The Draft Articles on *Responsibility of States for Internationally Wrongful Acts*[26] cover them in articles 4 and 8. Article 4 which deals with state actors in para.1 stipulates as follows: 'The conduct of any State organ shall be considered an act of that State under international law ... whatever position it holds in the organisation of the State, and whatever its character as an organ of the central Government or of a territorial unit of the State.' And then para. 2 goes on to clarify: 'An organ includes any person or entity which has that status in accordance with the internal law of that State.' Article 8, on the other hand, spells out the position of international law on non-state actors: 'The conduct of a person or group of persons shall be considered an act of a State under international law if the person or group of persons is in fact acting on the instructions of, or under the control of, that State in carrying out the conduct.'

The International Law Commission in its commentary on the Draft Articles points out[27] that as a general rule, the conduct of private persons or non-state actors is not attributable to the state under international law. However, in its opinion, it becomes attributable to the state if there is specific factual relationship between the private persons or non-state actors engaged in the wrongful act and the state.

It envisages two situations in this regard: a) when private persons or non-state actors are carrying out a wrongful act on the instructions of the state; b) when they are acting under the direction or control of the state. The degree of control that the state exercises over the private persons or non-state actors is a critical factor in determining whether their conduct can be attributed to the state. The World Court took up this issue in the *Military and Paramilitary Activities in and against Nicaragua (merits)* case when it examined the question whether the conduct of *contras*, the rebel group which was fighting against the Nicaraguan government, could be attributed to the US in order to hold it responsible for violations of humanitarian law by the former. It looked at the issue in terms of the 'control' that the US allegedly exercised over the *contras*. On the basis of the evidence produced before it, it held the US responsible for 'planning, direction and support' of the *contras* but rejected Nicaragua's broader claim that all the conduct of the *contras* was attributable to the US on account of its control over them in these words:[28`]

> The Court has taken the view ... that the United States participation, even if preponderant or decisive, in the financing, organizing, training, supplying and equipping of *contras*, the selection of its military or paramilitary targets, and the planning of the whole of its operation, is still insufficient in itself, on the basis of the evidence in the possession of the Court, for the purpose of attributing to the United States the acts committed by the *contras* in the course of their military and paramilitary operations in Nicaragua. All the forms of United States participation mentioned above, and even the general control by the respondent State over a force with a high degree of dependency on it, would not in themselves mean, without further evidence, that the United States directed or enforced the perpetration of the acts contrary to human rights and humanitarian law alleged by the applicant State. Such acts could well be committed by members of the *contras* without the control of the United States. For this conduct to give rise to legal responsibility of the United States, it could in principle have to be proved that the United States had effective control of the military and paramilitary operations, in the course of which the alleged violations were committed.

In other words, the Court laid down the principle of *effective control* by the state over the acts of private persons or non-state actors as the criterion to hold the state responsible for the latter's conduct. It thus rejected the criterion of state's preponderant or decisive role over the latter as being insufficient. When we examine the militancy that raged in the Indian-administered Jammu and Kashmir during the 1990s, we note that independent commentators and serious Indian analysts are all agreed that it was originally an indigenous movement which the Pakistani military authorities exploited, as this happens in such situations. How much control did the latter exercise over the insurgents, particularly the LeT, which was originally a Pakistani outfit but had established itself in the Indian-administered Jammu and Kashmir for what it termed a freedom struggle? The Pakistan government completely denies any role except for moral and diplomatic support that it acknowledged it extended to what it termed freedom fighters. The disclaimer is understandable because no government admits its involvement in wrongful acts, let alone exercising control over an armed group on an alien territory.

The Indian government, however, holds an entirely different view. It is of the opinion that the entire struggle was an ISI (Inter-Services Intelligence, the name of Pakistan's premier military intelligence agency) affair in which the Pakistani military authorities enjoyed, in the Court's language, *effective control* over the militants. Who is right and who is wrong? There may be evidence to suggest that Pakistan exercised more than what it terms moral and diplomatic support. However, in the absence of the kind of data or evidence that the Court had while adjudging the Nicaraguan complaint against the United States, it is hard to conclusively say that the latter enjoyed effective control over the militants.

Voicing the Indian viewpoint, one may argue that when Musharraf concluded the Islamabad Declaration in 2004 with Prime Minister Vajpayee pledging not to allow the use of Pakistani territory or the territory under its control for purposes of 'cross-border terrorism', the militancy in Jammu and Kashmir came to a halt. It shows that Pakistan enjoyed effective control over the LeT militants in Jammu and Kashmir. It is true that though Musharraf's sealing of the LoC (the Line of Control that separates the Indian and Pakistan-administered Jammu and Kashmir) largely put an end to the infiltration of militants into the Indian-administered Jammu and Kashmir, it does not prove that Pakistan enjoyed effective control over the acts of the LeT. This is because closing the tap of infiltration by the Pakistani authorities evoked a strong reaction against Musharraf by the militants who, sore about this decision, made several attempts on his life. Given this backdrop it is hard to maintain that the Pakistani authorities enjoyed effective control over the militants.

Similarly, to prove that Pakistan enjoyed effective control over the militants one would have to prove that it controlled a whole gamut of their activities or that the latter were completely dependent on Pakistan. As pointed out above, in the absence of the kind of evidence that the World Court had in the *Nicaraguan* case, it would be hard to prove it. However, in order to take this line of thought to its logical end, let us assume for a moment that Pakistan enjoyed such control over the militants. Then the question arises whether militancy in the Indian-administered Jammu and Kashmir constituted an 'armed attack' against India. This is an important question for India in order to claim the right of self-defence under article 51 of the UN Charter. But here lies the problem: irrespective of what India may say on this point, the State of Jammu and Kashmir, as a whole, is an internationally recognized disputed territory and there cannot be, as shown above, aggression in a disputed territory.

B) *REBUS SIC STANTIBUS* AS GROUNDS FOR TERMINATION

Indian analysts have also advocated repudiation of the Treaty on the grounds that it is obsolete. One Indian writer has argued that the Treaty was concluded in response to the water problems that Pakistan and India faced in the twentieth century whereas the issues that they confront in the twenty-first century are of an entirely different character. Elaborating this idea, he has pointed out that since 1960, when the Treaty was concluded, lots of changes have occurred in the domain of science and technology, development and management of waters, administration, political boundaries, etc. which have rendered it redundant.[29] Simply put, he has justified the denunciation of the Treaty on the grounds of the principle of *rebus sic stantibus* or the fundamental change of circumstances. Can this be a valid basis for termination of the Treaty? To address this question, we propose to examine the principle of fundamental change of circumstances which the Vienna Convention on the Law of Treaties of 1969 enshrines in its article 62, para. 1 in these words:

> A fundamental change of circumstances which has occurred with regard to those existing at the time of the conclusion of a treaty, and which was not foreseen by the parties, may not be invoked as a ground for terminating or withdrawing from the treaty unless:
> a) the existence of those circumstances constituted an essential basis of the consent of the parties to be bound by the treaty; and
> b) the effect of the change is radically to transform the extent of obligations still to be performed under the treaty ....

When we carefully analyse the above provision, we discover that it lays down a very rigorous test for invoking this ground which, according to the International Law Commission, can be reduced to the following five conditions:[30]

1. the change must be of circumstances which existed at the time the treaty was concluded;
2. the change must be a fundamental one;
3. the change must also be one not foreseen by the parties at the time of the conclusion of the treaty;
4. the existence of those circumstances must have constituted an essential basis of the consent of the parties to be bound by the treaty;
5. the effect of the change must be radically to transform the scope of obligations still to be performed under the treaty.

To find out whether or not India has a case for denouncing the Treaty on the grounds of it being outdated, we begin by examining the first condition, namely whether or not there is a change in the circumstances that existed at the time the treaty was concluded. To answer this question we refer back to 1948 when India, the upper riparian, decided to appropriate the waters of the three Eastern Rivers for its exclusive use and deny to Pakistan (the lower riparian) the quantity which it had previously used on the basis of 'historic withdrawals'. It was so ruthless in the pursuit of this objective that it refused to make a compromise and instead bullied Pakistan by stopping the flow of water to Pakistan's canals. Subsequently, it made Pakistan sign a Joint Statement on 4 May 1948, by virtue of which the latter was forced to surrender its proprietary rights over the waters of the Eastern Rivers. This engendered so much tension that a distinct possibility of armed conflict existed between the two countries, which Lilienthal described in his famous *Collier's* article:[31] 'India and Pakistan are today on the very razor's edge of a war that would directly involve more than 360,000,000 people, one sixth of the world's population .... That there is danger that war will break out any day is neither speculative not alarmist. In Kashmir's high mountains two armies face each other in anger.' Henceforth, the Bank intervened and successfully helped the two antagonists bring about an amicable settlement of the dispute in the shape of the Indus Waters Treaty.

Here, we ask whether the circumstances that led to the conclusion of the Treaty have undergone any change. Our conclusion based on facts on the ground is that they remain totally unchanged. Pakistanis believe that India is as predatory a state today as it was when the water dispute erupted in 1948. They think that after controlling the three Eastern Rivers, it is now keen to deny Pakistan the exclusive use of the three Western Rivers that the latter is entitled to under the Treaty. It is true that India has not formally staked a claim to a share in these waters but the kind of measures it is taking on the Western Rivers is cause for concern for Pakistan. No wonder that the latter regards all attempts to deprive it of its exclusive rights to the Western Rivers as a *casus belli*. In fact, it has reportedly stated that it would be prepared to use nuclear weapons over the water issue.[32] Incidentally, outsiders appear to share Pakistan's perception that India's behaviour on the waters of the Western Rivers could lead to a war between the two countries. For instance, in March 2009, a group of more than 20 UN bodies warned that given the rising tension over the water issue between Pakistan and India, the world would be perilously close to its water war.[33] This analysis shows that the circumstances under which the Treaty was concluded have not changed. Given the fact that the first condition is not fulfilled, it is not necessary to investigate whether or not other conditions are fulfilled.

The foregoing was a juridical analysis. As for the political dimension of the case, India is hesitant to denounce the Treaty as being outdated because, as underlined by many Indian writers, it would stir up a hornet's nest in the international community. For example, Ramaswamy Iyer contends that since the Bank midwifed the Treaty and is a party to it, its reaction would be adverse as would that of the international community in case India jettisoned the Treaty. He further adds that, '[f]rom the position of being praised for the steadfast maintenance of the Treaty, India would have fallen into that of incurring international displeasure for violating a solemn covenant. Other countries with which India has similar treaties (Nepal, Bangladesh) might have found their faith in India shaken and become apprehensive about their own treaties with us.'[34] He also opposes the Treaty's termination on the grounds that it would not work because a treaty on water is different. He holds that even without an agreement like the Indus Waters Treaty, the upper riparian is not entitled to withhold the flow of water towards the territory of the lower riparian because of the obligations that it has towards the latter under various international conventions.[35] Similarly, some critics who oppose the abrogation of the Treaty have pointed out that western countries which financed dams and reservoirs in Pakistan under the Treaty such as the US, Britain, Germany, Australia, and Canada would also be unhappy with this development and may intervene to save it.[36]

Before closing the debate relating to the *rebus sic stantibus* principle, there is another aspect that deserves our attention. India in another attempt to jettison the Treaty has now found a Chinese dimension to the issue. Thus, one writer has advanced the view that since China has constructed the Zada Gorge Barrage[37] on the upper Sutlej and the Senge Ali Dam[38] on the upper Indus in western Tibet, it has staked its rights to the Treaty. He contends that given this development, India should invoke article 62 of the Vienna Convention to seek a review of the Treaty.[39] Is this demand justified? We do not deem it so, for the simple reason that the Treaty is a partitioning construct which, of the six rivers of the Indus Basin, allocates three each to Pakistan and India. With this backdrop, each country is responsible to sort out the problems, internal or external, with which it may be confronted regarding the rivers allocated to it. In other words, if China is building a barrage on the Sutlej River, it is for India to decide how to take care of this matter. Similarly, if China is building a dam on the Indus River, it is for Pakistan to decide how to deal with it. The Chinese factor cannot be a justification to seek a review of the Treaty. If we were to accept India's logic, then Pakistan would be entitled to seek a review of the Treaty and a share in the Eastern Rivers if tomorrow Afghanistan builds dams on the Kabul River (which alone contributes 16.8 million acres feet of water to the Indus River) reducing the flow of water towards Pakistan. We are doubtful that India would be agreeable to this kind of proposition. Incidentally, if there is a review of the Treaty then both the Eastern as well as the Western Rivers would be a subject of negotiation, which most of the Indian writers, writing on the subject, conveniently fail to mention. Their focus is entirely on the Western Rivers as if the Eastern Rivers were never a part of the Indus Basin. India cannot have it all. Therefore, the Indian government should be careful before advancing the argument based on the principle of *rebus sic stantibus.*

C) KASHMIRI CONCERNS AS GROUNDS FOR TERMINATION

Besides the Indian government, the Legislative Assembly of Jammu and Kashmir has also threatened to terminate the Treaty. Thus, in 2002 the Kashmir Assembly members, arguing[40] that the state was suffering a loss of Rs. 6,500 crore per year, urged the Indian government to scrap the Treaty. They contended that the Kashmir state could not use waters of its own rivers and that the Treaty was the greatest bottleneck in its progress and prosperity. They adopted a resolution to abrogate the Treaty in the best interests of Kashmiris and submitted a memorandum to that effect to the government in New Delhi. If the Kashmir State Assembly moved the resolution for the termination of the Treaty, Farooq Abdullah, the former chief minister of Jammu and Kashmir and son of the legendary leader, Sheikh Abdullah, provided justification by arguing that the State was not consulted during the negotiations that led to the Treaty.[41] In other words, he is suggesting that the Treaty is unacceptable to Kashmiris because despite the existence of article 370 in the Indian Constitution which confers special status on Jammu and Kashmir, New Delhi never consulted it on the Treaty before, or after, ratifying it. It is noteworthy here that one member (Harshdev Singh of the Panthers Party) made a statement in the assembly, which later on became a catchy slogan, 'We export water to Pakistan, they export terrorism to Jammu and Kashmir'.[42] The statement is mostly false because according to the terms of the Treaty, the waters of the Western Rivers belong to Pakistan except for the quantum the Treaty allocates to Jammu and Kashmir for local needs and that the so-called terrorism is the direct result of India's refusal to allow the Kashmiris the right to self-determination.

The Kashmir Assembly and Kashmiri politicians are not alone in voicing their discontentment with the Treaty as presently drafted. They are joined by people in academia as well. One Indian writer has said:[43] 'But already opinions have been expressed that the Indus Waters Treaty, which decided India's share from the two rivers

Jhelum and Chenab, ignored interests of the constituent Jammu and Kashmir state. The Government should therefore open the Indus Waters Treaty and/or look into the allocations between different states of India once more and correct the injustice. There may be need if not now, [then] later for Pakistan to resolve such internal issues. How concerns of within country-country sharing of allocated waters for both can be met with a new treaty has not been studied, nor voiced so far.'

How far are these concerns justified by law? To answer this question, we refer to the Vienna Convention on the Law of Treaties 1969 whose articles 27 and 46 are relevant here. According to article 27, a party is barred from invoking the provision of its internal law as a justification to perform a treaty. However, this provision operates subject to article 46 which specifies that a state is not bound by a treaty that violates a provision which is 'manifest' and 'concerned a rule of its internal law of fundamental importance.' It is true that the Kashmir Assembly did not give its consent to the Treaty before or after its ratification and consequently it constitutes a violation of the provision of the Indian Constitution which is 'manifest' and of 'fundamental importance.' There are two problems with this argument. First, since the Vienna Convention, which was concluded in 1969 and came into force in 1980, does not have retrospective effective and so, this argument does not apply to the Treaty. Second, Farooq Abdullah's plea for the denunciation of the Treaty for non-consultation of Kashmiris during the negotiations cannot be entertained 50 years after its conclusion.

Critics have rejected the Kashmir Assembly resolution on other grounds as well. For example, Ramaswamy Iyer has contended that '[t]here is no reason to believe that the interests of Jammu and Kashmir were not kept in view while negotiating the Treaty. The negotiations were long and hard, and the Indian team did its best under whatever briefing it had from the government. The Treaty was signed and approved at the highest level. Therefore it is not open to us that it is disadvantageous to India.'[44] Similarly, the former chief minister of the Indian-administered Jammu and Kashmir, Mufti Sayeed, ruled out the scrapping of the Treaty on the grounds that India was bound to observe international obligations. Furthermore, advising the members of the Kashmir Assembly to take a holistic view of the issue, he rejected the demand stating that 'we don't want to further increase the tension between India and Pakistan in view of the already strained relations.'[45] India has not used even half of the amount of water allocated to it under the Treaty for agricultural use in Jammu and Kashmir. Similarly, as far as the generation of hydropower in Jammu and Kashmir is concerned, the Pakistan government has no issue with it as long as the projects that the Indian government puts up are in consonance with the provisions of the Treaty, i.e. they should be run-of-river projects rather than dams and reservoirs for storage purposes which the Treaty prohibits.

D) UNFAIRNESS AS GROUNDS FOR RENEGOTIATIONS

Indian detractors also object to the Treaty on the grounds that it was an unfair deal *ab initio* because, in their opinion, while Pakistan got 80 per cent of water from the Indus Basin system India barely got 20 per cent.[46] They blame the Indian leadership of the time for this 'foolishness' and contend that time has now come to remove this injustice by making Pakistan share waters of the Western Rivers with India.[47] One critic, S. K. Garg, is of the view that India should have 42.8 per cent of the water from the Indus Basin as opposed to the 20 to 25 per cent that the Treaty allocates to it. He thinks that Pakistan was unjustifiably allocated 75 to 80 per cent of the waters. He justifies his calculation for the entitlement of India on the basis of Pakistan's and India's share of population, drainage areas, length of rivers, and culturable areas of the Indus Basin.[48] He invokes the Helsinki Rules on the Uses of the Waters of International Rivers in favour of his computation.[49] Is the Indian argument based on the 'unfairness' of the Treaty justified? Why didn't India raise this objection in the course of the Indus waters negotiations?

The Indian objection is utterly devoid of fairness for a number of reasons. First, India accepted the Treaty's scheme of river-partition voluntarily and after protracted negotiations. Nobody forced this legislation. On the contrary, it may not be wrong to say that it was virtually imposed on Pakistan. As demonstrated by the *travaux préparatoires* of the negotiations (see chapter 3) Pakistan was not at all happy with the Bank proposal to allocate the entire flow of the Eastern Rivers to India, which was *grosso modo* based on the Indian plan, but had to accept it because the Bank put its entire weight behind it and left Pakistan virtually no choice but to reject it. As for the reason why India favoured it, Ramaswamy Iyer reveals the logic that drove it in these words:[50]

> [I]f Pakistan got the near-exclusive allocation of the three western rivers, India for its part got the eastern rivers. This was important from the viewpoint of the Indian negotiators, because the water needs of Punjab and Rajasthan weighed heavily with them in seeking an adequate allocation of Indus waters from India .... As early as in 1953, many years before the treaty, thinking had begun on the transfer of waters from the eastern rivers to Rajasthan through a canal. In the 1950s again, Bhakra Nangal was already under construction. If the Ravi, Beas and Sutlej had not been allocated to India, Pakistan would have had the usual lower riparian rights over these rivers, and would have had to be consulted about these projects and, would surely have raised objections. The projects might not have come up at all, or might have had to be substantially smaller. In a sense, one might say that the allocation of the eastern rivers to India under the Indus treaty removed Pakistan from the picture in relation to these rivers, and retrospectively legitimised the Bhakra-Nangal and Rajasthan canal projects. The price paid for this was the sacrifice of rights over the western rivers.

Secondly, the Indian argument does not hold because out of the three Western Rivers that went to Pakistan, the Indus and the Jhelum are of no use to India in agricultural terms except in Jammu and Kashmir for which the Treaty has allocated 3.6 million acre feet of water which is more than enough for its present and future agricultural needs.

The only river that it can make use of in proper India (meaning India excluding the disputed territory of Jammu and Kashmir) is the Chenab River. But if it relinquished its claim to it, it got in return exclusive proprietary rights to the three Eastern Rivers in perpetuity. Incidentally, this made eminent sense because these are the rivers that matter the most to it. Michel agrees with this assessment in these words:[51]

> In agreeing to recognize Pakistan's right in perpetuity to virtually all the waters of the three western rivers (Indus, Jhelum, and Chenab), India was really giving away only one stream, the Chenab that she could really use herself (by diversion into the Ravi or Beas). She was gaining undisputed possession of the waters of the three eastern rivers (Ravi, Beas, and Sutlej) in perpetuity after the Transition Period ends in 1970 or at the latest in 1973. These are the rivers that are really useful to India, and the Indus Waters Treaty gives her the right to dry them up entirely if she so chooses.

Thirdly, the Indian contention is deemed unjustified because it was argued that the territories that went to India as a result of partition were historically 'using less than 10 per cent of the Indus waters and that the Treaty was generous to India in giving 20 per cent of the waters.' During the Bank-sponsored negotiations[52] there were two conflicting principles on the table, namely 'no appreciable harm' and 'equitable utilization'. The Indians, which favoured the 'equitable utilization' formula viewed the allocation of 75 per cent of the water as a fundamental violation of this principle whereas the Pakistanis, which supported the 'no appreciable harm' principle, criticized the allocation of only 75 per cent of the water on the ground that it possessed 90 per cent of the irrigated land. Neutral observers have tended to support Pakistan's contention on the grounds that, among others, it had prior use of these waters which was greater than India's.[53] As noted above, the Treaty nonetheless allocated to India 3.6 million acre feet of water from the Western Rivers for agricultural use in Jammu and Kashmir. It also allowed India to use the waters of these rivers for hydropower generation by establishing run-of-river projects. It is noteworthy that if India is unhappy with the surrender of the Western Rivers to Pakistan, Pakistan is equally unhappy with the surrender of the three Eastern Rivers to India in perpetuity which has adversely affected it in a variety of ways.

Before concluding this point, we need to examine Garg's claim that India should have gotten 42.8 per cent of the water from the Indus Basin rather than the 20 or 25 per cent that it was actually allocated. We have seen above that he reached this conclusion on the basis of the Helsinki Rules. The entire exercise by Garg is meaningless because the International Law Association adopted these Rules in 1966 whereas the Treaty was concluded and came into force in 1960. In other words, the Helsinki Rules post-date the Treaty which signifies that *ipso facto* they do not apply to the Treaty. Even if they were to apply, the situation would not change because they are exhortatory in character hence, not binding. There is another aspect that we need to keep in mind in this respect. The Rules prescribe the principle of 'equitable utilization' for sharing the common international watercourses for which purpose they identify eleven factors to determine what is reasonable and equitable. Of these factors, one relates to the existing utilization of waters or the principle of prior appropriation. Despite Garg's mention of other factors to substantiate his thesis, he conveniently forgets the principle of prior appropriation which Pakistan emphasized for its claim to waters of the Eastern Rivers. This shows the arbitrary and partisan character of Garg's exercise. Perhaps, cognizant of the foregoing facts, Ramaswamy Iyer made the following observation on Garg's plea:[54] 'All that one can say is that when prolonged inter-country negotiations by teams acting under governmental briefings lead to a treaty, and the treaty is approved and signed at the highest levels, it must be presumed that it was the best possible outcome that could have been negotiated under the circumstances, either side is then precluded from saying that it was unfair, unequal, poorly negotiated, etc.' In light of these observations, it is obvious that Garg's claim is utterly unfounded.

E) INDIA'S MOTIVE FOR REVISITING THE TREATY

There are two reasons why India would like to scuttle the Treaty in favour of another agreement. First, according to international and Indian reports, India is increasingly becoming a water-deficit country. The 1999 World Bank report reveals that the per capita water availability in India, which was 5,000 cubic metres per annum in 1947, dropped to less than 2,000 cubic metres per annum in 1997 and is expected to further decline to 1,500 cubic metres per annum by 2025. It further estimates that the overall water demand in India would almost double from 552 BCM (Billion Cubic Metres) in 1999 to 1050 BCM in 2025. It concludes that India would become 'water stressed' by 2025.[55] Similarly, the McKinsey report, published ten years later in 2009, makes projections in the same direction. According to it, India's water needs which stood at 740 billion m^3 in 2009 would approximately double to 1.5 trillion m^3 by 2030, resulting in severe water shortages in most of its river basins.[56] Regarding national reports, the National Commission for Integrated Water Resource Development (NCIWRD) of India makes projections which are similar to those made by international reports. For example, it estimates that by 2050 India's total water requirements would be approximately 1180 BCM—though it links these figures with population growth—whereas the ministry of water resources puts the estimates for 2025 at 1093 BCM and for 2050 at 1447 BCM.[57]

Ultimately, all indications are that the water resources in India will fall short of its needs in the future. With this perspective, it is not surprising that the Indian government has accorded high priority to water security as evidenced by the speech of the former Indian prime minister, Manmohan Singh, made on Independence Day in 2004 in which he described it as one of the *saat sutras* requiring special attention.[58] Meanwhile, the Indian government has devised different strategies to implement this policy, no document suggests that India has its sights set on the Western Rivers. However, that does not signify that it has no interest in them. In fact the Indians covet them. For

example, one Indian analyst advised his government to tempt Pakistan
the prospect of increased water supplies from the Eastern Rivers to
the province of Sindh in return for an amendment to the Treaty.
Reproduced below is the proposal *in extenso*.[59]

If India wishes to stop the dominant Pakistani province from drying up
downriver Sindh and Baluchistan through usurper policies, it will have
to look at innovative means. Such an upstream appropriation of waters
even though intercountry, does violate the spirit of the Treaty, founded
on the concept of open flow of river waters. Given the cross-border
implications of irreversible damage to the natural ecosystems India
needs to explore prudent and feasible options. One possible option,
dependent on Pakistani agreement to amend the treaty, would be for
India—instead of continuing to release all the waters to Pakistani
Punjab directly or through Pakistan-held Kashmir—to build southward
canals to offer substantial waters at the Rajasthan-Sindh border for the
benefit of Sindhis and to save the deltaic ecosystems. Essentially, this
would mean constructing a southward spur from the 949-kilometre
Rajasthan canal, also known as the Indira Gandhi Canal, which
originates close to the confluence of the Beas and the Sutlej rivers and
terminates near the Rajasthan city of Jaisalmer.

The water delivered at the Sindh-Rajasthan border will be from the
Indus system's eastern rivers, which are reserved exclusively for India's
use. But to compensate users in its water-stressed northern states for
the water diversion to Sindh, the Indian Government would have to
make equivalent water withdrawals from the western rivers—which are
set aside for Pakistan's use but in practice, have been meant for use
mainly by Pakistani Punjab. Such Indian withdrawals could be made
by reopening old canals and building new ones between the western
and eastern rivers. *The proposal, in essence, would seek to force Pakistan's hand
to renegotiate the terms of the treaty.*

Secondly, India is keen to keep Pakistan under pressure for security
reasons. The least expensive and the most inoffensive way to go about
this business is through control of the waters that are Pakistan's lifeline.
The ideal way for the former to do so is to build structures on the
Western Rivers in Kashmir which it can use at the opportune moment

to stop the flow of water towards Pakistan. Lilienthal pointed to this potential danger in his article in *Collier's*:[60] 'Pakistan includes some of the most productive food-growing lands in the world in western Punjab (the Kipling country) and Sindh. But *without water for irrigation* this would be desert, 20,000,000 acres would dry up in a week, tens of millions would starve. No army, with bombs and shellfire, could devastate a land as thoroughly as Pakistan could be devastated by the simple expedient of India's permanently shutting off the sources of water that keeps the fields and the people of Pakistan alive. India has never threatened such a drastic step and indeed denies any such intention—but the power is there nonetheless.'

Is India harbouring any such intentions towards Pakistan? There is no indication of it in current Indian behaviour. But this is perhaps because it does not possess the means to do so. However, that does not mean that things would remain the same in the future. They can change in response to a given political or security situation. A country does not formulate its defence strategy against its adversary on the basis of the latter's intentions because those can change at any moment. India may not be currently harbouring any desire to harm Pakistan but is building capacity to control the flow of water as demonstrated by the number of dams and reservoirs it is feverishly constructing and planning to construct on the Western Rivers. Once it has built a sufficient number of such structures, it would be in a position to effect a change in its intentions.

Independent observers have also pointed towards this danger. John Briscoe, the former senior water advisor to the World Bank and currently a Harvard University professor, has this to say about the matter:[61] 'If Baglihar was the only dam being built by India on the Chenab and Jhelum, this would be a limited problem. But following Baglihar is a veritable caravan of Indian projects—Kishenganga, Sawalkot, Pakuldul, Bursur, Dul Hasti, Gypsa. The cumulative storage will be large, giving India an unquestioned capacity to have major impact on the timing of flows into Pakistan. Using Baglihar as a reference, simple back-of-the-envelope calculations suggest that once it has constructed all of the planned hydropower plants on the Chenab, India will have the ability to effect major damage on Pakistan.' Similarly, the US Senate Foreign Relations Committee's Report, 2011, seconds this assessment in these words:[62] 'The number of dams under construction and their management is a source of significant bilateral tension …. While studies show that no single dam along the waters controlled by the Indus Waters Treaty will affect Pakistan's access to water, the cumulative effect of these projects could give India the ability to store enough water to limit the supply to Pakistan at crucial moments in the growing season.' In addition to posing a threat to Pakistan's agriculture, India can use water as a weapon for military purposes at the time of possible or actual hostilities between the two countries by releasing or withholding water. Incidentally, this apocalyptic scenario is not imaginary because let us not forget that India demonstrated this in 1948 when it closed canals supplying water to Pakistan.

Pakistan believes that India's intentions to use water as a weapon against it are not limited to building storages on the Western Rivers alone but extend to Afghanistan whose rivers bring a large quantity of water to Pakistan. Before examining the soundness of this thesis, we propose to detail the degree of Pakistan's dependence on waters flowing from the territory of its neighbour in the north. There are nine rivers that flow from Afghanistan into Pakistan with an annual flow of about 18.7 MAF. Out of these, the Kabul River is the largest, it joins the Indus River after crossing into Pakistan, and it alone accounts for 16.8 MAF. The Chitral River, a tributary of the Kabul River, originates from Pakistan, enters into Afghanistan where it is called the Kunar River, and joins the Kabul River near Jalalabad before re-entering Pakistan. It contributes 8.5 MAF or about 60 per cent of water to the Kabul River.[63] Pakistan gets about 17 per cent of its water supplies from the Kabul River during the winter season. In 2011, Prime Minister Manmohan Singh made an official visit to Kabul during which he pledged a sum of $1.5 billion in development assistance to Afghanistan in various sectors of its economy, especially for the construction of 14 medium and small dams with a total storage capacity of 7.4 MAF.[64]

The Pakistan government views the Indian offer to build dams and reservoirs on the rivers that Pakistan and Afghanistan share with deep suspicion as, in its opinion, it is motivated by the desire to deny to Pakistan as much water as possible. Is Pakistan's perception well-founded, particularly when we take into the account the fact that Afghanistan is not only a water-stressed country but also figures amongst the least developed countries (LDCs) in the world and that construction of dams and reservoirs could go a long way in bringing vast tracts of its barren land under irrigation? The question becomes all the more pertinent when we realize that not only India, but international financial institutions including the World Bank have offered huge financial assistance to Afghanistan (reportedly $7.079 billion) to build dams and reservoirs. Pakistan's view becomes all the more problematic when we realize that Pakistan raises objections to the Indian offer but keeps mum over that made by international financial institutions.

Pakistan's objections to India's offer of help to Afghanistan, on the face of it, look not only utterly unjustified but also paranoid. However, upon closer scrutiny it may not be as indefensible as it appears to be, especially in the framework of Pakistan-India relations. Consider the following: Pakistan and India, ever since their emergence as independent states, have rarely enjoyed friendly relations with each other. They have constantly engaged in a political tug of war with no holds barred when it comes to achieving their objectives. This power struggle has manifested itself in different forms and at different places. One place of special significance where they have jockeyed for influence is Afghanistan. Pakistan dreads Indian influence in Kabul because it fears that it enables India to encircle it as well as foment trouble in Balochistan and Pakistan's frontier regions. Critics have accused Pakistan of seeking strategic depth in Afghanistan to take care of the Indian threat, though the latter denies it.[65] India, on its part, seeks a friendly government in Kabul and wants to exclude Pakistan's clout there in order to eliminate the Taliban factor. It also aims to use its presence in Kabul to project power in the region and demonstrate global interests.[66] It is feared that the two are likely to get involved in a proxy war in Afghanistan in the wake of the complete withdrawal of American troops from there.

Pakistan enjoyed very close relations with the government in Kabul during the second half of the 1990s when the Taliban ruled there. However, after their remover from power by the Americans following 9/11, and replacement by the government of Hamid Karzai, India has found a niche for itself in Kabul which is a cause for grave concern for Pakistan. Pakistan accuses India of pursuing an anti-Pakistan agenda in Afghanistan. It alleges that the Indian consulates in Afghanistan, located near the Pakistan border, in cahoots with Afghan intelligence are supporting Baloch insurgents and fomenting trouble in FATA (Federally Administered Tribal Areas). It objects to the training of the Afghan national army by India on the grounds that it would permeate the Afghan soldiers with its anti-Pakistan worldview.[67] It regards the Indian offer to build dams and reservoirs in Afghanistan as part of its agenda to harm Pakistan.

However, India rejects Pakistan's charges as utterly unfounded and denies that its consulates are involved in any objectionable activity. It justifies the training of the Afghan national army personnel as an element which would contribute towards building Afghanistan's defence capability. It claims that its presence is completely benign and points out that its companies have made commercial investments of more than $10 billion there. It denies any nefarious design in extending help to Afghanistan, including the offer to assist it in harnessing its water resources. It justifies its stance as a positive step for the economic development of Afghanistan and regional peace. It considers Pakistan's reaction in the matter as nothing short of paranoia. Are the Indian denials credible? Is Pakistan paranoid?

Independent evidence does not seem to support India's contention that its presence in Afghanistan is utterly benign. For example, the American academic, Christine Fair, representing the Rand Corporation in a roundtable discussion organized by the *Foreign Affairs* magazine in March 2009, observed that, '[h]aving visited the Indian mission in Zahedan, Iran, I can assure you they are not issuing visas as the main activity. Moreover, India has run operations from its mission in Mazar (through which it supported the National Alliance) and is likely doing so from the other consulates it has reopened in Jalalabad and Qandhar along the border. *Indian officials have told me privately that they are pumping money into Balochistan.*'[68] There is the damning testimony of the former US secretary of defence, Chuck Hagel, a known friend of India, who has seconded Pakistan's allegations. In a talk that he delivered at the University of Oklahoma in 2011, he made the following startling revelation: '*India has over the years financed problems for Pakistan on that side of the border, and you can carry that into many dimensions.*'[69] One such dimension could be the building of dams and reservoirs in order to deprive Pakistan of as much water as possible. It is true that these structures would ultimately help Afghanistan harness its water resources but given the character of India's relations with Pakistan its offer is perhaps motivated less with the intention to help the latter and more to destabilize Pakistan.

F) INDIA'S WATER DIPLOMACY WITH ITS SOUTH ASIAN NEIGHBOURS

We have seen that India is a predatory state insofar as Pakistan's water resources are concerned. Despite the considerable evidence that we have presented in support of our thesis, many may still have reservations. Consequently, the matter needs to be probed further. We propose to undertake this exercise by examining the Indian attitude towards its South Asian neighbours, Nepal and Bangladesh, with whom it shares many rivers. This may shed light on whether or not Pakistan's apprehensions about India are justified.

a) Nepal

(i) Kosi Agreement

We begin by looking at Nepal's relations with India. Nepal is the upper and India the lower riparian. The two countries have, over the years, entered many water-related agreements that include Kosi (1954), Gandak (1959), Tanakpur (1991), and Mahakali (1996). Nepal and India concluded an agreement on 25 April 1954, regarding the river Kosi which envisaged a multipurpose project covering flood control, hydropower generation, and irrigation.[70] Pursuant to this agreement, a 1150-metre barrage was built at a place called Bhimnagar which is about 8 kilometres inside the India-Nepal border. Two canals take off from the barrage, the Eastern Canal which irrigates 612,000 hectares of Indian land and the Western Canal which irrigates 356,610 hectares of Indian and 11,300 hectares of Nepalese land. The project also has a powerhouse with an installed capacity of 20,000 kW of electricity. The Nepalese opposition political parties subjected the agreement to severe criticism on the grounds that it was one-sided and was not at all beneficial to Nepal.[71] It also contended that it had granted extra-territorial rights to India for an indefinite period without obtaining adequate compensation in return; that Nepal lost fertile land without equivalent gains in exchange; and that the project was designed to further India's interest at the expense of the Nepalese people.[72]

It was also alleged that India would gain more from the hydropower resources and that Nepal would receive only a negligible part of the total irrigated land. During 1960–61 when relations between Nepal and India dipped, criticism against the agreement intensified. However, the character of criticism this time was different from the past because the Nepal government, which had been defending the agreement previously, changed its stance by attacking it. The Nepal government put pressure on India to change the terms of the agreement. Later on, the Indian minister for power and irrigation, K. L. Rao, visited Nepal in 1962–63 when he showed readiness to effect changes in the agreement. The Nepal government was not satisfied with the assurances as it wanted immediate revision which, however, did not come. Then in 1965, Prime Minister Lal Bahadur Shastri visited Nepal and assured the Nepal government they would amend the agreement. Subsequently, the agreement was revised and signed on 19 December 1966.[73]

Notwithstanding the revision, the agreement is still tilted in favour of India based on the responsibilities that the agreement lays down for its implementation by the two countries. For example, according to article 2 of the agreement, the Nepal government is bound to provide necessary facilities to the officers or those acting under them when they undertake any survey or investigation for the execution of the agreement. Similarly, article 3 binds the Nepal government to grant approval to any construction work, not envisaged in the revised text, as and when the Indian government sanctions any project or part of it and issues a notice to the Nepal government. Commenting on it, it has been rightly observed that:[74] 'it is clear that the Project did not give priority to consensus in decision-making. Practically all the decisions regarding the Project execution can be undertaken by India, a *modus operandi* criticised by the Nepalese .... It should also be noted that while each country is required to give to the other the data resulting from its surveys and investigations on the "Kosi in Nepal," there is no provision about the exchange of data on the "Kosi in India".'

Secondly, given the nomenclature that the revised text uses for the two governments by referring to India as 'the Union' and Nepal as 'the Government', it has been argued that the agreement undermines the sovereign status of Nepal.[75] Is this contention justified? Observers have agreed with this assessment:[76] 'However weak this argument may be in treaty law, it certainly comes out as a political stain for Nepal.' They have also noted that, in opposition to this agreement, other agreements between the two countries concluded during the same period have used the terms 'Government of India' and 'Government of Nepal.' Thirdly, one notices that the conflict resolution mechanism is tilted towards India. Consider the following: the contracting parties are to settle their disputes through discussion and if they fail to do so, they can resort to arbitration. However, the arbitration mechanism adumbrated in the agreement is flawed. This is for two reasons. First, there is no provision to resolve the matter if one party frustrates the process by failing to nominate its arbitrator. Secondly, if the arbitrators fail to resolve a dispute, the two parties are to appoint an umpire whose award would be final and binding. However, there is no binding obligation to appoint an umpire. Commenting on this lacuna, it has been observed that, 'a desire to submit an issue to arbitration by one party to the agreement can easily be frustrated by the other.'[77]

(ii) Gandak Agreement

India and Nepal next concluded an agreement on 4 December 1959, on Gandaki or Gandak (as it is called in India) River.[78] It is also a multipurpose project focusing on irrigation, hydropower, and flood control. Pursuant to the agreement, a barrage was built at Bhaisalotan on the reaches of Gandaki River located on the India-Nepal boundary. Two canals issue from the barrage, which irrigate 1,850,000 hectares of Indian and 57,000 hectares of Nepalese land. The project has a power house of 15,000 kW of electricity. Like the Kosi agreement, the opposition political parties subjected it to strong criticism because, in their opinion, it was contrary to Nepal's interests. They challenged the Nepal government's authority to conclude it.

Again, as in the case of the Kosi agreement, they mounted protests against it which led to its review in 1964.[79] However, despite the revision, the Nepalese observers regard it as flawed on different grounds. Firstly, they are of the view that though article 9 entitles Nepal to withdraw as much water as it requires from the river and its tributaries for irrigation and other purposes, it is nothing but an eye-wash. This is so, they contend, because the article in question forbids the latter from withdrawing water from February through April. The criticism is absolutely well-founded because the agreement allows Nepal to withdraw water when it does not need it (because it is in abundance) and disallows it when it is in dire need (when it is in short supply).[80]

Secondly, critics regard the conflict resolution mechanism, which is an exact replica of the one contained in the Kosi agreement, as defective. Since we have explained the reasoning before, we desist from repeating it here except to say that it is entirely tilted towards India. Thirdly, critics find the provision on the land acquisition defective. The agreement obliges the Nepal government to acquire land for the barrage and then transfer it to India in return for compensation. The Government of India, on payment of the compensation, acquires the title to the land in perpetuity. It is true that the agreement also provides that the land will return to Nepal free of cost in case the Government of India does not need it. However, such a situation ever arising seems far-fetched. It goes without saying that this provision is not in the interests of Nepal. We would like to point out here that this provision is totally at variance with the one contained in the Kosi agreement which grants land to India on lease (though for an extended period of 199 years).[81] Fourthly and finally, critics point out that though the Gandaki project was totally financed by the Indian government, the benefits accruing to Nepal have not matched the social cost that it has incurred. There is merit in this argument because as one report points out, the social cost of the land submerged behind the barrage, the rehabilitation of the displaced people, and the economic cost of erecting huge structures was quite significant.[82]

(iii) Tanakpur Agreement

In December 1991, India entered into a memorandum of understanding (MoU) with Nepal, commonly known as the Tanakpur agreement[83] by virtue of which it acquired the right to use 2.9 hectares of the Nepalese territory to construct the left afflux bund (or the retaining wall) for the purpose of building the Tanakpur Barrage located in Uttar Pradesh. As a quid pro quo for the concession, it agreed to supply Nepal with 10 megawatts of electricity and 150 cusecs of water for irrigation purposes. It also agreed to provide

Nepal with additional water if there was an increase in water supply at the Pancheshwar Reservoir. According to commentators, the Nepalese government, while entering into the agreement, either did not appreciate its implications or deliberately overlooked them to please India. Irrespective, the general impression in Nepal was that the Tanakpur agreement was not voluntary but a contrived one which India forced upon the Nepal government.[84]

Following its conclusion, the agreement became the subject of huge controversy as the Nepalese opposition accused Prime Minister Girija Koirala's government of compromising Nepal's territorial sovereignty. It also charged him with selling Nepal's water interests to India. Besides, it was of the view that since the Tanakpur agreement related to natural resources, it fell within the ambit of Nepal's constitution and hence needed to be ratified by Nepal's parliament to be effective. The Koirala government, on its part, was reluctant to submit the agreement to the parliament on the ground that it was no more than an 'understanding' between the two governments and hence a non-constitutional matter. Following the government's stonewalling in the matter, a writ petition was moved in Nepal's Supreme Court which challenged the validity of the agreement. The petitioner took the plea that the agreement required ratification before it could be enforced. The Supreme Court gave its ruling in December 1992, in which it adjudged that it was indeed a treaty which required ratification from the parliament. The Indian government had, right from the start, adopted a 'no negotiations' policy on the Tanakpur agreement. Despite the Supreme Court ruling, the Indian government refused to change its stance. The Nepalese government, on its part, never presented the Tanakpur agreement to the parliament. Therefore, the latter never ratified it.

(iv) Mahakali Treaty

Nepal and India signed the Mahakali agreement in 1996.[85] First, a word about the Mahakali River is in order. The Mahakali River, known as Sarada River in India, flows along the Nepal-India western border. It was the British who, during their rule in the subcontinent, fixed the border between the two countries. While doing so, they made the mid-stream of the river the boundary. It is necessary to underline here that the Sarada Treaty, which was concluded by the British, was the predecessor agreement that governed the Mahakali River Treaty. It took the two governments five years to negotiate the water-sharing arrangement for the integrated development of the water resources of the Mahakali River with a focus on Sarada Barrage, Tanakpur Barrage, and Pancheshwar multipurpose project. The outstanding feature of the Mahakali Treaty was that it not only enshrined the Sarada Treaty regime but also validated the highly controversial Tanakpur Treaty. The ratification issue became a rallying point for the people of Nepal who were utterly opposed to the Mahakali Treaty. Leading the movement against the Treaty, Nepalese students assaulted the parliamentary office of the United Marxist and Leninist Party (UML) and locked-up its leaders. Some opposition parties tried to besiege the parliament which led to their incarceration by the government.

Despite these protests, the Nepalese parliament in February 1996 ratified the Mahakali Treaty. There was considerable foreign pressure to do so. For example, the British minister of state for parliamentary affairs, Liam Fox, and the US assistant secretary of state for South Asia, Robin Raphael, during their visits to Nepal in August 1996, urged the Nepalese parliament to ratify the Treaty failing which, they warned it would send a wrong signal to foreign investors and thus drive them away from investing in Nepal.[86] The parliament succumbed to the pressure and made the ratification but subject to the four following conditions or strictures as the Nepalese call them: a) Nepal's electricity bought by India will be sold as per the 'avoided cost' principle; b) when the Mahakali Commission is constituted, it will be done only upon agreement by the main opposition party in parliament as well as by parties recognized as national parties; c) equal entitlement in the utilization of the waters of the Mahakali River without prejudice to their respective consumptive uses of the Mahakali

River means equal rights to all the waters of the Mahakali; d) saying that 'Mahakali is a boundary river on major stretches between the two countries' is the same as saying it is 'basically a border river.'[87] The two countries exchanged the instruments of ratification during Prime Minister Inder Kumar Gujral's visit to Nepal in June 1997. While undertaking this exercise, the Nepalese government utterly overlooked the mandatory conditions that the parliament had directed it to fulfil. Critics point out that the Mahakali Treaty was ratified conditionally and conditional ratification is no ratification. In their view, without the inclusion of the strictures in the Treaty, it is not a proper treaty.[88]

The foregoing analysis of the Kosi, Gandak, Tanakpur, and Mahakali agreements shows that they are heavily tilted towards India. Their terms in general, including the conflict resolution clauses and the way the Nepal government conducted itself with regard to some of them (e.g. Tanakpur and Mahakali in particular) indicate that Nepal was less than a free agent. If this is so, it is hard to resist concluding that the agreements were less than voluntary. If India imposed itself on Nepal, was its attitude towards Bangladesh any different? This is what we propose to examine below.

b) Bangladesh

(i) Farakka Barrage Agreement

India and Bangladesh share more than fifty rivers between them, particularly the large Himalayan river systems, namely the Ganges, Brahmaputra, and Meghna. They have a number of water-related disputes of which perhaps the most important one related to the Ganges River which arose in 1951 (Bangladesh at that time was still part of Pakistan and was known as East Pakistan) when India started to plan the construction of a barrage at a place called Farakka, 18 kilometres upstream of the East Pakistan border. The principal feature of the proposed project was the construction of a 38-kilometre long canal with a 40,000 cusecs capacity to divert the water of the Ganges River to the Bhagirathi-Hooghly distributary in order to flush out silt to keep the port of Calcutta (now Kolkata) navigable. Since the quantum of water available during the highly lean period was between 50,000 and 55,000 cusecs, East Pakistan was to receive barely 10,000 to 15,000 cusecs which was too minimal. India never informed Pakistan of the project which the latter learned about from the press. Pakistan protested against the project on the grounds that it was in violation of its historic rights over the Ganges Rivers, but to no avail. India rejected Pakistan's claim as it did to a similar claim that Pakistan made to waters of the Eastern Rivers in West Punjab. The work on the project began in 1961 and was completed in 1971. There were occasional talks between Pakistan and India during this period but no serious high level discussions or negotiations because India claimed that the Ganges was not an international river. It justified this stand on the ground that about 80 per cent of the river lay in its territory.[89]

Despite refusing to negotiate with Pakistan, India was conscious of the hollowness of its claim. To make its case convincing, on 26 March 1956, it denounced the Barcelona Convention and Statute concerning the Regime of Navigable Waterways of International Concern, 1921.[90] India looked upon this convention as a hurdle in its claim since article 1 of the convention laid down that a waterway that separates or traverses different states is a navigable waterway of international concern. Pakistan raised objections against the Indian denunciation, claiming that India did so in order to bypass the international obligations that the foregoing instruments imposed. India rejected this by arguing that 'the Barcelona Convention and Statute dealt with only some aspects of inland navigation and its purpose had been superseded by GATT [General Agreement on Tariffs and Trade].'[91] Pakistan rejected the Indian contention citing para. 1 of article 10 of the Barcelona Convention which states: 'Each riparian State is bound, on the one hand, to refrain from all measures likely to prejudice the navigability of the waterways, or to reduce the facilities for navigation, and the other hand, to take as rapidly as possible all necessary steps for removing any obstacles and dangers which may occur to navigation.' Commenting on the Indian justification, Salman M. A. Salman and Kishor Uprety have opined that '[t]he Indian contention was difficult to accept and the linkage of denunciation of the Barcelona Convention to the Farakka Barrage cannot be escaped.'[92]

In a bid to move things forward, Pakistan proposed to India that they jointly seek UN help for the cooperative development of rivers in East and West Bengal. As expected, India was not amenable to the proposal. Low level delegations of the two countries continued to meet occasionally but failed to make any progress. As a dilatory tactic, India continued to seek more and more data on the utilization of the Ganges River by Pakistan. In 1962, Pakistan proposed the construction of the Ganges Barrage in the area of the Hardinge Bridge in East Bengal, close to the Indian border for which it subsequently presented India with a feasibility report. The purpose of the project was to store the excess water during the monsoon season to irrigate southwestern parts of East Pakistan during the dry season. India rejected Pakistan's proposal terming it a retaliatory measure against the Farakka Barrage. It also opposed it on the ground that it would submerge some areas in the Indian State of West Bengal.

Following the disintegration of Pakistan and the emergence of Bangladesh, an independent state created with the help of Indian military intervention, it was thought that the *bonhomie* existing between India and Bangladesh would help the two countries approach the Farakka Barrage issue in a positive political framework. However, India did not make any concession to Bangladesh. The two countries took up the issue and held several rounds of high-level talks which resulted in the formation of an Indo-Bangladesh Joint River Commission. However, its deliberations did not lead to a settlement as India kept insisting on more and more studies. A breakthrough of sorts occurred during Sheikh Mujib-ur-Rahman's visit to India in 1974 when the prime ministers of the two countries issued a Joint Declaration in which they agreed on the allocation of water during the lean period and also on the need to augment the flow of water during the dry season. The second part of the relevant provision of the Joint Declaration had special significance as it stated that the augmentation would be undertaken through 'optimum utilization of the *water resources of the region* available to the two countries ...'.[93]

What India had in mind by this formulation was that the waters of the Brahmaputra River would be used to augment the flow of the Ganges River by interlinking the two which was in utter contradiction to Bangladesh's policy. In April 1975, India demanded and Bangladesh accepted the test running of the feeder canal through diversion of waters from the Ganges River. The two countries then announced through a joint press release a Partial Accord on the Farakka Barrage. It was partial because it lasted for forty-one days during the lean period in 1975 and after its expiry it was neither renewed nor replaced by another agreement.

The concession that Bangladesh made in favour of India was enormous on several counts. First, it accepted the Farakka Barrage through an agreement rather than let India complete it unilaterally and be tainted with illegitimacy. By virtue of the accord, it also accepted that the existence of the Farakka Barrage was no longer a contentious issue and that what remained to be settled was merely the allocation of water during the lean period. Secondly, it sent a delegation to India to participate in the inauguration ceremony of the Farakka Barrage which further put its stamp of approval on the project. Finally, it did not seek from India the release of a fixed quantity of water from the Ganges River until the settlement of the issue of augmentation which it should have done.

Why did Bangladesh make such a huge concession? The answer lies in the fact that India exerted political pressure and Sheikh Mujib-ur-Rahman thought that he would be able to persuade the Indian prime minister, Indira Gandhi, to make reciprocal concessions. India refused to oblige Bangladesh in the matter. In the face of the Indian hardline, Bangladesh took the plea that the Partial Accord was applicable to the test running of the feeder canal but to no avail. There was also a deadlock on the issue of augmentation of flow of the Ganges River. India wanted this to be done by diverting the Brahmaputra River through a canal linking it with the Ganges River whereas Bangladesh wanted it through storing of waters in reservoirs in the upper reaches of the Ganges River in India and Nepal.

These differences led to the deterioration of relations between the two countries. In August 1975, the Bangladesh armed forces staged a coup during which they killed Sheikh Mujib-ur-Rahman and some members of his family. There were a variety of reasons for his assassination, including perhaps 'the already growing tide of anti-Indianism in Bangladesh, symbolized by the disapproval of the commissioning of the Farakka Barrage [that] turned out to be one of the justifications for the brutal action.'[94] There was also an assassination attempt on the life of the Indian high commissioner in Dhaka in 1975. In the meantime, bilateral negotiations on the issue were totally deadlocked. Bangladesh was so frustrated by the Indian negotiating behaviour that on 21 August 1976, it decided to take the matter to the United Nations General Assembly. As expected, India opposed the move on the grounds that it was an attempt by Bangladesh to internationalize the issue which it felt was unwarranted. With the view of informing the members of the United Nations of its position in the matter, Bangladesh issued a white paper on the Farakka Barrage in which it gave details of the disastrous effects of the project. India retaliated by issuing its own white paper in which it rebutted the Bangladeshi claims. Bangladesh, however, succeeded in getting the matter inscribed on the agenda of the 31st session of the General Assembly. It was discussed in the Assembly's political committee where it presented a resolution which failed to muster enough support because of India's clout in the international community. However, it succeeded in getting a 'Consensus Statement' which marked a victory of sorts.

The adoption of the 'Consensus Statement' put considerable pressure on India to seek a settlement with Bangladesh. Consequently, it decided to resume negotiations with the latter in 1976. The two countries held a series of meetings at the ministerial level as a result of which they agreed to have a temporary arrangement on sharing the Ganges water, to be followed by a permanent settlement based on augmentation of the flow of the river. They signed an agreement in November 1977 on the above lines and also agreed on the share that each country would have in the water of the Ganges River. What is noteworthy here is that the Agreement used the word 'Ganga' which is the nomenclature used in India for the Ganges River rather than the expression 'Ganges' by which the river is internationally known.[95] The motive behind this semantic legerdemain was to drive home the point that the Ganges River was an Indian and not an international river. Salman M. A. Salman and Kishor Uprety in their analysis of the 1977 Agreement also subscribe to this view in these words: 'This was perhaps an assertion by India that the river is an overwhelmingly Indian river and as such the name given to it by India should be the name to be used in the Agreement.'[96] This point is further reinforced when we look at column 4 of the Schedule to the Agreement entitled 'Release to Bangladesh', which, in the opinion of Salman M. A. Salman and Kishor Uprety, implies that these releases to Bangladesh were 'something less than a right.'[97] They go on to add that '[p]erhaps those were concessions that Bangladesh had to make to get the Agreement.'[98]

The 1977 Agreement was meant to last for five years. However, it ceased to be operational in May 1982, or six months before the date of its expiry. The president of Bangladesh, Hussain Mohammed Ershad, visited India in October 1982, or less than a month before the termination of the Agreement during which he sought its extension under article XV. India refused, forcing Bangladesh to conclude a MoU with it in October 1982 which was valid for the next two years. The salient feature of the MoU was that it deleted the guarantee clause of the 1977 Agreement according to which Bangladesh was to receive a minimum of 80 per cent of the water that it was slated to receive and replaced it with a formula which was less favourable to Bangladesh. The latter felt the effect of the new clause in the 1983 dry season when out of the promised share of 35,000 cusecs, it received only 25,000 cusecs. As opposed to the 1977 Agreement,

India also forced Bangladesh to accept the use of the expression 'Ganga' rather than the Ganges in the MoU. In November 1985, the two countries signed a new MoU which was valid for three years and whose clauses relating to water sharing were similar to those of the 1982 MoU. The new agreement used both the expressions 'Ganga' and the Ganges which was a concession to Bangladesh. However, the Joint Expert Committee established under the 1985 MoU reproduced the augmentation clause of the 1972 Joint Declaration by putting emphasis on the optimal utilization 'of the surface water resources of the region' which meant that it supported the Indian proposal to link the Brahmaputra River with the Ganges River which, as seen above, was an anathema to Bangladesh.

In 1988, Bangladesh sought a permanent water sharing arrangement with India following the pattern of the 1985 MoU, along with a burden sharing clause in case of exceptionally low flow. Why did Bangladesh show readiness to seek an agreement which was unfavourable to its cause? Salman M. A. Salman and Kishor Uprety have explained: 'This concession from Bangladesh was understandable, with another dry season about to start. However, India was in no mood to cooperate, having seen its proposal for augmentation of the flow of the Ganges constantly rejected by Bangladesh. India kept insisting that any water sharing arrangement should be linked to a study of proposals to augment the flow of the Ganges during the dry season. With no agreement for sharing the dry season flow of the Ganges, India continued its diversions of the waters of the Ganges, and again Bangladesh began to complain about the adverse effects of India's withdrawals.'[99] A number of meetings at the ministerial, secretary, and Joint Rivers Commission level were held between the two countries to sort out the issue but without any result. The Indian attitude in the matter was so uncompromising that Bangladesh was forced to, once again, raise the issue at the 1993 session of United Nations General Assembly where it accused India of reneging on the promises that it had made on water sharing.

There was a change of governments both in Bangladesh and India in 1996. The Awami League, which enjoyed close ties with India, won the elections in Bangladesh and Inder Kumar Gujral of the Congress Party, who headed a collection of 13 regional parties, was successful in India. The two governments collaborated with each other to work out a Treaty which they signed on 12 December 1996.[100] The Treaty focused on sharing the waters of the Ganges River and this was valid for thirty years. Bangladesh 'gained' from the Treaty as, unlike the previous agreements, there was no linkage between the water sharing arrangements and the scheme for augmentation of the flow of waters of the Ganges River. Another 'gain' for Bangladesh was that, unlike the 1977 Agreement and the 1985 MoU, India did not insist on calling the Ganges an Indian river and agreed to call it Ganges/Ganga. However, Bangladesh had to pay a heavy price for these 'concessions'. First, it had to abandon its claim to the water of the Ganges River based on 'historic uses' just as Pakistan was made to do on the Eastern Rivers under the Indus Waters Treaty. Secondly, the Treaty reduced Bangladesh's share of water from 59 to 52 per cent. Thirdly, it did not incorporate the guarantee clause on the basis of which Bangladesh could claim a minimum of supplies from the Ganges River. Fourthly, it did not provide the latter with protection in terms of water-sharing arrangements. Thus, if the flow of water fell below 50,000 cusecs, Bangladesh was at the mercy of India for help. In other words, as noted by one commentary, 'Moreover, future reviews of the Treaty could result in a decrease of the share of Bangladesh to 90 per cent of the share specified under the Treaty.'[101] Commenting on the Indian negotiating behaviour, B. M. Abbas A. T., the leading Bangladeshi expert and the principal negotiator on water issues has passed the following verdict:[102]

The last 34 years of negotiations on Ganges water has shown the futility of trying to solve bilaterally a problem that is essentially multilateral in character. It has not brought about any improvements in the situation for Bangladesh; rather Bangladesh has lost the bulk of the dry season flow of the Ganges, as well as of other rivers. While Bangladesh has

been engaged in a never ending dialogue with India, India has gone
ahead with her schemes of building barrages and irrigation projects
on the Ganges one after another and withdrawing water progressively
reducing the flows in Bangladesh ... Bangladesh can only hope that
India will yet give evidence of its oft-repeated expressions of goodwill
by giving in Bangladesh its right full share in the waters of the Ganges
and other common rivers.

(ii) Teesta River Dispute

Another outstanding dispute between Bangladesh and India relates to
the sharing of the waters of the Teesta River. Originating in the Indian
state of Sikkim, the Teesta River is 414 kilometres long, of which 151
kilometres lie in Sikkim, 142 kilometres along the Sikkim-West Bengal
boundary and West Bengal and 121 kilometres in Bangladesh. After
traversing the Indian territory, it joins the Brahmaputra River in the
Rangpur district of Bangladesh. Its basin covers an area of 12,159
square kilometres of which 10,155 lie in India (which is more than
70 per cent hilly) and 2,004 in Bangladesh. With a maximum flow of
about 280,000 cusecs and a minimum flow of about 10,000 cusecs
of water, it is the fourth largest transboundary river of Bangladesh.
India has planned about 30 major hydropower projects involving large
dams on it. Of these, the largest one is the Teesta Barrage Project
(TBP) which is located in eastern India. When completed, it would
irrigate 922,000 hectares of land. India initiated it in 1976 and plans
to complete it in stages, some of which have already been achieved
like the Teesta Barrage at Gazaldoba and barrages on the Mahananda
and Dauk Rivers. It is also developing two low dams in the Darjeeling
district in West Bengal.[103]

The dispute between East and West Bengal goes back to the time
when the former was part of Pakistan. The negotiations between
Pakistan and India on the issue took place in the 1950s and 1960s
but proved inconclusive. After East Bengal became independent,
Bangladesh and India established a Joint Rivers Commission to deal
with the issue. In 1983, the Commission reached an ad hoc agreement

by dividing 75 per cent of the water, allocating 39 per cent to India and 36 per cent to Bangladesh and deferring the allocation of the remaining 25 per cent till after further study. It was not a formal agreement (hence made part of the minutes of the Commission's report) and was to last two years. In 1997, the two countries set up a Joint Commission of Experts which failed to achieve any results. In 2004, a Joint Technical Group replaced the Commission but it too failed to resolve the matter. In 2010, the prime ministers of the two countries met and issued a joint communiqué in which they called for a resolving of the issue expeditiously. Later on, they agreed on a draft agreement and a statement of principles which became the basis of a treaty that divided 80 per cent of the water with each country getting 40 per cent, 20 per cent left for the river flow. The treaty was ready for signature in 2011 when disagreements between the two countries cropped up. The signing was then scheduled for September 2013 but could not take place because of opposition from the West Bengal government which wanted 75 per cent of the water for itself and 25 per cent for Bangladesh. The deadlock persists to date because apparently the former is not prepared to budge from its position.[104]

Is the Indian claim to an enhanced share of water on the Teesta River justified? The answer is in the negative. Consider the following: according to the Asia Foundation study on the Teesta River, about 21 million Bangladeshis directly or indirectly depend on it for their livelihood. Additionally, the Teesta floodplain covers about 14 per cent of the total cropped area of Bangladesh and directly provides livelihood to about 9.15 million of its people or about 7.3 per cent of its population.[105] It has been estimated that merely 8 million Indians and ½ million Sikkimese depend on the Teesta River for their livelihood. In other words, the ratio of Bangladeshi and Indian populations dependent on it is 70 and 30 per cent respectively.[106] The Gazaldoba Dam renders the Dalia Barrage, which Bangladesh has built on its side of the border, almost useless during the dry season. It has affected more than 20 tributaries of the Teesta River as many of them have dried up. It has also been reported that people belonging to various professions dependent on the Teesta River are changing their jobs because the river stays dry for six months in a year. On the other hand, the release of excess water by India during the monsoon period causes floods, bank erosion, and extensive damage to crops. Accusing India of stealing Bangladesh's share of water, Bangladeshi political parties, including the Khalida Zia's BNP, have held long marches against it. They have blamed the government of Hasina Wajid for selling Bangladeshi water to India to stay in power.[107]

The matter does not end here. Bangladesh is quite concerned about India's Inter-Linking of Rivers Programme which also involves utilization of the Teesta River. Under the plan, India intends to divert water from the Himalayan and peninsular rivers through some thirty inter-basin canals and dams to areas of southern India which are prone to water scarcity and drought. Through this project, it proposes to irrigate almost 30 million hectares of land and generate 20,000 to 25,000 megawatts of electricity. Bangladesh fears that the project, if and when completed, would increase flooding in the monsoons and reduce availability of water during the dry season. Bangladesh is of the view that the project violates the 1966 Helsinki Rules on uses of water resources and the 2004 Berlin Rules on equitable sharing of river waters between co-riparians. It may not be out of place to mention here that in February 2012, the Supreme Court of India ordered the setting up of a special committee to expedite the implementation of the project.[108] In light of these observations, it is obvious that India is utterly unjustified in seeking an enhanced share of water from the Teesta River.

The foregoing analysis shows that India's water diplomacy with regards to Nepal and Bangladesh is predatory in character; it is not based on the equitable sharing of waters of the rivers which they have in common, but is driven by the desire to appropriate as much water as possible. It signifies that India's treatment of Nepal and Bangladesh is no different from that of Pakistan. We can therefore say that Pakistan's apprehensions about India's intentions are justified. Given this perspective, what should be Pakistan's response if India seeks redressal of the 'inherent injustice' that the Treaty supposedly inflicted on it by allocating more water to Pakistan or denounces it on any of the grounds discussed above? In such an eventuality, Pakistan should categorically refuse to go along with any such demand or denunciation because, as shown above, they are utterly unwarranted. Incidentally, there is a consensus among water experts, political circles, and the educated laymen of Pakistan that the country should under no circumstances reopen the Treaty. However, that does not mean that there is no room for improvement in the Treaty. On the contrary, there is a dire need to address the shortcomings or lacunae, particularly relating to pollution and issues arising out of climate change that the two countries have discovered in the course of its implementation during the last 50 odd years.

REFERENCES AND NOTES

1. For example, this is how one prominent Indian writer has commented on the functioning of the Treaty: 'The Indus Waters Treaty must rank among the triumphs of the United Nations system since it was signed in 1960. It has worked remarkably well in keeping the peace ...' (B. G. Verghese, 'The Indus, POK and the Peace Process: Building on the Foundations of the Indus Waters Treaty', in Virendra Gupta and Alok Bansal (eds.), *Pakistan Occupied Kashmir: The Untold Story* (New Delhi: Manas Publications, 2007), 195. The Neutral Expert in the *Baglihar* case also declared the Treaty a success story in these words: 'His opinion [Neutral Expert's] is that in fact, specific Parties emerge successfully from the treatment of this difference: the authors of the Treaty. The Treaty is the successful document'. *Baglihar Hydroelectric Plant, Expert Determination, Executive Summary* (Lausanne: 12 February 2007), 20.

2. Ramaswamy Iyer, 'Indus Treaty: A Different View', *Economic and Political Weekly* (16 July 2005), 3140. However, Iyer in a vein of realism has also observed that, '[t]he high praise of the IWT as a successful instance of conflict resolution seems somewhat exaggerated. Echoing E. M. Forster on democracy, one might say that two cheers are quite enough for the IWT and that three cheers are not called for.' Ibid. 3141. John Briscoe has made similar observations in these words: 'Whereas once the Indus Waters Treaty (IWT) could correctly be described as a beacon of light in an otherwise gloomy relationship, the situation has changed.' John Briscoe, 'Troubled Waters: Can a Bridge be Built over the Indus.' *Economic and Political Weekly* (11 December 2010), vol. XLV, no. 50, 28.

3. For example, a Pakistani group called Engineers Study Forum has revealed in a study that India is stealing 15–20 per cent of water from the Western Rivers causing a loss of $12 billion to Pakistan's agriculture, <http://Siyasipakistan. wordpress.com/2010/02/08/india-stealing-water-causing-12-bln-loss-to-Pakistan/>.

4. Sandeep Waslekar, *The Final Settlement: Restructuring India-Pakistan Relations* (Mumbai: Strategic Foresight Group, 2005), Executive Summary, 3, <http://strategicforesight.com/publications_inner.php?id=24#.Vqrp65p942w>. The study further adds: 'The statements made by Pakistan's military officials, Kashmiri leaders, and newspaper editorials describe Jammu and Kashmir as a supplier of crucial rivers, and project the bloodshed there as the sacrifice made by Kashmiri youth to ensure Pakistan's water security.' Ibid. 4.

5. Kayani on a visit to Washington reportedly argued that 'water had replaced Kashmir as the primary non-military concern with India.' (John Briscoe, 'Troubled Waters', supra note 2, 30). Sandeep Waslekar traces this change in Pakistan's attitude to a dissertation that the former President Musharraf wrote as a student at the Royal College of Defence Studies, London, in 1990. According him, Musharraf argued in it that water had come to replace the territory of Jammu and Kashmir as the core issue between Pakistan and India and that its resolution was the key to peace in the subcontinent. He also points out that Musharraf proposed division of Jammu and Kashmir between Pakistan and India on the basis of the Chenab Formula as a solution because, in his opinion, it would give Pakistan control of the catchment area of the three Western Rivers (*Final Settlement*, Part III: Water, Chapter 6: The Secret, supra note 4). Waslekar believes that Musharraf triggered the Kargil war in order to secure the three Western Rivers. Ibid.

6. We can only make an educated guess as to why the two countries did not show any enthusiasm for the celebration of the fiftieth anniversary of the Treaty. Pakistan did not do so perhaps because, given the largely shared view that Pakistan should not have surrendered the three Eastern Rivers to India, it might have raised an unnecessary controversy regarding the wisdom of concluding the Treaty. Given again the largely shared view that India is building dams on the Western Rivers in violation of the Treaty, it did not feel that there was anything to celebrate. India did not manifest much enthusiasm perhaps because it would have been incongruous with its desire to get rid of the Treaty.

7. Salman M. A. Salman, the former lead counsel of the Bank made this claim to the present writer in the course of an interview that he gave at his residence in Washington DC, on 22 August 2010.

8. There are others who belong to this school. For example, the authors of the IDSA Report who made the following statement: 'Given some stringent provisions in the Indus Waters Treaty that thwart India's plans of developing projects on the western rivers, a "modification" of the provisions of the treaty

should be called for. Whether it is done through renegotiations or through establishing an Indus II treaty, modifications of the provisions are crucial in case of the western rivers.' *Water Security for India: The External Dynamics,* Institute of Defence Studies and Analyses (IDSA), (New Delhi, September 2010). 43.

9. B. G. Verghese, 'It's Time for Indus-II: A New Treaty will Help the Peace Process too', *The Tribune,* 26 May 2007.

10. It provides as follows: 'The provisions of this Treaty may from time to time be modified by a duly ratified treaty concluded for that purpose between the two Governments.'

11. B. G. Verghese, 'Talk of Abrogating Indus Waters Treaty', *The Tribune* (29 April 2002).

12. Ramaswamy Iyer, 'Indus Treaty', supra note 2, 3144. Emphasis original.

13. Ibid.

14. Erin Blankenship, 'Kashmiri Water: Good Enough for Peace', *Pugwash on Line* (July 2010), <http://cc.bing.com/cache.aspx?q-http%3a%2fwww.pugwash. org%2freports%2frc%2f2fsa%[http://www.indiawaterportal.org/articles/ kashmiri-water-good-enough-peace]>; Robert Wirsing and Christopher Jasparro, *Spotlight on Indus River Diplomacy: India, Pakistan and the Baglihar Dam Dispute,* Asia-Pacific Centre for Security Studies, Department of Defense (May 2006), <http://www.apcss.org>.

15. Iyer admitted this fact in one of his statements: 'I have been writing extensively on India-Pakistan water relations and on the Indus Waters Treaty 1960. Whenever the water issue flares up between the two countries, the Ministries of External Affairs and Water Resources and even the Prime Minister call me in for consultations. I have a degree of readership in Pakistan too. I am an active participant in Track II initiatives between the two countries', Per Steineide Refseth, *The Indus River Basin, 1999–2008: An Intellectual History of Hydropolitics,* master's thesis submitted to the Institute of Archaeology, Conservation and History, University of Oslo, Spring 2013, 22. He also admitted that he had been asked to contribute by Pakistani newspapers and was from time to time consulted by India's high commissioner in Pakistan. Ibid.

16. Jawed Naqvi, 'New Delhi Planning Tougher Actions: Scrapping of Indus Treaty, Suspension of Overflights', *Dawn* (23 December 2001).

17. Kak, M. L., 'Indus Waters Treaty must go: PM', <http://www.tribuncindia. com/2002/200205 23/main3.htm>.

18. A. G. Noorani, 'A Treaty to Keep', *Frontline* (13–26 April 2002), vol. 19, issue 8.

19. Ibid.

20. *Water Security for India,* supra note 8, 43.

21. As opposed to the official line important Indian analysts have opposed starvation of Pakistanis as a means to punish Pakistan as being contrary to humanitarian law. See B. G. Verghese, 'Talk of Abrogating', supra note 11; A. G. Noorani, 'A Treaty to Keep', supra note 18.

22. That explains why the British prime minister, Harold Wilson, described the Indian attack on Lahore during the Pakistan-India war in 1965 as an 'act of aggression' but did not do so when Pakistan introduced its troops in Jammu and Kashmir.

23. Derek W. Bowett, *Self-Defence in International Law* (Manchester University Press, 1958), 34.

24. A. O. Cukwurah, *The Settlement of Boundary Disputes in International Law* (Manchester University Press, 1967), 7.

25. It is generally believed that the Kashmir resolutions are exhortatory in character and are not binding on the parties because they were adopted under chapter 6 of the UN Charter. However, both Pakistan and India affirm that there are two resolutions adopted by the United Nations Commission for India and Pakistan (UNCIP) on 13 August 1948, and 5 January 1949, which call for a plebiscite in Jammu and Kashmir and are binding on the parties because they were concluded with the consent of the parties involved. For a detailed discussion on the subject, see Ijaz Hussain, *Kashmir Dispute: An International Law Perspective* (Islamabad: Quaid-i-Azam Chair, National Institute of Pakistan Studies, QAU, 1998), 180–5.

26. The text of the 'Draft Articles' is available in the annex to the General Assembly resolution 56/83 of 12 December 2001. It is also available online, <http://untreaty.un.org/ilc/texts/instruments/english/draft%articles/9_6_2001.pdf>. On 12 December 2001, while adopting the resolution, the General Assembly 'commended [the Draft Articles] to the attention of Governments without prejudice to the question of their future or other appropriate action.' The Draft Articles have not been given the shape of a convention because whereas some countries favoured convening a diplomatic conference for the purpose others were in favour of maintaining it as an ILC text approved *ad referendum* by the UN General Assembly. Even though the Draft Articles has not been adopted as a convention, it has widely been cited by various bodies including the World Court which did, for example, in the *Gabcikovo-Nagymaros* case between Hungary and Slovakia.

27. Ibid. 47.

28. *ICJ Reports* 1986, 64–5, para. 115.

29. Chandrakant D. Thatte, 'Indus Water and the 1960 Treaty between India and Pakistan', in Olli Varis; C. Tortajada, and A. K. Biswas (eds.), *Management of Transboundary Rivers and Lakes* (Heidelberg: Springer-Verlag Berlin, 2008), 206.

30. Draft Articles on the Law of Treaties with Commentaries, *Yearbook of International Law Commission* (1966), vol. II, 259, para. 9.

31. David E. Lilienthal, 'Another Korea in the Making?', *Collier's* (4 August 1951).

32. Daniel Nelson, 'Water War Warning as Tension Escalates in Kashmir', *One World.net* (21 May 2002), accessed online at Common Dreams News Centre:

<http://www.commondreams.org/egi-bin/print.egi?file=headlines02/0521-07. htm> cited in Erin Blankenship, 'Kashmiri Water', supra note 14, 5.

33. Ashfak Bokhari, 'Water Dispute and War Risk', *Dawn (Economic and Business Review)* (18–24 January 2010), VI. Other bodies, like the American CIA, have warned of the possibility of an armed conflict between Pakistan and India over the water issue in the twenty-first century. For details, see the 'Conclusion' of this book.

34. Ramaswamy Iyer, 'Was the Indus Waters Treaty in Trouble?' *Economic and Political Weekly* (22–28 June 2002), vol. 37, no. 25, 2402.

35. Ibid. Iyer also terms abrogation talk as 'irresponsible and indefensible' on the ground that withdrawal from a treaty by a party because of dissatisfaction with it would deprive treaties of force and sanctity. Ibid.

36. Zahir-ud-Din, 'Seeking Abrogation of IWT is as Good as Seeking Independence', <http://www.defenceforumindia.com/forum/threads/abrogate-indus-water-treaty.12647/>.

37. It is a diversion structure (locally called Lang Gorge Khambab) which China is building across the Zada Gorge on the Sutlej River. The Indian government does not appear to be bothered about the possible reduction of water in the flow of the river on the grounds that the upstream uses in Tibet would be minimal due to the terrain and sparse population. S. M. Menon, 'Concerns over Chinese Barrage on Sutlej', *The Hindu* (30 July 2006).

38. It is a medium-sized dam that China has built on the Indus River near Demchok, Ladakh. It would initially generate 11 megawatts of electricity and has the capacity to double the output. According to Pakistani sources (on the authority of Idris Rajput, former secretary, Sindh Irrigation Department), the dam can be used for power generation only. Since Aksai Chin, Ngari, and Eastern Ladakh are located at a height of 13,000 feet above the mean sea level, the dam cannot be put to agricultural uses. Senge S. Sering, 'China Builds Dam on the Indus near Ladakh', *Journal of Defence Studies* (April 2010), vol. 4, no. 2, 136–9.

39. M. S. Menon, 'Indus Waters Treaty Needs Relook', *The Pioneer* (27 November 2013).

40. Excelsior Correspondent, 'Assembly's Unanimous Call for Scrapping Indus Treaty', *The Daily Excelsior* (4 April 2002), <http://www.hvk.org/ articles/0402/35.html>. There are voices in Kashmir which oppose the Kashmir Assembly's approach in the matter. For example, a well-known commentator, Dr Sheikh Showkat Hussain, advised his fellow Kashmiris to safeguard Kashmir's interests within the ambit of the Treaty rather than denounce it. He has contended that the state must seek its share within what India has got out of the Treaty. He specifically mentioned in this regard that it has not been compensated for the benefits which Punjab, Haryana, and Rajasthan got out of the Treaty. He went on to say that the green revolution in

India was entirely indebted to the Treaty. Zahir-ud-Din, 'Seeking Abrogation' supra note 36.

41. 'IWT must Go', supra note 17.

42. Excelsior Correspondent, 'Assembly's Unanimous Call', supra note 40.

43. Chandrakant D. Thatte, 'Indus Waters and the 1960 Treaty', supra note 29, 204.

44. Ramaswamy Iyer, 'Was the Indus Waters Treaty in Trouble?', supra note 34, 2402. Brahma Chellaney, the former member of the Indian National Security Council Advisory Board and the former member of the Indian Policy Advisory Group however disagrees with Iyer by contending that Nehru ignored the interests of Jammu and Kashmir while signing the Treaty. Chellaney, 'Water Treaties and Diplomacy: India Faces Difficult Choices over Water', *The Economic Times* (10 May 2012).

45. *The Asian Age* (12 March 2003).

46. Ramaswamy Iyer, 'Indus Treaty', supra note 2, 3141; *Water Security for India*, supra note 8, 39; Uttam Kumar Sinha, '50 Years of Indus Waters Treaty', *Strategic Analysis* (September–October 2010), 66–70.

47. Ramaswamy Iyer, 'Indus Treaty', supra note 2, 3141.

48. Garg who invokes articles IV and V of the Helsinki Rules bases his calculations as follows. i) On the basis of the population as at the time of Partition, India's entitlement 21 million/46 million= 45.65 per cent, Pakistan's 25 million/46 million= 54.35 per cent; ii) On the basis of the culturable area, India's 26 million acres/65 million acres=40 per cent, Pakistan's 39 million acres/65 million acres=60 per cent; iii) On the basis of the respective drainage areas, India's 3,21,289 sq. km/10,13,985 sq. km=31.7 per cent, Pakistan's 6,92,696 sq. km/10,13,985 sq. km=68.3 per cent; iv) On the basis of the length of the arms or river beds, India's 3500 km/6500 km=53.84 per cent, Pakistan's 3000km/6500km=46.16 per cent. He concludes from the foregoing that India deserved 45.65+40+31.7+53.84/4=42.8. He contends that the Bank should have proceeded on the basis of the quantum of water to be divided rather than the rivers. Besides, he contends that it should have included the Kabul River and not restricted itself to the six rivers. Santosh Kumar Garg, *International and Interstate River Water Disputes* (New Delhi: Laxmi Publications (p) Ltd., 1999), 85. See also K. Warikoo, 'Indus Waters Treaty: View from Kashmir', <http://www. jammu-kashmir.com/insights/insight20060601a.html>.

49. For the text, see <http://digi.library.tu.ac.th/thesis/la/1417/17APPENDIX_C. pdf>.

50. Ramaswamy Iyer, 'Indus Treaty', supra note 2, 3143.

51. Aloys Michel, *The Indus Rivers: A Study of the Effects of Partition* (New Haven and London: Yale University Press, 1967), 8.

52. *Pakistan's Water Resource Assistance Strategy* (Washington DC: World Bank, 2005-PK-34081), 7–8.

53. Trilochan Upreti, *International Watercourses Law and its Application in South Asia* (Kathmandu: Pairavi Prakashan, 2006), 63. The other arguments that Upreti invokes to justify Pakistan's contention is the dependence of its huge population and considerations spelled out in article 6 of the United Nations Convention on International Watercourses, 1997, which lays down as follows:

1) Utilization of an international watercourse in an equitable and reasonable manner within the meaning of article 5 requires taking into account all relevant factors and circumstances, including: a) Geographic, hydrographic, hydrological, climatic, ecological and other factors of a natural character; b) The social and economic needs of the watercourse States concerned: c) The population dependent on the watercourse in each watercourse State; d) The effects of the use or uses of the watercourses in one watercourse State on other watercourse States; e) Existing and potential uses of the watercourse; f) Conservation, protection, development and economy of use of the water resources of the watercourse and the costs of measures taken to that effect; g) The availability of alternatives, of comparable value, to a particular planned or existing use.

2) In the application of article 5 or paragraph 1 of this article, watercourse States concerned shall, when the need arises, enter into consultations in a spirit of cooperation.

3) The weight to be given to each factor is to be determined by its importance in comparison with that of other relevant factors. In determining what is reasonable and equitable use, all relevant factors are to be considered together and a conclusion reached on the basis of the whole.

54. Ramaswamy Iyer, 'Indus Treaty', supra note 2, 3141.

55. *Water Security for India*, supra note 8, 15. According to the Falkenmark index, which the Bank report uses, a 'water stressed' country is one where water availability ranges between 1,000 and 1,700 cubic metres per person per annum whereas 'water scarce' is one where it ranges between 500 and 1,000 cubic metres.

56. McKinsey Report, *Charting our Water Future* (November 2009), <http://www.mckinsey.com/business-functions/sustainability-and-resource-productivity/our-insights/charting-our-water-future>, Executive Summary, 10.

57. *Water Security for India*, supra note 8, 22.

58. Ibid. 20.

59. Brahma Chellaney, *Water: Asia's New Battleground* (Washington DC: George Town University, 2011), 226. Emphasis added.

60. David E. Lilienthal, 'Another Korea', supra note 31. Emphasis original.

61. John Briscoe, 'War or Peace on the Indus' (18 April 2010), <http://www.thefrontierpost.com/News.aspx?ncat_ar&nid+255>.

62. *Avoiding Water Wars: Water Scarcity and Central Asia's Growing Importance for Stability in Afghanistan and Pakistan*, A Majority Staff Report, Committee on Foreign Relations, US Senate, 12th Congress, 22 February 2011, 9.

63. Zulfiqar Halepoto, 'No Integrated Approach', *Dawn* (24 May 2011).

64. Ibid.

65. 'Strategic depth' is a military term which signifies the distance between enemy forces and the main centre of gravity of a country encompassing not only military bases but also economic and commercial hubs. Military planners are of the view that the more distance enemy forces have to cover to reach these bases, the greater the chances of successful defence as this strategy leads to a war of attrition (Gharanai Khwakhuzhi, 'Afpak: Strategic Depth', *Khaama Press* (20 December 2013), <http://www.khaama.com/afpak-the-strategic-depth-9876>). It has been alleged that Pakistan's military establishment has pursued the policy of 'strategic depth' to control Afghanistan. Mirza Aslam Beg, the former head of the ISI and the COAS, is credited with formulating the concept in the 1980s to deal with the dilemma of the Indian threat. Pakistan's military establishment, however, denies the allegation. Pakistan's former COAS, Gen. Ashfaq Pervaiz Kayani, on many occasions stated that the concept is not meant to control Afghanistan but to ensure a 'peaceful, friendly and stable' relationship with Afghanistan in order to ensure that Pakistan is not threatened with long-term security problems with Afghanistan. He also clarified that a talibanized Afghanistan is not in Pakistan's interests.

66. Larry Hanauer and Peter Chalk, *India's and Pakistan's Strategies in Afghanistan: Implications for the United States and the Region* (Centre for Asia Pacific Policy, International Programs of the Rand Corporation, 2012), 11–13.

67. When Hamid Karzai was president of Afghanistan, he developed very close ties with India and an antagonistic relationship with Pakistan. When it came to raising the Afghan national army, he decided to send his officers for training to India which was not to the liking of Pakistan. However, Ashraf Ghani, following his election as president, reversed this policy by mending fences with Pakistan and downgrading Afghanistan's relationship with India. Following this paradigm shift, he decided to send Afghan cadets to Pakistan rather than to India and not to seek military hardware from India which his predecessor had earlier requested. The change in Ghani's thinking is explained by Pakistan army's operation in North Waziristan and help in the Afghan peace process. What this new arrangement holds lies in the womb of time.

68. Emphasis added. See <http://www.outlookindia.com/article/pakistanis-have-blown-my-comments-out-of-proportion/261113>. It may be noted here that the statement in question was supposed to have figured in an article entitled 'What is the Problem with Pakistan?' written by Ms Fair for the *Foreign Affairs* magazine but never saw the light of the day. On 18 January 2016, we contacted Ms Fair through an email seeking her comments on the issue to

which she responded with the following observation 'Well … Foreign Affairs removed it when they reorganized their site.' What does that signify? Should we understand that *Foreign Affairs* did not publish the article because it did not want to embarrass Ms Fair for having exposed India regarding its role in Balochistan? It may be noted that following the Freudian slip in an unguarded moment during the roundtable discussion she subsequently maintained that she did not mean what the Pakistanis understood by that statement; and that they blew her statement out of proportion. The recent arrest and confessions of the Indian spy Kulbushan Yadav, a serving officer of the Indian Navy, have confirmed Pakistan's claims of India's subversive activities in Balochistan.

69. Chuck Hagel, 'India Financed Problems for Pakistan from Afghanistan: Chuck Hagel', *The Express Tribune* (26 February 2013). Emphasis added.

70. *Agreement on the Kosi Project 1954*, see appendix 2, <http://www.internationalrivers.org/files/attached-files/treaties_between_nepal-india.pdf>.

71. B. C. Upreti, *Politics of Himalayan River Water: An Analysis of the River Water Issues of Nepal, India and Bangladesh* (Jaipur, New Delhi: Nirala Publications, 1993), 98.

72. Ibid. 119, notes 7–9.

73. *Agreement on the Kosi Project 1966* (amended), <http://www.internationalwaterlaw.org/documents/regionaldocs/Amended_Kosi_Project_Agreement-1966.pdf>.

74. Salman M. A. Salman and Kishor Uprety, *Conflict and Cooperation in South Asia's International Rivers: A Legal Perspective* (Washington DC: The World Bank, 2002), 73.

75. Aditya Man Shrestha, *Bleeding Mountains of Nepal* (Kathmandu: Ekta Books, 1999), 157. Aditya also claims that the Nepal government wanted to build the dam in the foothills of the Siwalik Range but the Indian government rejected the idea for fear that it might lose control over its management and operation because of the extended distance. Ibid.

76. Salman M. A. Salman and Kishor Uprety, *Conflict and Cooperation*, supra note 74, 73–4, note 21.

77. Ibid. 79.

78. *Gandak Irrigation and Power Project 1959*, <http://www.moen.gov.np/pdf_files/gandak_treaty.pdf>.

79. *Gandak Irrigation and Power Project 1964* (amended), see appendix 3, <http://www.internationalrivers.org/files/attached-files/treaties_between_nepal-india.pdf>.

80. Khaga Nath Adhikari, 'Turning the Tide: Developing Cooperation on Water Resources in South Asia', *Regional Studies* (Institute of Regional Studies, Spring 2015) vol. XXXIII, no. 2, 94.

81. Ibid.

82. *A Review of Existing Nepalese Laws, Policies and Practices on Land Acquisition, Compensation, Resettlement and Rehabilitation*, Report of the Development Law Inc. Nepal for WECS, vol. 1, 24 July 1998, 25–6, cited in Salman M. A. Salman and Kishor Uprety, *Conflict and Cooperation*, supra note 74, 94.

83. *Tanakpur Agreement 1991*, see appendix 4, <http://www.internationalrivers.org/
    files/attached-files/treaties_between_nepal-india.pdf>.

84. ATS Ahmed, 'Challenges of Governance in Nepal: Politico-Economic and
    Ethno-Religious Dimensions', *Journal of Contemporary Asia* (1994) vol. 24, 360–2;
    see also, Dipak Gyawali and Othmar Schwank, 'Interstate Sharing of Water
    Rights: An Alps-Himalayan Comparison', in Ajaya Dixit (ed.), *Water Nepal*
    (1994), vol. 4, no. 1, 23–5.

85. *Treaty concerning the Integrated Development of the Mahakali Barrage including Sarada
    Barrage, Tanakpur Barrage and Pancheshwar Project 1996*, <http://www.international
    waterlaw.org/documents/regionaldocs/Mahakali_Treaty-1996.pdf>.

86. Dipak Gyawali and Ajaya Dixit, 'How Not to Do a South Asian Treaty', *Himal*
    (April 2001), 9. <http://www.himalmag.com/component/content/article/
    article/1975=How-not-to-do-a-South-Asia-Treaty>.

87. Farzana Noshab and Nadia Mushtaq, 'Water Disputes in South Asia', Institute
    of Strategic Studies, Islamabad, <http://www.issi.org.pk/journal/2001-files/
    no_3/article/4a.htm>.

88. 'Indo-Nepal Mahakali Treaty has not been Properly Ratified', <http://
    ekantipur.com>, 8 December 2009, cited in Pia Malhotra, *Water Issues between
    Nepal, India and Bangladesh*, IPCS Special Report, July 2010, note 20.

89. Ben Crow et al., *Sharing the Ganges: The Politics and Technology of River Development*
    (New Delhi: Sage Publications, 1995), 84, cited in Salman M. A. Salman and
    Kishor Uprety, *Conflict and Cooperation*, supra note 74, 136.

90. For the text of the treaty, see *League of Nations Treaty Series*, vol. 7, 35. For the
    Indian denunciation, see *Multilateral Treaties Deposited with the Secretary-General*,
    Status as at 31 December 1997, 982.

91. B. M. Abbas A. T., *The Ganges Waters Dispute* (Bangladesh: University Press
    Limited, 1982), 19.

92. Salman MA Salman and Kishor Uprety, *Conflict and Cooperation*, supra note 74,
    137.

93. Emphasis added. The full text of the relevant clause is as follows:

    17. The two Prime Ministers took note of the fact that the Farakka Barrage
        Project would be commissioned before the end of 1974. They recognised
        that during the periods of minimum flow in the Ganga, there might
        not be enough water to meet the needs of the Calcutta Port and the
        full requirements of Bangladesh and therefore, the fair weather flow
        of the Ganga in the lean months would have to be augmented to
        meet the requirements of the two countries. It was agreed that the
        problems should be approached with understanding so that the interests
        of both countries are reconciled and the difficulties removed in the
        spirit of friendship and cooperation. It was accordingly decided that
        the best means of such augmentation through the optimum utilization
        of the water resources of the region available to the two countries

should be studied by the Joint Rivers Commission. The Commission should make suitable recommendations to meet the requirements of both countries.

See Avtar Singh Bhasin, *India-Bangladesh Relations, 1971–1994: Documents,* vol. 1 (Delhi: Siba Exim Pvt., Ltd., 1996), 88–93.

94. *Keesings Contemporary Archives,* vol. XXI, 27381, cited in Iftekharuzzaman, *The Ganges Water Sharing Issue: Diplomacy and Domestic Politics in Bangladesh* (Bangladesh: Institute of International Strategic Studies, 1994), vol. 15, no. 3, 222.

95. The side letter to the 1977 Agreement, however, used the expression 'Ganges'.

96. Salman M. A. Salman and Kishor Uprety, *Conflict and Cooperation,* supra note 74, 154.

97. Ibid.

98. Ibid.

99. Ibid. 168–9.

100. *Ganges Treaty 1996,* <http://www.gov.bd/attachment/Ganges.Water_Sharing_treaty,1996.pdf>.

101. Salman M. A. Salman and Kishor Uprety, *Conflict and Cooperation,* supra note 74, 177.

102. B. M. Abbas A. T., 'Agreement on the Ganges' in M. Ali, G. E. Radosevich and A. A. Khan (eds.), *Water Resources Policy for Asia* (A. A. Balkema Publishers, 1987), 537–8.

103. *Political Economy Analysis of the Teesta River Basin* (The Asia Foundation, March 2013), 10, <http://asiafoundation.org/resources/pdfs/TheAsiaFoundation.Political EconomyAnalysisoftheTeestaRiverBasin.March20131.pdf>.

104. Ibid. 13.

105. Ibid. 8.

106. Sajjad Shaukat, 'India Suppresses Bangladesh on Teesta River Issue', *Writing for Godot* (20 May 2014), <http://readersupportednews.org/pm-section/21-21/23778-india-suppresses-bangladesh-on-teesta-river-issue>.

107. Ibid.

108. *Political Economy Analysis,* supra note 103, 11–12.

Conclusion

In 1960, when the Treaty was concluded, it was commonly believed that it had permanently taken care of water as an issue between Pakistan and India.[1] For the first four decades of its existence, the promise that it initially held appeared to be justified.[2] However, during the last decade or so it has come under considerable strain principally because India would like to get rid of it at the earliest opportunity either with a new treaty or a revised one. If India has not denounced the Treaty so far, it is because it fears that this would offend both the Bank and the Western powers that backed it politically and financially when it was signed. It equally fears that this would put a dent in its reputation as a law-abiding and responsible member of the international community in addition to scuttling the ongoing peace process between the two countries with serious consequences for the regional and global peace. Given this background, instead of outrightly denouncing it and earning the ire of the international community, it has decided to achieve its objective by endowing the dams and reservoirs that it is building on the Western Rivers[3] for the purpose of power generation with the capacity to withhold or release water at critical moments, which can not only ruin Pakistan's agriculture but also seriously affect its security.[4]

Pakistan views the new Indian dam-building spree as a threat to its security. India, on the other hand, rejects this charge and deems the large number of projects that it is undertaking on the Western Rivers strictly in accordance with the terms of the Treaty. Neutral observers, however, tend to agree with Pakistan's position. For example, John Briscoe, a Harvard professor, without mincing words has declared that, '[t]he cumulative storage of these dams will be large, giving India an unquestioned capacity to have major impact on the timing of flows into Pakistan. Using Baglihar as a reference, simple back-of-the-envelope calculations suggest that once it has constructed all of the planned hydropower plants on the Chenab, India will have the ability to effect major damage on Pakistan.'[5] Similarly, the US Senate Foreign Relations Committee's report issued in 2011 has endorsed this assessment in these words:[6] 'The number of dams under construction and their management is a source of significant bilateral tension .... While studies show that no single dam along the waters controlled by the Indus Waters Treaty will affect Pakistan's access to water, the cumulative effect of these projects could give India the ability to store enough water to limit the supply to Pakistan at crucial moments in the growing season.'

India's decision to build a series of dams on the Western Rivers has sounded alarm bells in Pakistan and made various leaders, civil and military, issue warnings of dire consequences. For example, the former COAS of the Pakistan army, General Ashfaq Pervaiz Kayani, justified Pakistan's India-centric approach to the water issue between the two countries.[7] The former finance minister of Zulfikar Ali Bhutto and current president of the Shaheed Bhutto faction of the PPP (Punjab), Dr Mubashir Hasan, has stated in a press conference:[8] 'It would be construed as an act of war by the other party if India or Pakistan make an attempt to violate the provisions of the Indus Basin Water Treaty between the two countries.' He did not stop there. Furthermore, he recommended military action 'to reduce India's capability to divert water from the rivers Ravi, Sutlej, and Beas for use in its Punjab, Haryana, and Rajasthan which will then naturally flow into Pakistan. A few suicide bombers can perpetrate such horrible acts causing widespread starvation in India.' Mansur Ejaz, an American-Pakistani businessman has also observed: 'Agriculture is not possible in Punjab and Sindh without river water. Therefore, unless Pakistan was assured the supply of water, it will never abandon the proxies that keep India on its toes by destabilizing Kashmir.'[9] Even non-state actors like the LeT have issued threats on the water issue. For example, Hafiz Saeed, during a protest rally on 7 March 2010 against the theft of Pakistan's water, accused India of 'water terrorism', and threatened that it could trigger a war between the two nuclear-armed neighbours if India did not stop this aggression.[10]

There are others in Pakistan who foresee a nuclear war if India deprives Pakistan of the water of the Western Rivers. For example, Majid Nizami, the late chief editor of the Urdu group of newspapers *Nawa-e-Waqt* has said: 'Pakistan can become a desert within the next ten or fifteen years. We should show upright posture or otherwise prepare for a nuclear war.'[11] Similarly, Palwasha Khan, a member of Pakistan's National Assembly, accusing India of 'water terrorism' against Pakistan, has said:[12] 'India's usurping of Pakistan's share of water may result in a confrontation. Experts foresee nuclear war over the issue in the future and any war in the region would be no less than a nuclear war.' Lt Gen. Khalid Kidwai, the former head of the Strategic Plan Division, made a statement about the circumstances that could make Pakistan resort to nuclear weapons which he identified as follows: 'India attacks and captures large parts of Pakistani territories; b) destroys large part of Pakistan's armed forces; c) India imposes an economic blockade; and d) India creates large-scale internal subversion in Pakistan.'[13] It has been said that the reference to 'economic blockade' incorporates India's strangulation of Pakistan by blocking water.[14]

Given these warnings, what options are available to effect a change in India's mindset in the matter? One option that John Briscoe has proposed consists of advice to India to show 'compassion' towards Pakistan on the ground that compared to India, Pakistan is a small and poor country. He pleaded that it should demonstrate leadership on the water issue. To illustrate his point, he made a comparison between the attitude of India and Brazil, the hegemons of South Asia and Latin America respectively, towards their neighbours. He pointed out that Paraguay, which has an agreement with Brazil on the huge Itaipu Binacional Hydropower Project, used earnings from it for an anti-Brazil campaign during the presidential election in that country. Similarly, he cited the example of Bolivia which mounted an anti-Brazil campaign during its presidential election by criticizing the gas agreement that it has with Brazil. The Brazilian media and public were outraged by the attitude of Paraguay and Bolivia. Consequently, they mounted pressure on their government to take action against them. However, swimming against the popular tide, the government of President Luis Inacio Lula decided to reject the demand by contending that[15] 'these are poor countries, and these are huge issues for them. They are our brothers. Yes, we are in our legal rights to be harsh with them, but we are going to show understanding and generosity, so I am unilaterally doubling (in the case of Paraguay) and tripling (in the case of Bolivia) the payments we make to them. Brazil is a big country and a relatively rich one, so this will do a lot for them and won't harm us much.' John Briscoe argued that taking a leaf from Brazil's book, India should try to understand Pakistan's perspective and act as a good neighbour and great power with a generosity of spirit towards the latter.

Briscoe's advice while well-intentioned, is utterly inappropriate because nations do not conduct their international relations on the basis of sentiments or feelings but national interest and realpolitik.[16] Therefore, there is no way that India would heed this advice. Conscious of this stark reality, Pakistan never bothered to contemplate it, let alone tender it. Instead, it tried to take care of the water issue by getting it included in the Composite Dialogue for which purpose in 2010 it submitted a non-paper[17] to the Indian government. In it, it called on India to strictly observe the terms of the Treaty by letting the waters of the Western Rivers flow to Pakistan uninterruptedly; provide complete and timely information on the run-of-river projects that it decides to construct; plan the construction of a project after having fully satisfied Pakistan on that count. Additionally, it emphasized the need for the two countries to undertake watershed management and joint environmental studies in order to address emerging concerns arising from reduced flows.[18] The Indian government, however, refused to consider it on the grounds that the Treaty spells out a mechanism to take care of water problems. Ramaswamy Iyer, an unofficial mouthpiece of the Indian

government, endorsed this viewpoint in these terms:[19] '[T]here is no
water issue between India and Pakistan. The sharing of the Indus
waters stands settled by the Indus Waters Treaty 1960, and the nature
of the sharing ... is such that no disputes can arise on this matter ....
Why has Pakistan suddenly decided to raise water as an issue for the
"Composite Dialogue" as and when resumed? The answer is very
simple. The intention is to give the people of Pakistan and the world
in general the impression that "water" is indeed an issue between the
two countries.'

Another and certainly more potent option relates to the putative
role of the Indian media. The media of a country plays the role of a
watch-dog by pointing out flaws in its government's policies on a given
matter and suggests ways and means to mend them. Unfortunately, the
situation in India does not look promising on this count as its media,
instead of taking the Indian government to task for jeopardizing peace
in the subcontinent by its attitude on the water issue with Pakistan,
seems to be hand in glove with it. In fact, it would not be wrong to say
that as far as Jammu and Kashmir and the water issue are concerned,
it has no thinking of its own and is no more than his master's voice.
John Briscoe, who spent forty years dealing with challenges of water
management in South Asia, talks about this point:[20]

> Living in Delhi and working in both India and Pakistan, I was struck by
> a paradox. One country was a vigorous democracy, the other a military
> regime. But whereas an important part of the Pakistani press regularly
> reported India's view on the water issue in an objective way, the Indian
> press never did the same. I never saw a report which gave Indian
> readers a factual description of the enormous vulnerability of Pakistan,
> of the way in which India had socked it to Pakistan when filling
> Baglihar. How could this be, I asked? Because, a journalist colleague
> in Delhi told me, 'when it comes to Kashmir—and the Indus Treaty
> is considered an integral part of Kashmir—the ministry of external
> affairs instructs newspapers on what they can and cannot say, and often
> tells them explicitly what it is they are to say.' This apparently remains
> the case. In the context of the recent talks between India and Pakistan
> I read, in Boston, the electronic reports on the disagreement about

'the water issue' in *The Times of India, The Hindustan Times, The Hindu, The Indian Express and The Economic Times.* Taken together, these reports make astounding reading. Not only was the message the same in each case ('no real issue, just Pakistani shenanigans'), but the arguments were the same, the numbers were the same and the phrases were the same. And in all cases the source was 'analysts' and 'experts'—in not one case was the reader informed that this was reporting an official position of the Government of India.

John Briscoe then goes on to narrate an eye-opening incident which confirms the general impression about the Indian media on the issue. He states that as part of the *Aman ki Asha* (desire for peace) programme which the *Jang* and *The Times of India,* two major media houses of Pakistan and India respectively, launched sometime back, he was asked to write a piece on the Indus Waters Treaty which they agreed to publish simultaneously in their newspapers. According to him, when he sent the article for publication the *Jang* and *The News* (the *Jang's* English language outlet) brought it out both in print form and electronically but *The Times of India* did not publish it. He also revealed that the article generated 'an avalanche of email'; and 'a cottage industry of comments on the internet' mostly from Indian readers in which he was called a *jihadi*, a closet ISI general, and a Paki-lover.[21]

If the Indian media appears to be complicit with the Indian government to sabotage the Indus Waters Treaty, Indian intellectuals are not far behind. Briscoe laments:[22] 'Equally depressing is my repeated experience—including at a recent major international meeting of security institutions in Delhi—that even the most liberal and enlightened analysts (many of whom are friends who I greatly respect) seem constitutionally incapable of seeing the (obvious to an outsider) great vulnerability and legitimate concern of Pakistan.' Incidentally, while commenting on the hate mail he received after the publication of his article in the *Jang* and *The News,* Briscoe disclosed that 'some of India's most prominent intellectuals excoriated me for knowing nothing, defending the indefensible Pakistan and putting my nose in something that was not my business.'[23]

Another possible option to effect a change in the Indian thinking is for Pakistan to ask the Bank, which was originally responsible for the settlement of the Indus waters dispute, to intervene in the matter and take care of the new irritants between the two countries. On the face of it, it looks an attractive proposition but upon closer scrutiny it is terribly flawed. There are two reasons why this is so. First, though Pakistan would indeed be receptive to it, India would not agree because of its aversion to any third party role in its disputes with Pakistan (and other South Asian neighbours). Let us not forget that if India accepted the Bank's role as an arbitrator in the 1950s, it was due to the peculiar circumstances of that time and which do not exist at present. Therefore, most probably the latter would not be amenable to change its position. And without India's consent, the Bank would be absolutely loath to intervene in the matter. Secondly, if the Bank has to intervene in the matter it must have a vested interest to do so. Last time around, when it offered its services as an 'honest broker', it did so because the US, on whose behalf it acted, wanted India on its side during the struggle against the communist world during the Cold War period and on cue from the US it pressurized Pakistan to surrender the three Eastern Rivers to India.[24] The situation has radically changed since then; the US has won the Cold War and in the new political configuration that has emerged in South Asia it has adopted an 'India First' policy which it is pursuing with great success through a close strategic relationship with India. Consequently, it has no interest in putting pressure on the latter to make a settlement with Pakistan on the water issue. It signifies that this option is not available to Pakistan either.

Still another option that needs to be explored is the suggestion based on a new concept termed IWRM or Integrated Water Resources Management which emerged from the International Conference on Water and the Environment held in Ireland in 1992 (the Dublin Principles) which aims to promoting 'the coordinated development and management of water, land and related resources in order to maximize economic and social welfare in an equitable manner without compromising the sustainability of vital ecosystems.'[25] In other words, what it proposes to do is to replace the traditional, fragmented, and sectoral approach to water resources and management with a cross-sectoral one and proceed on the assumption that water resources are an integral part of the ecosystem, a natural resource, and a social and economic good.[26] Inspired by this concept, India-born Professor Ashok Swain of the Uppsala University, Sweden, among others has contended that, '[t]he development of rivers occurs most optimally on the basin level. The whole international river basin needs to be regarded as an economic unit irrespective of state boundaries. Under an integrated water development programme dam and storage are to be located at the best possible places and the benefits are to be used by the riparian states in need of those benefits.'[27] Proceeding from the foregoing worldview, Professor Swain contends that the Treaty supplied only limited solution to the issue of water scarcity that Pakistan and India faced in 1950s; and that it has outlived its utility as new issues emerged in the 1990s and 2000s. Consequently, he pleads for renegotiations of the Treaty on the basis of the IWRM principles in order to get optimal results from water resources of the Indus Rivers.[28]

The genesis of the foregoing IWRM-based proposal can be traced to the proposal of joint management and control that Lilienthal mooted in the *Collier's* magazine and which became the basis of the Bank-sponsored negotiations in the 1950s. John Briscoe recently advanced a similar proposal as a way out of the water imbroglio between Pakistan and India. Believing that India's own security is dependent on a stable and cooperative Pakistan and that a fair and balanced regime on the Indus Rivers can be 'critical not only in its own rights but as a catalyst for a broader normalization of ties', he proposed joint projects by the two countries with, 'benefits flowing both ways and even operating rules on Indian dams that would benefit Pakistan.'[29] We concur with the foregoing suggestion in the belief that joint projects can not only help the two countries achieve optimal utilization of the water resources of the Indus Rivers and also take care of the issues that currently bedevil or are likely to strain relations between them. For example, it will at once remove Pakistan's apprehensions about the Indian intention to build storages under the guise of run-of-river structures[30] and India's constant grouse of Pakistan always trying to thwart its hydropower projects on the pretext of their being in conflict with the Treaty. Similarly, it will take care of the blame game that the two countries are likely to indulge in resulting from the adverse effect of the climate change and which has the potential to further aggravate relations between them. Waslekar is absolutely justified in observing that '[t]he integrated development approach may be Utopian, but all the other options will lead to destruction sooner or later.'[31]

To realize this objective, we need not replace the Treaty by another one because article VII (c) provides for such cooperation.[32] Notwithstanding this sound recipe, the question remains as to how to go about it. This is because the latest proposals for joint projects have as much chance of success as that of Lilienthal (which we need to remember, though the starting point of the Indus water negotiations, was soon discarded in favour of the division of the six Indus Rivers) for the simple reason that there is an unbridgeable trust deficit between the two countries. To take care of the culture of mistrust, Briscoe proposes the involvement of trusted third parties to act as interlocutors but realizing India's instinctive rejection of any such proposal on Jammu and Kashmir (which he knows in the Indian thinking covers the water issue as well) and emergence as a global player, he despairs of it.[33]

Mistrust is not the only hurdle in the way of the realization of this proposal. Unfortunately, India is not positively inclined to accept the idea of control and management of joint projects on Indian territory. In 1987, following the failure of talks on the Wullar barrage at the level of the Indus Commission, the matter was taken up by the Governments of Pakistan and India at the secretaries level. In 1989, each country submitted a draft agreement for its resolution. Since India was adamant on creating man-made storage on the

Jhelum River by building a barrage on the Wullar Lake, the Benazir government, in the spirit of compromise, incorporated an innovative provision in its draft agreement which stipulated joint control and management of the barrage by the two countries. India shot it down with utter disdain. According to a member of the Pakistan delegation who took part in the meeting, 'Dr Roy from the Ministry of Law, who was a member of the Indian delegation, was simply livid with indignation at this proposal and frothing at the mouth dismissed it peremptorily.'[34] Taking into account this incident, we can say that prospects for the success of the IWRM-based proposal do not look bright.

If the above mentioned options are not available, then what is the way out of the imbroglio? Ramaswamy Iyer believes that it will come about when Kashmir becomes irrelevant and relations between the two countries cease to be adversarial.[35] Similarly, Waslekar thinks that it will become possible only if there is 'a paradigm shift in the mindset' which he thinks will come about with the 'end of hostilities, both physical and psychological, from both sides. It will have to be a part of final settlement in letter and spirit.'[36] In other words, both have linked it to the success of the ongoing peace process between Pakistan and India. Is this a sound proposition? To address this question, we need to examine the history of the peace process and make a balance sheet.

The peace process began in February 1999, when Atal Bihari Vajpayee took a bus journey to Lahore and the prime ministers of Pakistan and India decided to start a new beginning in the two countries' relations by signing the Lahore Declaration. By virtue of this declaration, they agreed to start what came to be known as Composite Dialogue as a result of which they divided their contentious issues between them into eight baskets.[37] However, the process soon got derailed when the Kargil incident took place a few months after the declaration was signed. Subsequently, the situation further deteriorated when the LeT (Lashkar-e-Taiba) committed a terrorist attack on the Indian Parliament in December 2001. It took almost five years to put the process back on track when Prime Minister Vajpayee and President Musharraf signed the Islamabad Declaration on the sidelines of the SAARC summit held in January 2004. Following its conclusion, the two countries engaged in serious negotiations on various contentious issues between them. Nonetheless, the process was interrupted again in the wake of the Mumbai terrorist attack in November 2008. It remained suspended for three years till India agreed to its resumption in 2011. However, the latter tried to restrict its scope to two issues, namely terrorism and trade. After two rounds of talks, it was again interrupted in January 2013, because of incidents on the Line of Control (LoC). It remained suspended till 8 December 2015, when on the sidelines of the Heart of Asia conference the Indian foreign minister, Sushma Swaraj, gave the green light to resume it.

On this occasion, the two countries, at the behest of India, changed the nomenclature of the Composite Dialogue to 'comprehensive bilateral dialogue'. The reason for this change is not clear. Perhaps the Modi government wants to focus on the terrorism issue as the new nomenclature, in addition to the eight issues encompassed in the Composite Dialogue.[38] If this is its intention, it is not surprising because at the start of its term it had indicated that the Composite Dialogue had run its course and that it wanted to replace it with a 'new architecture' of talks. Pakistan is wary of this move as it could mean putting long-standing disputes, in whose resolution Pakistan is interested, on the backburner.

In a dramatic development on 25 December 2015, Prime Minister Modi, on his return from Kabul made a lightning visit to Raiwind in Lahore, the residence of Prime Minister Nawaz Sharif, where the two leaders agreed to take the peace process forward. They also indicated mid-January as a tentative date for the foreign secretaries to meet for the purpose. In the meantime, on 2 January 2016, a terrorist attack took place on the Pathankot airbase. India blamed Pakistan-based Jaish-e-Muhammad for the heinous act but for a change did not hold the Pakistan government, its army, or the ISI responsible for it.

Pakistan on its part agreed to investigate the matter and punish those responsible if they turned out to be Pakistani nationals. Despite the fact that two countries showed maturity in handling the issue, it is too early to predict the fate of the peace process.

When we make a balance sheet of what the peace process has achieved so far, we find the result quite disappointing. As far as Jammu and Kashmir, the most contentious of all Pakistan-India disputes, is concerned, the two countries conducted talks mostly through back-channel diplomacy. They reportedly worked out a settlement which sought to make the existing LoC in Kashmir 'soft' by allowing people and goods on both sides of the divide to move freely; grant greater autonomy and self-government to the Kashmiris within the existing boundaries; effect staggered demilitarization of the state; and put in place joint supervisory mechanism over selected issues across the LoC.[39] It was said that the two countries were a signature away from giving the deal a formal shape when the judicial crisis erupted in Pakistan in March 2007, after President Musharraf dismissed 60 judges of the higher judiciary including the CJP. This prevented Manmohan Singh from travelling to Islamabad for the purpose. Subsequently, the Zardari government, which replaced the Musharraf government, could not pursue the matter because of the Mumbai terrorist attack in November 2008.

As for other issues in the peace process, the Siachen Glacier is considered a relatively easy matter to handle, a kind of low-hanging fruit. Prime Ministers Benazir Bhutto and Rajiv Gandhi reached an agreement on it as far back as 1989 but progress on it was stalled because the Indian government backed out of it, demanding authentication of the existing position of the troops of the two countries on the map before their redeployment. This demand was an after-thought as India did not raise it in the course of negotiations. Pakistan rejected it because, in its opinion, it amounted to legitimizing an illegal act of occupation of the glacier by the Indian troops and because of its fear that India could use it as a legal claim in subsequent negotiations to delineate the undemarcated area beyond NJ 9842 on the LoC. However, in a spirit of accommodation, it offered to accept it by agreeing to place the authenticating document as an annexure to the agreement on condition that it would not become a basis for a subsequent legal claim to the glacier but India refused to accept it. Subsequently, India hardened its position further on the matter on the grounds that the Saltoro range of mountains which is part of the Siachen Glacier is indispensable for Ladakh's security which, in its estimation, a Pakistan-China axis could threaten as a result of the Karakoram Pass coming under Pakistan's control.

As far as the Sir Creek issue is concerned, it is also considered a low-hanging fruit, easy to be plucked. Starting from 1969 till January 2016, twelve rounds of talks have taken place between the two countries but without result. Given the impasse in talks, Pakistan proposed arbitration on the issue but, as expected, India spurned the offer. Instead of resolving it bilaterally, India is now contemplating unilaterally and arbitrarily erecting a 'floating fence' anchored by submerged metallic meshes along the body of water to which it lays its claims. It was thought that the Law of the Sea (LoS) Convention 1982, which stipulates that the signatories who have claims to the extended Continental Shelf (CS) and the Exclusive Economic Zone (EEZ) must fix their maritime boundaries by 2009 failing which they would lose their rights to the International Sea Bed Authority (ISBA), would force the two countries into a bilateral settlement. However, notwithstanding this Damocles' sword, they failed to do so even though they have expansive claims.[40] Regarding trade and commerce there was some progress during the last days of the Manmohan government. However, the Modi government has gone back on it on the plea that it would pick up the thread from where the two countries left it in September 2012, rather than what was agreed in January/February 2014, because it suits India. Resultantly the progress on trade and commerce has stalled as well.

Can we expect any movement on the water issue divorced from the failing peace process? We propose to examine the period between 2004–07 when relations between the two countries had vastly improved because of the large number of confidence-building measures (CBMs) that the two countries had taken. During this period the Baglihar Dam issue was making headlines and had the potential to vitiate the political atmosphere. In an ambience of *bonhomie* between the two countries, Pakistan made an appeal to India to temporarily suspend work on it but the latter refused to accede to Pakistan's request. There was a feeling in Pakistan that India was going ahead with the project because it wanted to present the latter with a fait accompli. It is also noteworthy that for quite some time during this period, India did not allow Pakistan's commissioner to make an on-the-spot inspection of the dam site which due to the Treaty, it was under legal obligation to do so. This occurred on the pretext that there was a security threat to the Pakistan delegation from militants. This incident debunks the proposition that the removal of the trust deficit would make India observe the Treaty because in 2004 India proposed 72 CBMs but did not make the one that could have transformed the atmospherics of the peace process, namely the temporary suspension of work on the Baglihar Dam. In fact, instead of suspending the work, it reportedly expedited it. It is also noteworthy that the water issue is not part of the peace process except for the Tulbul Navigational Project/Wullar Barrage which figures as one of the issues in the 'comprehensive bilateral dialogue'. With this backdrop, it is hard to be Micawberish in the matter.

Why does India adopt such a rigid and uncompromising attitude on the water issue? Why does it want a resolution of disputes with Pakistan on its own terms rather than through a compromise? The answers to these questions lie in two parts. First, it is the place that the Indian leadership assigns to itself and to Pakistan in its strategic worldview. To understand the nature of this worldview, we need to keep in mind that India is of subcontinental size; houses a population which is greater than the combined populations of the three major global powers (namely the US, the European Union, and Russia); is strategically located; is currently an economic power which is slated to grow exponentially; is nuclear-armed; and is home to a great and ancient civilization. Conscious of this reality, it firmly believes that its 'manifest destiny' is to play a major role in the comity of nations. As far as the subcontinent is concerned, Leo Rose, the late editor of *Asian Survey* and professor at the University of California, Berkley, opines that India views itself, 'as the "dominant", "paramount", "hegemonic", "preeminent", or just plain "major" (kindly select the adjective preferred) power in South Asia and the rights and privileges that accompany such status.'[41] Continuing this line of thought, he further observes that, '[t]here is now a broad consensus among most Indian leaders and commentators that a primary objective is to obtain the acceptance of India's status as *the* major regional power in South Asia from both regional states and from major external powers.'[42] Indian strategic thinkers also share this view. For example, Raja Mohan, an Indian academic and journalist, while observing that India's grand strategy divides the world into three concentric circles, acknowledges that *'India has sought primacy and veto over the actions of the outside powers'*[43] in the immediate neighbourhood which he places in the first circle.[44]

India has had, over the years, more or less no problem in making most South Asian countries accept the reality of Indian pre-eminence in the region. The only exception has been Pakistan which, right from its inception, has refused to kowtow before it. India got a chance to make Pakistan accept its pre-eminence in 1971 when East Pakistan revolted against the domination of West Pakistan. Availing itself of this opportunity (which the Indian strategic thinkers termed the opportunity of the century) to dismember Pakistan, it intervened in the conflict militarily and apparently achieved its objective by helping to create the independent state of Bangladesh. Following Pakistan's dismemberment, it must have expected the latter to accept its pre-eminent status. However, to India's chagrin, Pakistan soon started to reassert itself again as a fully sovereign state. Not only that, in due course it acquired nuclear weapons which gave it parity with India and thus nullified the superiority that the latter enjoyed in conventional weapons over it.

Despite this development, India has not abandoned its dream of making Pakistan accept its pre-eminence in the region as admitted by Jaswant Singh, the former foreign minister of India and one of the stalwarts of the BJP:[45] *'Unless Pakistan accepts India's dominant role in Asia and its readiness to live in friendly neighbbourly manner there can be no peace.'* To achieve this objective, India is prepared to go to any extent. That explains why it is reluctant to solve any dispute with Pakistan (including water) except on its own terms.

The second explanation why India is reluctant to make concessions to Pakistan, both in the peace process and the water talks, is the balance of power that lies between them. The kind of diplomacy that a country conducts with another country and the concessions that it is prepared to make to it is largely a function of the balance of power that exists between them.[46] A country in whose favour the balance is tilted is not prone to make concessions to the country against whom it is tilted and tries to seek a settlement over a contentious issue on its own terms. As far as Pakistan and India are concerned, the balance is heavily tilted in favour of the latter. Consider the following: with an average annual growth rate of 8 per cent during the last decade or so, India is currently the third and tenth largest economy of Asia and the world respectively. It is an economic power house which has attracted major investors from Japan, China, and the US.

As far as global politics is concerned, there too India appears to be favoured. For instance, the P5 favour India's candidature as a permanent member of the restructured UN Security Council (the four permanent members have publicly committed whereas China has privately conveyed it to New Delhi that when the time comes it would not oppose its candidature). Similarly, it is a de facto member of the exclusive nuclear club even without signing the Non-Proliferation Treaty (NPT) as a result of the 123 Agreement (or the 2005 Civil Nuclear Deal with the US). Again, the US, the sole superpower, accepts it as the 'indispensable partner' in global politics and is committed to make it a 'major world power in the 21st century'.[47]

Compared to India, Pakistan is weak and vulnerable. On the economic front, according to the latest World Bank report (covering the fiscal year ending 30 June 2015) with a 4.4 per cent growth rate Pakistan figures at the bottom of the eight economies in South Asia, the second lowest after Afghanistan.[48] Besides, its exports are falling and with the exception of the CPEC (China-Pakistan Economic Corridor) it has miserably failed to attract foreign investments. Foreign debt has risen enormously and the country cannot stay economically afloat without further foreign capital inflows. According to the finance minister, Ishaq Dar's own confession, with $2 as the minimum daily wage, more than half of Pakistan's population lives below the poverty line.[49] The energy crisis with widespread load shedding has further compounded the already dire economic situation. Pakistan is also burdened with a large number of Internally Displaced Persons (IDPs) and frequent devastating floods have wreaked havoc on its economy.

Furthermore, on the political front, the picture is far from rosy. Though the country has made remarkable gains in controlling terrorism, it has failed to strike at its source. If it fails to do so, which is the case at present, it may not succeed in rooting it out. The country also suffers from lack of political stability due to constant bickering of the political class. These developments have tilted the balance of power in South Asia squarely in India's favour which has emboldened it to take an inflexible and uncompromising attitude with Pakistan in negotiations.

Given the current state of affairs between the two countries, there are two possible scenarios that could emerge out of the possible standoff over the water issue. One is that Pakistan knuckles under Indian pressure with or without war and accepts *Pax Indica*. In that eventuality, the matter would stand resolved as Pakistan would accept Indian supremacy. The second is that Pakistan refuses to toe the Indian line and remains defiant. In that case, the prospects for peace do not look promising because, deprived of water, Pakistan may not take things lying down. The reason is very simple. It can live without Siachen Glacier or Sir Creek, even Jammu and Kashmir, but not without water. If this happens, it could trigger armed hostilities between the two countries with unknown consequences. This is not an alarmist view or mere kite-flying but a realistic assessment. A similar situation happened between the two countries in the 1950s when the water issue had the potential to become a *casus belli* as testified by Lilienthal's following observation:[50] 'It [water] is pure dynamite, a Punjab powder keg. Peace in the Indo-Pakistan subcontinent is not in sight with these inflammables lying around. Unless a better answer on water is soon forthcoming, *even if the Kashmir plebiscite could be held,* peace would not come.' The two countries were fortunate that the Bank intervened and helped them successfully negotiate the Indus Waters Treaty.

International bodies have predicted that the two countries could go to war over water. For example, the 2009 American CIA report, observing that, 'water issues likely will continue to be a major source of conflict between India and neighboring Pakistan ... in coming decades',[51] warned that, '*if shortages are acute, Pakistan may feel compelled to threaten India or even to utilize military force. While general discord is not uncommon in the India-Pakistan relationship, Pakistan would actually have a potential pathway for successful military action to secure additional water resources. With water resources already scarce and key dams and reservoirs just inside Indian-controlled Kashmir, limited military action could be a rewarding proposition for Pakistan. Given both nations' nuclear arsenal, a direct offensive seems unlikely due to potential for a catastrophic conflict. However, the campaign could shift to encompass the objective of threatening, capturing or even destroying key dams and reservoirs.*'[52] Similarly, the opinion of American diplomats posted in the region is no different. According to a number of secret cables sent by them to the US State Department as revealed in Wikileaks, they were not very sanguine of the long-term prospects of Pakistan and India resolving their disagreement over the 'emotional issues' of water, especially given 'Pakistani anxiety over access to water'. Of particular importance is the secret cable sent by Mulford, the US ambassador in New Delhi on 5 February 2005, in which it stated that 'there are several hydrological dams planned for Indian Kashmir that might be questioned under the IWT', he expressed the fear that 'India's dams in Jammu and Kashmir have the potential to *destroy the peace process and even lead to war.*'[53] Again, the Intelligence Community Assessment report issued in 2012 has put on notice that, *'[p]hysical infrastructure, including dams has been used as a convenient and high publicity targets by extremists, terrorists, and rogue states threatening substantial harm and will become more likely beyond the next 10 years.*'[54] Finally, a group of more than twenty UN bodies in March 2009 warned that given the rising tension over the water issue between Pakistan and India the world would be *perilously close to its water war.*[55]

This is indeed a very scary scenario. However, there may be 'doubting Thomases' who would not agree with it because it goes against the contemporary thinking which denies a causal relationship between water and war and hence discounts the possibility of nations going to war over water. There are a number of writers who fall in this category. Aaron T. Wolf, professor at the Oregon State University, is one of them. He has tried to establish, on the basis of extensive data, that he has marshalled that the relationship between water scarcity and war is nothing but anecdotal and based on selective evidence.[56] Relying on a fresh water transboundary database that he created involving 6,400 cases of water-related conflicts covering the period 1948–2005, he asserts that cooperation overrides confrontation between states.[57] He claims that only seven skirmishes have taken place between states in this century and that no war has ever been fought over water.[58] He contends that 145 water-related treaties were signed during this period. He discounts the possibility of war in the future because, in his opinion, it is 'neither strategically rational, hydrologically effective, nor economically viable.'[59] Pursuing this line of thought he argues that, '[s]hared interests along a waterway seem to constantly outweigh water's conflict-inducing characteristics. Furthermore, once cooperative water regimes are established, through treaty, they turn out to be impressively resilient over time, even between otherwise hostile riparians, and even as conflict is waged over other issues.'[60]

On the basis of this thesis, one can say that the predictions of a water war between Pakistan and India are utterly unfounded. However, there are serious problems with this proposition. First, even if we accept that there has been no water war in history between states, it does not mean that the future would necessarily follow the past; that instead of war there would be cooperation between the contending parties. To definitively deny the possibility of such wars between states in the future is highly dicey. The situation can be compared to the one which existed on the eve of World War I. Between the Congress of Vienna held in 1815 and World War I which started in 1914, the Europeans became so indifferent to the prospect of a war that they had come to believe that they had touched the acme of civilization and that they were immune to the destructive forces of armed conflict. When the war came, they were utterly shocked. We are afraid that if we continue to believe in the impossibility of war between states over water in the future we may be succumbing to the same complacent mindset that plagued Europe on the eve of World War I.

Secondly, there is a chink in the armour of this thesis. During the last decade of the twentieth century it was commonly held that water scarcity could lead to wars in the international community. However, at the beginning of the twenty-first century there occurred a shift in this view as it was commonly believed that the relationship between water scarcity and war was nothing but a myth; that states, as opposed to communities, have never fought a war over water in history; and that whatever water conflicts have taken place in the past have been mild in character. It was further claimed, supposedly on the basis of empirical evidence, that water promotes interdependence and cooperation and that 'water and river basins are pathways to peace'.[61] It is this view which Wolf et al. have referred.

However, in the second half of the first decade of the twenty-first century, this view came under attack. The Swedish writer, Terje Tvedt, has tried to prove, on the basis of two case studies related to River Nile, that water wars are no myth and that they have been fought in the past with far-reaching geopolitical consequences.[62] Similarly, the current UN Secretary General, Ban Ki-moon, on two occasions in 2007 (once at the Davos summit and then while commenting on war in Darfur) sounded an alarm bell of impending wars between states on the water issue. He was not alone in doing this. A number of high profile politicians and activists made claims that water scarcity led to war in Darfur.[63]

We believe that the last word on the question of a causal relationship between water and wars has not been said and that if we deny the possibility of such a war between Pakistan and India, we would be doing so at our own peril.

REFERENCES AND NOTES

1. The president of the Bank, Eugene Black, justified the Treaty on the grounds that it would be a harbinger of peace between Pakistan and India:

   This is story of how a bank—the World Bank—helped to resolve the dispute that might well have caused war between India and Pakistan .... Had war broken out, it would have involved the almost 500,000,000 people, who live in the two countries—nearly a fifth of the world's population—and would have had unforeseeable consequences for the rest of the world.

   So ended a fateful chapter, and so began a hopeful chapter, in the relations between India and Pakistan ... as long as the two countries are working on the concrete tasks of developing the Indus, there is every hope that they will live together in tolerance and peace.

   The billion dollars or so that will be spent would not cover the cost of a single week of modern war. To devote it to peace along the Indus is to make an investment that we and the generations to come may consider well worth the price.

   Eugene Black, 'A Moral for Nations', *New York Times*, 11 December 1960.

   Black was certain that the Treaty had had secured the peace forever as attested by the following statement that he made many years after it was concluded: 'And that we hope has put the Indus water dispute to bed ... in fact, more than put it to bed, because that rather suggests it may waken again tomorrow morning. I hope it's buried it.'

2. It is true that the Treaty has worked smoothly over the years. However, there are instances where one party made complaints against the other for not acting in conformity with the terms of the Treaty. For example, in the wake of the 1965 war, Pakistan sent a letter to India in which it accused the latter of withholding

almost the entire flow of the waters of the Sutlej and Beas Rivers, causing the supplies to fall below the level stipulated by the Treaty; and stopping all supplies to the Central Bari Doab Channels. See 'The Note of the Ministry of Foreign Affairs of Pakistan to the Indian High Commissioner', annexed to the letter of the Permanent Representative of Pakistan to the UN Secretary-General, 29 November 1965. *UN Document No, S/6985, 1965.*

3. Pakistan estimates that the overall number of structures that India is planning on the Western Rivers runs into more than a hundred (For the exact location of Indian dam projects, see Appendix 3). According to the US Senate Report, India is building 33 dams in Jammu and Kashmir which are at various stages of completion. *Avoiding Water Wars: Water Scarcity and Central Asia's Growing Importance for Stability in Afghanistan and Pakistan*, A Majority Staff Report, Committee on Foreign Relations, US Senate, 12th Congress, 22 February 2011, 9.

4. John Briscoe has estimated that with these dams in place India 'will create something like 40 days of live storage on the Chenab alone.' John Briscoe, 'Troubled Waters: Can a Bridge be Built over the Indus', *Economic and Political Weekly* (11 December 2010), vol. XLV, no. 50, 30. He has observed that at one time the Indians were claiming that India would never use water as a weapon against Pakistan but the LeT's terrorist attack on Mumbai changed their perspective as, in its wake, they started contending that it 'should use anything and everything, including water.' Ibid. 31.

5. John Briscoe, 'War or Peace on the Indus' (18 April 2010), <http://www.thefrontierpost.com/News.aspx?ncat_ar&nid+255>. See also ibid. It is noteworthy that Bashir Ahmad, a Srinagar geologist, commenting on India's intentions on the Western Rivers has warned: 'They will switch the Indus off to make Pakistan solely dependent on India. It's going to be a water bomb' ('South Asia's Water: Unquenchable Thirst', *The Economist* (19 November 2011), <www.economist.com/node/21538687>). Similarly, Michael Kugelman, South Asia associate at the Woodrow Wilson International Centre for Scholars in Washington DC, has observed: 'There is definitely potential for conflict based on water, particularly if we are looking to the year 2050 when there would be considerable water scarcity in India and Pakistan.' Nita Bhalla, 'Thirsty South Asia's River Drifts Threaten "Water Wars"', *Reuters* (23 July 2012), <http://www.reuters.com/article/2012/07/23/us-water-southasia-idUSBRE86MOC820120723>.

6. *Avoiding Water Wars*, supra note 3, 9.

7. 'South Asia's Water', supra note 5.

8. Ashraf Mumtaz, 'India can't Scrap IWT: Experts', <http://www.dawn.com/2001/12/24/local48.htm>.

9. 'Washington Diary: India-Pakistan Conundrum', *Daily Times* (3 March 2010).

10. B. G. Verghese, 'Ideology Threatens Indus Treaty', *The South Asian Journal* (25 March 2010).

11. 18 January 2010, *Dawn*, cited in *Water Security for India: The External Dynamics*, Report of the Institute for Defence Studies and Analysis (IDSA), 2010, 38. <http://www.indiaenvironmentalportal.org/in/files/book_WaterSecurity.pdf>. Similarly, an editorial in *Nawa-e-Waqt* published in April 2011, warned that: 'Pakistan should convey to India that a war is possible on the issue of water and this time war will be a nuclear one.' 'South Asia's Water', supra note 5.

12. 'MNAs Lambast India for "Water Terrorism"', *Daily Times* (17 February 2010).

13. Zia Mian and M. V. Ramana, 'Going MAD: Ten Years of the Bomb in South Asia', *Economic and Political Weekly* (28 June 2008), 203.

14. *Water Security for India*, supra note 11, 38, note 47.

15. John Briscoe, 'War or Peace', supra note 5. See also the article by the same writer 'Troubled Waters', supra note 4, 30.

16. It is true nations conduct their relations with each other on the basis of realpolitik. However, sometimes they depart from this norm. For example, Indonesia which is huge in size does not follow it vis-à-vis its neighbours. Gen. Zia ul-Haq, Pakistan's military ruler, narrates that when he asked Gen. Suharto, Indonesia's military ruler, to explain the secret of the success of the Asean, he replied that it was due to his country's decision not to act as a big brother towards other members of the organization. Unfortunately, the Indian rulers do not think on these lines and there does not appear to be any possibility that they would change their present mindset towards other states of South Asia.

17. The term 'non-paper' is used in diplomacy to connote documents which are not formal in character but are exchanged for informal discussion.

18. Ramaswamy Iyer, 'Pakistan's Questionable Move on Water', *Economic and Political Weekly* (27 March 2010), vol. XLV, no. 13, 11.

19. Ibid. 10.

20. John Briscoe, 'Troubled Waters', supra note 4, 30. Briscoe started his South Asian career in the early 1970s with a project on the Ganga and Narmada Rivers which was a collaborative arrangement between Harvard University and the Government of India. He lived in Bangladesh in the 1970s and Delhi from 2002 to 2005 as a senior water advisor to the Bank. He has published two books with his Indian and Pakistani colleagues on water issues of the two countries which are entitled as follows: *India's Water Economy: Facing a Turbulent Future*; and *Pakistan's Water Economy: Running Dry*. In short, he is well-qualified to make pronouncements on the water issue that troubles the two countries. As regards Briscoe's observations on how the Indians conduct themselves when it comes to Kashmir, incidentally my personal experience is no different from his. My book entitled *Issues in Pakistan's Foreign Policy: An International Law Perspective* focused on different foreign policy issues but not on Kashmir. My publisher told me that when the book was published he got a big order from India. When my next book entitled *Kashmir Dispute: An International Law Perspective* came out, my publisher told me that he contacted many Indian book stores but did not receive

even a single order. The explanation for this lies in the fact that the Indians do not want to hear anything on Kashmir which challenges their version of it. It was obvious to them on the basis of the flyer which was sent to them and the fact that the book was written by a Pakistani and was published in Pakistan that it presented the Pakistani viewpoint. Therefore it was not welcome in India. It would not be out of place to mention here that the Indian authorities in the 1990s not only banned David Sugarman's pictorial survey book but also declared him *persona non grata* because it covered human rights violations by the Indian security forces in Jammu and Kashmir.

21. Ibid. 30.
22. Ibid.
23. Ibid. 30-1.
24. For a detailed discussion on this point, see chapter 3, 'World Bank: The Honest Broker', under 'Bank's Motive to Side with India.'
25. *What is IRWM Global Water Partnership, The Challenge,* <http://www.gwp.org/The-Challenge/What-is-IWRM/>.
26. Ibid.
27. Ashok Swain, 'Water Wars: Fact or Fiction', *Futures* (October 2001), vol. 33, no. 8, 777.
28. Per Steineide Refseth, *The Indus River Basin 1999–2008: An Intellectual History in Hydropolitics,* master's thesis submitted to the Institute of Archaeology, Conservation and History, University of Oslo, Sweden, Spring 2013, 40.
29. John Briscoe, 'Troubled Waters', supra note 4, 32. Briscoe believes that it would be much easier for India to adopt a fair and balanced regime on the Indus waters if Pakistan were to end its too-cute-by-half dance with organizations that commit acts of terrorism in India (ibid.). We are of the view that Briscoe has a very simplistic view of the situation because he ignores the fact that currently Pakistan has adopted this policy in response to India's use of local terrorist proxies in Balochistan, Karachi, and FATA. Consequently, it is highly unlikely that Pakistan would agree to take such a step as long as India does not reciprocate. We believe that both countries should move to snap their ties with terrorist organizations.
30. One Indian academic has cited these grounds as justification for a joint management proposal:

> Finally, because the design of the treaty only requires a one-time quid pro quo—the division of rivers at the time of signing—and not a continuing compromise, the status quo in any given dispute lacks incentive to cooperate. For example, because India no longer has any stake in pleasing Pakistan beyond the minimal threshold needed to avoid abrogation, it has an incentive to constantly push the boundaries of acceptable behavior allowed by the treaty because Pakistan has no real leverage besides the invocation of a slow dispute resolution process. As a result, Pakistani officials have often

encountered an Indian negotiating strategy whereby India 'proceed[s] with construction plans, even when aware that the plans might well violate the treaty, so that Pakistan, confronted eventually with a fait accompli, would have no choice but to cut its losses and accept an unfavorable compromise settlement.' This treaty design is easily contrasted with one in which India and Pakistan jointly share and administer a river basin based on their specialized expertise, each dependent on the other for good-faith use of water and a share in energy production. In this hypothetical case ... both countries have a stake in, and leverage over, the water practices of the other.

Manav Bhatnagar, 'Reconsidering the Indus Waters Treaty', *Tulane Environmental Law Journal* (summer 2009), vol. 22, issue 2, 288. Briscoe makes a similar proposal when he suggests: 'What is needed is to use the resetting of the terms by the ICA [International Court of Arbitration] for India and Pakistan to start out in a new direction. This should be one in which there is a search for joint benefits (such as hydropower plants built in the best possible sites, with power sold both ways, and, with operating rules which benefit both parties built into the project). 'Winning the Battle but Losing the War', *The Hindu* (22 February 2013).

31. Sandeep Waslekar, *The Final Settlement: Restructuring India-Pakistan Relations* (Mumbai: Strategic Foresight Group, 2005), Executive Summary.

32. The article is couched in these terms: 'At the request of either Party, the two Parties may, by mutual agreement, co-operate in undertaking engineering works on the Rivers.'

33. John Briscoe, 'Troubled Waters', supra note 4, 32. We know that renegotiations are out of the question because Pakistan would not agree to it. However, if by some miracle the latter was to do so and the parties settle for joint management, would it encompass all the six rivers of the Indus Basin or only the Western Rivers? We have raised this question because the writings of the Indian water experts indicate that India wants to restrict the water-sharing arrangement with Pakistan to the Western Rivers (Ramaswamy Iyer, 'Indus Treaty: A Different View', *Economic and Political Weekly*, 18 July 2005, 3144). We understand that as far as Pakistan is concerned this proposition is a non-starter. Iyer contends that India cannot restrict sharing of waters to the Western Rivers alone and will have to extend it to the Eastern Rivers as well. However, he does not think that it is an easy task because it would, in his opinion, open a Pandora's Box. He is therefore of the opinion that the existing arrangement as enshrined in the Treaty should continue and that we should wait till such time 'the Kashmir issue has become a thing of the past, and the relations between India and Pakistan have ceased to be adversarial' (Ibid).

34. M. H. Siddiqui, consultant to the Punjab Irrigation Department who participated in the meeting as a member of the delegation, narrated this incident to the present writer during an interview that he accorded him on 9 August 2013, at his residence in Lahore, Pakistan.

35. Ramaswamy Iyer, 'Indus Treaty', supra note 33, 3140.

36. Waslekar, *Final Settlement*, supra note 31.

37. The Composite Dialogue owes its origin to the meeting between Prime Minister Nawaz Sharif and Prime Minister IK Gujral during the SAARC summit held at Malé, Maldives, in 1997. The meeting resulted in what came to be known as Malé Declaration by virtue of which the two leaders decided to adopt a new approach to Pakistan India relations by agreeing to focus negotiations on various issues simultaneously rather than debate which issue is to be accorded priority as was the case previously. This came to be known as Composite Dialogue process which covers eight baskets of issues which are Jammu and Kashmir, Siachen, Wullar Barrage/Tulbul Navigation Project, Sir Creek, Terrorism and Drug Trafficking, Economic and Commercial Cooperation, Peace and Security, and Promotion of Friendly Exchanges in various fields (later termed as people to people contact).

38. Pakistan's former diplomat, Munir Akram, opines that the change from 'composite' to 'comprehensive' connotes that whereas the previous nomenclature was meant to convey the interlinkages between the two plus six agenda, the new one loosens it, in addition to enabling the inclusion of 'terrorism' as a new item on the agenda. See 'Engaging India', *Dawn* (10 January 2016).

39. Khurshid Mahmud Kasuri, *Neither a Hawk Nor a Dove* (Karachi: Oxford University Press, 2015), 323–50.

40. The two countries submitted their claims for extended Continental Shelf (CS) and the Exclusive Economic Zone (EEZ) to the UN in 2009. Their claims would not be considered until they resolved their boundary claims.

41. Leo Rose, 'India's Regional Policy: Non-Military Dimensions' in Stephen P. Cohen (ed.), *The Security of South Asia: American and Asian Perspectives* (Illinois: University of Illinois Press, 1987), 3.

42. Ibid. 4. Emphasis original.

43. What Mohan has in mind can be described in these terms: 'India strongly opposes outside intervention in the domestic affairs of other South Asian nations, especially by outside powers whose goals are perceived to be inimical to Indian interests. Therefore, no South Asian government should ask for outside assistance from any country, rather, if a South Asian nation genuinely needs external assistance, it should seek from India. A failure to do so will be considered anti-Indian.' Devin T. Hagerty, 'India's Regional Security Doctrine', *Asian Survey* (April 1991), vol. 31, no. 4, 352.

44. As to the second circle which covers the area stretching across Asia and the Indian Ocean littoral, India seeks to balance the influence of other powers and prevent them from undercutting its interests. As to the third circle which includes the world at large, it tries to occupy a place among the great powers. C. Raja Mohan, 'India and the Balance of Power', *Foreign Affairs* (July–August 2006), 18. Emphasis added.

45. Jaswant Singh, 'Against the Current', *Seminar* (January 1989), 41, in Javed Hassan, Lt. Col., *India: A Study in Profile* (Rawalpindi: Army Education Press, 1990), 228. Emphasis added.

46. See Hans J. Morgenthau, *Politics among Nations: The Struggle for Power and Peace* (New York, etc.: McGraw Hill International Editions (Political Science Series), 1997), 187–97.

47. Condoleezza Rice, the US secretary of state under the George Bush administration made this statement. In order to realize this goal, she also unveiled a plan under which Washington offered to establish a strategic dialogue with India to boost missile defence and other security initiatives as well as cooperation in technology, economy, and energy.

48. Shahbaz Rana, 'World Bank Projects: Pakistan's Economic Growth at 4.4 per cent', *The Express Tribune* (15 April 2015). The WB's projected growth rate is in line with the IMF and ADB projections which put it at 4.3 per cent for the same period. The official growth rate figure stands at 5.1 per cent. Ibid.

49. 'Over Half of Pakistan Lives under Poverty Line: Dar', *Dawn* (3 June 2014).

50. David E. Lilienthal, 'Another 'Korea' in the Making?', *Collier's* (4 August 1951). Emphasis original.

51. *Resource Dispute in South Asia: Water Scarcity and the Potential for Interstate Conflict*, CIA Report, 1 June 2009, 50.

52. Ibid. 39. Emphasis added.

53. 'Water Issues could Sweep Away Indo-Pak Peace Process', <http://beta.dawn.com/news/638350/secret-us-cables-accessed-by-dawn-through-wikileaks-water-issues-could-sweep-away-indo-pak-peace-pr>. Emphasis added. According to *Dawn*, the former secretary of state, Henry Kissinger, in an opinion piece that he penned in 2013 predicted a nuclear war over the next couple of decades between Pakistan and India. He also reminded us that the nuclear factor was in play in four major and one minor Pakistan-India crises, namely 1987, 1990, 1998, 1999, and 2002 (*Dawn*, 26 October 2014). The well-respected Canadian syndicated journalist, Gwynne Dyer, in his book published in the recent past, has depicted a scenario where Pakistan and India will exchange nuclear-tipped missiles with each other over water rights. *Climate Wars* (Toronto: Random House (Canada), 2008), (Scenario 4: Northern India, 2036), 113–24.

54. *Global Water Security*, Intelligence Community Assessment (ICA), 2 February 2012, 4. Emphasis added. Briscoe has warned that if India continues to consider Pakistan's water concerns as merely a trifle and disregard what appears to an outsider an existential threat, it will certainly become a powerful recruiting tool for Pakistan's jihadis against India. John Briscoe, 'Troubled Waters', supra note 4, 32.

55. Ashfak Bokhari, 'Water Dispute and War Risk', *Dawn (Economic and Business Review)*, (18–24 January 2010), VI. One well-known Norwegian writer, Terje

Tvedt, has opined that the struggle for the water of the Indus Rivers will determine whether or not the nuclear-armed Pakistan and India will wage war or peace. Per Steineide Refseth, *The Indus River Basin*, supra note 28, 63, note 255.

56. See Jannick Boesen and Helle Munk Ravnborg (eds.), *From Water 'Wars' to Water Riots?: Lessons from Water Transboundary Management* (Copenhagen: Danish Institute for International Studies, 2004), 6.

57. Aaron T. Wolf, *International Water Event Database: 1950–2005, Program in Water Conflict, Management and Transformation*, Institute for Water and Watersheds, Oregon State University, <http://transboundarywaters.orst.edu/database/interwatereventdata.html>.

58. Wolf, however, mentions an exception which took place in 2,500 BCE between the Sumerian city-states of Lagash and Umma over the right to boundary channels along the Tigris as the only documented conflict between states. Aaron T. Wolf, 'Conflict and Cooperation along International Waterways', *Water Policy* (1998), vol. 1, no. 2, note 5.

59. Ibid. 251. The IPCC Working Group II in its 2014 report shares this assessment as revealed by the following statement: 'However, there is high scientific agreement that this increased rivalry [over access to the resources of the transboundary river basins] is unlikely to lead to warfare between states. The evidence to date shows that the nature of resources such as transboundary water and a range of conflict resolution institutions have been able to resolve rivalries in ways that avoid conflict.' *ARSWGII*, chapter 12, 'Human Security', 772.

60. Ibid.

61. Terje Tvedt, 'Water: A Source of Wars or a Pathway to Peace? An Empirical Critique of two Dominant Schools of Thought on Water and International Politics', in Terje Tvedt et al. (eds.), *Water Geopolitics and the New World Order* (A History of Water Series 2), 78–9.

62. Ibid. 78–108.

63. Ibid. 78–9. Researchers have claimed that an extreme drought in Syria between 2006 and 2009 was a factor for the violent uprising that erupted there in 2011. Henry Fountain, 'Researchers Link Syrian Conflict to a Drought Made Worse by Climate Change' (2 March 2015), <http://www.nytimes.com/2015/03/03/science/earth/study-links-syria-conflict-to-drought-caused-by-climate-change.html/p_r=0>.

Bibliography

A) OFFICIAL DOCUMENTS

a) General

Agreement on the Kosi Project, 1954, <http://www.internationalrivers.org/files/attached-files/treaties_between_nepal-india.pdf>.

Agreement on the Kosi Project, 1966 (amended), <http://www.international waterlaw.org/documents/regionaldocs/Amended_Kosi_Project_Agreement-1966.pdf>.

Agreement on the Salal Hydroelectric Plant between the Governments of Pakistan and India, 1978. The text is available at the Creative Commons Attribution-Share-Alike Licence website under certain conditions.

Barcelona Convention and Statute concerning the Regime of Navigable Waterways of International Concern, 20 April 1921, League of Nations Treaty Series, vol. 7, <http://www.internationalwaterlaw.org/documents/intldocs/barcelona_conv.html>.

Convention dated 21 May 1906 between the US and Mexico, <http://www.ibwc.gov/Files/1906Conv.pdf>.

Convention on the Law of Non-Navigational Uses of International Watercourses, 21 May 1997, <http://waterwiki.net/index.php?title=Convention_on_the_Law_of_the_Non-Navigational_Uses_of_International_Watercourses>.

Denunciation of the Joint Statement (Delhi Agreement) dated 4 May 1948 by India, United Nations Treaty Series, vol. 85.

Draft Articles on the Law of Treaties with Commentaries, Yearbook of International Law Commission, vol. II, 1966, <http://legal.un.org/ilc/texts/instruments/english/commentaries/1_1_1966.pdf>.

Exchange of Notes on the Use of the Waters of the River Nile for Irrigation Purposes, 7 May, 1929, Treaty Series 17, Cmd. 3348, House of Commons—Parliamentary Papers online, London: His Majesty's Stationery Office.

Express Postal Telegram dated 26 April 1948 from Chief Secretary, East Punjab, to Chief Secretary, West Punjab.

Gandak Irrigation and Power Project, 1959, <http://www.moen.gov.np/pdf_files/gandak_treaty.pdf>.

Gandak Irrigation and Power Project, 1964 (amended), <http://www.international rivers.org/files/attached-files/treaties_between_nepal-india.pdf>.

Ganges Treaty, 1996, <http://www.jrcb.gov.bd/attachment/Ganges_Water_ Sharing_treaty,1996.pdf>.

Hearings on Treaty with Mexico relating to the Utilization of the Waters of Certain Rivers, Committee on Foreign Relations, US Senate 79th Congress, 1st session, Part 5.

Helsinki Rules on Uses of the Waters of International Rivers, 1966, <http://digi. library.tu.ac.th/thesis/la/1417/17APPENDIX_C.pdf>.

ICOLD Bulletin no. 50 (Dams and the Environment: Notes on Regional Influences), <http://www.icold-cibg.org/GB/publications/bulletins.asp>.

ICOLD Definition of a Large Dam, <https://en.wikipedia.org/wiki/International _Commission_on_Large_Dams>.

India's Rejection of Pakistan's Note of Denunciation of the Joint Statement (Delhi Agreement) of 4 May 1948, United Nations Treaty Series, vol. 128.

Indus Basin Irrigation Water Dispute: The Lilienthal Proposal, Government of Pakistan Publication (Press Release no. 139), 1953.

Indus Basin Dispute, Government of Pakistan Publication, 1958.

Indus Basin Water Dispute, Government of Pakistan Publication, 1958.

Indus Waters Treaty, <http://siteresources.worldbank.org/INTSOUTHASIA Resources/223497-1105737253588/IndusWatersTreaty1960.pdf>.

IPCC Climate Change, 2007: Working Group II: Impacts, Adaptation, and Vulnerability, <https://www.ipcc.ch/pdf/assessment-report/ar4/wg2/ar4_wg2_full_ report.pdf>.

IPCC Climate Change, 2014: Working Group II: Impacts, Adaptation, and Vulnerability, <http://www.ipcc.ch/report/ar5/wg2/>.

IPCC Fifth Assessment Synthesis Report, 2014, <http://www.ipcc.ch/report/ ar5/syr/>.

IPCC Statement on the Melting of Himalayan Glaciers, 20 January 2010, Geneva, <http://www.ipcc.ch/pdf/presentations/himalaya-statement-20january 2010.pdf>.

IPCC Synthesis Report, 2007 (adopted section by section at the IPCC plenary XXII (Valencia, Spain, 12–17 November 2007), <https://www.ipcc.ch/ pdf/assessment-report/ar4/syr/ar4_syr.pdf>.

IPCC Synthesis Report, 2007, Summary for Policymakers, <http://www.ipcc.ch/ pdf/assessment-report/ar4/syr/ar4_syr.pdf>.

Joint Statement (Delhi Agreement), 4 May 1948, United Nations Treaty Series, vol. 54.

Quaid-i-Azam Muhammad Ali Jinnah: Speeches as Governor General, Karachi: Pakistan Publications, 1963.

Let the Reader Judge, Government of Pakistan Publication, 1958.

Letter dated 12 September 1950 from the Prime Minister of India to the Prime Minister of Pakistan.

Letter dated 8 October 1950 from the Prime Minister of India to the Prime Minister of Pakistan.

Letter dated 21 November 1950 from the Prime Minister of India to the Prime Minister of Pakistan.

Letter dated 27 November 1950 from the Prime Minister of Pakistan to the Prime Minister of India.

Letter dated 11 December 1950 from the Prime Minister of India to the Prime Minister of Pakistan.

Multilateral Treaties Deposited with the Secretary General (Status as of 1 April 2009), New York: UN, 2009.

Pakistan: The Struggle for Irrigation Water-and Existence, Government of Pakistan Publication, 1953.

The Partition of the Punjab 1947: A Compilation of Official Documents, vols. I and III, Lahore: National Documentation Centre, 1983.

Protocol to Treaty of Friendship between Turkey and Iraq, 29 March 1947.

Register of International Rivers Water Supply Management, vol. 2, no. 1, New York: United Nations (Pergamon Press), 1978.

Report of the Indian Irrigation Commission, vol. II, Calcutta: Office of the Superintendent of Government Printing, India, 1903.

Responsibility of States for Internationally Wrongful Acts (Draft Articles), International Law Commission, 2001, available in the Annex to the General Assembly Resolution 26/83 of December 2001, <http://legal.un.org/ilc/texts/instruments/english/draft_articles/9_6_2001.pdf>.

A Review of the Efforts made to Settle the Irrigation Water Dispute between Pakistan and India, undated, Government of Pakistan Publication, File no. 1787276.

Sabharwal, Sharat, 'The Indus Water Treaty', *Speech of the Indian High Commissioner to the Karachi Council on Foreign Relations and Pakistan-India Citizen Friendship Forum*, 3 April 2010, <http://www.india.org.pk/docs/SpeechbyMr.SharatSabharwalHighIndusWatersTreaty3April2010.pdf>.

Tanakpur Agreement, 1991, <http://www.internationalrivers.org/files/attached-files/treaties_between_nepal-india.pdf>.

Telegram no. 1681 PRIMIN dated 18 October 1948 from the Government of India to the Government of Pakistan.

Treaty concerning the Integrated Development of the Mahakali Barrage including Sarada Barrage, Tanakpur Barrage and Pancheshwar Project, 1996, <http://www.inter nationalwaterlaw.org/documents/regionaldocs/Mahakali_Treaty-1996. pdf>.

UN Economic Commission for Europe: Legal Aspects of the Hydroelectric Development of Rivers and Lakes of Common Interest, Geneva, January 1952, File no. 1787277.

UNFCCC Rules of Business, FCCC/KP/CMP/2005/8/Add.1, dated 30 March 2006.

US and World Population Clock-census.gov, <http://www.census.gov/popclock/>.

Vienna Convention on the Law of Treaties, 1969, <https://treaties.un.org/doc/ Publication/UNTS/Volume%201155/volume-1155-I-18232-English. pdf>.

Wire no. 1006 dated 21 April 1948 from Chief Secretary, East Punjab, to Chief Secretary, West Punjab.

b) World Bank-Related Documents

Aide-Memoire dated 21 May 1956, World Bank.

Iliff's Letter to Joseph Rucinski, File no. 1787277.

Iliff's Message dated 12 September 1960 to Manzur Qadir, Pakistan's Foreign Minister, File no. 1787284.

Iliff's Message (undated) to Shoaib, File no. 1787284.

Letter dated 6 September 1951 from Black to Liaquat Ali Khan, the Prime Minister of Pakistan.

Letter dated 6 September 1951 from Black to Jawaharlal Nehru, the Prime Minister of India.

Letter dated 8 November 1951 from Black to Khawaja Nazimuddin, the Prime Minister of Pakistan.

Letter dated 8 November 1951 from Black to Jawaharlal Nehru, the Prime Minister of India.

Letter dated 13 March 1952 from Black to Khawaja Nazimuddin, the Prime Minister of Pakistan.

Letter dated 13 March 1952 from Black to Jawaharlal Nehru, the Prime Minister of India.

Letter dated 8 April 1953 from Jawaharlal Nehru, the Prime Minister of India to Black.

Letter dated 11 August 1953 from Black to the Government of Pakistan.

Letter entitled 'Fact and Fiction of the Indus Basin Irrigation Water Dispute', by L Shaffi, Pakistan's Consul General in New York, File no. 1787278.

Letter dated 15 March 1954 covering an American Engineer's Appraisal of the Bank's Proposal, Indus Basin Irrigation Water Problem: Working Draft of Introductory Memorandum on Preparation of Workable Plan under the Good Offices of the International Bank, December, 1954, File no. 1787280.

Letter dated 24 August 1954 of Sir Zafrulla Khan, the Foreign Minister of Pakistan to Black, Indus Basin Irrigation Water Problem: Working Draft of the Introductory Memorandum on Preparation of Workable Plan under the Good Offices of the International Bank, December, 1954, File no. 1787280.

Letter dated 20 October 1954 from the Chief Minister of Punjab (Firoz Khan Noon) to the Minister of Interior (Mushtaq Ahmed Gurmani), File no. 1787280.

Letter dated 28 June 1956 from Wheeler to Black, File no. 1787281.

Letter dated 20 May 1957 from Mueenuddin to Iliff, File no. 1782282.

Letter dated 25 July 1957 from Mueenuddin to Iliff, File no. 1787282.

Letter dated 10 September 1957 from Mueenuddin to Iliff, File no. 1787282.

Letter dated 22 December 1958 from Mueenuddin to Iliff, File no. 1787282.

Letter (termed 'Flier') dated 21 August 1959 from Mueenuddin to CRO London, File no. 1787283.

Letter dated 7 July 1960 from Black to Nehru, File no. 1787284.

Letter dated 8 September 1960 (relating to Court of Arbitration), File no. 1787797.

Lilienthal's letter dated 13 December 1951 to Khosla, File no. 1787276.

Memo dated 22 May 1953 from Sommers to Black, File no. 1787278.

Memo dated 20 July 1959 from Iliff to Wheeler, File no. 1787283.

Memo dated 3 May 1960 by Iliff of the Meeting between Black, Nehru and Himself, File no. 1787284.

Memo dated 3 May 1960 by Iliff of a talk with Pakistan Foreign Minister, Manzur Qadir, File no. 1787284.

Notes dated 30 June 1960 from Black's Conversation with Gulhati, File no. 1787284.

Report dated 28 June 1956 from Wheeler to Black, File no. 1787281.

Report dated 9 December 1956 of Iliff's Meeting with Nehru, File no. 1782281.

Report dated 11 June 1957 of Iliff's Meeting with Chaudhri Mohammad Ali, the Pakistan Prime Minister, File no. 1787282.

Report dated 13 May 1959 of a Meeting of Messrs Black and Iliff with Nehru, File no. 1787283.

Report of Washington Working Party 3rd Meeting, 4 December 1959, File no. 1787283.

Report dated 9 December 1959 of Meeting of Pakistan's Ambassador in Washington and Mueenuddin with Black and Iliff, File no. 1787283.

Text of Shoaib's Draft of 12 July 1954 and Iliff's Correction, File no. 1787280.

World Bank Operational Manual, March 2012, <http://go.worldbank.org/ 2G5SSZAET0>.

World Bank Proposal for a Plan for the Development and Use of the Indus Basin Waters, 5 February 1954, World Bank.

World Bank Press Releases (26 June 1952; 5 February, 1954; and 10 December 1954).

• File No.	Title	Start Date
1787276	*Indus Basin Dispute-General Negotiations-Corr. 01*	4/1/49
1787277	*Indus Basin Dispute-General Negotiations-Corr. 02*	3/1/52
1787278	*Indus Basin Dispute-General Negotiations-Corr. 03*	2/1/53
1787279	*Indus Basin Dispute-General Negotiations-Corr. 04*	11/1/53
1787280	*Indus Basin Dispute-General Negotiations-Corr. 05*	8/1/54
1787281	*Indus Basin Dispute-General Negotiations-Corr. 06*	1/1/56
1787282	*Indus Basin Dispute-General Negotiations-Corr. 07*	8/1/57
1787283	*Indus Basin Dispute-General Negotiations-Corr. 08*	1/1/59
1787284	*Indus Basin Dispute-General Negotiations-Corr. 09*	1/1/60
1787766	*Indus Basin Dispute-Bank Analysis-Pak Plans-Corr. 01*	4/1/57
1787267	*Working Party-Corr. 01*	1/1/52
1787268	*Working Party-Corr. 02*	11/1/52
1787269	*Working Party-Corr. 03*	5/1/53
1787270	*Working Party-Corr. 04*	7/1/53
1787294–96	*India Comprehensive Plan*	1/2/03
1787312	*Pakistan Comprehensive Plan-Corr. 01*	12/1/53
1787759	*Bank Proposal-Corr. 01*	11/1/54
1788098–100	*Tipton Study-Documents 01–02–03*	11/1/54
1787391–92	*Pak Alternate London Plan-Corr. 01, Doc. 01*	1/1/58
1787393–96	*Pak Alternate London Plan-Docs. 02–05*	10/1/58
1787310	*Indian Comments on Pak London Plan, Corr. 01*	11/1/58
1787399–402	*Pak Comments on London Plan-Corr. 01–02, Docs. 01–02*	1/1/59
1787307–09	*Marhu Plan- Corr., 01-Docs. 01–02*	10/1/58
1787793	*Agriculture Use by India on Western Rivers-Corr. 01*	4/1/59
1787794	*Generation of Hydropower by India on W. Rivers-Corr. 01*	8/1/59
1787796	*Storage of Waters by India on W. Rivers-Corr. 01*	2/1/60
1787797	*Court of Arbitration Corr. 01*	3/1/60

B) BOOKS AND REPORTS

Abbas A. T., B. M., *The Ganges Water Dispute* (Bangladesh: University Press Ltd., 1982).

Abbas A. T., B. M., 'Agreement on the Ganges', in Ali, M., Radosevich, G. E., and Khan, A. A. (eds), *Water Resources Policy for Asia* (Boston: Balkema, 1987), 517–38.

Abdullah, Sheikh M., *Aatish-e-Chinar* (Autobiography in Urdu) (Lahore: Chaudhry Academy, nd).

Abdullah, Sheikh M., *Flames of the Chinar* (Autobiography in English abridged and translated from Urdu by Khushwant Singh) (Viking, Penguin India, 1993).

Akhtar, Shaheen, *Emerging Challenges to Indus Waters Treaty: Issues of Compliance and Transboundary Impacts of Indian Hydroprojects on the Western Rivers,* Focus on the Regional Issues, Institute of Regional Studies, vol. XXIII, no. 3, 2010.

Alam, Undala Z., *Water Rationality: Mediating the Indus Waters Treaty* (Doctoral Thesis, Geography Department, University of Durham, 1998), <http://etheses.dur.ac.uk/1053/1/1053.pdf>.

Alam, Undala Z., *Notes from a Conversation with Sir William Iliff* (10 June 1970), <http://siteresources.worldbank.org/EXTARCHIVES/Resources/William_Iliff_Summarized_Interview_1970.pdf>.

Ali, Chaudhri M., *The Emergence of Pakistan* (Lahore: The Research Society of Pakistan, University of the Punjab, 1973).

Ali, Imran, *The Punjab under Imperialism (1885–1947)* (Karachi: Oxford University Press, 1989).

Analyzing Kishanganga Award, Report of the Institute of Policy Studies (IPS) Seminar (January 2014), <http://www.ips.org.pk/whats-new/92-seeminar/1748-ips-seminar-reviews-kishanganga-award>.

Arora, R. K., *The Indus Water Treaty Regime* (New Delhi: Mohit Publications, 2007).

Avoiding Water Wars: Water Scarcity and Central Asia's Growing Importance for Stability in Afghanistan and Pakistan, A Majority Staff Report, Committee on Foreign Relations, US Senate, 12th Congress (22 February 2011).

Bashin, Avtar Singh, *India-Bangladesh Relations, 1971–94: Documents,* vol. 1 (Delhi: Siba Exim (Pvt.), Ltd., 1996).

Baxter, R. R., 'The Indus Basin', in Garretson, A. H., Hayton, R. C., and Olmstead, C. J. (eds), *The Law of International Drainage Basins* (Dobb's Ferry, New York: Oceana Publications, Inc., 1967), 443–85.

Boesen, Jannick, and Ravnborg, Helle Munk (eds), *From Water Wars to Water Riots? Lessons from Water Transboundary Management* (Copenhagen: Danish Institute for International Studies, 2004).

Bowett, Derek W., *Self-Defence in International Law* (Manchester: Manchester University Press, 1958).

Briscoe, John, and Qamar, Usman, *Pakistan's Water Economy: Running Dry* (Karachi: Oxford University Press, 2005).

Briscoe, John, and Malik, R. P. S., *India's Water Economy Bracing for a Turbulent Future* (New Delhi: Oxford University Press, 2006).

Burke, S. M., *Pakistan's Foreign Policy: An Historical Analysis* (Karachi: Oxford University Press, 1973).

Campbell-Johnson, Alan, *Mission with Mountbatten* (London: Robert Hale, 1953).

Caponera, D., 'International Water Resources Law in the Indus Basin', in Ali, M., et al. (eds), *Water Resources Policy for Asia* (Boston: Balkema, 1987), 509–15.

Changing Glaciers and Hydrology in Asia: Addressing Vulnerabilities to Glacier Melt Impacts (USAID Publication, 20 November 2010).

The Changing Himalayas—Impact of Climate Change on Water Resources and Livelihood in the Greater Himalayas (Kathmandu: International Centre for Integrated Mountain Development (ICIMOD), 2009).

Chellaney, Brahma, *Water: Asia's New Battleground* (Washington D. C.: George Town University, 2011).

Chemillier-Gendreau, Monique, 'Equity', in Bedjaoui, M. (ed.), *International Law: Achievements and Prospects* (Paris, Dordrecht, etc.: UNESCO, Martinus Nijhoff Publishers, 1991).

Climate Depot Special Report: More than 1000 Scientists Dissent over Man-Made Global Warming Claims, <http://www.climatedepot.com/2011/05/31/special-report-more-than-1000-international-scientists-dissent-over-manmade-global-warming-claims-challenge-un-ipcc-gore/>.

Collins, Larry, and Lapierre, Dominique, *Freedom at Midnight* (Pan Books, 1977).

Crow, Ben, and Singh, Nirvikan, *Impediments and Innovation in International Rivers: The Waters of South Asia* (Santa Cruz: Department of Sociology and Economics, University of California, August, 1999).

Cukwurah, A. O., *The Settlement of Boundary Disputes in International Law* (Manchester: Manchester University Press, 1967).

Dixit, J. N., *India-Pakistan: In War and Peace* (London: Routledge, 2002).

Dyer, Gwynne, *Climate Wars* (Toronto: Random House (Canada), 2008).

Elias, T. O., *The Modern Law of Treaties* (Dobbs Ferry, New York: Oceana Publications Inc; Leiden: AW Sijthoff, 1974).

Engineers Study Forum Report, <https://siyasipakistan.wordpress.com/2010/02/08/india-stealing-water-causing-12-bln-loss-to-pakistan/>.

Fairley, Jean, *The Lion River: The Indus* (London: Penguin Books, 1975).

Garg, Santosh Kumar, *International and Interstate River Water Disputes* (New Delhi: Laxmi Publications Ltd., 1999).

Gleick, P. H. (ed.), *Water in Crisis: A Guide to the World's Fresh Water Resources* (New York: Oxford University Press, 1993).

Global Water Security, Intelligence Community Assessment, ICA 2012(08) (2 February 2012), <https://fas.org/irp/nic/water.pdf>.

Goodrich, Hambro, and Simmons, *Charter of the United Nations: Commentary and Documents* (New York and London: Columbia University Press, 1969).

Graves Jr., H., 'The Bank as International Mediator: Three Episodes', in Mason, E. S. and Asher, R. E. (eds), *The World Bank Since Bretton Woods* (Washington, D. C.: Brookings Institution Press, 1973), 595–643.

Gulhati, Niranjan D., *Indus Waters Treaty: An Exercise in International Mediation* (Bombay, Calcutta, etc.: Allied Publishers, 1973).

Hanauer, Larry and Chalk, Peter, *India's and Pakistan's Strategies in Afghanistan: Implications for the United States and the Region* (Centre for Asia Pacific Policy, International Programs of the Rand Corporation, 2012), <http://www.rand.org/pubs/occasional_papers/OP387.html>.

Hassan, (Lt. Col.) Javed, *India: A Study in Profile* (Rawalpindi: Army Education Press, 1990).

Hayat, Sirdar Shaukat, *The Nation that Lost its Soul: Memoirs of Sirdar Shaukat Hayat Khan* (Lahore: Jang Publishers, 1995).

Hodson, H. V., *The Great Divide* (Karachi: Oxford University Press, 1985).

Hussain, Ijaz, 'Pakistani and Indian attitudes towards the World Court', in Hussain, Ijaz, *Issues in Pakistan's Foreign Policy: An International Law Perspective* (Lahore: Progressive Publishers, 1988), 201–21.

Hussain, Ijaz, 'Sir Zafrulla Khan – The Silent Judge', in Hussain, Ijaz, *Issues in Pakistan's Foreign Policy: An International Law Perspective* (Lahore: Progressive Publishers, 1988), 118–53.

Hussain, Ijaz, *Kashmir Dispute: An International Law Perspective*, Islamabad: Quaid-i-Azam Chair, National Institute of Pakistan Studies, Quaid-i-Azam University, Islamabad, 1998.

Hussain, Ijaz, 'Baglihar must be Depoliticized' in Hussain, Ijaz, *Dimensions of Pakistan-India Relations* (Lahore: Heritage Publications, 2006), 237–42.

Hussain, Ijaz, 'Baglihar under Spotlight Again', in Hussain, Ijaz, *Dimensions of Pakistan-India Relations* (Lahore: Heritage Publications, 2006), 243–48.

Hussain, Ijaz, 'Pakistan's Baglihar Dilemma', in Hussain, Ijaz, *Dimensions of Pakistan-India Relations* (Lahore: Heritage Publications, 2006), 254–58.

Hussain, Ijaz, 'What Does Baglihar Portend?', in Hussain, Ijaz, *Dimensions of Pakistan-India Relations* (Lahore: Heritage Publications, 2006), 232–36.

Hussain, Ijaz, 'Who is Responsible for the Baglihar Debacle?', in Hussain, Ijaz, *Dimensions of Pakistan-India Relations* (Lahore: Heritage Publications, 2006), 259–64.

Hussain, Ijaz, 'Wolfensohn Statement and the Baglihar Dam', in Hussain, Ijaz, *Dimensions of Pakistan-India Relations* (Lahore: Heritage Publications, 2006), 264–68.

Iyer, Ramaswamy, *India's Water Relations with her Neighbours* (USI, National Security Studies, 2007, KW Publishers (Pvt.) Ltd., 2008).

Jennings, Sir Robert, 'Treaties', in Bedjaoui, M (ed.), *International Law: Achievements and Prospects* (Paris, Dordrecht, etc., UNESCO, Martinus Nijhoff Publishers, 1991).

Jennings, Sir Robert, and Watts, Sir Arthur, *Oppenheim's International Law (Peace)*, vol. I, (parts 2 to 4) (Longman, 1992).

Kasuri, Khurshid Mahmud, 'India: "A Road Less Travelled"', in Kasuri, Khurshid Mahmud, *Neither a Hawk Nor a Dove* (Karachi: Oxford University Press, 2015).

Kelsen, Hans, *Peace through Law* (Chapel Hill: University of California, 1944).

Khan, Mohammad Ayub, *Friends not Masters: A Political Biography* (Lahore, Karachi, etc.: Oxford University Press, 1967).

Khan, Sir Muhammad Zafrulla, *Tehdis-e-Naimat,* [online presentation] <http://www.slideshare.net/muzaffertahir9/tehdis-enaimat>.

Khan, Sir Muhammad Zafrulla, *The Agony of Pakistan* (1974). No other publishing details are available.

Khwakhuzhi, Gharanai, *Afpak: the Strategic Depth,* (Khaama Press, 20 December 2013), <http://www.khaama.com/afpak-the-strategic-depth-9876>.

Kirmani, Syed, and Le Moigne, Guy, *Fostering Riparian Cooperation in International River Basins* (Washington, D. C.: World Bank, 1997).

Kulshreshtka, S. N., *World Water Resources and Regional Vulnerability: Impact of Future Changes,* RR-93–10. (IIASA Laxenbury, Austria).

Kux, Dennis, *The United States and Pakistan: Disenchanted Allies* (Karachi: Oxford University Press, 2001).

Kux, Dennis, *India-Pakistan Negotiations: Is Past Still Prologue?* (Washington, D. C.: US Institute of Peace, 2006).

Lamb, Alastair, *Kashmir: A Disputed Legacy 1846–1990* (Karachi, etc.: Oxford University Press, 1992).

Lauterpacht, Hersch, *The Function of Law in the International Community* (Hamden-Connecticut: Archon Books, 1966).

Lilienthal, David E., *The Journals of David E Lilienthal: Venturesome Years 1950–55,* vol. 3 (New York, Evanston and London: Harper and Row, 1966).

Lilienthal, David E., *The Journals of David E Lilienthal: Creativity and Conflict*

1964–67, vol. 6 (New York, Evanston and London: Harper and Row, 1976).

McMahon, Robert J., *The Cold War on the Periphery: The United States, India and Pakistan* (New York: Columbia University Press, 1994).

Malik, Bashir A., *Indus Waters Treaty in Retrospect* (Lahore: Brite Books, 2005).

Malik, Ramiz Ahmad, *Teen Darya Kaisay Khoe, Sutlej, Beas aur Ravi: Muhaida-e-Sindh Taas Ke Andruni Kahani* (in Urdu) (Lahore: Takhleqat, 2008).

Malhotra, Pia, *Water Issues between Nepal, India and Bangladesh* (New Delhi: Institute of Peace and Conflict Studies (IPCS), Special Report, July, 2010).

Mansergh, Nicholas, and Moon, Penderel (eds), *Constitutional Relations between Britain and India: The Transfer of Power 1942–7, The Mountbatten Viceroyalty: Formulation of a Plan, 22 March-30 May, 1947*, vol. X (London: Her Majesty's Stationary Office, 1981).

Mansergh, Nicholas, and Moon, Penderel (eds), *Constitutional Relations between Britain and India: The Transfer of Power 1942–7, The Mountbatten Viceroyalty: Princes, Partition and Independence, 8 July–15 August 1947*, vol. XII (London: Her Majesty's Stationary Office, 1983).

Mckinsey Report, *Charting our Water Future* (November 2009), <http://www.mckinsey.com/business-functions/sustainability-and-resource-productivity/our-insights/charting-our-water-future>, Executive Summary.

Mehta, Jagat S., 'The Indus Waters Treaty', in Vlachos, E., Webb, A., and Murphy, I. L. (eds), *The Management of International River Basin Conflicts* (Proceedings of a workshop held at the Institute for Applied System Analysis, Laxenberg, Austria, 22–25 September 1986), 1–24.

Memon, Altaf A., *An Overview of the History and Impacts of the Water Issue in Pakistan* (Adelphi, Maryland: University of Maryland, November, 2002).

Michel, Aloys Arthur, *The Indus Rivers: A Study of the Effects of Partition* (New Haven and London: Yale University Press, 1967).

Mirza, Abdul Latif, and Ali, Ch. Mazhar, 'Canal Irrigation', in Hasan, Mubashir (ed.), *Hundred Years of PWD* (Lahore: Publications Committee of PWD (Centennial), October 1963), 3–51.

Morgenthau, Hans J., *Politics Among Nations: The Struggle for Power and Peace* (New York, etc.: McGraw-Hill International Editions (Political Science Series), 1997).

Mosley, Leonard, *The Last Days of the British Raj* (London: Weidenfeld & Nicolson, 1961).

Mukerjee, Amitabha, 'The Ganga Basin', <https://www.cs.albany.edu/~amit/ganges.html>.

Murty, B. S., 'Settlement of Disputes', in Sorenson (ed.), *Manual of International Law* (London, Toronto: Macmillan; New York: St. Martin's Press, 1968), 673–737.

Munir, Muhammad, *From Jinnah to Zia* (Lahore: Vanguard Books Ltd., 1979).

Naqvi, Saiyid Ali, *Indus Waters and Social Change: The Evolution and Transition of Agrarian Society in Pakistan,* (Karachi: Oxford University Press, 2013).

NASA – What is in a Name? Global Warming v. Climate Change, <http://www.nasa.gov/topics/earth/features/climate_by_any_other_name.html>.

Nehru, B. K., *Nice Guys Finish Second: Memoirs* (New Delhi: Viking (India), 1997).

Noon, Firoz Khan, *From Memory* (National Book Foundation: 1993).

Noorani, A. G., *Article 370: A Constitutional History of Jammu and Kashmir* (Karachi: Oxford University Press, 2011).

Origin of the Phrase 'Climate Change'? Blame it on Bush-Gore Lie, <http://algorelied.com/?p=3760>.

'Pakistanis have Blown my Comments out of Proportion', *Outlook,* (10 August 2009), <http://www.outlookindia.com/magazine/story/pakistanis-have-blown-my-comments-out-of-proportion/261113>.

Palejo, Rasul Bux, *Sindh-Punjab Water Dispute 1859–2003,* <http://www.panhwar.com/>.

Pirzada, Sharifuddin, 'Radcliffe Award', in Sadullah, Mian Muhammad, and Mujahid, Sharif-al, et al., (eds), *The Partition of the Punjab, A Compilation of Official Documents,* vol. I (Lahore: National Documentation Centre, 1983), vii–xl.

Pitman, G. T. Keith, 'The Role of the World Bank in Enhancing Cooperation and Resolving Conflict on International Watercourses: The Case Study of the Indus Basin', in Salman, Salman M. A., and Chazournes, Laurence Boisson de, *International Watercourses: Enhancing Cooperation and Managing Conflict* (Washington D. C.: World Bank, 1998), 155–66.

Political Economy Analysis of the Teesta River Basin (The Asia Foundation, March 2013), <https://asiafoundation.org/resources/pdfs/TheAsiaFoundation.PoliticalEconomyAnalysisoftheTeestaRiverBasin.March20131.pdf>.

Qureshi, M. Aslam, *Anglo-Pakistan Relations 1947–76* (Lahore: Research Society of Pakistan, University of the Punjab, 1976).

Rausching, Dietrich, 'Indus Water Dispute', in Bernhardt, Rudolf (ed.), *The Encyclopaedia of International Law,* vol. 2 (Elsevier (North-Holland), 1995), 962–64.

Refseth, Per Steineide, *The Indus River Basin, 1999–2008: An Intellectual History in Hydropolitics* (master's thesis, Institute of Archaeology, Conservation and History, University of Oslo, Spring, 2013).

Resource Dispute in South Asia: Water Scarcity and the Potential for Interstate Conflict (CIA Report of a Workshop in International Public Affairs, Office of South Asia, 1 June, 2009).

Roberts, Andrew, 'Lord Mountbatten and the Perils of Adrenalin', in Roberts, Andrew, *Eminent Churchillians* (London: Weidenfeld and Nicolson, 1994), 55–136.

Rose, Leo E., 'India's Regional Policy: Non-Military Dimensions', in Cohen, Stephen P. (ed.), *Security of South Asia: American and Asian Perspectives* (Urbana-Champagne, Illinois: University of Illinois Press, 1987), 3–23.

Sadullah, Mian Muhammad, Mujahid, Sharif-al, et al. (eds), *The Partition of the Punjab, A Compilation of Official Documents*, vol. III (Lahore: National Documentation Centre, 1983).

Salman, Salman M. A., and Uprety, Kishor, *Conflict and Cooperation on South Asia's International Rivers: A Legal Perspective* (Washington, D. C.: World Bank, 2002).

Saraf, Muhammad Yusuf, *Kashmiris Fight for Freedom*, vol. II (Lahore: Ferozsons Ltd., 1979).

'Scientists Predict Warmer and Wetter Himalayas', (5 August 2013), <http://www.icimod.org/?q=11326>.

Sette-Camara, Jose, 'Methods of Obligatory Settlement of Disputes', in Bedjaoui, M. (ed.), *International Law: Achievements and Prospects* (Paris, Dordrecht, etc.: UNESCO, Martinus Nijhoff Publishers, 1991), 519–44.

Sherwani, Latif Ahmed, *The Partition of India and Mountbatten* (Karachi: Council for Pakistan Studies, 1986).

Shreshta, Aditya Man, *Bleeding Mountains of Nepal* (Kathmandu: Ekta Books, 1999).

Singer, Fred (ed.), *Nature, Not Human Activity Rules the Climate: Summary for Policymakers of the Report of the Nongovernmental International Panel on Climate Change* (Chicago: The Heartland Institute, 2008).

Singh, Khushwant, *A History of the Sikhs*, vol. 2 (New Delhi: Oxford University Press, 1999).

Starke, J. G., *Introduction to International Law* (London: Butterworths (10th Edition), 1989).

Tabassum, Shaista, *River Water Sharing between India and Pakistan: Case Study of Indus Waters* (Policy Studies no. 24) (Colombo: Regional Centre for Strategic Studies, 2004).

Thatte, Chandrakant D., 'Indus Waters and the 1960 Treaty between India and Pakistan', in Varis, Olli, Tortajada, C., and Biswas, A. K. (eds), *Management of Transboundary Rivers and Lakes* (Heidelberg: Springer-Verlag Berlin, 2008), 165–206.

Transboundary Fresh Water Dispute Database (TFDD) (Oregon State University: 2007), <http://www.transboundarywaters.orst.edu/>.

Transcript of Interview with Eugene R. Black, Oral History Program (Columbia University) (World Bank/IFC Archives, Brookings Institution, 6 August 1961).

Transcript of Interview with Robert Garner, Oral History Program (Columbia University) (World Bank/IFC Archives, Brookings Institution, 19 July 1961).

Transcript of Interview with Sir William Iliff, Oral History Program (Columbia University) (World Bank/IFC Archives, Brookings Institution, 12–16 August 1961).

Transcript of Interview with Davidson Sommers, Oral History Program (Columbia University) (World Bank/IFC Archives, Brookings Institution, 2 August 1961).

Tvedt, Terje, 'Water: A Source of Wars or a Pathway to Peace? An Empirical Critique of two Dominant Schools of Thought and the New World Order', in Tvedt, Terje, et al. (eds), *Water Geopolitics and the New World Order* (A History of Water Series 2) (2010), 78–108.

Twain's Whiskey/Water Quote Appears Greatly Exaggerated, <http://www.mcclatchydc.com/news/politics-government/article24609343.html>.

ul Haq, Noor (ed.), *Pakistan's Water Concerns*, Islamabad Policy Research Institute (IPRI) Factfile, vol. XII, no. 10, (October 2010).

Upreti, B. C., *Politics of Himalayan River Waters: An Analysis of the River Water Issues of Nepal, India and Bangladesh* (Jaipur, New Delhi: Nirala Publication, 1993),

Upreti, Trilochan, *International Watercourses Law and its Application in South Asia* (Kathmandu: Pairavi Prakashan, 2006).

Verghese, B. G., 'The Indus, POK and the Peace Process: Building on the Foundations of the Indus Waters Treaty', in Gupta, Virendra and Bansal, Alok (eds), *Pakistan Occupied Kashmir: The Untold Story* (New Delhi: Manas Publications, 2007), 195–211.

Washington, Haydn, and Cook, John, *Climate Change Denial: Heads in the Sand* (London, New York: Earth Scan, 2011).

Waslekar, Sandeep, *The Final Settlement: Restructuring India-Pakistan Relations* (Mumbai: Strategic Foresight Group, 2005).

Water Security for India: The External Dynamics, Report of the Institute for Defence Studies and Analysis (IDSA) (2010), <http://www.idsa.in/sites/default/files/book_WaterSecurity.pdf>.

Water – Ismail Serageldin, <http://www.serageldin.com/water.htm>.

What is Carbon Credit? CBS News, <http://www.cbsnews.com/news/what-is-carbon-credit/>.

What is IWRM? Global Water Partnership, The Challenge, <http://www.gwp.org/The-Challenge/What-is-IWRM/>.

Wirsing, Robert G., *India, Pakistan and the Kashmir Dispute* (New York: St. Martin's Press, 1994).

Wirsing, Robert G., *Rivers in Contention: Is there a Water War in South Asia's Future?,* Working Paper no. 4 (South Asia Institute, Department of Political Science, University of Heidelberg, October, 2008).

Wirsing, Robert G., and Jasparro, Christopher, *Spotlight on Indus River Diplomacy: India Pakistan and the Baglihar Dam Dispute* (Asia-Pacific Centre for Security Studies: May 2006), <http://apcss.org/>.

Wolf, Aaron T., *International Water Event Database: 1950–2005* Program in Water Conflict Management and Transformation, Institute for Water and Watersheds, Oregon State University, <http://www.transboundarywaters.orst.edu/database/interwatereventdata.html>.

Wolf, Aaron T., and Newton, Joshua, *Case Study of Transboundary Dispute Resolution: The Indus Waters Treaty,* Institute for Water and Watersheds, Oregon State University, <http://www.transboundarywaters.orst.edu/research/case_studies/Indus_New.htm>.

Wolpert, Stanley, *Jinnah of Pakistan* (Karachi, etc.: Oxford University Press, 1984).

Zeigler, Philip, *Mountbatten: The Official Biography* (London: Book Club Associates, 1985).

C) PERIODICAL LITERATURE

Adhikari, Khaga Nath, 'Turning the Tide: Developing Cooperation on Water Resources in South Asia', *Regional Studies,* Institute of Regional Studies, vol. XXXIII, no. 2, Spring (2015), 67–83.

Ahmed, A. T. S., 'Challenges of Governance in Nepal: Politico-Economic and Ethno-Religious Dimensions', *Journal of Contemporary Asia*, vol. 24, no. 3 (1994), 360–62.

Akhtar, Shaheen, 'Quest for Re-Interpretation of the Indus Waters Treaty: Pakistan's Dilemma', *Margalla Papers (Pakistan's Water Security Dilemma: Re-Visiting the Efficacy of Indus Waters Treaty)*, National Defence University, Special Edition, vol. XV, no. I (2011), 15–46.

Alam, Undala Z., 'Questioning the Water Wars Rationale: A Case Study of the Indus Waters Treaty', *The Geographical Journal*, vol. 168, no. 4 (December 2002), 341–53.

Bhatangar, Manav, 'Reconsidering the Indus Waters Treaty', *Tulane Environmental Law Journal*, vol. 22, no. 2, Summer (2009), 271–313.

Biswas, Asit K., 'Indus Waters Treaty: The Negotiating Process', *Water International*, vol. 11, no. 4 (1992), 201–09.

Blankenship, Erin, 'Kashmiri Water: Good Enough for Peace?', *Pugwash Online*, July, 2010.

Briscoe, John, 'Troubled Waters: Can a Bridge be Built over the Indus', *Economic and Political Weekly*, vol. XLV, no. 50 (11–17 December 2010), 28–32.

Bukhari, M. H., and Sayal, Ejaz Ahmad, 'Emerging Climate Changes and Water Resource Situation in Pakistan', *Pakistan Vision*, vol. 12, no. 2, 236–54.

Burki, Shahid Javed, 'The Indus River Plain', *Encyclopaedia Britannica*, <http://www.britannica.com/place/Pakistan/The-Indus-River-plain>.

Chakraborty, Roshni, and Nasir, Sadia, 'Indus Basin Treaty: Its Relevance to Indo-Pak Relations', *Pakistan Horizon*, no. 4 (2002), 53–62.

Chellaney, Brahma, 'Climate Change and Security in South Asia', *The RUSI Journal*, vol. 152, no. 2, 62–09.

Concannon, Brian E., 'The Indus Waters Treaty: Three Decades of Success, yet, Will it Endure?' *The Georgetown International Environmental Law Review*, vol. 2 (1989), 55–79.

Cully, McPatrick, 'Nepal, India Sign Deal to Build World's Highest Dam', *World Rivers Review*, vol. 11, no. 4, (September 1996), 32.

Deen, T., 'Climate Change Deepening World Water Crisis', (19 March 2008), <https://www.globalpolicy.org/social-and-economic-policy/global-public-goods-1-101/45362-climate-change-deepening-world-water-crisis.html>.

Fitzmaurice, Sir Gerald, 'The Law and Procedure of the International Court of Justice 1951–4: Treaty Interpretation and Other Treaty Points', *British Yearbook of International Law* (1957), vol. 33, 203–93.

Ghali, Boutros Boutros, 'I Support the Algerian Government', *Middle East Quarterly*, vol. 4, no. 3 (September 2007), 59–66, <http://www.meforum.org/364/boutros-boutros-ghali-i-support-the-algerian>.

Gilmartin, D., 'Scientific Empire and Imperial Science: Colonialism and Irrigation in the Indus Basin', *The British Journal of Asian Studies*, vol. 53, no. 4 (1994), 1127–49.

Giordano, Meredith A., 'Managing the Quality of International Rivers: Global Principles and Basin Practice', *Natural Resources Journal*, no. 1, Winter (2003), 111–36.

Gyawali, Dipak and Dixit, Ajaya, 'How Not to Do a South Asian Treaty', *Himal*, (April 2001), <http://old.himalmag.com/component/content/article/1975-How-not-to-do-a-South-Asian-Treaty....html>.

Gyawali, Dipak, and Schwank, Othmar, 'Interstate Sharing of Water Rights: An Alps-Himalayan Comparison', in Dixit, Ajaya (ed.), *Water Nepal*, vol. 4, no. 1 (1994), 233–5.

Hagerty, Devin T., 'India's Regional Security Doctrine', *Asian Survey*, vol. 31, no. 4, (April 1991), 351–63.

Husain, Mahe Zehra, 'The Indus Waters Treaty in Light of Climate Change', *Transboundary Water Resources*, Spring (2010), <http://www.ce.utexas.edu/prof/mckinney/ce397/Topics/Indus/Indus_2010.pdf>.

Hussain, Ijaz, 'Pakistan and the Wullar Barrage', *Regional Studies*, vol. VI. no. 2, Spring (1988), 47–62.

Iftikharuzzaman, 'The Ganges Water Sharing Issue: Diplomacy and Domestic Politics in Bangladesh', *Bangladesh Institute of Strategic Studies Journal*, vol. 15, no. 3 (1994), 215–35.

'India Informs Pakistan before Releasing Water, Baig', *Pakistan Defence* (2 July 2014), <http://defence.pk/threads/pak-to-ask-india-to-inform-before-releasing-dam-water.321938/>.

Iyer, Ramaswamy, 'Was the Indus Waters Treaty in Trouble?', *Economic and Political Weekly*, vol. 37, no. 25 (22–28 June 2002), 2401–02.

Iyer, Ramaswamy, 'Indus Treaty: A Different View', *Economic and Political Weekly* (16 July 2005), 3140–44.

Iyer, Ramaswamy, 'Pakistan's Questionable Move on Water', *Economic and Political Weekly* (27 March 2013), 10–12.

Jaitly, Ashok, 'South Asian Perspectives on Climate Change', in Michel, David, and Pandya, Amit (eds), *Troubled Waters: Climate Change, Hydropolitics, and Transboundary Resources,* Washington, D. C.: Stimson Centre (2009), 17–31.

Kakakhel, Shafqat, 'The Indus Waters Treaty: Negotiation, Implementation, New Challenges, and Future Prospects', *Criterion Quarterly,* vol. 9, no. 2, (10 May 2014), 44–60.

Kirmani, Syed, 'Water, Peace and Conflict Management: The Experience of the Indus and Mekong River Basins', *Water International,* vol. 15, no. 4 (1990), 200–05.

Kulz, Helmut R., 'Further Water Disputes between India and Pakistan', *International and Comparative Law Quarterly,* vol. 18, no. 3 (1969), 718–38.

Laylin, John G., 'Principles of Law Governing the Uses of International Rivers: Contributions from the Indus Basin', *Proceedings of the American Society of International Law,* vol. 51, (25–27 April 1957), 20–36.

Lilienthal, David E., 'Are we Losing India?', *Collier's* (23 June 1951).

Lilienthal, David E., 'Another "Korea" in the Making', *Collier's* (4 August 1951).

Mccaffrey, Stephen C., 'The Harmon Doctrine, One Hundred Years Later: Buried, Not Praised', *Natural Resources Journal,* vol. 36, no. 3, Summer (1996), 549–90.

Mehta, Jagat S., 'The Indus Waters Treaty: A Case Study in the Resolution of an International River Basin Conflict', *Natural Resources Forum,* vol. 12, no. 1 (1988), 69–77.

Mian, Zia, and Ramana, M. V., 'Going MAD: Ten Years of the Bomb in South Asia', *Economic and Political Weekly,* vol. 43, nos. 26/27 (28 June–11 July 2008), 201–08.

Miner, M., Patankar, G., Gamkhar, S., and Eaton, D. J., 'Water Sharing between India and Pakistan: A Critical Evaluation of the Indus Water Treaty', *Water International,* vol. 34, no. 2 (2009), 204–16.

Mohan, Raja C., 'India and the Balance of Power', *Foreign Affairs* (July–August 2006), 17–32.

Noorani, A. G., 'How and Why Nehru and Abdullah Fell Out', *Economic and Political Weekly,* vol. XXXIV, no. 5, (30 January 1999), 268–72.

Noorani, A. G., 'A Treaty to Keep', *Frontline,* vol. 19, issue 8 (13–26 April 2002), <http://www.frontline.in/static/html/fl1908/19080830.htm>.

Noshab, Farzana, and Mushtaq, Nadia, 'Water Disputes in South Asia', *Strategic Studies,* Islamabad Institute of Strategic Studies, vol. 21, no. 3, Summer (2001).

Paukert, M., 'The Indus Umbilical', *Himal* (July 2002), 28–30.

'Population Seven Billion: UN Sets out Challenges', BBC News, <http://www.bbc.com/news/world-15459643>.

Salman, Salman M. A., 'The Baglihar Difference and its Resolution Process: A Triumph for the Indus Waters Treaty', *Water Policy*, vol. 10, issue 2 (April 2008), 105–17, <http://wp.iwaponline.com/content/10/2/105.full.pdf>.

Salman, Salman M. A. and Uprety, Kishor, 'Hydro-Politics in South Asia: A Comparative Analysis of Mahakali and the Ganges Treaty', *Natural Resources Journal*, vol. 39 (1999), 295–343.

Sarfraz, Hamid, 'Revisiting the 1960 Indus Waters Treaty', *Water International*, vol. 38, no. 2 (2013), 304–16.

Sering, Senge H., 'China Builds Dam on Indus near Ladakh', *Journal of Defence Studies* (India), vol. 4, no. 2, (April 2010), 136–9.

Shah, Jamaat Ali, 'Indus Waters Treaty under Stress: Imperatives of Climate Change or Political Manipulation', *Margalla Papers* (Pakistan's Water Security Dilemma: Re-Visiting the Efficacy of Indus Waters Treaty), National Defence University, Special Edition, vol. XV, no. I, (2011), 1–14.

Siyad, A. C. Mohammed, 'Indus Water Treaty and Baglihar Project: Relevance of International Watercourse Law', *Economic and Political Weekly*, vol. 40, no. 29 (16–22 July 2005), 3145–54.

Spens, Sir Patrick, 'The Arbitral Tribunal in India 1947–48', *Transactions of the Grotius Society*, vol. 36 (1950), 61–74.

Swain, Ashok, 'Water Wars: Fact or Fiction?', *Futures*, vol. 33, no. 8 (2001), 769–81.

Swain, Ashok, 'The Indus II and Siachen Peace Park: Pushing the India-Pakistan Peace Process Forward', *The Round Table*, vol. 98, no. 4 (October 2009), 569–82.

Verghese, B. G., 'Ideology Threatens Indus Treaty', *The South Asian Journal* (25 March 2010), <http://www.bgverghese.com/WaterSharing.htm>.

Warikoo, K., 'Indus Waters Treaty: View from Kashmir', *Himalayan and Central Asian States*, vol. 9, no. 3 (July–September 2005), <http://www.jammu-kashmir.com/insights/insight20060601a.html>.

'Water Wars?' *Geographical Perspectives* (December 2002), 293–312, <http://www.jstor.org/stable/i368330>.

Wolf, Aaron T., 'Conflict and Cooperation along International Waterways', *Water Policy*, vol.1, no. 2 (1998), 251–65, <http://www.ce.utexas.edu/prof/mckinney/ce397/Topics/conflict/Conflictandcooperation.pdf>.

Wolf, Aaron T., et al., 'International River Basins of the World', *International*

Journal of Water Resources Development, vol. 15, no. 4 (1999), 387–427, <http://www.tandfonline.com/doi/abs/10.1080/07900629948682>.

D) NEWSPAPER ARTICLES

Abbasi, Ansar, 'Pakistan to Move World Court if India Pursues Ratlé Dam Project', *The News* (20 July 2013).

Abbasi, Ansar, 'Marginal, not 'Big Victory' on Kishenganga', *The News* (24 December 2013).

Abbasi, Ansar, 'Charge Sheet against Pakistan', *The News* (26 December 2013).

Abbasi, Arshad H., 'Lethal Change of Course', *Dawn* (9 February 2010).

Aftab, Noor, 'Pakistan Authorities in a Fix as India Secures Carbon Credit', *The News* (14 January 2011).

Ahmed, Khaled, 'Getting Ready for a 'Water War?', *The Friday Times* (8 March 2010).

Ahmed, Khaled, 'Water War' Pakistani Style', *The Friday Times* (30 November 2012).

Ahmed, Khaled, 'Target: Jamaat Ali Shah', *The Express Tribune* (2 December 2012).

Akhtar, Majed, 'The Indus and the 'Territorial Trap'', *Dawn* (14 February 2012).

Alam, Ahmad Rafay, Kugelman, Michael, and Bakshi, Gitanjali, 'A Kabul River Treaty', *The News* (1 December 2011).

Ali, Kalbe, 'India Assuming Aggressive Posture: Pakistan', *Dawn* (16 March 2013).

Analyst, 'Indian Design to Divert Pakistan's Water', *The Muslim* (Islamabad, 19 March 1987).

Aziz, Mir Abdul, 'Wullar and the Proposed Barrage', *The Muslim* (Islamabad, 24 October 1986).

Bhalla, Nita, 'Thirsty South Asia's River Drifts Threaten "Water Wars"', *Reuters* (23 July 2012), <http://www.reuters.com/article/us-water-southasia-idUSBRE86M0C820120723>.

Bhutta, Zafar, 'Pakistan-India Water Dispute: No Headway in Wullar Barrage Negotiations', *The Express Tribune* (13 May 2011).

Bhutta, Zafar, 'Cold-Feet: Setbacks in Diamer-Basha Funding', *The Express Tribune* (3 August 2012).

Bidwai, Praful, 'The Himalayan Glacier Controversy', *The News* (22 February 2010).

Black, Eugene, 'The Indus: A Moral for Nations', *The New York Times* (11 December 1960).

Bokhari, Ashfak, 'Water Dispute and War Risk', *Dawn (Economic and Business Review)* (18 January 2010).

Bokhari, Ashfak, 'Water Talks', *Dawn (Economic and Business Review)* (5 April 2010).

Bokhari, Ashfak, 'Threat from Melting Glaciers', *Dawn (Economic and Business Review)* (20 September 2010).

Bokhari, Ashfak, 'Growing Risk of Water Wars', *Dawn (Economic and Business Review)* (9 April 2012).

Bokhari, Ashfak, 'Failure to Defend Riparian Rights', *Dawn (Economic and Business Review)* (3 December 2012).

Bokhari, Ashfak, 'Kishanganga Verdict a Tilt in India's Favour', *Dawn (Economic and Business Review)* (25 February 2013).

Bokhari, Ashfak, 'Issues in Kishanganga Hydropower Project', *Dawn (Economic and Business Review)* (8 April 2013).

Briscoe, John, 'War or Peace on the Indus' (18 April 2010), <http://johnbriscoe.seas.harvard.edu/files/johnbriscoe/files/108._john_briscoe_war_or_peace_on_the_indus_201004.pdf>?m=1393430744.

Briscoe, John, 'Winning the Battle but Losing the War', *The Hindu* (22 February 2013).

Briscoe, John, 'Peace, not War, on the Indus', *The News* (1 January 2014).

Chellaney, Brahma, 'Water Treaties and Diplomacy: India Faces Difficult Choices over Water', *The Economic Times* (10 May 2012).

'Claims Himalayan Glaciers could Melt by 2035 were False, say UN Scientists', *The Guardian* (20 January 2010).

Economic Correspondent, 'India Manipulating to Build 7 Dams on 3 Pak Rivers', *The News* (17 May 2012).

Excelsior Correspondent, 'Assembly's Unanimous Call for Scrapping Indus Treaty', *The Daily Excelsior* (4 April 2002), <http://www.hindunet.org/hvk/articles/0402/35.html>.

'British Fair Play' (Editorial), *The New York Times* (5 March 1890).

Fazl-e-Haider, Syed, 'Pak Dam Dealt Funding Blow by India', *Asia Times* (online) (18 August 2012).

Fountain, Henry, 'Researchers Link Syrian Conflict to a Drought Made Worse by Climate Change' (2 March 2015), <http://www.nytimes.com/

2015/03/03/science/earth/study-links-syria-conflict-to-drought-caused-by-climate-change.html>.

Gandapur, Fateh Ullah Khan, 'Implications of Water issue', *Dawn* (8 April 2010).

Gilani, Iftikhar, 'Pakistan on Brink of 'Water Disaster', *The Daily Times* (15 January 2009).

Gilani, Syed Nazir, 'Water Resources in Kashmir', *The Nation* (18 May 2006).

Halepoto, Zulfiqar, 'No Integrated Approach', *Dawn* (24 May 2011).

Haq, M. Anwarul, '30-year Historic Water Treaty', *Daily Star* (Dhaka) (13 December 1996).

Hashmey, N. H., 'Men Behind Salal Accord', *The Pakistan Times (Magazine Section)* (28 April 1978).

'Himalayan Glacier Controversy', *The Hindustan Times* (12 February 2013), <http://www.hindustantimes.com/india/himalayan-glacier-controversy/story-AnFurSg9aW4SrMinMYfXZJ.html>.

'Himalayan Melting: How a Climate Panel Got it Wrong', *Time (Science and Space)*, <http://content.time.com/time/health/article/0,8599, 1955405,00.html>.

Husain, Irfan, 'Edwina Mountbatten, Nehru and Radcliffe', *Dawn.com* (20 January 2014).

Hussain, Ijaz, 'Not Treated According to the Treaty', *The Daily Times* (28 February 2007).

Hussain, Ijaz, 'Can Pakistan Challenge Baglihar Verdict?', *The Daily Times* (14 March 2007).

Hussain, Masood, 'Baglihar Dam: Spillways a Blow to Islamabad', *The Economic Times* (14 February 2007).

'IFIs Refuse to Fund Bhasha Dam under Indian Pressure, says WAPDA', *The News* (4 January 2013).

'India Allays Pakistan Fears on Tulbal Project', *The Times of India* (27 September 1986).

'India Defends Jhelum Project', *The Telegraph* (Calcutta) (October 1986).

'India Financed Problems for Pakistan from Afghanistan: Chuck Hagel', *The Express Tribune* (26 February 2013).

'India Offers to Amend Design of Wullar Barrage', *Dawn.com* (13 May 2011), <http://www.dawn.com/news/628533/india-offers-to-amend-design-of-wullar-barragg>.

'India Resumes Construction Work on Wullar Barrage, International Arbitration…', *The Lahore Times* (14 December 2012).

'Indus Waters Treaty not in Pak Interest: Asif', *The News* (26 October 2013).

Iyer, Ramaswamy, 'What Water Wars?', *The Indian Express* (23 March 2010).

Iyer, Ramaswamy, 'Dealing with Pakistan's Fears on Water', *The Hindu* (28 January 2012).

Iyer, Ramaswamy, 'Troubled Waters', *The Friday Times* (22–28 June 2012).

Iyer, Ramaswamy, 'Jubilation at Indus Win is Premature', *The Hindu* (28 February 2013).

Jan, Mohammad Aslam, 'Salal Dam: Pus Manzar Aur Muzmairat', *Nawa-e-Waqt* (Urdu newspaper, Lahore) (6 February 1978).

Kak, M. L., 'Indus Waters Treaty May Go: PM', *The Tribune* (23 May 2002), <http://www.tribuneindia.com/2002/20020523/main3.htm>.

Kazi, Asif H., 'Misusing the Indus Treaty', *The News* (1 July 2011).

Khalil, Tahir, and Yasin, Asim, 'Shujaat Says Aziz was Fully Aware of Lal Masjid Operation', *The News* (20 March 2013).

Khan, Mubarak Zeb, 'India can Divert only Minimum Water from Kishanganga: Tribunal', *Dawn* (19 February 2013).

'Khawaja Asif Terms PCA Award on Kishenganga Dam a "Big Victory" for Pakistan', *The Express Tribune* (21 December 2013).

Khurshedi, Nusrat, 'Sharing Water Resources with Afghanistan', *Dawn (Economic and Business Review)* (14 November 2011).

Kiani, Khaleeq, 'Agencies Seize Record of two Ministries', *Dawn* (16 April 2011).

Kiani, Khaleeq, 'Joint Management of Water Proposed with Afghanistan', *Dawn* (14 June 2011).

Kiani, Khaleeq, 'Grant of Carbon Credits to India: Pakistan to Challenge UN Decision in World Court', *Dawn* (2 January 2012).

Kiani, Khaleeq, 'Pakistan's Institutional Lapse Helps India Complete Water Projects', *Dawn* (30 July 2012).

Kiani, Khaleeq, 'Objections to Four More Indian Projects Raised', *The News* (28 September 2013).

'Kishanganga Project: Arbitration Court's Decision not a Legal Defeat for Pakistan says Farhatullah Babar', *The Lahore Times* (27 February 2012).

'Kishenganga Verdict at the Hague Court Absolute Victory for India', *The Times of India* (25 December 2013).

Lodhi, Maleeha, 'Talks with no Solutions', *The News* (20 April 2010).

Lodhi, Maleeha, 'Can the Ice Melt on Siachen?', *The News* (24 April 2012).

Mansur, Ijaz, 'Washington Diary: India-Pakistan Conundrum', The *Daily Times* (3 March 2010).

Menon, M. S., 'Concerns over Chinese Barrage on Sutlej', *The Hindu* (30 July 2006).

Menon, M. S., 'A Redundant Treaty', *The Hindu* (8 April 2007).

Menon, M. S., 'Indus Waters Treaty Needs Relook', *The Pioneer* (27 November 2013).

Miah, M. Maniruzzaman, 'How Much Water in the Ganges', *Daily Star* (Dhaka) (9 April 1997).

'Miss Jinnah's Question to President Ayub on Indus Waters Treaty', *The Nation* (16 December 2003).

'MNAs Lambast India for 'Water Terrorism', *The Daily Times* (17 February 2010).

Mountbatten Interview, *Time*, Golden Jubilee Issue (Asia Edition) (11 August 1997).

Mumtaz, Ashraf, 'India can't Scrap IWT: Experts', <http://www.dawn.com/news/11864/lahore-india-can-t-scrap-indus-water-treaty-experts>.

'Mush Weakened Baglihar Dam Case: Jamaat Shah', *The News* (19 March 2012).

Mustafa, Khalid, 'India Starts Construction of Another Mega Dam on Chenab', *The News* (26 February 2011).

Mustafa, Khalid, 'India Plans to Build 190 Dams on Pak Rivers in Six Years', *The News* (28 February 2011).

Mustafa, Khalid, 'India Set to Squeeze Pak Waters while Officials Sleep', *The News* (12 May 2011).

Mustafa, Khalid, 'Pakistan to Convey US Concerns over Kabul River Projects', *The News* (16 May 2011).

Mustafa, Khalid, 'Who "Cleared" India to Bag Multi-Million Dollar Carbon Credits', *The News* (30 May 2011).

Mustafa, Khalid, 'Pakistan Virtually Loses Out on Indian Dam', *The News* (7 July 2011).

Mustafa, Khalid, 'Kishanganga to Cause Pakistan Rs. 12b Annual Loss', *The News* (18 August 2011).

Mustafa, Khalid, 'Pak Govt Helped India Build Dam on Indus River', *The News* (24 September 2011).

Mustafa, Khalid, 'Nation Misled on Kishanganga "Victory"', *The News* (26 September 2011).

Mustafa, Khalid, 'India Rejects Pakistan's Objections over Another Power Project', *The News* (9 October 2011).

Mustafa, Khalid, 'FIA to Contact Interpol for Jamaat Ali Shah's Arrest', *The News* (16 January 2012).

Mustafa, Khalid, 'Kishanganga in The Hague: Flawed Strategy led to Pak Defeat, say Experts', *The News (Business News)* (20 February 2013).

Mustafa, Khalid, 'Another Blow to Pakistan's Water Interests', *The News* (10 July 2013).

Mustafa, Khalid, 'India Emerges Winner in Kishenganga Case', *The News* (22 December 2013).

Mustafa, Khalid, 'Pakistan to Face $145 Million Yearly Loss', *The News* (23 December 2013).

Mustafa, Khalid, 'Establishment Starts Looking into Kishenganga Fiasco', *The News* (29 December 2013).

Naqvi, Jawad, 'New Delhi Planning Tougher Actions: Scrapping of Indus Treaty, Suspension of Overflights', *Dawn* (23 December 2001).

Nizami, Arif, 'Salal Moahida Aur Us Ke Mauharaqat', *Nawa-e-Waqt* (Urdu newspaper) (Lahore, 10 February 1978).

Nizami, Majid, 'The Water Bomb', *The Nation* (7 May 2008).

Nizami's (Majid) Statement on Nuclear War between Pakistan and India, *Dawn* (18 January 2010).

Noorani, A. G., 'The Two Punjabs', *Dawn* (26 October 2010).

'Over Half of Pakistan Lives under Poverty Line: Dar', *Dawn* (3 June 2014).

'Pakistan Seeks Indian Consent for Neutral Expert', *The News* (18 November 2015), <http://www.thenews.com.pk/print/15720-pakistan-seeks-indian-consent-for-neutral-expert>.

'Pakistan Stand on Barrage Issue Surprises India', *The Times of India* (2 October 1986).

Praveen, Swami, 'A Treaty Questioned', *Frontline* (27 April–10 May 2002).

Rahman, Atiqur, and Chakrabarty, Swapan, 'Water Accord: Opposition Charges Sell Out', *Dhaka Courier* (20 December 1996).

Rana, Shahbaz, 'World Bank Projects: Pakistan's Economic Growth at 4.4 per cent', *The Express Tribune* (15 April 2015).

Rose, David, 'Glacier Scientist: I Knew Data hadn't been Verified', *Mail Online* (24 January 2010), <http://www.dailymail.co.uk/news/article-1245636/Glacier-scientists-says-knew-data-verified.html>.

Sayeed's (Mufti) Statement on the IWT, *The Asian Age* (12 March 2003).

Shariff, Maqbul, 'Pakistan, India Sign Agreement on Salal Dam Design', *The Pakistan Times* (14 April 1978).

Shaukat, Sajjad, 'India Suppresses Bangladesh on Teesta River Issue', *Writing for Godot* (20 May 2014), <http://readersupportednews.org/pm-section/21%E2%80%9321/23778-india-suppresses-bangladesh-on-teesta-river-issue>.

Sheikh, Aslam, 'India Told to Stop Work on Jhelum Barrage', *The Muslim* (Islamabad, 29 September 1986).

Singh, S. K., 'India's Threat of Ending the Indus Waters Treaty', *Dawn* (23 December 2001).

'South Asia's Water: Unquenchable Thirst', *The Economist* (19 November 2001), <http://www.economist.com/node/21538687>.

Special Correspondent, 'Indian Barrage Project: A Violation of Treaty', *The Muslim* (Islamabad, 25 October 1986).

Srinath, M. G., 'India Denies Pak Charge on Jhelum Barrage', *The Hindustan Times* (27 September 1987).

Talib, Arjimand Hussain, 'New Fantasy on Jhelum River', *Jang* (September 2007), <http://jang.com.pk/thenews/sep2007-weekly/nos-30-09-2007/pol1.htm#8>.

Verghese, B. G., 'Talk of Abrogating Indus Waters Treaty', *The Tribune* (Chandigarh, India, 29 April 2002).

Verghese, B. G., 'It's Time for Indus–II: A New Treaty will Help the Peace Process too', *The Tribune* (Chandigarh, India, 26 May 2007).

Wasim, Amir, 'Kishanganga Dispute with India: Govt Criticised in Senate over Hague Court Verdict', *The News* (27 February 2013).

'Water for Water', *The Post* (Islamabad, 19 September 2008).

'Water Issues could Sweep Away Indo-Pak Peace Process', <http://www.dawn.com/news/638350/secret-us-cables-accessed-by-dawn-through-wikileaks-water-issues-could-sweep-away-indo-pak-peace-process-feared-us>.

'World Bank Accused of Dithering over Basha Dam', *Dawn.com* (24 October 2011).

'World Bank has Agreed to Finance Diamer-Basha Dam–Dar', *Pakistan Defence* (21 August 2013).

Zahir-ud-Din, 'Seeking Abrogation of IWT is as Good as Seeking Independence', [online forum] <http://defenceforumindia.com/forum/threads/abrogate-indus-water-treaty.12647/>.

E) CASE LAW

a) General

Aerial Incident of 10 August, 1999 (Pakistan v. India), Jurisdiction of the Court, ICJ Reports 2000.

Arbitration Regarding the Iron Rhine (Ijzeren Rijn) Railway (Belgium v. Netherlands), PCA, 24 May 2005.

Case Concerning Gabcikovo-Nagymaros Project (Hungary v. Slovakia), ICJ Reports 1997.

Case Concerning the Arbitral Award made by the King of Spain, ICJ Reports 1966.

Competence of the General Assembly for the Admission of a State to the United Nations, ICJ Reports 1950.

Certain Expenses of the United Nations, ICJ Reports 1962.

Eastern Carelia, PCIJ, series B, no. 5 [1923].

Fisheries Jurisdiction (Federal Republic of Germany v. Iceland), Jurisdiction of the Court, ICJ Reports 1973.

Fisheries Jurisdiction (United Kingdom v. Iceland), Jurisdiction of the Court, ICJ Reports 1973.

Interpretation of Greco-Turkish Agreement of 1ˢᵗ December 1926 (Final Protocol, Article IV), PCIJ, series B, no. 16 [1928].

Interpretation of Peace Treaties with Bulgaria, Hungary and Rumania, ICJ Reports 1950.

Kansas v. Colorado, 185 US 125 [1902], and 206 US 46 [1907].

Kasikili/Sedudu Island (Botswana v. Namibia), ICJ Reports 1999.

Lake Lanoux Affair (France v. Spain), 12 RIAA (Reports of International Arbitration Awards), 16 November 1957.

Legal Consequences for States of the Continued Presence of South Africa in Namibia (South West Africa) notwithstanding Security Council Resolution 276 (1970), ICJ Reports 1971.

Legality of the Use by a State of Nuclear Weapons in Armed Conflict, ICJ Reports 1996.

Legal Status of Eastern Greenland (Denmark v. Norway), PCIJ, series A/B, no. 53 [1933].

Military and Paramilitary Activities in and against Nicaragua (Nicaragua v. US), Jurisdiction and Admissibility, ICJ Reports 1984.

Military and Paramilitary Activities in and against Nicaragua (Nicaragua v. US), Merits, ICJ Reports 1986.

Missouri v. Illinois, 200 US 496 [1906].

North Dakota v. Minnesota, 263 US 365 [1923].

Oil Platforms (Iran v. US), Preliminary Objections, ICJ Reports 1996.

Pulp Mills on the River Uruguay (Argentina v. Uruguay), ICJ Reports 2010.

Rann of Kutch Arbitration Award (Pakistan v. India), PCA, 19 February 1968.

South West Africa (Liberia and Ethiopia v. South Africa), Preliminary Objections, ICJ Reports 1962.

South West Africa (Liberia and Ethiopia v. South Africa), 2nd Phase, ICJ Reports 1966.

Sovereignty over Pulau Ligitan and Pulau Sipadan (Indonesia v. Malaysia), ICJ Reports 2002.

Territorial Dispute (Libya v. Chad), ICJ Reports 1994.

Trail Smelter Case (US v. Canada), 3 RIAA (Reports of International Arbitration Awards), 11 March 1941.

United States Diplomatic and Consular Staff in Tehran (US v. Iran), ICJ Reports 1980.

United States Nationals in Morocco (France v. US), ICJ Reports 1952.

Wyoming v. Colorado, 259 US 419 [1922].

b) Baglihar and Kishenganga

Baglihar Hydroelectric Plant, Memorial of the Pakistan Government, 14 August 2005.

Baglihar Hydroelectric Plant, Counter Memorial of India, 23 September 2005.

Baglihar Hydroelectric Plant, Pakistan's Reply, 31 January 2006.

Baglihar Hydroelectric Plant, India's Rejoinder, 20 March 2006.

Baglihar Hydroelectric Plant, Final Draft, 30 October 2006.

Baglihar Hydroelectric Plant, Transcript of Meeting No. 5 with the Neutral Expert and the Delegations of India and Pakistan, vols. 1–3, World Bank, Washington, D. C., November 2006.

Baglihar Hydroelectric Plant, Expert Determination, Lausanne, 12 February 2007.

Baglihar Hydroelectric Plant, Expert Determination, Executive Summary, Lausanne, 12 February 2007.

Kishenganga Arbitration, Memorial of the Government of Pakistan, vol. 1, 27 May 2011.

Kishenganga Arbitration, Counter Memorial of the Government of India, vol. 1, 2011.

Kishenganga Arbitration, Order, Interim Measures, PCA, 23 September 2011.

Kishenganga Arbitration, Press Release, PCA, 1 September 2012.

Kishenganga Arbitration, Partial Award, PCA, 18 February 2013.

Kishenganga Arbitration, Press Release, PCA, 19 February 2013.

Kishenganga Arbitration, Final Award, PCA, 20 December 2013.

Kishenganga Arbitration, India's Request for Clarification or Interpretation dated 20 May 2013, PCA, 20 December, 2013.

Kishenganga Arbitration, Press Release, PCA, 21 December 2013.

F) PAKISTAN'S TV TALK SHOWS (AVAILABLE ON THE INTERNET)

Kishanganga Dam Case (Sochta Pakistan, PTV), 10 March 2012; and 7 June 2012.

Kishanganga Dam Award (Tonight with Moeed Pirzada, Waqt TV), 22 February 2013.

Water Issue (Tonight @8 with Malik, Dunya TV), 25 July 2013.

Indo-Pakistan Relations (Tonight with Moeed Pirzada, Waqt TV), 26 July 2013.

G) INTERVIEWS OR CORRESPONDENCE WITH WATER EXPERTS

Asif H. Kazi, on 18 August 2012, at his residence in Lahore.

Bashir Ahmad, on 9 March 2014, at his residence in Lahore.

Jamaat Ali Shah, on 13 September 2010, in his office in Lahore and on 2 September 2011; 21 August 2012; and 10 July 2014, at his residence in Lahore.

M. H. Siddiqui, on 9 August 2013, and 9 July 2014, at his residence in Lahore.

Mirza Asif Baig, on 5 April 2013; 22 March 2014, in his office in Lahore and on 20 July 2014, in his office in Islamabad.

Prof. Raymond Lafitte, email correspondence which, starting on 15 March 2014, continued till the end of the year during which seventeen communications were exchanged.

Salman M. A. Salman, on 22 August 2010, at his residence in Washington, D. C.

Shams-ul Mulk, on 5 March 2012, in his office in Islamabad.

APPENDICES

Appendix 1

Inter-Dominion Agreement

1. A dispute has arisen between the East and West Punjab Government regarding the supply by East Punjab of water to the Central Bari Doab and the Depalpur canals in West Punjab. The contention of the East Punjab Government is that under the Punjab Partition (Apportionment of assets and liabilities) Order, 1947, and the Arbitral Award the proprietary rights in the waters of the rivers in East Punjab vest wholly in the East Punjab Government and that the West Punjab Government cannot claim any share of these waters as a right. The West Punjab Government disputes this contention, its view being that the point has conclusively been decided in its favour by implication by the Arbitral Award and that in accordance with international law and equity, West Punjab has a right to the waters of the East Punjab rivers.

2. The East Punjab Government has revived the flow of water into these canals on certain conditions of which two are disputed by West Punjab. One, which arises out of the contention in paragraph 1, is the right to the levy of seigniorage charges for water and the other is the question of the capital cost of the Madhopur Head Works and carrier channels to be taken into account.

3. The East and West Punjab Governments are anxious that this question should be settled in a spirit of goodwill and friendship. Without prejudice to its legal rights in the matter the East Punjab Government assured the West Punjab Government that it has no intention to withhold water from West Punjab without giving it time to tap alternative sources. The West Punjab Government on its part recognize the natural anxiety of the East Punjab Government to discharge the obligations to develop areas where water is scarce and which were under-developed in relation to parts of West Punjab.

4. Apart, therefore, from the question of law involved the Governments are anxious to approach the problem in a practical spirit of the East Punjab

Government progressively diminishing it supply to these canals in order to give reasonable time to enable the West Punjab Government to tap alternative sources.

5. The West Punjab Government has agreed to deposit immediately in the Reserve Bank such ad hoc sum as may be specified by the Prime Minister of India. Out of this sum, that Government agrees to the immediate transfer to East Punjab Government of sums over which there are no dispute.

6. After an examination by each party of the legal issues, of the method of estimating the cost of water to be supplied by the East Punjab Government and of the technical survey of water resources and the means of using them for supply to these canals, the two Governments agree that further meetings between their representatives should take place.

7. The Dominion Governments of India and Pakistan accept above terms and express the hope that a friendly solution will be reached.

(Signed)

JAWAHARLAL NEHRU, GHULAM MOHD, SWARAN SINGH, SHAUKAT HYAT KHAN, N. V. GADGIL, MUMTAZ DAULTANA.

New Delhi May 4, 1948

Appendix 2

Indus Waters Treaty
PREAMBLE

The Government of India and the Government of Pakistan, being equally desirous of attaining the most complete and satisfactory utilisation of the waters of the Indus system of rivers and recognizing the need, therefore, of fixing and delimiting, in a spirit of goodwill and friendship, the rights and obligations of each in relation to the other concerning the use of these waters and of making provision for the settlement, in a cooperative spirit, of all such questions as may hereafter arise in regard to the interpretation or application of the provisions agreed upon herein, have resolved to conclude a Treaty in furtherance of these objectives, and for this purpose named as their plenipotentiaries:

The Government of India:
  Shri Jawaharlal Nehru, *Prime Minister of India,* and

The Government of Pakistan:
  Field Marshal Mohammad Ayub Khan, H.P., H.J., *President of Pakistan,*

who, having communicated to each other their respective Full Powers and found them in good and due form, have agreed upon the following Articles and Annexures:

Article I

DEFINITIONS

As used in this Treaty:
…

(2) The term "Tributary" of a river means any surface channel, whether in continuous or intermittent flow and by whatever name called, whose waters in the natural course would fall into that river, e.g. a tributary, a torrent, a

natural drainage an artificial drainage, a *nadi*, a *nallah*, a *nai*, a *khad*, a *cho*. The term also includes any sub-tributary or branch or subsidiary channel, by whatever name called, whose waters, in the natural course, would directly or otherwise flow into that surface channel.

...

(4) The term "Main" added after Indus, Jhelum, Chenab, Sutlej, Beas or Ravi means the main stem of the named river excluding its Tributaries, but including all channels and creeks of the main stem of that river and such Connecting Lakes as form part of the main stem itself. The Jhelum Main shall be deemed to extend up to Verinag, and the Chenab Main up to the confluence of the river Chandra and the river Bhaga.

(5) The term "Eastern Rivers" means The Sutlej, The Beas and The Ravi taken together.

(6) The term "Western Rivers" means The Indus, The Jhelum and The Chenab taken together.

(7) The term "the Rivers" means all the rivers, The Sutlej, The Beas, The Ravi, The Indus, The Jhelum and The Chenab.

...

(9) The term "Agricultural Use" means the use of water for irrigation, except for irrigation of household gardens and public recreational gardens.

(10) The term "Domestic Use" means the use of water for:

(a) drinking, washing, bathing, recreation, sanitation (including the conveyance and dilution of sewage and of industrial and other wastes), stock and poultry, and other like purposes;

(b) household and municipal purposes (including use for household gardens and public recreational gardens); and

(c) industrial purposes (including mining, milling and other like purposes);

but the term does not include Agricultural Use or use for the generation of hydro-electric power.

(11) The term "Non-Consumptive Use" means any control or use of water for navigation, floating of timber or other property, flood protection or flood control, fishing or fish culture, wild life or other like beneficial purposes,

provided that, exclusive of seepage and evaporation of water incidental to the control or use, the water (undiminished in volume within the practical range of measurement) remains in, or is returned to, the same river or its Tributaries; but the term does not include Agricultural Use or use for the generation of hydro-electric power.

(12) The term "Transition Period" means the period beginning and ending as provided in Article 11(6).

(13) The term "Bank" means the International Bank for Reconstruction and Development.

(14) The term "Commissioner" means either of the Commissioners appointed under the provisions of Article VIII (1) and the term "Commission" means the Permanent Indus Commission constituted in accordance with Article VIII (3).

(15) The term "interference with the waters" means

    (a) Any act of withdrawal therefrom; or

    (b) Any man-made obstruction to their flow which causes a change in the volume (within the practical range of measurement) of the daily flow of the waters: Provided however that an obstruction which involves only an insignificant and incidental change in the volume of the daily flow, for example, fluctuations due to afflux caused by bridge piers or a temporary by-pass, etc., shall not be deemed to be an interference with the waters.

(16) The term "Effective Date" means the date on which this Treaty takes effect in accordance with the provisions of Article XII, that is, the first of April 1960.

Article II

Provisions Regarding Eastern Rivers

(1) All the waters of the Eastern Rivers shall be available for the unrestricted use of India, except as otherwise expressly provided in this Article.

(2) Except for Domestic Use and Non-Consumptive Use, Pakistan shall be under an obligation to let flow, and shall not permit any interference with,

the waters of the Sutlej Main and the Ravi Main in the reaches where these rivers flow in Pakistan and have not yet finally crossed into Pakistan. The points of final crossing are the following: (a) near the new Hasta Bund upstream of Suleimanke in the case of the Sutlej Main, and (b) about one and a half miles upstream of the syphon for the B-R B-D Link in the case of the Ravi Main.

(3) Except for Domestic Use, Non-Consumptive Use and Agricultural Use (as specified in Annexure B), Pakistan shall be under an obligation to let flow, and shall not permit any interference with, the waters (while flowing in Pakistan) of any Tributary which in its natural course joins the Sutlej Main or the Ravi Main before these rivers have finally crossed into Pakistan.

(4) All the waters, while flowing in Pakistan, of any Tributary which, in its natural course, joins the Sutlej Main or the Ravi Main after these rivers have finally crossed into Pakistan shall be available for the unrestricted use of Pakistan: Provided however that this provision shall not be construed as giving Pakistan any claim or right to any releases by India in any such Tributary. If Pakistan should deliver any of the waters of any such Tributary, which on the Effective Date joins the Ravi Main after this river has finally crossed into Pakistan, into a reach of the Ravi Main upstream of this crossing, India shall not make use of these waters; each Party agrees to establish such discharge observation stations and make such observations as may be necessary for the determination of the component of water available for the use of Pakistan on account of the aforesaid deliveries by Pakistan, and Pakistan agrees to meet the cost of establishing the aforesaid discharge observation stations and making the aforesaid observations.

(5) There shall be a Transition Period during which, to the extent specified in Annexure H, India shall

(i) limit its withdrawals for Agricultural Use,

(ii) limit abstractions for storages, and

(iii) make deliveries to Pakistan from the Eastern Rivers.

(6) The Transition Period shall begin on 1st April 1960 and it shall end on 31st March 1970, or, if extended under the provisions of Part 8 of Annexure H, on the date up to which it has been extended. In any event, whether or not the replacement referred to in Article IV (1) has been accomplished, the Transition Period shall end not later than 31st March 1973.

(7) If the Transition Period is extended beyond 31st March 1970, the provisions of Article V (5) shall apply.

(8) If the Transition Period is extended beyond 31st March 1970, the provisions of Paragraph (5) shall apply during the period of extension beyond 31st March 1970.

(9) During the Transition Period, Pakistan shall receive for unrestricted use the waters of the Eastern Rivers which are to be released by India in accordance with the provisions of Annexure H. After the end of the Transition Period, Pakistan shall have no claim or right to releases by India of any of the waters of the Eastern Rivers. In case there are any releases, Pakistan shall enjoy the unrestricted use of the waters so released after they have finally crossed into Pakistan: Provided that in the event that Pakistan makes any use of these waters, Pakistan shall not acquire any right whatsoever, by prescription or otherwise, to a continuance of such releases or such use.

Article III

PROVISIONS REGARDING WESTERN RIVERS

(1) Pakistan shall receive for unrestricted use all those waters of the Western Rivers which India is under obligation to let flow under the provisions of Paragraph (2).

(2) India shall be under an obligation to let flow all the waters of the Western Rivers, and shall not permit any interference with these waters, except for the following uses, restricted (except as provided in item (c) (ii) of Paragraph 5 of Annexure C) in the case of each of the rivers, The Indus, The Jhelum and the Chenab, to the drainage basin thereof:

(a) Domestic Use;

(b) Non-Consumptive Use;

(c) Agricultural Use, as set out in Annexure C; and

(d) Generation of hydroelectric power, as set out in Annexure D.

(3) Pakistan shall have the unrestricted use of all waters originating from the source other than the Eastern Rivers which are delivered by Pakistan

into The Ravi or The Sutlej, and India shall not make use of these waters. Each Party agrees to establish such discharge observation stations and make such observations as may be considered necessary by the Commission for the determination of the component of water available for the use of Pakistan on account of the aforesaid deliveries by Pakistan.

(4) Except as provided in Annexures D and E, India shall not store any water of, or construct any storage works on, the Western Rivers.

Article IV

PROVISIONS REGARDING EASTERN RIVERS AND WESTERN RIVERS

(1) Pakistan shall use its best endeavours to construct and bring into operation, with due regard to expedition and economy, that part of a system of works which will accomplish the replacement, from the Western Rivers and other sources, of water supplies for irrigation canals in Pakistan which, on 15th August 1947, were dependent on water supplies from the Eastern Rivers.

(2) Each Party agrees that any Non-Consumptive Use made by it shall be so made as not to materially change, on account of such use, the flow in any channel to the prejudice of the uses on that channel by the other Party under the provisions of this Treaty. In executing any scheme of flood protection or flood control each Party will avoid, as far as practicable, any material damage to the other Party, and any such scheme carried out by India on the Western Rivers shall not involve any use of water or any storage in addition to that provided under Article III.

(3) Nothing in this Treaty shall be construed as having the effect of preventing either Party from undertaking schemes of drainage, river training, conservation of soil against erosion and dredging, or from removal of stones, gravel or sand from the beds of the Rivers: Provided that

(a) in executing any of the schemes mentioned above, each Party will avoid, as far as practicable, any material damage to the other Party;

(b) any such scheme carried out by India on the Western Rivers shall not involve any use of water or any storage in addition to that provided under Article III;

(c) except as provided in Paragraph (5) and Article VII (1) (b), India shall not take any action to increase the catchment area, beyond the area on the Effective Date, of any natural or artificial drainage or drain which crosses into Pakistan, and shall not undertake such construction or remodelling of any drainage or drain which so crosses or falls into a drainage or drain which so crosses as might cause material damage in Pakistan or entail the construction of a new drain or enlargement of an existing drainage or drain in Pakistan; and

(d) should Pakistan desire to increase the catchment area, beyond the area on the Effective Date, of any natural or artificial drainage or drain, which receives drainage waters from India, or, except in an emergency, to pour any waters into it in excess of the quantities received by it as on the Effective Date, Pakistan shall, before undertaking any work for these purposes, increase the capacity of that drainage or drain to the extent necessary so as not to impair its efficacy for dealing with drainage waters received from India as on the Effective Date.

(4) Pakistan shall maintain in good order its portions of the drainages mentioned below with capacities not less than the capacities as on the Effective Date:

(i)   Hudiara Drain

(ii)   Kasur Nala

(iii)   Salimshah Drain

(iv)   Fazilka Drain.

(5) If India finds it necessary that any of the drainages mentioned in Paragraph (4) should be deepened or widened in Pakistan, Pakistan agrees to undertake to do so as a work of public interest, provided India agrees to pay the cost of the deepening or widening.

(6) Each Party will use its best endeavours to maintain the natural channels of the Rivers, as on the Effective Date, in such condition as will avoid, as far as practicable, any obstruction to the flow in these channels likely to cause material damage to the other Party.

(7) Neither Party will take any action which would have the effect of diverting the Ravi Main between Madhopur and Lahore, or the Sutlej Main between Harike and Suleimanke, from its natural channel between high banks.

(8) The use of the natural channels of the Rivers for the discharge of flood or other excess waters shall be free and not subject to limitation by either Party, and neither Party shall have any claim against the other in respect of any damage caused by such use. Each Party agrees to communicate to the other Party, as far in advance as practicable, any information it may have in regard to such extraordinary discharges of water from reservoirs and flood flows as may affect the other Party.

(9) Each Party declares its intention to operate its storage dams, barrages and irrigation canals in such manner, consistent with the normal operations of its hydraulic systems, as to avoid, as far as feasible, material damage to the other Party.

(10) Each Party declares its intention to prevent, as far as practicable, undue pollution of the waters of the Rivers which might affect adversely uses similar in nature to those to which the waters were put on the Effective Date, and agrees to take all reasonable measures to ensure that, before any sewage or industrial waste is allowed to flow into the Rivers, it will be treated, where necessary, in such manner as not materially to affect those uses: Provided that the criterion of reasonableness shall be the customary practice in similar situations on the Rivers.

(11) The Parties agree to adopt, as far as feasible, appropriate measures for the recovery, and restoration to owners, of timber and other property floated or floating down the Rivers, subject to appropriate charges being paid by the owners.

(12) The use of water for industrial purposes under Articles II (2), II (3) and III (2) shall not exceed:

(a) in the case of an industrial process known on the Effective Date, such quantum of use as was customary in that process on the Effective Date;

(b) in the case of an industrial process not known on the Effective Date:

(i) such quantum of use as was customary on the Effective Date in similar or in any way comparable industrial processes; or

(ii) if there was no industrial process on the Effective Date similar or in any way comparable to the new process, such quantum of use as would not have a substantially adverse effect on the other Party.

(13) Such part of any water withdrawn for Domestic Use under the provisions of Articles11(3) and III (2) as is subsequently applied to Agricultural Use shall be accounted for as part of the Agricultural Use specified in Annexure B and Annexure C respectively; each Party will use its best endeavours to return to the same river (directly or through one of its Tributaries) all water withdrawn therefrom for industrial purposes and not consumed either in the industrial processes for which it was withdrawn or in some other Domestic Use.

(14) In the event that either Party should develop a use of the waters of the Rivers which is not in accordance with the provisions of this Treaty, that Party shall not acquire by reason of such use any right, by prescription or otherwise, to a continuance of such use.

(15) Except as otherwise required by the express provisions of this Treaty, nothing in this Treaty shall be construed as affecting existing territorial rights over the waters of any of the Rivers or the beds or banks thereof, or as affecting existing property rights under municipal law over such waters or beds or banks.

Article V

FINANCIAL PROVISIONS

(1) In consideration of the fact that the purpose of part of the system of works referred to in Article IV (1) is the replacement, from the Western Rivers and other sources, of water supplies for irrigation canals in Pakistan which, on 15th August 1947, were dependent on water supplies from the Eastern Rivers, India agrees to make a fixed contribution of Pounds Sterling 62,060,000 towards the costs of these works. The amount in Pounds Sterling of this contribution shall remain unchanged irrespective of any alteration in the par value of any currency.

(2) The sum of Pounds Sterling 62,060,000 specified in Paragraph (1) shall be paid in ten equal annual installments on the 1st of November of each year. The first of such annual installments shall be paid on 1st November 1960, or if the Treaty has not entered into force by that date, then within one month after the Treaty enters into force.

(3) Each of the instalments specified in Paragraph (2) shall be paid to the Bank for the credit of the Indus Basin Development Fund to be established and administered by the Bank, and payment shall be made in Pounds Sterling, or in such other currency or currencies as may from time to time be agreed between India and the Bank.

(4) The payments provided for under the provisions of Paragraph (3) shall be made without deduction or set-off on account of any financial claims of India on Pakistan arising otherwise than under the provisions of this Treaty: Provided that this provision shall in no way absolve Pakistan from the necessity of paying in other ways debts to India which may be outstanding against Pakistan.

(5) If, at the request of Pakistan, the Transition Period is extended in accordance with the provisions of Article II (6) and of Part 8 of Annexure H, the Bank shall thereupon pay to India out of the Indus Basin Development Fund the appropriate amount specified in the Table below:

Table

Period of Aggregate Extension of Transition Period	Paynet to India £Stg.
One year	3,125,000
Two years	6,406,250
Three years	9,850,000

(6) The provisions of Article IV (1) and Article V (1) shall not be construed as conferring upon India any right to participate in the decisions as to the system of works which Pakistan constructs pursuant to Article IV (1) or as constituting an assumption of any responsibility by India or as an agreement by India in regard to such works.

(7) Except for such payments as are specifically provided for in this Treaty neither Party shall be entitled to claim any payment for observance of the provisions of this Treaty or to make any charge for water received from it by the other Party.

Article VI

EXCHANGE OF DATA

(1) The following data with respect to the flow in, and utilisation of the waters of, the Rivers shall be exchanged regularly between the Parties:

  (a) Daily (or as observed or estimated less frequently) gauge and discharge data relating to flow of the Rivers at all observation sites.

  (b) Daily extractions for or releases from reservoirs.

  (c) Daily withdrawals at the heads of all canals operated by government or by a government agency (hereinafter in this Article called canals), including link canals.

  (d) Daily escapages from all canals, including link canals.

  (e) Daily deliveries from link canals.

These data shall be transmitted monthly by each Party to the other as soon as the data for a calendar month have been collected and tabulated, but not later than three months after the end of the month to which they relate: Provided that such of the data specified above as are considered by either Party to be necessary for operational purposes shall be supplied daily or at less frequent intervals, as may be requested. Should one Party request the supply of any of these data by telegram, telephone, or wireless, it shall reimburse the other Party for the cost of transmission.

(2) If, in addition to the data specified in Paragraph (1) of this Article, either Party requests the supply of any data relating to the hydrology of the Rivers, or to canal or reservoir operation connected with the Rivers, or to any provision of this Treaty, such data shall be supplied by the other Party to the extent that these are available.

Article VII

FUTURE CO-OPERATION

(1) The two Parties recognize that they have a common interest in the optimum development of the Rivers, and, to that end, they declare their intention to co-operate, by mutual agreement, to the fullest possible extent. In particular:

(a) Each Party, to the extent it considers practicable and on agreement by the other Party to pay the costs to be incurred, will, at the request of the other Party, set up or install such hydrologic observation stations within the drainage basins of the Rivers, and set up or install such meteorological observation stations relating thereto and carry out such observations thereat, as may be requested, and will supply the data so obtained.

(b) Each Party, to the extent it considers practicable and on agreement by the other Party to pay the costs to be incurred, will, at the request of the other Party, carry out such new drainage works as may be required in connection with new drainage works of the other Party.

(c) At the request of either Party, the two Parties may, by mutual agreement, co-operate in undertaking engineering works on the Rivers.

The formal arrangements, in each case, shall be as agreed upon between the Parties.

(2) If either Party plans to construct any engineering work which would cause interference with the waters of any of the Rivers and which, in its opinion, would affect the other Party materially, it shall notify the other Party of its plans and shall supply such data relating to the work as may be available and as would enable the other Party to inform itself of the nature, magnitude and effect of the work. If a work would cause interference with the waters of any of the Rivers but would not, in the opinion of the Party planning it, affect the other Party materially, nevertheless the Party planning the work shall, on request, supply the other Party with such data regarding the nature, magnitude and effect, if any, of the work as may be available.

Article VIII

PERMANENT INDUS COMMISSION

(1) India and Pakistan shall each create a permanent post of Commissioner for Indus Waters, and shall appoint to this post, as often as a vacancy occurs, a person who should ordinarily be a high-ranking engineer competent in the field of hydrology and water-use. Unless either Government should decide to take up any particular question directly with the other Government,

each Commissioner will be the representative of his Government for all matters arising out of this Treaty, and will serve as the regular channel of communication on all matters relating to the implementation of the Treaty, and, in particular, with respect to

(a) the furnishing or exchange of information or data provided for in the Treaty; and

(b) the giving of any notice or response to any notice provided for in the Treaty.

(2) The status of each Commissioner and his duties and responsibilities towards his Government will be determined by that Government.

(3) The two Commissioners shall together form the Permanent Indus Commission.

(4) The purpose and functions of the Commission shall be to establish and maintain co-operation between the Parties in the development of the waters of the Rivers and, in particular,

(a) to study and report to the two Governments on any problem relating to the development of the waters of the Rivers which may be jointly referred to the Commission by the two Governments: in the event that a reference is made by one Government alone, the Commissioner of the other Government shall obtain the authorization of his Government before he proceeds to act on the reference;

(b) to make every effort to settle promptly, in accordance with the provisions of Article IX (1), any question arising thereunder;

(c) to undertake, once in every five years, a general tour of inspection of the Rivers for ascertaining the facts connected with various developments and works on the Rivers;

(d) to undertake promptly, at the request of either Commissioner, a tour of inspection of such works or sites on the Rivers as may be considered necessary by him for ascertaining the facts connected with those works or sites; and

(e) to take, during the Transition Period, such steps as may be necessary for the implementation of the provisions of Annexure H.

(5) The Commission shall meet regularly at least once a year, alternately in India and Pakistan. This regular annual meeting shall be held in November or in such other month as may be agreed upon between the Commissioners. The Commission shall also meet when requested by either Commissioner.

(6) To enable the Commissioners to perform their functions in the Commission, each Government agrees to accord to the Commissioner of the other Government the same privileges and immunities as are accorded to representatives of member States to the principal and subsidiary organs of the United Nations under Sections 11, 12 and 13 of Article IV of the Convention on the Privileges and Immunities of the United Nations (dated 13th February, 1946) during the periods specified in those Sections. It is understood and agreed that these privileges and immunities are accorded to the Commissioners not for the personal benefit of the individuals themselves but in order to safeguard the independent exercise of their functions in connection with the Commission; consequently, the Government appointing the Commissioner not only has the right but is under a duty to waive the immunity of its Commissioner in any case where, in the opinion of the appointing Government, the immunity would impede the course of justice and can be waived without prejudice to the purpose for which immunity is accorded.

(7) For the purposes of the inspections specified in Paragraph (4) (c) and (d) each Commissioner may be accompanied by two advisers or assistants to whom appropriate facilities will be accorded.

(8) The Commission shall submit to the Government of India and to the Government of Pakistan, before the first of June of every year, a report on its work for the year ended on the preceding 31st of March, and may submit to the two Governments other reports at such times as it may think desirable.

(9) Each Government shall bear the expenses of its Commissioner and his ordinary staff. The cost of any special staff required in connection with the work mentioned in Article VII (1) shall be borne as provided therein.

(10) The Commission shall determine its own procedures.

Article IX

SETTLEMENT OF DIFFERENCES AND DISPUTES

(1) Any question which arises between the Parties concerning the interpretation or application of this Treaty or the existence of any fact which, if established, might constitute a breach of this Treaty shall first be examined by the Commission, which will endeavour to resolve the question by agreement.

(2) If the Commission does not reach agreement on any of the questions mentioned in Paragraph (1), then a difference will be deemed to have arisen, which shall be dealt with as follows:

    (a) Any difference which, in the opinion of either Commissioner, falls within the provisions of Part I of Annexure F shall, at the request of either Commissioner, be dealt with by a Neutral Expert in accordance with the provisions of Part 2 of Annexure F;

    (b) If the difference does not come within the provisions of Paragraph (2) (a), or if a Neutral Expert, in accordance with the provisions of Paragraph 7 of Annexure F, has informed the Commission that, in his opinion, the difference, or a part thereof, should be treated as a dispute, then a dispute will be deemed to have arisen which shall be settled in accordance with the provisions of Paragraphs (3), (4) and (5);

Provided that, at the discretion of the Commission, any difference may either be dealt with by a Neutral Expert in accordance with the provisions of Part 2 of Annexure F or be deemed to be a dispute to be settled in accordance with the provisions of Paragraphs (3), (4) and (5), or may be settled in any other way agreed upon by the Commission.

(3) As soon as a dispute to be settled in accordance with this and the succeeding paragraphs of this Article has arisen, the Commission shall, at the request of either Commissioner, report the fact to the two Governments, as early as practicable, stating in its report the points on which the Commission is in agreement and the issues in dispute, the views of each Commissioner on these issues and his reasons therefor.

(4) Either Government may, following receipt of the report referred to in Paragraph (3), or if it comes to the conclusion that this report is being unduly

delayed in the Commission, invite the other Government to resolve the dispute by agreement. In doing so it shall state the names of its negotiators and their readiness to meet with the negotiators to be appointed by the other Government at a time and place to be indicated by the other Government. To assist in these negotiations, the two Governments may agree to enlist the services of one or more mediators acceptable to them.

(5) A Court of Arbitration shall be established to resolve the dispute in the manner provided by Annexure G.

    (a) upon agreement between the Parties to do so; or

    (b) at the request of either Party, if, after negotiations have begun pursuant to Paragraph (4), in its opinion the dispute is not likely to be resolved by negotiation or mediation; or

    (c) at the request of either Party, if, after the expiry of one month following receipt by the other Government of the invitation referred to in Paragraph (4), that Party comes to the conclusion that the other Government is unduly delaying the negotiations.

(6) The provisions of Paragraphs (3), (4) and (5) shall not apply to any difference while it is being dealt with by a Neutral Expert.

Article X

EMERGENCY PROVISION

If, at any time prior to 31st March 1965, Pakistan should represent to the Bank that, because of the outbreak of large-scale international hostilities arising out of causes beyond the control of Pakistan, it is unable to obtain from abroad the materials and equipment necessary for the completion, by 31st March 1973, of that part of the system of works referred to in Article IV (1) which relates to the replacement referred to therein, (hereinafter referred to as the "replacement element") and if, after consideration of this representation in consultation with India, the Bank is of the opinion that

    (a) these hostilities are on a scale of which the consequence is that Pakistan is unable to obtain in time such materials and equipment as must be procured from abroad for the completion, by 31st March 1973, of the replacement element, and

(b) since the Effective Date, Pakistan has taken all reasonable steps to obtain the said materials and equipment and, with such resources of materials and equipment as have been available to Pakistan both from within Pakistan and from abroad, has carried forward the construction of the replacement element with due diligence and all reasonable expedition, the Bank shall immediately notify each of the Parties accordingly. The Parties undertake, without prejudice to the provisions of Article XII (3) and (4), that, on being so notified, they will forthwith consult together and enlist the good offices of the Bank in their consultation, with a view to reaching mutual agreement as to whether or not, in the light of all the circumstances then prevailing, any modifications of the provisions of this Treaty are appropriate and advisable and, if so, the nature and the extent of the modifications.

Article XI

GENERAL PROVISIONS

(1) It is expressly understood that

(a) this Treaty governs the rights and obligations of each Party in relation to the other with respect only to the use of the waters of the Rivers and matters incidental thereto; and

(b) nothing contained in this Treaty, and nothing arising out of the execution thereof, shall be construed as constituting a recognition or waiver (whether tacit, by implication or otherwise) of any rights or claims whatsoever of either of the Parties other than those rights or claims which are expressly recognized or waived in this Treaty.

Each of the Parties agrees that it will not invoke this Treaty, anything contained therein, or anything arising out of the execution thereof, in support of any of its own rights or claims whatsoever or in disputing any of the rights or claims whatsoever of the other Party, other than those rights or claims which are expressly recognized or waived in this Treaty.

(2) Nothing in this Treaty shall be construed by the Parties as in any way establishing any general principle of law or any precedent.

(3) The rights and obligations of each Party under this Treaty shall remain unaffected by any provisions contained in, or by anything arising out of the execution of, any agreement establishing the Indus Basin Development Fund.

Article XII

FINAL PROVISIONS

(1) This Treaty consists of the Preamble, the Articles hereof and Annexures A to H hereto, and may be cited as "The Indus Waters Treaty 1960".

(2) This Treaty shall be ratified and the ratifications thereof shall be exchanged in New Delhi. It shall enter into force upon the exchange of ratifications, and will then take effect retrospectively from the first of April 1960.

(3) The provisions of this Treaty may from time to time be modified by a duly ratified treaty concluded for that purpose between the two Governments.

(4) The provisions of this Treaty, or the provisions of this Treaty as modified under the provisions of Paragraph (3), shall continue in force until terminated by a duly ratified treaty concluded for that purpose between the two Governments.

IN WITNESS WHEREOF the respective Plenipotentiaries have signed this Treaty and have hereunto affixed their seals.

DONE in triplicate in English at Karachi on this Nineteenth day of September 1960.

For the Government of India:
(Signed) Jawaharlal NEHRU

For the Government of Pakistan:
(Signed) Mohammad Ayub Khan
Field Marshal, H.P., H.J.

For the International Bank for Reconstruction and Development,
for the purposes specified in Articles V and X and Annexures F, G and H:

(Signed) W. A. B. ILIFF

ANNEXURE A – EXCHANGE OF NOTES BETWEEN GOVERNMENT OF INDIA AND GOVERNMENT OF PAKISTAN

I. NOTE DATED 19TH SEPTEMBER 1960, FROM THE HIGH COMMISSIONER FOR INDIA IN PAKISTAN, KARACHI, TO THE MINISTER FOR FOREIGN AFFAIRS AND COMMONWEALTH RELATIONS, GOVERNMENT OF PAKISTAN

and

II. NOTE DATED 19TH SEPTEMBER 1960, FROM THE MINISTER FOR FOREIGN AFFAIRS AND COMMONWEALTH RELATIONS, GOVERNMENT OF PAKISTAN, TO THE HIGH COMMISSIONER FOR INDIA IN PAKISTAN, KARACHI.

By virtue of these notes, the two governments agreed that on the ratification of the Treaty, the agreement/document of "May, 1948, and the rights and obligations of the either party thereto claimed under, or arising out of, that" agreement/document "shall be without effect as from 1st April, 1960".

ANNEXURE B – AGRICULTURAL USE BY PAKISTAN FROM CERTAIN TRIBUTARIES OF THE RAVI

(Article II (3))

...

2. Pakistan may withdraw from the Basantar Tributary of the Ravi such waters as may be available and necessary for the irrigation of not more than 100 acres annually.

3. In addition to the area specified in Paragraph 2, Pakistan may also withdraw such waters from each of the following Tributaries of The Ravi as may be available and as may be necessary for the irrigation of that part of the following areas cultivated on *sailab* as on the Effective Date which cannot be so cultivated after that date: Provided that the total area whether irrigated or cultivated on *sailab* shall not exceed the limits specified below, except during a year of exceptionally heavy floods when *sailab* may extend to areas which were not cultivated on *sailab* as on the Effective Date and when such areas may be cultivated in addition to the limits specified:

Name of Tributary	Maximum Annual Cultivation (acres)
Basantar	14,000
Bein	26,600
Tarnah	1,800
Ujh	3,000

4. The provisions of Paragraphs 2 and 3 shall not be construed as giving Pakistan any claim or right to any releases by India in the Tributaries mentioned in these paragraphs.

5. Not later than 31st March 1961, Pakistan shall furnish to India a statement by Districts and Tehsils showing (i) the area irrigated and (ii) the area cultivated on *sailab*, as on the Effective Date, from the waters of each of the Tributaries specified in Paragraphs 2 and 3.

6. As soon as the statistics for each crop year (commencing with the beginning of *kharif* and ending with the end of the following *rabi*) have been compiled at the District Headquarters, but not later than the 30th November following the end of that crop year, Pakistan shall furnish to India a statement arranged by Tributaries and showing for each of the Districts and Tehsils irrigated or cultivated on *sailab* from the Tributaries mentioned in Paragraphs 2 and 3:

(i) the area irrigated, and

(ii) the area cultivated on *sailab*.

ANNEXURE C – AGRICULTURAL USE BY INDIA FROM THE WESTERN RIVERS

(Article III (2) (c))

...

2. As used in this Annexure, the term "Irrigated Cropped Area" means the total area under irrigated crops in a year, the same area being counted twice if it bears different crops in *kharif and rabi* ....

3. India may withdraw from the Chenab Main such waters as India may need for Agricultural Use on the following canals limited to the maximum withdrawals noted against each:

Name of Canal	Maximum Withdrawals for Agricultural Use
(a) Ranbir Canal ...	1,000 cusecs from 15th April to 14th October, and 350 cusecs from 15th October to 14th April.
(b) Pratap Canal ...	400 cusecs from 15th April to 14th October, and 100 cusecs from 15th October to 14th April.

Provided that

...

4. Apart from the irrigation from the Ranbir and Pratap Canals under the provisions of Paragraph 3, India may continue to irrigate from the Western Rivers those areas which were so irrigated as on the Effective Date.

5. In addition to such withdrawals as may be made in accordance with the provisions of Paragraphs 3 and 4, India may, subject to the provisions of Paragraphs 6, 7, 8 and 9, make further withdrawals from the Western Rivers to the extent India may consider necessary to meet the irrigation needs of the areas specified below:

Particulars	Maximum Irrigated Cropped Area (over and above the cropped area irrigated under the provisions of Paragraphs 3 and 4) acres)
a) From The Indus, in its drainage basin ...	70,000
b) From The Jhelum, in its drainage basin ...	400,000
c) From The Chenab,	
i) in its drainage basin ...	225,000 of which not more than 100,000 acres will be in the Jammu District
ii) outside its drainage basin in the area west of the Deg Nadi (also called Devak River), the aggregate capacity of irrigating channels leading out of the drainage basin of the Chenab to this area not to exceed 120 cusecs ...	6,000

Provided that

...

    (ii) the maximum Irrigated Cropped Area shown against items (a), (b) and (c) (i) above shall be deemed to include cropped areas, if any, irrigated from an open well, a tube-well, a spring, a lake (other than a Connecting Lake) or a tank, in excess of the areas so irrigated as on the Effective Date; and

    (iii) the aggregate of the areas specified against items (a), (b) and (c) (i) above may be redistributed among the three drainage basins in such manner as may be agreed upon between the Commissioners.

6   (a) ...

    (b) Within the limits of the maximum Irrigated Cropped Areas specified against items (b) and (c) in Paragraph 5, there shall be no restriction on the development of such of these areas as may be irrigated from General Storage (as defined in Annexure E); the areas irrigated from General Storage may, however, receive irrigation from river flow but, unless the Commissioners otherwise agree, only in the following periods:

        (i)   from The Jhelum: 21st June to 20th August

        (ii) from The Chenab: 21st June to 31st August

Provided that withdrawals for such irrigation, whether from General Storage or from river flow, are controlled by Government.

7. Within the limits of the maximum Irrigated Cropped Areas specified against items (b) and (c) in Paragraph 5, the development of these areas by withdrawals from river flow (as distinct from withdrawals from General Storage *cum* river flow in accordance with Paragraph 6 (b)) shall be regulated as follows:

    (a) Until India can release water from Conservation Storage (as defined in Annexure E) in accordance with sub-paragraphs (b) and (c) below, the new area developed shall not exceed the following:

        (i)   from The Jhelum: 150,000 acres

(ii) from The Chenab: 25,000 acres during the Transition Period and 50,000 acres after the end of the Transition Period.

(b) In addition to the areas specified in (a) above, there may be developed from The Jhelum or The Chenab an aggregate area of 150,000 acres if there is released annually from Conservation Storage, in accordance with Paragraph 8, a volume of 0.2 MAF into The Jhelum and a volume of 0.1 MAF into The Chenab; provided that India shall have the option to store on and release into The Chenab the whole or a part of the volume of 0.2 MAF specified above for release into The Jhelum.

(c) Any additional areas over and above those specified in (a) and (b) above may be developed if there is released annually from Conservation Storage a volume of 0.2 MAF into The Jhelum or The Chenab, in accordance with Paragraph 8, in addition to the releases specified in (b) above.

...

9. On those Tributaries of The Jhelum on which there is any Agricultural Use or hydro-electric use by Pakistan, any new Agricultural Use by India shall be so made as not to affect adversely the then existing Agricultural Use or hydroelectric use by Pakistan on those Tributaries.

...

11. (a) As soon as the statistics for each crop year (commencing with the beginning of *kharif* and ending with the end of the following *rabi*) have been compiled at the District Headquarters, but not later than the 30th November following the end of that crop year, India shall furnish to Pakistan a statement showing for each of the Districts and Tehsils irrigated from the Western Rivers, the total Irrigated Cropped Areas (excluding the area irrigated under the provisions of Paragraph 3) arranged in accordance with items (a), (b), (c) (i) and (c) (ii) of Paragraph 5: Provided that, in the case of areas in the Punjab, the 30th November date specified above may be extended to the following 30th June in the event of failure of communications.

...

ANNEXURE D – GENERATION OF HYDRO-ELECTRIC POWER BY INDIA ON THE WESTERN RIVERS

(Article III (2) (d))

1. The provisions of this Annexure shall apply with respect to the use by India of the waters of the Western Rivers for the generation of hydroelectric power under the provisions of Article III (2) (d) and, subject to the provisions of this Annexure, such use shall be unrestricted: Provided that the design, construction and operation of new hydroelectric plants which are incorporated in a Storage Work (as defined in Annexure E) shall be governed by the relevant provisions of Annexure E.

PART I—DEFINITIONS

2. As used in this Annexure

(a) "Dead Storage" means that portion of the storage which is not used for operational purposes and "Dead Storage Level" means the level corresponding to Dead Storage.

(b) "Live Storage" means all storage above Dead Storage.

(c) "Pondage" means Live Storage of only sufficient magnitude to meet fluctuations in the discharge of the turbines arising from variations in the daily and the weekly loads of the plant.

(d) "Full Pondage Level" means the level corresponding to the maximum Pondage provided in the design in accordance with Paragraph 8 (c).

(e) "Surcharge Storage" means uncontrollable storage occupying space above the Full Pondage Level.

(f) "Operating Pool" means the storage capacity between Dead Storage level and Full Pondage Level.

(g) "Run-of-River Plant" means a hydroelectric that develops power without Live Storage as an integral part of the plant, except for Pondage and Surcharge Storage.

(h) "Regulating Basin" means the basin whose only purpose is to even out fluctuations in the discharge from the turbines arising from variations in the daily and weekly loads of the plant.

(i) "Firm Power" means the hydroelectric power corresponding to the minimum mean discharge at the site of the plant, the minimum mean discharge being calculated as follows:

...

(j) "Secondary Power" means the power, other than Power Firm, available only during certain periods of the year.

PART 2—HYDRO-ELECTRIC PLANTS IN OPERATION, OR UNDER CONSTITUTION, AS ON THE EFFECTIVE DATE

3. There shall be no restriction on the operation of the following hydroelectric plants which were in operation as on the Effective Date:

...

4. There shall be no restriction on the completion by India, in accordance with the design adopted prior to the Effective Date, or, on the operation by India, of the following hydroelectric plants which were actually under construction on the Effective Date, whether or not the plant was on that date in partial operation.

5. As soon as India finds it possible to do so, but not later than 31st March 1961, India shall communicate to Pakistan the Information specified in Appendix I to this Annexure for each of the plants specified in Paragraphs 3 and 4. If any such information is not available or is not pertinent to the design of the plant or to the conditions at the site, it will be so stated.

6. (a) If any alteration proposed in the design of any of the plants specified in Paragraphs 3 and 4 would result in a material change in the information furnished to Pakistan under the provisions of Paragraph 5, India shall, at least 4 months in advance of making the alteration, communicate particulars of the change to Pakistan in writing and the provisions of Paragraph 7 shall then apply.

(b) In the event of an emergency arising which requires repairs to be undertaken to protect the integrity of any of the plants specified in Paragraphs 3 and 4, India may undertake immediately the necessary repairs or alterations and, if these repairs or alterations result in a change in the information furnished to Pakistan under the provisions of Paragraph 5, India shall as soon as possible communicate particulars of the change to Pakistan in writing. The provisions of Paragraph 7 shall then apply.

7. Within three months of the receipt of the particulars specified in Paragraph 6 Pakistan shall communicate to India in writing any objection it may have with regard to the proposed change on the ground that the change involves a material departure from the criteria set out in Paragraph 8 or 18 of this Annexure or Paragraph II of Annexure E as the case may be. If no objection is received by India from Pakistan within the specified period of three months, then Pakistan shall be deemed to have no objection. If a question arises as to whether or not the change involves a material departure from such of the criteria mentioned above as may be applicable, then either Party may proceed to have the question resolved in accordance with the provisions of Article IX (1) and (2).

PART 3—NEW RUN-OF-RIVER PLANTS

8. Except as provided in Paragraph 18, the design of any new Run-of-River Plant (hereinafter in this Part referred to as a Plant) shall conform to the following criteria:

(a) The works themselves shall not be capable of raising artificially the water level in the Operating Pool above the Full Pondage Level specified in the design.

(b) The design of the works shall take due account of the requirements of Surcharge Storage and of Secondary Power.

(c) The maximum Pondage in the Operating Pool shall not exceed twice the Pond required for Firm Power.

(d) There shall be no outlets below the Dead Storage Level, unless necessary for sediment control or any other technical purpose; any such outlet shall be of the minimum size, and located at the highest level, consistent with sound and economical design and with satisfactory operation of the works.

(e) If the conditions at the site of a Plant make a gated spillway necessary, the bottom level of the gates in normal closed position shall be located at the highest level consistent with sound and economical design and satisfactory construction and operation of the works.

(f) The intakes for the turbines shall be located at the highest level consistent with satisfactory and economical construction and operation of the Plant as a Run-of-River Plant and with customary and accepted practice of design for the designated range of the Plant's operation.

(g) If any Plant is constructed on the Chenab Main at a site below Kotru (Longitude 74° – 59'East and Latitude 33° – 09'North), a Regulating Basin shall be incorporated.

9. To enable Pakistan to satisfy itself that the design of a Plant conforms to the criteria mentioned in Paragraph 8, India shall, at least six months in advance of the beginning of construction of river works connected with the Plant, communicate to Pakistan, in writing, the information specified in Appendix II to this Annexure. If any such information is not available or is not pertinent to the design of the Plant or to conditions at the site, it will be so stated.

10. Within three months of the receipt by Pakistan of the information specified in Paragraph 9, Pakistan shall communicate to India, in writing, any objection that it may have with regard to the proposed design on the ground that it does not conform to the criteria mentioned in Paragraph 8. If no objection is received by India from Pakistan within the specified period of three months, then Pakistan shall be deemed to have no objection.

11. If a question arises as to whether or not the design of a Plant conforms to the criteria set out in Paragraph 8, then either Party may proceed to have the question resolved in accordance with the provisions of Article IX (1) and (2).

12. (a) If any alteration proposed in the design of a Plant before it comes into operation would result in a material change in the information furnished to Pakistan under the provisions of Paragraph 9, India shall immediately communicate particulars of the change to Pakistan in writing and the provisions of Paragraphs 10 and 11 shall

then apply but the period of three months specified in Paragraph 10 shall be reduced to two months.

(b) If any alteration proposed in the design of a Plant after it comes into operation would result in a material change in the information furnished to Pakistan under the provisions of Paragraph 9, India shall, at least four months in advance of making the alteration, communicate particulars of the change to Pakistan in writing and the provisions of Paragraphs 10 and 11 shall then apply, but the period of three months specified in Paragraph 10 shall be reduced to two months.

13. In the event of an emergency arising which requires repairs to be undertaken to protect the integrity of a Plant, India may undertake immediately the necessary repairs or alterations; if these repairs or alterations result in a change in the information furnished to Pakistan under the provisions of Paragraph 9, India shall, as soon as possible, communicate particulars of the change to Pakistan in writing to enable Pakistan to satisfy itself that after such change the design of the Plant conforms to the criteria specified in Paragraph 8. The provisions of Paragraphs 10 and 11 shall then apply.

14. The filling of Dead Storage shall be carried out in accordance with the provisions of Paragraph 18 or 19 of Annexure E.

15. Subject to the provisions of Paragraph 17, the works connected with a Plant shall be so operated that

(a) the volume of water received in the river upstream of the Plant, during any period of seven consecutive days, shall be delivered into the river below the Plant during the same seven-day period, and

(b) in any one period of 24 hours within that seven-day period, the volume delivered into the river below the Plant shall be not less than 30%, and not more than 130%, of the volume received in the river above the Plant during the same 24-hour period: Provided however that:

(i) where a Plant is located at a site on the Chenab Main below Ramban, the volume of water received in the river upstream of the Plant in any one period of 24 hours shall be delivered into the river below the Plant within the same period of 24 hours;

(ii) where a Plant is located at a site on the Chenab Main above Ramban, the volume of water delivered into the river below the Plant in any one period of 24 hours shall not be less than 50% and not more than 130%, of the volume received above the Plant during the same 24-hour period; and

(iii) where a Plant is located on a Tributary of The Jhelum on which Pakistan has any Agricultural use or hydroelectric use, the water released below the Plant may be delivered, if necessary, into another Tributary but only to the extent that the then existing Agricultural Use or hydro-electric use by Pakistan on the former Tributary would not be adversely affected.

...

18. The provisions of paragraphs 8, 9, 10, 11, 12 and 13 shall not apply to a new Run-of-River Plant which is located on a Tributary and which conforms to the following criteria (hereinafter referred to as a Small Plant):

(a) the aggregate designed maximum discharge through the turbines does not exceed 300 cusecs;

(b) no storage is involved in connection with the Small Plant, except the Pondage and the storage incidental to the diversion structure; and

(c) the crest of the diversion structure across the Tributary, or the top level of the gates, if any, shall not be higher than 20 feet above the mean bed of the Tributary at the site of the structure.

19. The information specified in Appendix III to this Annexure shall be communicated to Pakistan by India at least two months in advance of the beginning of construction of the river works connected with a Small Plant. If any such information is not available or is not pertinent to the design of the Small Plant or to the conditions at the site, it will be so stated.

20. Within two months of the receipt by Pakistan of the information specified in Appendix III, Pakistan shall communicate to India, in writing, any objection that it may have with regard to the proposed design on the ground that it does not conform to the criteria mentioned in Paragraph 18.

If no objection is received by India from Pakistan within the specified period of two months, then Pakistan shall be deemed to have no objection.

21. If a question arises as to whether or not the design of a Small Plant conforms to the criteria set out in Paragraph 18, then either Party may proceed to have the question resolved in accordance with the provisions of Article IX (1) and (2).

22. If any alteration in the design of a Small Plant, whether during the construction period or subsequently, results in a change in the information furnished to Pakistan under the provisions of Paragraph 19, then India shall immediately communicate the change in writing to Pakistan.

23. If, with any alteration proposed in the design of a Small Plant, the design would cease to comply with the criteria set out in Paragraph 18, then the provisions of Paragraphs 18 to 22 inclusive shall no longer apply and, in lieu thereof, the provisions of Paragraphs 8 to 13 inclusive shall apply.

PART 4—NEW PLANTS ON IRRIGATION CHANNELS

24. Notwithstanding the foregoing provisions of this Annexure, there shall be no restriction on the construction and operation by India of new hydroelectric plants on any irrigation channel taking off the Western Rivers, provided that

(a) the works incorporate no storage other than Pondage and the Dead Storage incidental to the diversion structure, and

(b) no additional supplies are run in the irrigation channel for the purpose of generating hydroelectric power.

PART 5—GENERAL

25. If the change referred to in Paragraphs 6 (a) and 12 is not material, India shall communicate particulars of the change to Pakistan, in writing, as soon as the alteration has been made or the repairs have been undertaken. The provisions of Paragraph 7 or Paragraph 23, as the case may be, shall then apply.

...

ANNEXURE E – STORAGE OF WATERS
BY INDIA ON THE WESTERN RIVERS

(Article III (4))

...

2. As used in this Annexure:

(a) "Storage Work" means a work constructed for the purpose of impounding the waters of a stream; but excludes

   (i) a Small Tank,

   (ii) the works specified in Paragraphs 3 and 4 of Annexure D, and

   (iii) a new work constructed in accordance with the provisions of Annexure D.

(b) "Reservoir Capacity" means the gross volume of water which can be stored in the reservoir.

(c) "Dead Storage Capacity" means that portion of the Reservoir Capacity which is not used for operational purposes, and "Dead Storage" means the corresponding volume of water.

(d) "Live Storage Capacity" means the Reservoir Capacity excluding Dead Storage Capacity, and "Live Storage" means the corresponding volume of water.

(e) "Flood Storage Capacity" means that portion of the Reservoir Capacity which is reserved for the temporary storage of flood waters in order to regulate downstream flows, and "Flood Storage" means the corresponding volume of water.

(f) "Surcharge Storage Capacity" means the Reservoir Capacity between the crest of an uncontrolled spillway or the top of the crest gates in normal closed position and the maximum water elevation above this level for which the dam is designed, and "Surcharge Storage" means the corresponding volume of water.

(g) "Conservation Storage Capacity" means the Reservoir Capacity excluding Flood Storage Capacity, Dead Storage Capacity and Surcharge Storage Capacity, and "Conservation Storage" means the corresponding volume of water.

(h) "Power Storage Capacity" means that portion of the Conservation Storage Capacity which is designated to be used for generating electric energy, and "Power Storage" means the corresponding volume of water.

(i) "General Storage Capacity" means the Conservation Storage Capacity excluding Power Storage Capacity, and "General Storage" means the corresponding volume of water.

(j) "Dead Storage Level" means the level of water in a reservoir corresponding to Dead Storage Capacity, below which level the reservoir does not operate.

(k) "Full Reservoir Level" means the level of water in a reservoir corresponding to Conservation Storage Capacity.

(1) "Multi-purpose Reservoir" means a reservoir capable of and intended for use for more than one purpose.

(m) "Single-purpose Reservoir" means a reservoir capable of and intended for use for only one purpose.

(n) "Small Tank" means a tank having a Live Storage of less than 700 acre-feet and fed only from a non-perennial small stream: Provided that the Dead Storage does not exceed 50 acre-feet.

3. There shall be no restriction on the operation as heretofore by India of those Storage Works which were in operation as on the Effective Date or on the construction and operation of Small Tanks.

4. As soon as India finds it possible to do so, but not later than 31st March 1961, India shall communicate to Pakistan in writing the information specified in the Appendix to this Annexure for such Storage Works as were in operation as on the Effective Date. If any such information is not available or is not pertinent to the design of the Storage Work or to the conditions at the site, it will be so stated.

5. (a) If any alteration proposed in the design of any of the Storage Works referred to in Paragraph 3 would result in a material change in the information furnished to Pakistan under the provisions of Paragraph 4, India shall, at least 4 months in advance of making the alteration, communicate particulars of the change to Pakistan in writing and the provisions of Paragraph 6 shall then apply.

(b) In the event of an emergency arising which requires repairs to be undertaken to protect the integrity of any of the Storage Works referred to in Paragraph 3, India may undertake immediately the necessary repairs or alterations and, if these repairs or alterations result in a change in the information furnished to Pakistan under the provisions of Paragraph 4, India shall as soon as possible communicate particulars of the change to Pakistan in writing. The provisions of Paragraph 6 shall then apply.

6. Within three months of the receipt of the particulars specified in Paragraph 5, Pakistan shall communicate to India in writing any objection it may have with regard to the proposed change on the ground that the change involves a material departure from the criteria set out in Paragraph 11. If no objection is received by India from Pakistan within the specified period of three months, then Pakistan shall be deemed to have no objection. If a question arises as to whether or not the change involves a material departure from such of the criteria mentioned above as may be applicable, then either Party may proceed to have the question resolved in accordance with the provision of Article IX (1) and (2).

7. The aggregate storage capacity of all Single-purpose and Multi-purpose Reservoirs which may be constructed by India after the Effective Date on each of the River Systems specified in Column (2) of the following table shall not exceed, for each of the categories shown in Columns (3), (4) and (5), the quantities specified therein:

		Conservation Storage Capacity		
River System		General Storage Capacity	Power Storage Capacity	Flood Storage Capacity
(1)	(2)	(3)	(4)	(5)
		million acre-feet		
(a) The Indus		0.25	0.15	Nil
(b) The Jhelum (excluding the Jhelum Main)		0.50	0.25	0.75
(c) The Jhelum Main		Nil	Nil	As provided in Paragraph 9
(d) The Chenab (excluding the Chenab Main)		0.50	0.60	Nil
(e) The Chenab Main		Nil	0.60	Nil

Provided that

   (i) the storage specified in Column (3) above may be used for any purpose whatever including the generation of electric energy;

   (ii) the storage specified in Column (4) above may also be put to Non-Consumptive Use (other than flood protection or flood control) or to Domestic Use;

   (iii) India shall have the option to increase the Power Storage Capacity specified against item (d) above by making a reduction by an equal amount in the Power Storage Capacity specified against items (b) or (e) above; and

   (iv) Storage Works to provide the Power Storage Capacity on the Chenab Main specified against item (e) above shall not be constructed at a point below Naunut (Latitude 33° 19' N. and Longitude 75 59' E.).

8. The figures specified in Paragraph 7 shall be exclusive of the following:

   (a) Storage in any Small Tank.

   (b) Any natural storage in a Connecting Lake, that is to say, storage not resulting from any man-made works.

   (c) Waters which, without any man-made channel or works, spill into natural depressions or borrow-pits during floods.

   (d) Dead Storage.

   (e) The volume of Pondage for hydroelectric plants under Annexure D and under Paragraph 21(a).

   (f) Surcharge Storage.

   (g) Storage in a Regulating Basin (as defined in Annexure D).

   (h) Storage incidental to a barrage on the Jhelum Main or on the Chenab Main not exceeding 10,000 acre-feet.

9. India may construct on the Jhelum Main such works as it may consider necessary for flood control of the Jhelum Main and may complete any such works as were under construction on the Effective Date: Provided that

(i) any storage which may be effected by such works shall be confined to off-channel storage in side valleys, depressions or lakes and will not involve any storage in the Jhelum Main itself; and

(ii) except for the part held in takes, borrow-pits or natural depressions, the stored waters shall be released as quickly as possible after the flood recedes and returned to the Jhelum Main lower down.

These works shall be constructed in accordance with the provisions of Paragraph 11(d).

10. Notwithstanding the provisions of Paragraph 7, any Storage Work to be constructed on a Tributary of The Jhelum on which Pakistan has any Agricultural Use or hydro-electric use shall be so designed and operated as not to adversely affect the then existing Agricultural Use or hydro-electric use on that Tributary.

11. The design of any Storage Work (other than a Storage Work falling under paragraph 3) shall conform to the following criteria:

(a) The Storage Work shall not be capable of raising artificially the water level in the reservoir higher than the designed Full Reservoir Level except to the extent necessary for Flood Storage, if any, specified in the design.

(b) The design of the works shall take due account of the requirements of Surcharge Storage.

(c) The volume between the Full Reservoir Level and the Dead Storage Level of any reservoir shall not exceed the Conservation Storage Capacity specified in the design.

(d) With respect to the Flood Storage mentioned in Paragraph 9, the design of the works on the Jhelum Main shall be such that no water can spill from the Jhelum Main into the off-channel storage except when the water level in the Jhelum Main rises above the low flood stage.

(e) Outlets or other works of sufficient capacity shall be provided to deliver into the river downstream the flow of the river received upstream of the Storage Work, except during freshets or floods. These outlets or works shall be located at the highest level consistent with sound and economical design and with satisfactory operation of the Storage Work.

(f) Any outlets below the Dead Storage Level necessary for sediment control or any other technical purpose shall be of the minimum size, and located at the highest level, consistent with sound and economical design and with satisfactory operation of the Storage Work.

(g) If a power plant is incorporated in the Storage Work, the intakes for the turbines shall be located at the highest level consistent with satisfactory and economical construction and operation of the plant and with customary and accepted practice of design for the designated range of the plant's operation.

12. To enable Pakistan to satisfy itself that the design of a Storage Work (other than a Storage Work falling under Paragraph 3) conforms to the criteria mentioned in Paragraph 11, India shall, at least six months in advance of the beginning of construction of the Storage Work, communicate to Pakistan in writing the information specified in the Appendix to this Annexure; if any such information is not available or is not pertinent to the design of the Storage Work or to the conditions at the site, it will be so stated:

Provided that, in the case of a Storage Work falling under Paragraph 9,

(i) if the work is a new work, the period of six months shall be reduced to four months; and

(ii) if the work is a work under construction on the Effective Date, the information shall be furnished not later than 31st December 1960.

13. Within three months (or two months, in the case of a Storage Work specified in Paragraph 9) of the receipt by Pakistan of the information specified in Paragraph 12, Pakistan shall communicate to India in writing any objection that it may have with regard to the proposed design on the ground that the design does not conform to the criteria mentioned in Paragraph 11. If no objection is received by India from Pakistan within the specified period of three months (or two months, in the case of a Storage Work specified in Paragraph 9), then Pakistan shall be deemed to have no objection.

14. If a question arises as to whether or not the design of a Storage Work (other than a Storage Work falling under Paragraph 3) conforms to the

criteria set out in Paragraph 11, then either Party may proceed to have the question resolved in accordance with the provisions of Article IX (1) and (2).

15. (a) If any alteration proposed in the design of a Storage Work (other than a Storage Work falling under Paragraph 3) before it comes into operation would result in a material change in the information furnished to Pakistan under the provisions of Paragraph 12, India shall immediately communicate particulars of the change to Pakistan in writing and the provisions of Paragraphs 13 and 14 shall then apply, but where a period of three months is specified in Paragraph 13, that period shall be reduced to two months.

(b) If any alteration proposed in the design of a Storage Work (other than a Storage Work falling under Paragraph 3), after it comes into operation would result in a material change in the information furnished to Pakistan under the provisions of Paragraph 12, India shall, at least four months in advance of making the alteration, communicate particulars of the change to Pakistan in writing and the provisions of Paragraphs 13 and 14 shall then apply, but where a period of three months is specified in Paragraph 13, that period shall be reduced to two months.

16. In the event of an emergency arising which requires repairs to be undertaken to protect the integrity of a Storage Work (other than a Storage Work falling under Paragraph 3), India may undertake immediately the necessary repairs or alterations; if these repairs or alterations result in a change in the information furnished to Pakistan under the provisions of Paragraph 12, India shall, as soon as possible, communicate particulars of the change to Pakistan in writing to enable Pakistan to satisfy itself that after such change the design of the work conforms to the criteria specified in Paragraph 11. The provisions of Paragraphs 13 and 14 shall then apply.

17. The Flood Storage specified against item (b) in Paragraph 7 may be effected only during floods when the discharge of the river exceeds the amount specified for this purpose in the design of the work; the storage above Full Reservoir Level shall be released as quickly as possible after the flood recedes.

18. The annual filling of Conservation Storage and the initial filling below the Dead Storage Level, at any site, shall be carried out at such times

and in accordance with such rules as may be agreed upon between the Commissioners. In case the Commissioners are unable to reach agreement, India may carry out the filling as follows:

(a) if the site is on The Indus, between 1st July and 20th August;

(b) if the site is on The Jhelum, between 21st June and 20th August; and

(c) if the site is on The Chenab, between 21st June and 31st August at such rate as not to reduce, on account of this filling, the flow in the Chenab Main above Merala to less than 55,000 cusecs.

19. The Dead Storage shall not be depleted except in an unforeseen emergency. If so depleted, it will be refilled in accordance with the conditions of its initial filling.

20. Subject to the provisions of Paragraph 8 of Annexure C, India may make releases from Conservation Storage in any manner it may determine.

21. If a hydroelectric power plant is incorporated in a Storage Work (other than a Storage Work falling under Paragraph 3), the plant shall be so operated that:

(a) the maximum Pondage (as defined in Annexure D) shall not exceed the Pondage required for the firm power of the plant, and the water-level in the reservoir corresponding to maximum Pondage shall not, on account of this Pondage, exceed the Full Reservoir Level at any time; and

(b) except during the period in which a filling is being carried out in accordance with the provisions of Paragraph 18 or 19, the volume of water delivered into the river below the work during any period of seven consecutive days shall not be less than the volume of water received in the river upstream of the work in that seven-day period.

...

23. When the Live Storage Capacity of a Storage Work is reduced by sedimentation, India may, in accordance with the relevant provisions of this Annexure, construct new Storage Works or modify existing Storage Works so as to make up the storage capacity lost by sedimentation.

24. If a power plant incorporated in a Storage Work (other than a Storage Work falling under Paragraph 3) is used to operate a peak power plant and lies on any Tributary of The Jhelum on which there is any Agricultural Use by Pakistan, a Regulating Basin (as defined in Annexure D) shall be incorporated.

25. If the change referred to in Paragraph 5(a) or 15 is not material, India shall communicate particulars of the change to Pakistan, in writing, as soon as the alteration has been made or the repairs have been undertaken. The provisions of Paragraph 6 or Paragraphs 13 and 14, as the case may be, shall then apply.

ANNEXURE F – NEUTRAL EXPERT

(Article IX (2))

PART 1—QUESTIONS TO BE REFERRED TO A NEUTRAL EXPERT

1. Subject to the provisions of Paragraph 2, either Commissioner may, under the provisions of Article IX (2) (a), refer to a Neutral Expert any of the following questions:

    … [The Treaty lists 23 types of questions of technical nature].

2. If a claim for financial compensation has been raised with respect to any question specified in Paragraph 1, that question shall not be referred to a Neutral Expert unless the two Commissioners are agreed that it should be so referred.

3. Either Commissioner may refer to a Neutral Expert under the provisions of Article IX (2) (a) any question arising with regard to the determination of costs under Article IV (5), Article IV (11), Article VII (1) (a) or Article VII (1) (b).

PART 2—APPOINTMENT AND PROCEDURE

4. A Neutral Expert shall be a highly qualified engineer, …

5. If a difference arises and has to be dealt with in accordance with the provisions of Article IX (2) (a), the following procedure will be followed:

(a) The Commissioner who is of the opinion that the difference falls within the provisions of Part 1 of this Annexure (hereinafter in this paragraph referred to as "the first Commissioner") shall notify the other Commissioner of his intention to ask for the appointment of a Neutral Expert. Such notification shall clearly state the paragraph or paragraphs of Part I of this Annexure under which the difference falls and shall also contain a statement of the point or points of difference.

(b) Within two weeks of the receipt by the other Commissioner of the notification specified in (a) above, the two Commissioners will endeavour to prepare a joint statement of the point or points of difference.

(c) After expiry of the period of two weeks specified in (b) above, the first Commissioner may request the appropriate authority specified in Paragraph 4 to appoint a Neutral Expert; a copy of the request shall be sent at the same time to the other Commissioner.

(d) The request under (c) above shall be accompanied by the joint statement specified in (b) above; failing this, either Commissioner may send a separate statement to the appointing authority and, if he does so, he shall at the same time send a copy of the separate statement to the other Commissioner.

6. The procedure with respect to each reference to a Neutral Expert shall be determined by him, provided that:

(a) he shall afford to each Party an adequate hearing;

(b) in making his decision, he shall be governed by the provisions of this Treaty and by the *compromis*, if any, presented to him by the Commission; and

(c) without prejudice to the provisions of Paragraph 3, unless both Parties so request, he shall not deal with any issue of financial compensation.

7. Should the Commission be unable to agree that any particular difference falls within Part 1 of this Annexure, the Neutral Expert shall, after hearing both Parties, decide whether or not it so falls. Should he decide that the difference so falls, he shall proceed to render a decision on the merits; should

he decide otherwise, he shall inform the Commission that, in his opinion, the difference should be treated as a dispute. Should the Neutral Expert decide that only a part of the difference so falls, he shall, at his discretion, either:

(a) proceed to render a decision on the part which so falls, and inform the Commission that, in his opinion, the part which does not so fall should be treated as a dispute, or

(b) inform the Commission that, in his opinion, the entire difference should be treated as a dispute.

...

9. The Neutral Expert shall, as soon as possible, render a decision on the question or questions referred to him, giving his reasons. A copy of such decision, duly signed by the Neutral Expert, shall be forwarded by him to each of the Commissioners and to the Bank.

10. Each Party shall bear its own costs. The remuneration and the expense of the Neutral Expert and of any assistance that he may need shall be borne initially as provided in Part 3 of this Annexure and eventually by the Party against which his decision is rendered, except as, in special circumstances, and for reasons to be stated by him, he may otherwise direct. He shall include in his decision a direction concerning the extent to which the costs of such remuneration and expenses are to be borne by either Party.

11. The decision of the Neutral Expert on all matters within his competence shall be final and binding, in respect of the particular matter on which the decision is made, upon the Parties and upon any Court of Arbitration established under the provisions of Article IX (5).

12. The Neutral Expert may, at the request of the Commission, suggest for the consideration of the Parties such measures as are, in his opinion, appropriate to compose a difference or to implement his decision.

13. Without prejudice to the finality of the Neutral Expert's decision, if any question (including a claim to financial compensation) which is not within the competence of a Neutral Expert should arise out of his decision, that question shall, if it cannot be resolved by agreement, be settled in accordance with the provisions of Article IX (3), (4) and (5).

Part 3—Expenses

...

ANNEXURE G – COURT OF ARBITRATION

(Article IX (5))

1. If the necessity arises to establish a Court of Arbitration under the provisions of Article IX, the provisions of this Annexure shall apply.

...

4. Unless otherwise agreed between the Parties, a Court of Arbitration shall consist of seven arbitrators appointed as follows:

(a) Two arbitrators to be appointed by each Party in accordance with Paragraph 6; and

(b) Three arbitrators (hereinafter sometimes called the umpires) to be appointed in accordance with Paragraph 7, one from each of the following categories:

  (i) Persons qualified by status and reputation to be Chairman of the Court of Arbitration who may, but need not, be engineers or lawyers.

  (ii) Highly qualified engineers.

  (iii) Persons well versed in international law.

The Chairman of the Court shall be a person from category (b) (i) above.

...

6. The arbitrators referred to in Paragraph 4 (a) shall be appointed as follows:

The Party instituting the proceeding shall appoint two arbitrators at the time it makes a request to the other Party under Paragraph 2 (b). Within 30 days of the receipt of this request, the other Party shall notify the names of the arbitrators appointed by it.

...

12. Each Party shall be represented before the Court by an Agent and may have the assistance of Counsel.

...

15. At its first meeting the Court shall

    (a) establish its secretariat and appoint a Treasurer;

    (b) make an estimate of the likely expenses of the Court and call upon each Party to pay to the Treasurer half of the expenses so estimated: Provided that, if either Party should fail to make such payment, the other Party may initially pay the whole of the estimated expenses;

    (c) specify the issues in dispute;

    (d) lay down a programme for submission by each side of legal pleadings and rejoinders; and

    (e) determine the time and place of reconvening the Court.

Unless special circumstances arise, the Court shall not reconvene until the pleadings and rejoinders have been closed. During the intervening period, at the request of either Party, the Chairman of the Court may, for sufficient reason, make changes in arrangements made under (d) and (e) above.

16. Subject to the provisions of this Treaty and except as the Parties may otherwise agree, the Court shall decide all questions relating to its competence and shall determine its procedure, including the time within which each Party must present and conclude its arguments. All such decisions of the Court shall be by a majority of those present and voting. Each arbitrator, including the Chairman, shall have one vote. In the event of an equality of votes, the Chairman shall have a casting vote.

17. The proceedings of the Court shall be in English.

...

23. The Court shall render its Award, in writing, on the issues in dispute and on such relief, including financial compensation, as may have been claimed. The Award shall be accompanied by a statement of reasons. An Award signed by four or more members of the Court shall constitute the Award of the Court. A signed counterpart of the Award shall be delivered by the Court to each Party. Any such Award rendered in accordance with the provisions of this Annexure in regard to a particular dispute shall be final and binding upon the Parties with respect to that dispute.

...

26. In its Award, the Court shall also award the costs of the proceedings, including those initially borne by the Parties and those paid by the Treasurer.

27. At the request of either Party, made within three months of the date of the Award, the Court shall reassemble to clarify or interpret its Award. Pending such clarification or interpretation the Court may, at the request of either Party and if in the opinion of the Court circumstances so require, grant a stay of execution of its Award. After furnishing this clarification or interpretation, or if no request for such clarification or interpretation is made within three months of the date of the Award, the Court shall be deemed to have been dissolved.

28. Either Party may request the Court at its first meeting to lay down, pending its Award, such interim measures as, in the opinion of that Party, are necessary to safeguard its interests under the Treaty with respect to the matter in dispute, or to avoid prejudice to the final solution or aggravation or extension of the dispute. The Court shall, thereupon, after having afforded an adequate hearing to each Party, decide, by a majority consisting of at least four members of the Court, whether any interim measures are necessary for the reasons hereinbefore stated and, if so, shall specify such measures: Provided that

    (a) the Court shall lay down such interim measures only for such specified period as, in its opinion, will be necessary to render the Award: this period may, if necessary. be extended unless the delay in rendering the Award is due to any delay on the part of the Party which requested the interim measures in supplying such information as may be required by the other Party or by the Court in connection with the dispute, and

    (b) the specification of such interim measures shall not be construed as an indication of any view of the Court on the merits of the dispute.

29. Except as the Parties may otherwise agree, the law to be applied by the Court shall be this Treaty and, whenever necessary for its interpretation or application, but only to the extent necessary for that purpose, the following in the order in which they are listed:

    (a) International conventions establishing rules which are expressly recognized by the Parties.

    (b) Customary international law.

...

ANNEXURE H – TRANSITIONAL ARRANGEMENTS

(Article II (5))

PART 1—PRELIMINARY

[It specifies the duration of phases 1 and 2 of the transition period, gives definition of the terms used in the Annexures and lays down that "the provisions of this annexure shall lapse" soon after the end of the transition period.]

PART 2—DISTRIBUTION OF THE WATERS OF THE RAVI

[It specifies the detailed rules about the deliveries to be made by India to the CBDC generally in order to maintain the *status quo ante* from September 21 to April 10 during the year.]

PART 3—DISTRIBUTION OF THE WATERS OF THE SUTLEJ AND THE BEAS IN *KHARIF* DURING PHASE I

...

PART 4—DISTRIBUTION OF THE WATERS OF THE SUTLEJ AND THE BEAS IN *KHARIF* DURING PHASE II

...

PART 5—DISTRIBUTION OF THE WATERS OF THE SUTLEJ AND THE BEAS IN *RABI*

[These three parts specify in detail the rules for calculating the supplies that India was to let flow below Ferozpur (including those to be delivered to Dipalpur Canal), for each 10-day period of the year, under different supply conditions in the Sutlej, the Beas and also in the Chenab. It also includes incidental regulations.]

PART 6—WATER-ACCOUNTS AT FEROZPUR

[It requires the two Commissioners, separately to maintain, in accordance with specified rules and forms, an account of the distribution of, as at Ferozpur, under the provisions of Part 3, 4, and 5, for each 10-day period of the year. It specifies the manner in which the two Commissioners will exchange the water-accounts and in which the differences between the calculated supplies and those actually delivered have to be adjusted.]

PART 7—FINANCIAL PROVISIONS

[It specifies the yearly payments that Pakistan was to make to India for the proportionate working expenses on the Madhopur headworks, and matters incidental to these payments.]

PART 8—EXTENSION OF TRANSITION PERIOD

[It specifies the manner in which Pakistan could ask for an extension of the transition period by one, two or three years, beyond 31 March, 1970.]

PART 9—GENERAL

[It recognizes the uses in India, as on the Effective Date, from the Sutlej, the Beas and the Ravi, other than those specified in Parts 3, 4, and 5, and matters incidental to safeguarding these uses. It lays down the regulations that India and Pakistan are to enforce in order not to disturb the gains and losses in the rivers as assumed in Parts 3, 4, and 5. It also specifies action to be taken if there is damage to the link canals by floods.]

PART 10—SPECIAL PROVISIONS FOR 1960 AND 1961

[These became necessary because the Treaty could not be concluded until 19 September, 1960.]

APPENDICES TO ANNEXURE H ARE OMITTED

Appendix 3

Details of Projects by India on Western Rivers

A) RIVER CHENAB

a) Commissioned Projects

S. No.	Name of Plant	Location	Status	Power Capacity	Aggregate Discharge through turbines	G.T. Sheet No.
1	Salal (Phase-I & II) H/E Plant	45 miles U/S Marala Barrage Longitude: 74°-50' East Latitude: 33°-08' North	In operation	690 MW	14550	43-O
2	Dul-Hasti (I & II) H/E Plant	Near Kishtwar on Chenab River Longitude: 750-46' Latitude: 330-0'	Completed	780 MW	7522	43-O
3	Baglihar H/E Plant	On Chenab Main about 147 km U/S of Marala Headworks	Completed	450 MW	15473	43-O
4	Rajouri H/E Plant	On Darhali Nallah, a sub-tributary of Chenab	Completed	3 MW	87	43-K
5	Killer H/E Plant	On Mohal Nullah, a tributary of Chenab	Completed	0.3 MW	43	52-C/8

6	Thirot H/E Plant	On Thirot Nullah, a tributary of Chandra Bhaga	Completed	4.5 MW	81	52-D
7	Shansha H/E Plant	On Shansha Nullah, a tributary of Chenab	Completed	0.2 MW	50	52-D
8	Billing H/E Plant	On Billing Nullah, a tributary of Bhaga	Completed	0.1 MW	25	52-H
9	Sissu H/E Plant	Near Sissue, a tributary of Chenab	Completed	0.10 MW	25	52-H
10	Chinani-II H/E Plant	On Jammu Tawi River, a tributary of Chenab	Completed	2 MW	251	43-O
11	Bhadarwah (Remodling) H/E Plant	On Haloon Nullah, a tributary of Neeru Nullah near Dist. Bhadarwah		1 MW	300	43-P/13

b) Location

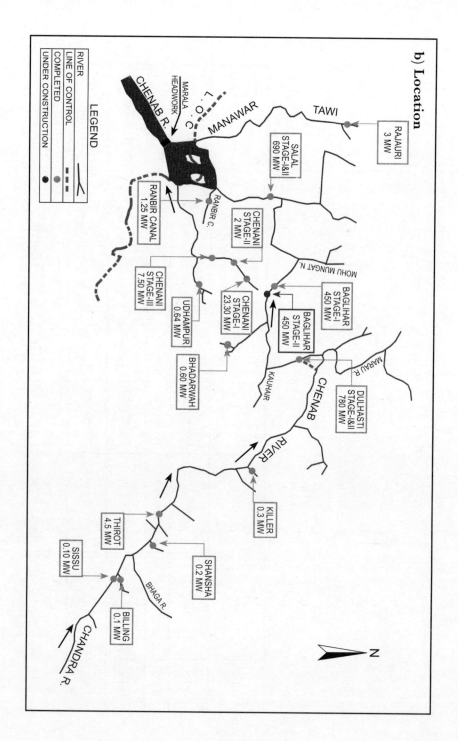

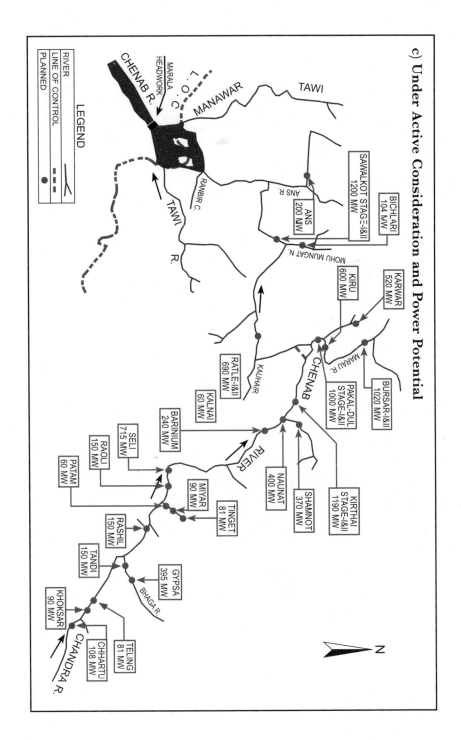

c) **Under Active Consideration and Power Potential**

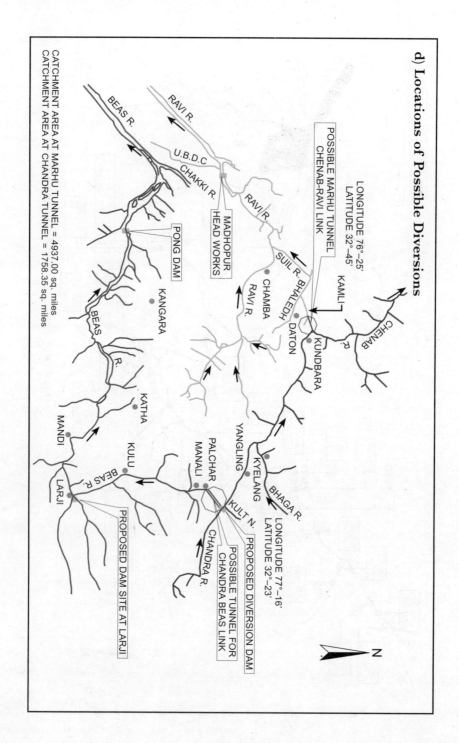

d) Locations of Possible Diversions

POSSIBLE MARHU TUNNEL
CHENAB-RAVI LINK

LONGITUDE 76°-25'
LATITUDE 32°-45'

LONGITUDE 77°-16'
LATITUDE 32°-23'

PROPOSED DIVERSION DAM

POSSIBLE TUNNEL FOR
CHANDRA BEAS LINK

PROPOSED DAM SITE AT LARJI

PONG DAM

MADHOPUR
HEAD WORKS

CHENAB R.

KAMLI

KUNDBARA

DATON

BHALEDH

CHAMBA
RAVI R.

SUIL R.

RAVI R.

RAVI R.

U.B.D.C

CHAKKI R.

BEAS R.

KANGARA

BEAS R.

KATHA

MANDI

LARJI

BEAS R.

KULU

MANALI

PALCHAR

CHANDRA R.

KULT N.

YANGLING

KYELANG

BHAGA R.

N

CATCHMENT AREA AT MARHU TUNNEL = 4937.00 sq. miles
CATCHMENT AREA AT CHANDRA TUNNEL = 1758.35 sq. miles

B) RIVER JHELUM

a) Commissioned Projects

S. No.	Name of Plant	Location	Status	Power Capacity	Aggregate Discharge through turbines in Cusecs
1	Asthan H/E Plant	On Asthan Nullah, a tributary of Kishenganga River	Completed	0.75 MW	39
2	Bandipura H/E Plant	On Madmati Nullah, a tributary of the Jhelum	Completed	0.03 MW	18
3	Dachhigam H/E Plant	On Degwan Nullah, a tributary of the Jhelum	Completed	0.04 MW	20
4	Karan H/E Plant	On Keshar Katta Nullah, a tributary of River Jhelum	Completed	0.70 MW	60
5	Karnah H/E Plant	On Quazi Nag Nullah, a tributary of Kishenganga River	Completed	2 MW	250
6	Lower Jhelum H/E Plant	8 miles D/S Baramula on the Jhelum	Completed	105 MW	7000
7	Matchil H/E Plant	On Dadhi Nullah, a tributary of Kishenganga	Completed	0.35 MW	141
8	Pahalgam H/E Plant	Corfluence of East Lidder and West Lidder, sub-tributaries of Jhelum River in Anantnag Distt.	Completed	4.5 MW	816
9	Parnai H/E Plant	On Suran River, a tributary of River Punch	Completed	37.5 MW	530
10	Poonch H/E Plant	On Betar Nullah, a tributary of the Jhelum	Completed	0.16 MW	10
11	Sambal H/E Plant	Near Village Sumbal on Sind River, a tributary of Jhelum	Completed	22 MW	615
12	Upper Sind (Phase-II) H/E Plant	On Wangat Nullah near Village Wangat, a tributary of Sind River	Completed	105 MW	1850
13	Uri I H/E Plant	Located at Uri about 16 miles D/S of Baramula on the Jhelum Main	Completed	480 MW	8000

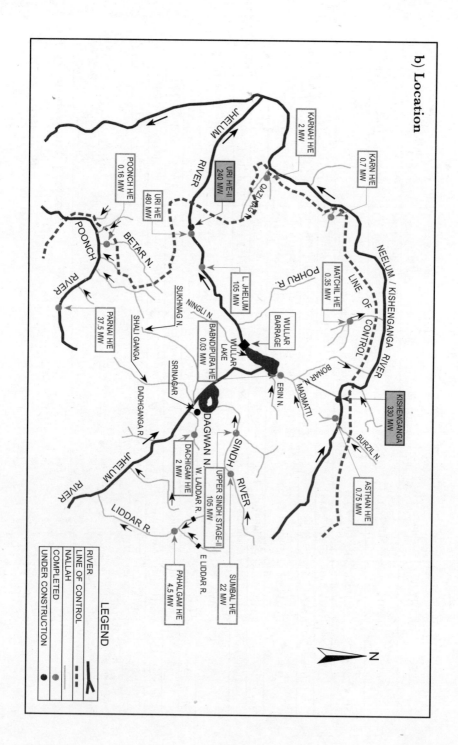

c) Power Potential

S. No.	Name of Scheme	District Location	River/Nallah	Anticipated installed Capacity MWs	Position of		Year of Preparation of DPR/PFR	Preparation of DPR/ Updation
					PFR	DPR		
1	Baderwah SHP	Doda	Haloon Nallah	1.5	Prepared	Prepared	2004	1 MW Commissioned
2	Shutkari Kullan HEP	Srinagar	Sindh Nallah	84	Prepared	Prepared	1989	Updation
3	Sonamarg Storage	--	Sindh Nallah	165	Prepared	Prepared	1987	Revision of DPR
4	Gangabal Storage	--	Gangabal Lake	100	Prepared		1978	-do-
5	Batkote Sakrus HEP	Anantnag	Lidder Nallah	36	Prepared	Prepared	1988	Updation
6	Lower Kalnai HEP	Doda	Kalnai Nallah	50	Prepared	Prepared	2006	-do-
7	Ujh Storage Multipurpose	Kathua	Ujh River	280	Prepared		2005	PFR Prepared by WAPCOS
8	Nunwan Batkote HEP	Anantnag	Lidder Nallah	22.6	Prepared	Prepared	Mar-87	Updation
9	Laripora SHP	Anantnag	Lidder Nallah	6	Prepared	Prepared	1987	Updation/Revision
10	Naigadh SHP	Doda	Naigadh Nallah	6	Prepared	Prepared	May-87	Updation
11	Hanswar SHP	Doda	Hanswar Nallah	1.3	Prepared	-	2004	Preparation of DPR

S. No.	Name of Scheme	District Location	River/Nallah	Anticipated installed Capacity MWs	Position of PFR	Position of DPR	Year of Preparation of DPR/PFR	Preparation of DPR/ Updation
12	New Rajouri SHP	Rajouri	Darhal Tawi	3	Prepared	Prepared	1997	Updation
13	Ravi Canal at Samba	Jammu	Ravi	1.5	Prepared	Prepared	1983	Updation
14	Modernization of Poonch	Poonch	Betar	0.5	Prepared	Prepared	1985	Updation
15	Athwathoo	Baramulla	Madhumati	10	Prepared	Prepared	2005	Allotted to IPP under Phase I
16	Tangmarg	Baramullah	Farozpur	10	Prepared	Prepared	2005	Allotted to IPP under Phase I
17	Boniyar-I	Baramullah	Hapatkhai	12	Prepared	Prepared	2005	Allotted to IPP under Phase I
18	Kahmil	Kupwara	Kahmil	4	Prepared	Prepared	2005	-do-
19	Aharbal	Anantnag	Vishow	22.5	Prepared	Prepared	2005	-do-
20	Brenwar	Budgam	Doodganga	5	Prepared	Prepared	2005	-do-
21	Hirapura	Pulwalma	Rambiara	12	Prepared	Prepared	2005	-do-
22	Mandi	Poonch	Mandi	12.5	Prepared	Prepared	2005	-do-
23	Ranjala Du nadi	Doda	Kalnai/ Upper Chenab	15	Prepared	Prepared	2005	-do-
24	Drung SHP	Kuthwa	Ujh	5	Prepared	Prepared	2007	-do-

S. No.	Name of Scheme	District Location	River/Nallah	Anticipated installed Capacity MWs	Position of PFR	Position of DPR	Year of Preparation of DPR/PFR	Preparation of DPR/Updation
25	Mawar (Nawgam)	Kupwara	Mawar Nallah	4.5	Prepared		2003	Under process for allotment to IPP in Phase II
26	Boniyar-II	Baramullah	Boniyar Nallah	2.6	Prepared		2003	-do-
27	Boniyar-III	Baramullah	Boniyar	1.2	Prepared		2003	-do-
28	Erin	Baramullah	Erin Nallah	6	Prepared		2003	Under process for allotment to IPP in Phase II
29	Chandenwari-Uri	Baramullah	Chandenwari Nallah	3	Prepared		2003	-do-
30	Kanzil Wangath	Srinagar	Wangath	12	Prepared		2003	-do-
31	Sranz Ningli	Baramulah	Ningli Nallah	2.3	Prepared		2003	-do-
32	Nihama (Kulgam)	Anantnag	Vishow Nallah	6.09	Prepared		2003	-do-
33	Aru (Pahalgam)	Anantnag	Lidder	2.5	Prepared		2003	-do-
34	Bringi	Anantnag	Bringi Nallah	3.5	Prepared		2004	-do-
35	Sukhnag	Budgam	Sukhnag Nallah	16	Prepared		2004	-do-
36	Shaliganga	Budgam	Shaliganga Nallah	10.5	Prepared		2004	-do-
37	Gulabgrah	Udhampur	Ans Nallah	3	Prepared		2004	Under process for allotment to IPP in Phase II

S. No.	Name of Scheme	District Location	River/Nallah	Anticipated installed Capacity MWs	Position of		Year of Preparation of DPR/PFR	Preparation of DPR/ Updation
					PFR	DPR		
38	Atalgrah	Doda	Neeru Nallah	2.5	Prepared		2004	Under process for allotment to IPP in Phase II
39	Girjan ki Gali	Poonch	Suran	15	Prepared		2004	-do-
40	Ans Stage-Udhampur		Ans River	22	Prepared		2004	-do-
41	Phagla	Poonch	Suran River	12	Prepared		2004	-do-
42	Bhalla	Doda	Binkud/Neru Nallah	1.5	Prepared		2004	-do-
43	Chingus-I	Rajouri	Nawshara Tawi	2.5	Prepared	Prepared	2005	Advertised under IPP Phase III
44	Chingus-II	Rajouri	Nawshara Tawi	2.5		Prepared	2005	-do-
45	Thana Mandi	Rajouri	Mangota Nallah	4.05		Prepared	2005	-do-
46	Nachlana	Banihal	Nachlana	1		Prepared	2005	Advertised under IPP Phase III. RFQ received.
47	Pogalgrah	Doda	Pogalgarh Nallah	1		Prepared	2005	-do-
48	Tulail SHP	Baramullah	Kishanganga	1.05	Prepared	Prepared	1987	Advertised under IPP Phase III
49	Mohara P/H	Baramullah	Boniyar/Jhelum	9		Prepared	2004	Advertised under IPP Phase III, RFQ received.

S. No.	Name of Scheme	District Location	River/Nallah	Anticipated installed Capacity MWs	Position of PFR	Position of DPR	Year of Preparation of DPR/PFR	Preparation of DPR/ Updation
50	Thanda Pani Wali	Rajouri	Thanda Pani Wali	0.75		Prepared	2004	Advertised under IPP Phase but there was no response.
51	Dachigam SHP	Srinagar	Dachigam Nallah	2	Prepared	Prepared	1988	Updation
52	Loran SHP	Poonch	Loran Nallah	6	Prepared		2006	Proposed for IPP Phase IV
53	Gund SHP	Srinagar	Sindh Nallah	0.73	Prepared		2005	-do-
54	Thajwas SHP	Srinagar	Sindh Nallah	1	Prepared		2005	-do-
55	Chittergul	Srinagar	Sindh Nallah	0.73	Prepared		2005	-do-
56	Kulan	Srinagar	Sindh Nallah	0.73	Prepared		2005	-do-
57	Lidder Stage-I (Shesnag-Chandanwari)	Anantnag	Lidder Nalah	50	Prepared		1981	Preparation of DPR
58	Lidder Stage-II (Chandanwari-Laripora)	Anantnag	Lidder Nalah	45	Prepared		1981	-do-
59	Martand Canal	Anantnag	Lidder Nallah Martand Canal	3	Prepared		2006	-do-
60	Vishow Stage-Anantnag		Vishow Nallah	15	Prepared		1981	-do-

S. No.	Name of Scheme	District Location	River/Nallah	Anticipated installed Capacity MWs	Position of		Year of Preparation of DPR/PFR	Preparation of DPR/ Updation
					PFR	DPR		
61	Ans Stage-II	Udhampur	Ans Nallah	15	Prepared		2006	Preparation of DPR
62	Jaglano	Udhampur	Upper Ans Nallah	4.5	Prepared		2006	-do-
63	Sultan Pathri (Loran)	Poonch	Androi/Nandi Chhul Nalah	3	Prepared		2006	-do-
64	Jori Budhal	Rajouri	Jori Nallah	0.6	Prepared		2006	-do-
65	Khari wali Darhal	Rajouri	Khariwali	0.3	Prepared		2007	-do-
66	Bhari Darhal	Rajouri	Darhali Nallah	1	Prepared		2007	-do-
67	Chenani IV	Udhampur	Tawi	11.9	Prepared		2005	Preparation of DPR
68	Chenani V	Udhampur	Tawi	5.4	Prepared		2005	Preparation of DPR
69	Anji Nallah SHP	Udhampur	Anji Nallah	0.3	Prepared	Prepared	1981	Updation
70	Chilli SHP	Doda	Chilli Nallah	1.6	Prepared		2005	Preparation of DPR
71	Gagal-Gwari	Doda	Gangal Nallah	2.2	Prepared		2005	-do-
72	Gawari Gundo	Doda	Gawari Nallah	2.5	Prepared		2005	-do-
73	Chewdara SHP	Budgam	Sukhnag N	0.126	Prepared		2005	-do-
74	Sai Ellahi Bakhsi (Loran)	Poonch	Chapper N	2.6	Prepared		2005	-do-

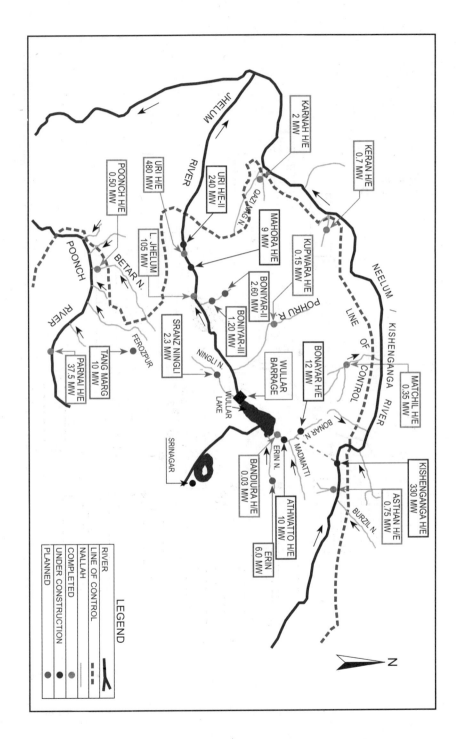

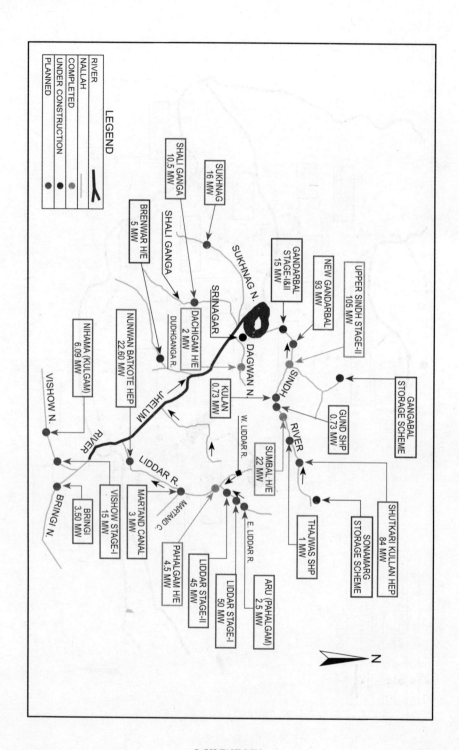

C) RIVER INDUS

a) Commissioned/Under Construction Projects

S. No.	Name of Scheme	District Location	River/ Nallah	Anticipated installed Capacity MWs	Position of PFR	Position of DPR	Year of Preparation of DPR/ PFR	Preparation of DPR/Updation
1	Kargil/Iqbal MHP	Kargil	Indus	3.75				Commissioned (State sector)
2	Sumoor, Hunder, Bazgo HEP	Kargil	Indus	0.8				Commissioned (State sector)
3	Stakna	Leh	Indus	3.24				Commissioned (State Sector). With PDD
4	Igo-Marcelong	Leh	Indus	3				Commissioned (State sector)
5	Marpachoo	Kargil	Indus	0.75				Commissioned (State sector)
6	Haftal MHP	Kargil	Indus	1				Commissioned (State sector)
7	Sanjak MHP	Kargil	Indus	1.26	Prepared	Prepared		Under Construction
8	Nimoo Bazgo HEP	Leh	Indus	45	Prepared	Prepared		Under Construction
9	Chutak HEP	Kargil	Indus	44	Prepared	Prepared		Under Consruction

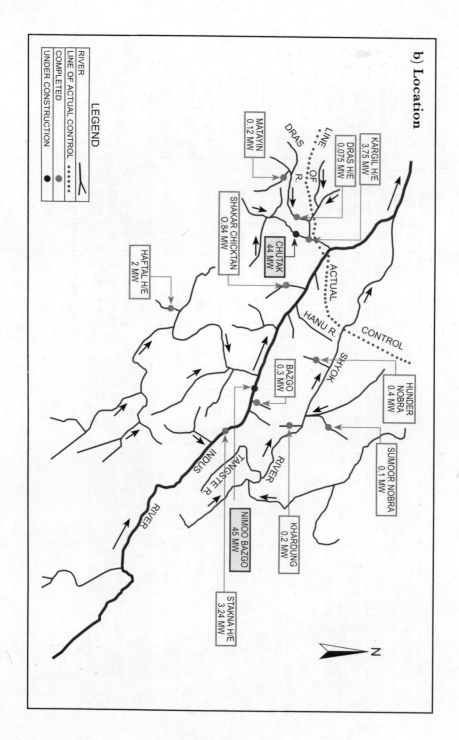

b) Location

LEGEND

RIVER	
LINE OF ACTUAL CONTROL	
COMPLETED	
UNDER CONSTRUCTION	

KARGIL H/E
3.75 MW

DRAS H/E
0.075 MW

MATAYIN
0.12 MW

DRAS R.

LINE OF

SHAKAR CHICKTAN
0.84 MW

CHUTAK
44 MW

ACTUAL

HAFTAL H/E
2 MW

HANU R.

CONTROL

BAZGO
0.3 MW

SHYOK

HUNDER
NOBRA
0.4 MW

SUMOOR NOBRA
0.1 MW

RIVER

INDUS

TANGSTE R.

KHARDUNG
0.2 MW

NIMOO BAZGO
45 MW

STAKNA H/E
3.24 MW

RIVER

N

c) Power Potential

S. No.	Name of Scheme	District Location	River/ Nallah	Anticipated installed Capacity MWs	Position of PFR	Position of DPR	Year of Preparation of DPR/ PFR	Preparation of DPR/ Updation
1	Karkit HEP	Leh	Indus	30	Prepared			PFR by WAPCOS
2	Khalsi HEP	Leh	Indus	60	Prepared			PFR by WAPCOS
3	Takmachik HEP	Leh	Indus	30	Prepared			PFR by WAPCOS
4	Kanyunchie HEP	Leh	Indus	45	Prepared			PFR by WAPCOS
5	Parkhachik – Panikhar HEP	Kargil	Indus	60	Prepared			PFR by CWC
6	Dah MHP	Drass Block, Leh	Indus	3	Prepared	Prepared	2003	DPR by AHEC
7	Hanu MHP	Drass Block, Leh	Indus	3	Prepared	Prepared	2003	DPR by AHEC; Tenders finalized under State sector
8	Tangste	Drass Block, Leh	Indus	3	Prepared	Prepared	2003	DPR by AHEC
9	Thusgam	Kargil	Indus	0.5	Prepared	Prepared	2003	DPR by AHEC
10	Chelloung	Kargil	Indus	0.4	Prepared	Prepared	2003	DPR by AHEC
11	Umbollong	Kargil	Indus	2	Prepared	Prepared	2003	DPR by AHEC
12	Bogdong	Leh	Indus	0.9	Prepared	Prepared	2003	DPR by AHEC
13	Sasome	Leh	Indus	0.75	Prepared	Prepared	2003	DPR by AHEC
14	Dumkhar HEP on Indus	Leh	Indus	45	Prepared			PFR by WAPCOS
15	Kumdok	Leh	Indus	0.5	Prepared			Revision of DPR by AHEC

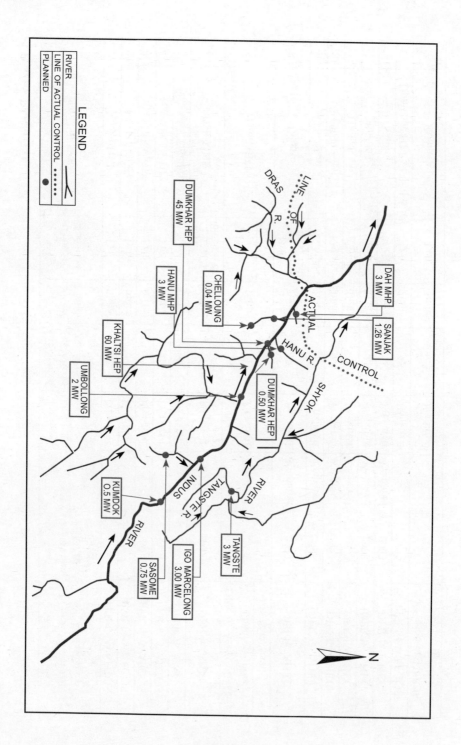

LEGEND

RIVER	
LINE OF ACTUAL CONTROL	
PLANNED	●●●●●

DUMKHAR HEP
45 MW

HANU MHP
3 MW

CHELLOUNG
0.04 MW

KHALTSI HEP
60 MW

UMBOLLONG
2 MW

DUMKHAR HEP
0.50 MW

KUMDOK
0.5 MW

SASOME
0.75 MW

IGO MARCELONG
3.00 MW

TANGSTE
3 MW

DAH MHP
3 MW

SANJAK
1.26 MW

DRAS R.

LINE OF

ACTUAL

HANU R.

CONTROL

SHYOK

INDUS

TANGSTE R.

RIVER

RIVER

N

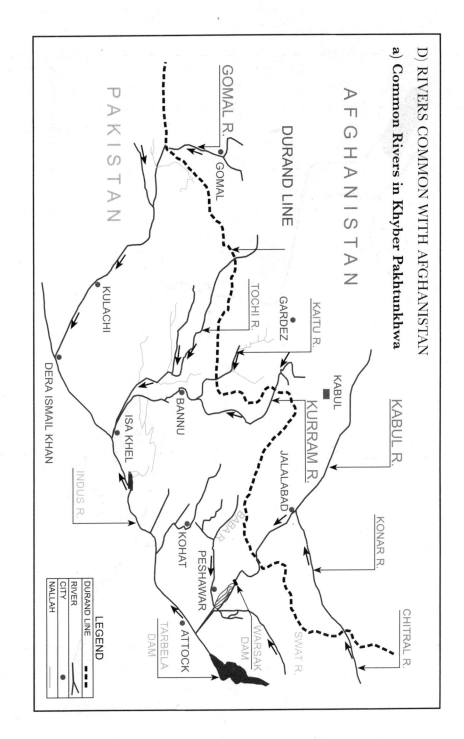

D) RIVERS COMMON WITH AFGHANISTAN

a) Common Rivers in Khyber Pakhtunkhwa

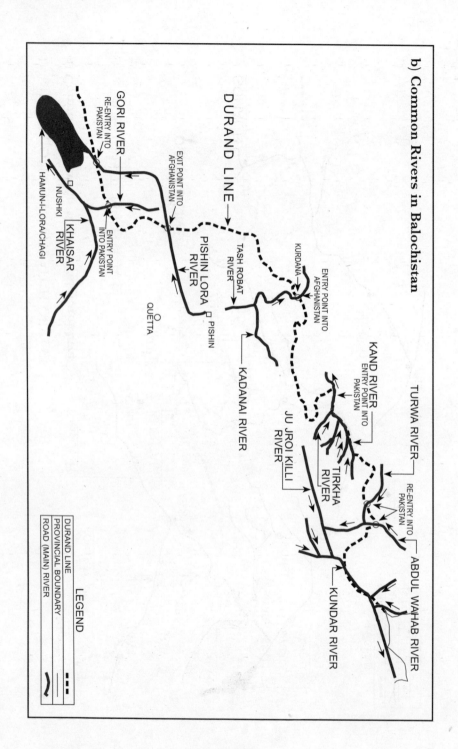

b) Common Rivers in Balochistan

c) Average Annual Inflows

S. No.	River	Inflow
NWFP		
1	Kabul	16.44
2	Khurram, Kaitu, Tochi	0.89
3	Gomal	0.79
BALOCHISTAN		
4	Abdul Wahab	0.0039
5	Kandar	0.0212
6	Kand	0.00293
7	Kadanai	0.0058
8	Pishin Lora	0.095
9	Kaisar	0.0175
	Total	18.266

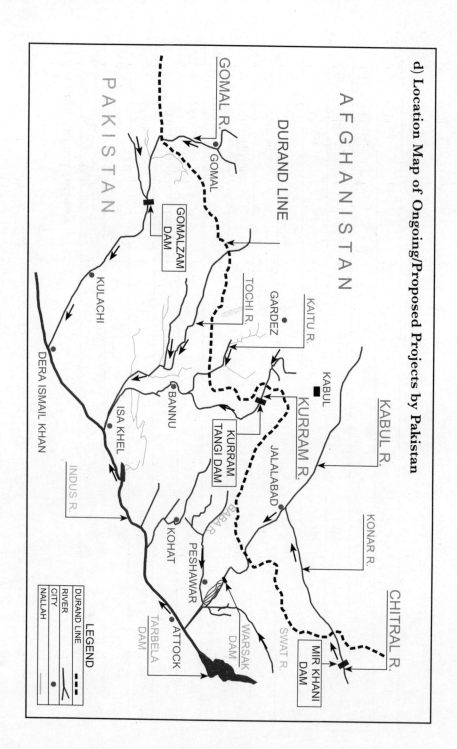

d) Location Map of Ongoing/Proposed Projects by Pakistan

e) Projects in Afghanistan

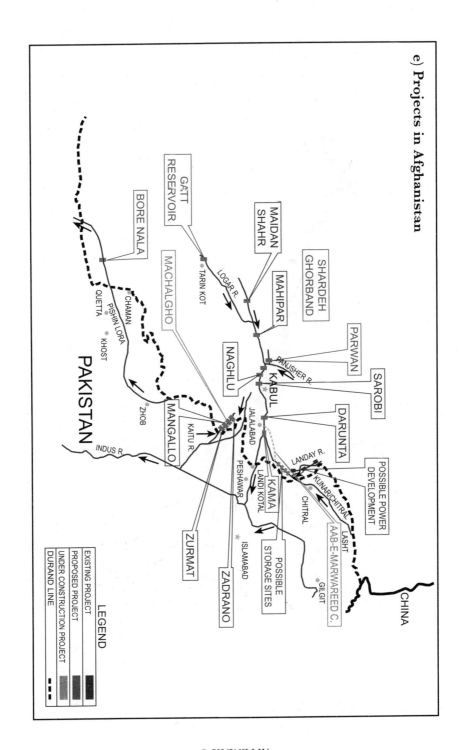

E) WULLAR BARRAGE AND STORAGE PROJECT/TULBUL NAVIGATION PROJECT

a) Location

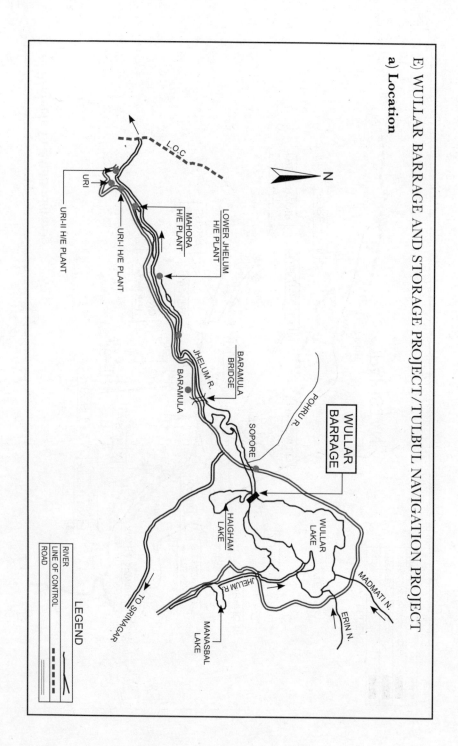

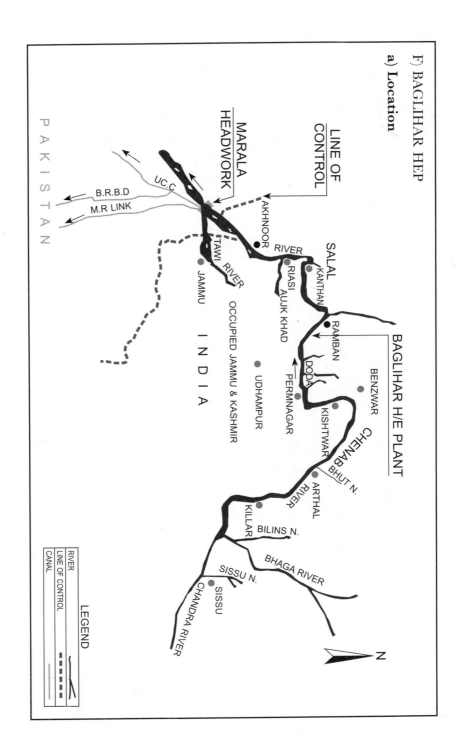

F) BAGLIHAR HEP

a) **Location**

LEGEND

RIVER
LINE OF CONTROL
CANAL

b) Salient Features

Location of Plant	Located on Chenab Main in District Doda of Occupied Jammu & Kashmir about 147 km upstream of Marala Headworks
Height and Length of Dam	143.0 M & 336.492 M long at top
Type	Concrete Graviy Dam
Gross Capacity Storage	321,048 Acre Feet
Storage Capacity upto Crest Level of Spillway	157,958 Acre Feet
Live Storage/Pondage	26,400 Acre Feet
Storage/Manipulatable Capacity	164,024 Acre Feet
Spillway gates	5 Nos. Sluice Spillway (10.0 m x 10.5 m) – El. 808 m 3 Nos. Chute Spillway (12.0 m x 19.0 m) – El. 821 m 1 No. Auxiliary Spillway (6.0 m x 3.0 m) – El. 837 m
Discharge through turbines	430.0 Cumecs
Installed Power Capacity	3 x 150 MW – 450 MW (Stage-I), 450 MW (Stage-II)

G) KISHENGANGA HYDROELECTRIC PLANT

a) Kishenganga Hydroelectric Plant on River Neelum

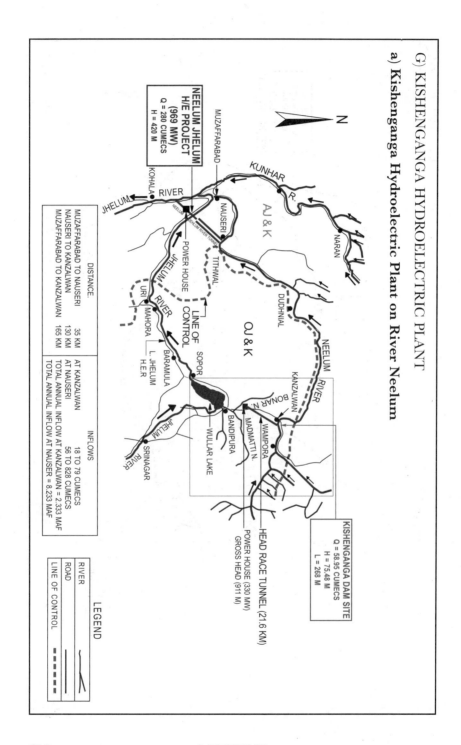

NEELUM JHELUM H/E PROJECT (969 MW)
Q = 280 CUMECS
H = 420 M

KISHENGANGA DAM SITE
Q = 58.95 CUMECS
H = 75.48 M
L = 268 M

HEAD RACE TUNNEL (21.6 KM)
POWER HOUSE (330 MW)
GROSS HEAD (911 M)

DISTANCE.

MUZAFFARABAD TO NAUSERI	35 KM	AT KANZALWAN
NAUSERI TO KANZALWAN	130 KM	AT NAUSERI
MUZAFFARABAD TO KANZALWAN	165 KM	

INFLOWS

AT KANZALWAN	18 TO 79 CUMECS
AT NAUSERI	56 TO 828 CUMECS
TOTAL ANNUAL INFLOW AT KANZALWAN = 2,333 MAF	
TOTAL ANNUAL INFLOW AT NAUSER = 8.233 MAF	

LEGEND
RIVER
ROAD
LINE OF CONTROL

b) Composite Map of Kishenganga, Wullar, LJHP, URI-I, and URI-II

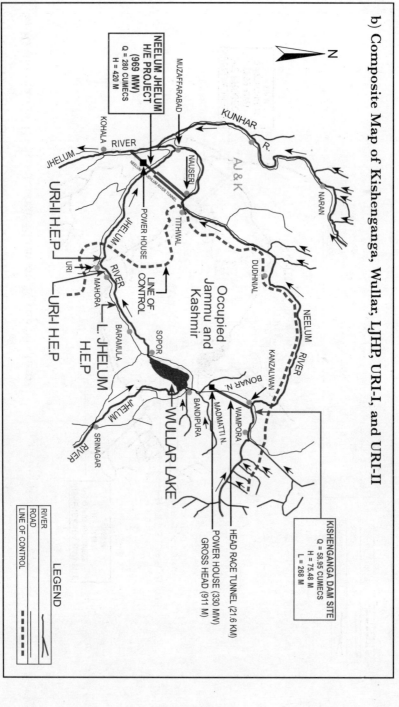

NEELUM JHELUM
H/E PROJECT
(969 MW)
Q = 280 CUMECS
H = 420 M

KISHENGANGA DAM SITE
Q = 58.95 CUMECS
H = 75.48 M
L = 268 M

HEAD RACE TUNNEL (21.6 KM)
POWER HOUSE (330 MW)
GROSS HEAD (911 M)

LEGEND
RIVER
ROAD
LINE OF CONTROL

Source: Pakistan Indus Waters Commission

Appendix 4

Western Rivers Inflow at Rim-Stations (MAF)

Years	Indus at Kalabagh			Jhelum at Mangla			Chenab at Marala			Total		
	Kharif	Rabi	Total	Kharif	Rabi	Total	Kharif	Rabi	Total	Kharif	Rabi	Total
22–23	81.65	16.04	97.68	20.39	05.37	25.76	19.44	04.55	23.99	121.47	25.96	147.43
23–24	95.09	14.95	110.04	18.14	04.79	22.93	17.23	03.81	21.04	130.46	23.55	154.01
24–25	69.62	13.08	82.70	22.71	03.75	26.45	17.23	03.30	20.52	109.55	20.12	129.67
25–26	66.36	11.40	77.76	17.03	03.71	20.74	17.12	03.11	20.22	100.50	18.21	118.72
26–27	61.15	11.71	72.85	19.03	03.36	22.39	18.98	03.08	22.06	99.16	18.14	117.30
2–28	57.07	12.66	69.74	16.18	04.51	20.69	17.17	03.24	20.41	90.42	20.42	110.84
28–29	68.37	12.75	81.12	21.80	05.42	27.22	18.05	03.92	21.96	108.22	22.09	130.31
29–30	62.45	14.22	76.67	15.83	07.74	23.58	18.92	04.98	23.90	97.21	26.95	124.15
30–31	74.09	12.32	86.42	21.16	04.42	25.58	21.88	02.99	24.86	117.13	19.73	136.86
31–32	63.27	14.79	78.06	20.83	04.47	25.30	17.00	03.04	20.04	101.10	22.30	123.40
32–33	71.08	10.97	82.05	17.83	08.51	21.34	18.71	03.16	21.87	107.62	17.64	125.26
33–34	80.25	11.60	91.84	22.52	03.66	26.18	22.91	03.51	26.42	125.68	18.77	144.45
34–35	74.46	11.61	86.07	14.36	03.66	18.02	19.37	03.40	22.77	108.19	18.67	126.87
35–36	74.61	12.82	87.43	20.23	05.57	25.80	21.97	03.89	25.86	116.81	22.28	139.09
36–37	82.49	13.20	95.70	19.90	04.41	24.31	22.52	03.30	25.82	124.92	20.91	145.83
37–38	75.54	11.99	87.53	16.25	04.78	21.03	18.31	04.57	22.88	110.10	21.34	131.44

Years	Indus at Kalabagh			Jhelum at Mangla			Chenab at Marala			Total		
	Kharif	Rabi	Total	Kharif	Rabi	Total	Kharif	Rabi	Total	Kharif	Rabi	Total
38–39	81.81	13.85	95.65	18.65	04.95	23.61	24.90	03.79	28.69	125.36	22.59	147.95
39–40	88.28	11.77	100.06	18.95	03.14	22.09	20.01	02.63	22.64	127.24	17.55	144.79
40–41	74.66	10.25	84.91	13.50	03.04	16.54	16.36	02.29	18.64	104.52	15.58	120.09
41–42	75.97	15.32	91.29	13.94	05.83	19.77	17.84	04.77	22.61	107.75	25.91	133.66
42–43	99.78	12.84	112.63	19.98	05.65	25.63	23.81	05.02	28.84	143.58	23.51	167.09
43–44	83.62	11.74	95.36	19.62	03.64	23.26	24.15	04.23	28.37	127.39	19.61	146.99
44–45	80.01	12.32	92.33	15.42	04.01	19.43	20.63	03.77	24.40	116.06	20.10	136.16
45–46	92.34	12.77	105.11	17.52	03.15	20.67	21.78	02.94	24.71	131.64	18.85	150.49
46–47	73.92	11.34	90.26	11.81	03.54			03.54	23.25	110.44	18.42	128.86
47–48	67.20	11.12	78.32	12.33	05.49	17.82	21.83	06.70	28.53	101.36	23.31	124.67
48–49	82.07	13.03	95.10	22.58	05.22	27.80	27.50	05.32	32.82	132.15	23.58	155.73
49–50	90.73	13.37	104.10	19.70	05.04	24.74	21.86	05.30	27.16	132.29	23.70	155.99
50–51	95.31	11.03	106.34	25.27	04.92	30.19	30.70	04.43	35.13	151.27	20.39	171.66
51–52	59.87	12.06	71.93	16.37	04.20	20.57	17.36	03.95	21.31	93.60	20.21	113.81
52–53	74.86	11.59	86.45	16.31	03.26	19.57	21.16	03.12	24.28	112.33	17.96	130.30
53–54	79.61	13.96	93.57	15.29	07.39	22.68	21.41	05.42	26.83	116.31	26.77	143.07
54–55	78.79	12.02	90.81	19.28	04.42	23.70	21.91	03.83	25.74	119.98	20.28	140.26
55–56	70.69	13.58	84.27	14.92	04.40	19.32	21.90	07.04	28.94	107.51	25.02	132.53

Years	Indus at Kalabagh			Jhelum at Mangla			Chenab at Marala			Total		
	Kharif	Rabi	Total	Kharif	Rabi	Total	Kharif	Rabi	Total	Kharif	Rabi	Total
56–57	85.10	13.69	98.79	19.28	05.74	25.02	27.54	06.03	33.57	131.92	25.46	157.38
57–58	71.04	14.83	85.37	25.07	07.67	32.74	26.89	05.60	32.49	123.00	28.09	151.09
58–59	81.92	17.55	99.47	19.25	08.15	27.40	23.30	08.39	31.69	124.47	34.09	158.55
59–60	99.84	20.25	120.09	25.60	06.05	31.65	29.30	05.75	35.05	154.74	32.05	186.79
60–61	90.99	13.52	104.51	13.00	03.26	16.26	20.98	03.96	24.94	124.97	20.75	145.72
61–62	80.79	13.06	93.85	14.24	03.55	17.79	24.55	04.32	28.87	119.58	20.92	140.51
62–63	60.23	11.10	71.33	11.91	04.26	16.17	17.82	04.49	22.31	89.96	19.85	109.81
63–64	76.41	12.95	89.36	17.63	04.38	22.01	19.36	04.33	23.69	113.40	21.66	135.06
64–65	75.15	13.58	88.73	19.30	04.30	23.60	21.66	04.44	26.10	116.11	22.32	138.43
65–66	76.83	12.91	89.74	22.37	04.23	26.60	18.69	03.95	22.64	117.89	21.08	138.97
66–67	77.54	13.93	91.47	17.58	05.52	23.10	21.52	04.38	25.90	116.64	23.83	140.47
67–68	81.92	15.06	96.98	18.41	05.49	23.90	20.10	05.21	25.31	120.43	25.75	146.19
68–69	78.83	14.46	93.29	16.40	05.24	21.64	20.40	03.51	23.91	115.63	23.20	138.84
69–70	74.47	13.03	87.50	20.11	04.10	24.21	19.91	02.64	22.55	114.49	19.77	134.26
70–71	61.36	10.16	71.52	12.33	03.02	15.35	16.58	02.72	19.30	90.27	15.90	106.17
71–72	62.38	09.36	71.74	10.19	03.36	13.55	15.83	03.02	18.85	88.40	15.74	104.14
72–73	67.00	12.58	79.58	17.81	07.15	24.96	16.81	04.72	21.53	101.62	24.45	126.07
73–74	95.27	11.42	105.69	22.13	04.31	26.44	27.57	03.39	30.96	144.97	19.11	164.09

Years	Indus at Kalabagh			Jhelum at Mangla			Chenab at Marala			Total		
	Kharif	Rabi	Total	Kharif	Rabi	Total	Kharif	Rabi	Total	Kharif	Rabi	Total
74–75	52.32	10.87	63.19	12.75	03.57	16.32	14.40	03.83	18.23	79.47	18.27	97.74
75–76	68.24	13.41	81.65	20.30	05.09	25.39	27.76	05.08	32.84	116.30	23.58	139.88
76–77	70.88	10.66	81.54	20.62	04.03	24.65	25.44	03.74	29.18	116.94	18.43	135.37
77–78	68.20	13.03	81.23	14.54	05.09	19.63	21.62	04.98	26.60	104.37	23.09	127.46
78–79	90.81	15.77	106.58	19.73	04.89	24.62	26.91	05.37	32.28	137.45	26.03	163.48
79–80	73.01	13.98	86.99	15.51	05.20	20.71	20.32	03.96	24.28	108.84	23.15	131.99
80–81	71.60	15.16	86.76	15.73	05.71	23.44	20.48	05.71	26.19	109.81	26.58	136.39
81–82	75.87	14.07	89.94	18.37	04.22	22.59	23.45	04.64	28.09	117.68	22.93	140.61
82–83	58.57	14.67	73.24	15.65	05.68	21.33	22.88	04.92	27.80	97.10	25.26	122.37
83–84	79.36	14.55	93.91	22.72	03.50	26.22	26.20	03.62	29.82	128.28	21.67	149.95
84–85	79.05	13.12	92.17	15.66	03.01	18.67	21.28	02.80	24.08	115.99	18.94	134.93
85–86	60.22	15.61	75.83	12.07	05.57	17.64	19.37	04.86	24.23	91.66	26.03	117.69
86–87	73.57	17.54	91.11	20.62	07.22	27.84	22.19	05.51	27.70	116.38	30.27	146.65
87–88	70.00	18.03	88.03	21.38	06.45	27.83	20.41	04.80	25.21	111.79	29.28	141.07
88–89	89.36	15.37	104.73	19.74	04.24	23.98	27.46	05.23	32.69	136.56	24.85	161.41
89–90	64.26	16.94	81.20	18.01	06.70	24.71	19.74	05.67	25.41	102.01	29.31	131.32
90–91	87.85	20.89	108.74	19.71	07.69	27.40	23.42	06.56	29.98	130.98	35.14	166.12
91–92	93.14	19.04	112.18	25.13	05.98	31.11	23.26	05.55	28.81	141.53	30.57	172.10

Years	Indus at Kalabagh			Jhelum at Mangla			Chenab at Marala			Total		
	Kharif	Rabi	Total	Kharif	Rabi	Total	Kharif	Rabi	Total	Kharif	Rabi	Total
92–93	90.84	19.06	109.90	25.18	06.82	32.00	22.60	05.18	27.78	138.62	31.07	169.69
93–94	66.46	15.33	81.79	18.69	04.01	22.70	19.53	03.45	22.98	104.69	22.78	127.46
94–95	92.65	16.47	109.12	20.82	05.67	26.49	24.55	05.65	30.20	138.03	27.78	165.81
95–96	81.49	17.42	98.91	21.91	00.26	28.08	26.40	05.47	31.87	129.80	29.06	158.86
96–97	85.08	15.23	100.31	24.93	04.11	29.04	27.48	04.41	31.89	137.49	23.75	161.24
97–98	71.45	18.48	89.93	16.96	07.06	24.02	21.74	06.55	28.29	110.15	32.10	142.25
98–99	83.71	16.18	99.89	18.11	03.61	21.72	23.16	04.78	27.94	124.98	24.57	149.55
99–00	77.51	14.57	92.08	11.24	03.19	14.43	18.70	04.35	23.05	107.45	22.11	129.56
00–01	58.86	11.55	70.41	10.27	02.28	12.75	16.00	02.73	18.90	86.33	16.56	102.89
01–02	55.65	10.72	66.37	08.23	03.66	11.89	17.20	02.90	19.93	79.88	17.28	97.16
02–03	64.64	12.49	77.13	12.31	05.10	17.41	18.02	05.47	23.49	94.97	23.06	118.03
03–04	76.61	12.79	89.40	17.67	05.00	22.67	21.50	04.36	25.86	115.78	22.15	137.93
04–05	55.43	17.32	72.75	11.74	06.72	18.46	14.90	06.41	21.31	82.07	30.45	112.52
05–06	82.37	14.45	96.82	17.71	05.46	23.17	21.12	04.02	25.14	121.20	23.93	145.13
06–07	74.02	17.73	91.75	16.43	06.78	23.21	21.38	06.33	27.71	111.83	30.84	142.67
07–08	75.40	12.39	87.79	13.51	04.18	17.69	16.98	03.62	20.60	105.89	20.19	126.08
08–09	65.89	13.51	79.40	13.38	05.88	19.26	16.21	03.61	19.82	95.48	23.00	118.48
09–10	68.18	13.22	81.40	16.48	04.57	21.05	14.46	03.39	17.85	99.12	21.18	120.30

Average Western Rivers Inflow at Rim Stations (MAF)

Years	Indus At Kalabagh			Jhelum At Mangla			Chenab At Marala			Total		
	Kharif	Rabi	Total	Kharif	Rabi	Total	Kharif	Rabi	Total	Kharif	Rabi	Total
(1922–47)	76.52	12.73	89.25	18.14	4.40	22.55 (Pre-Independence)	19.84	3.63	23.47	114.50	20.77	135.27
(1947–61)	80.57	13.69	94.26	18.87	5.37	24.25 (Pre-Treaty)	23.83	5.35	29.18	123.28	24.40	147.68
(1961–67)	74.49	12.92	87.41	17.17	4.37	21.54 (Pre-Mangla)	20.60	4.32	24.92	112.26	21.61	133.88
(1967–76)	71.31	12.26	83.57	16.71	4.59	21.31 (Pre-Tarbela)	19.93	3.79	23.72	107.95	20.64	128.60
(1976–2010)	74.47	15.22	89.69	17.43	5.16	22.59 (Post-Tarbela)	21.36	4.72	26.09	113.27	25.10	138.37
(1922–2010)	75.70	13.81	89.51	17.77	4.87	22.64 (Long Term)	21.12	4.39	25.51	114.60	23.06	137.66
MAX	99.84	20.89	120.09	25.60	8.15	32.74	30.70	8.39	35.13	154.74	35.14	186.79
MIN	52.32	9.36	63.19	8.23	2.28	11.89	14.40	2.29	17.85	79.47	15.58	97.16

Source: Pakistan Indus Waters Commission

Index

A

Abadan Oil Refinery Nationalization 170

Abasin 14

Abbas A. T., B. M. 395

Abbasia Canal 17

Abdullah, Farooq 369–70

Abdullah, Sheikh 67, 93, 95, 369

Abell, Sir George 48–50

Acheson, Dean 129, 165, 168, 173

Additional Protocol to Geneva Conventions (1977) 359

Afghan National Army 380, 406

Afghanistan 13–15, 19–20, 27, 38, 368, 378–81, 406, 426, 431

Agra Famine 25

Ahmad, Bashir 265, 278–9, 319, 326

Ahmad, Zubair 304

Ahmedpur Sial 17–18

Akram, S. M. 87

Alam, Undala Z. 7, 170, 172

Alberuni 16–19

Ali, Chaudhri Muhammad 48–9, 57, 60, 63, 65, 66, 70, 76, 97, 100, 141, 161

Aman ki Asha 415

American Constitution 130

Amritsar 19, 23, 27, 32, 41–2, 48, 76

Anderson Committee 116–17

Angles, Sir Claude 35

Annan, Kofi 1

Annandale, George 261, 264

Arabian Sea 20, 102

Arora, Rakesh Kumar 8

Asia 27, 79, 128, 130, 165, 346–7, 420, 425

Asian Development Bank (ADB) 206, 436

Asif, Khawaja 320, 332

Australia 24, 157, 242, 319, 328, 367

Awami League 394

Ayyangar, Gopalaswamy 72

Azad Kashmir 5, 11, 15–17, 20, 57–60, 67–8, 77, 93–5, 100–3, 105, 107, 113, 118, 123, 150–1, 153–5, 166, 169–70, 173–4, 184, 194, 202–7, 213–14, 221, 223, 228, 231–3, 273, 278, 281–3, 289, 294, 311, 325, 350, 355, 358–61, 363–4, 369–73, 376, 399–400, 402–4, 411, 414, 418–19, 421, 426–8, 431–5

Aziz, Shaukat 239, 319, 331

B

Baglihar Case 10, 82, 200, 210, 265, 270, 275, 284, 288, 293, 307–10, 315, 318, 322, 326, 399: reasons for loss of 257–70

Bahawalpur 19–21, 23, 31, 36, 42, 62, 70, 172

Bajaj, A. K. 304

Balloki 17–18: Balloki-Suleimanke Link Canals 18; Barrage 18

Balochistan 16, 376, 379, 381, 406–7, 433

Bambanwala-Ravi-Bedian-Dipalpur Link Canal (BRBD) 18

Bangladesh 186, 233, 346, 348, 367, 381, 388, 390–8, 424, 432: Agreement on Ganges River (1977) 392–5, 409; Bangladesh-India 186, 367, 381, 388, 390, 392, 394–8, 424; Farakka

Barrage Agreement (1996) 388–96; Farakka Barrage Partial Accord (1975) 391; Joint Declaration (1974) 390; Memorandum of Understanding (1982) 393–4; Memorandum of Understanding (1985) 394–5; National Party (BNP)

Baramulla 221, 223, 281

Bara River 20

Barcelona Convention and Statute on Navigable Waterways (1921) 389–90

Bari: Central Bari Doab Canal (CBDC) 62, 70, 74, 76, 208; Doab 18, 20–1, 23, 25, 30, 42–3, 62, 208, 223, 431; Lower Bari Doab Canal 18, 30, 223; Upper Bari Doab Canal (UBDC) 25

Bass, Neil 172–3

Batala 39, 57, 88

Beas River 8, 14–15, 17–19, 28, 40, 76–7, 97, 110, 116, 118, 145, 372–3, 376, 411, 431

Beaumont, Christopher 6, 48–9, 52–3, 56, 91: on change of Punjab Map 6–7, 48–9, 52–3, 56–7, 91; on Jinnah 53; on Mountbatten 53; on Nehru 53, on Radcliffe 6–7, 53, 57, 91

Beg, Mirza Aslam 406

Begari Canal 15

Bengal 27, 32, 46, 57, 67, 88, 94, 176, 390, 396–7

Bengal Boundary Commission 40, 57, 87

Benton, John 30

Bhagga River 17

Bhagirathi-Hooghly River 388

Bhakra 19, 372: Bhakra-Beas-Rajasthan Project 114; Canal 114, 132; Dam 19, 32–4, 75–6, 131

Bhawal Canal 19

Bhutto, Benazir 419, 421

Bihar 32, 106

Bikaner 31, 36, 42, 51–3

Black, Eugene 3, 9, 81, 98, 102–3, 105, 108, 110–13, 120, 122, 133–4, 138, 141, 143–4, 146, 149, 152, 155, 157, 159–63, 168–9, 172, 179–80, 184, 213, 430

Blackmore, David 319

Bogra, Muhammad Ali 122

Bombay Presidency 30–1, 35

Bonar Nallah 281–4, 290, 296

Brahmaputra River 19, 345, 347–8, 388, 391, 394, 396

Briscoe, John 12, 307, 377, 399, 410, 412–15, 417–18, 431–4, 436

British 7, 9, 22, 24–9, 32, 36, 38, 40–1, 44–5, 47–9, 54, 57–61, 71, 88–92, 96, 131–2, 143, 170, 182, 233, 387, 401: Commonwealth 57, 71, 96; Empire 25, 27; fair play 90; House of Commons 41; Raj 58–9

Bursur Dam 377

C

Calcutta 57, 88, 92, 94, 389, 408

Calcutta High Court 34

Canada 157, 177, 193, 314, 331, 367

carbon credits 4, 312–13, 315, 319

Central Treaty Organization (CENTO) 167–8

Chaj Doab 20–1

Chandra River 17, 145

Chandrabhagga River 17

Chashma: Barrage 15; Jhelum Link Canal 15; Right Bank Link Canal15

Chattha, Saifullah 331

Chelmsford, Lord 33

Chenab: Canal 17, 28, 30, 223; Formula 400; Lower Chenab Canal 17, 28; River 4, 17–20, 76, 123–4, 132, 145, 215, 217, 221, 230, 270, 272, 317, 373; Upper Chenab Canal 17, 30, 223

China 14, 84, 164–5, 368, 403, 425

China-Pakistan Economic Corridor (CPEC) 426
Chitral River 20, 378
Chittagong Hill Tracts 50, 88
Cholistan Desert 21
Churchill, Winston 75, 253
Chutak Dam 311–14, 319, 322
climate change 1, 333–6, 340, 344–5, 399: Climate Depot Special Report 337; deniers of 337–9; human induced 334–40, 346; and Indus Rivers 347–51; Non-Governmental International Panel on Climate Change (NIPCC) 336–9; and South Asia 344–7
Cold War 3, 164, 166, 328, 416
Colorado River 128–9, 131
Commonwealth Relations Office 143, 152
Composite Dialogue 4, 191, 200, 224, 236, 413–14, 419–20, 435
Comprehensive Bilateral Dialogue (see Composite Dialogue) 420, 423
Confidence-Building Measures (CBMs) 236, 423
Congress Party 39–41, 44–7, 54–5, 57, 60–1, 87, 95, 182, 395
Cotton Committee 33
Court of Arbitration 187, 195–7, 199–201, 308–9, 314, 320, 323, 331, 434
Covington and Burling 102, 169
Crawford, James 256, 260–1, 329
cross-border terrorism 11, 359, 361, 363
Curzon, Lord 29

D
Dadu Canal 16
Dal Lake 16
Dalia Barrage 397
Damodar Valley: Corporation 168; Project 106
Danube 17, 329

De Chazournes, Madame Laurence Boisson 268–70
Delhi 2, 23–4, 28, 30–2, 48–9, 51, 67–8, 74, 76, 81, 91, 116, 144, 158, 203, 213, 216, 218, 228, 236, 239, 267, 272–3, 282, 312, 353, 356–9, 369, 414–15, 425, 427, 432
Delhi Agreement (see Joint Statement) 2, 71, 73–4, 77, 82, 175
Dera Ghazi Khan Canal 15
Desert Canal 16
Diamer-Basha Dam 206, 207
Dipalpur Canal Systems 42–3, 62, 70, 74
disputes: justiciable or judicial 84–6, 275; non-justiciable or political 84–6, 275
Dixon, Sir Owen 92–3
Dublin Principles 416
Dul Hasti Dam 231, 316, 377
Dulles, John Foster 184–5
Durrani, Shakeel 206
Dyas, Captain J. H. 26–7

E
East Bengal 67, 94, 176, 390, 396
East Pakistan 176, 388–90, 424
East Punjab 9, 40, 60, 62–9, 71–2, 74–5, 81, 94–5, 97
Eastern Rivers 2–3, 8, 10, 45, 71–3, 75, 81, 95, 106–7, 115–16, 118–19, 121–3, 125, 131, 138, 140–1, 147, 149, 151–2, 156, 158–9, 161–3, 166, 168, 178, 183, 188, 190, 209, 215–16, 227, 246, 292, 347, 350, 366, 368, 372–4, 376, 389, 395, 400, 416, 434
Eastern Sadiqia Canal 19
Egypt 3, 114, 128, 129
Ejaz, Mansur 411
Environmental Impact Assessment (EIA) 5, 189, 287, 293, 312–13, 315, 319
Espoo Convention (see UNECE Convention on Environmental Impact Assessment) 189

exaequo et bono (*see* World Court) 201, 267, 280

existing uses of water 76, 109–11, 115–17, 119, 125, 132, 138, 151, 167, 176, 286, 290–1, 405

Exxon-Mobil 339, 352

F

Faisalabad 22, 42

Farakka Barrage Agreement (*see* Bangladesh) 388–96, 408

Ferozpur 2, 17, 39–40, 43–4, 48–53, 56, 88, 90, 92, 145: headworks 2, 9, 40, 42–5, 51, 53, 89, 227

Fisher, Roger 169

Fordwah Canal 19

Foster, Sir John 204–5

Fox, Liam 387

Fuleli Canal 16, 23

G

Gandak Agreement (*see* Nepal-India) 384–5

Gandaki River 384–5

Gandhi, Indira 391

Gandhi, Mahatma 45, 54, 89

Gandhi, Rajiv 421

Gandhi, Sonia 316

Ganga/Ganges: Ganges-Hooghly Rivers 176; Barrage 390; River 19, 75, 186, 345–8, 388–96, 408, 432

Garg, S. K. 371, 374, 404

Garner, Robert 152, 173–6

Gazaldoba Dam 397

General Agreement on Tariffs and Trade (GATT) 389

Germany 130, 157, 323, 367

Ghali, Boutros 1, 11

Ghazi: Ghazi-Barotha Power Project 15; Barrage 15

Ghulam Muhammad Barrage 16

Gilani, Yousaf Raza 273

Gilgit Baltistan 15

global warming 334, 337, 339

Gokhale, B. K. 97

Government of India Act: (1919) 31; (1935) 31, 34

Grand Anicut of Madras 23

Great Britain 24, 46, 48, 71, 90, 114, 128, 130, 367

Guddu Barrage 15

Guha, Ashok 182

Gujral, Inder Kumar 388, 395, 435

Gulhati, Niranjan D. 7, 63, 95, 97, 111, 121, 124, 143, 146–8, 171–2, 178, 181–2, 203

Gurdaspur 39, 41–3, 48, 50, 56–7, 59–60, 76, 88–9, 94, 360

Gurmani, Mushtaq Ahmed 173

Gypsa Dam 377

H

Hagel, Chuck 381

Hamid, S. A. 61

Hardinge Bridge 390

Harmon Doctrine 2, 129, 162, 177

Hasan, Ahmad 172

Hasan, Dr Mubashir 411

Haveli Canal 17

Hayat, Shaukat 64, 67–9, 94

Heads of Agreement 149–51, 154, 156, 175, 180, 286

Helsinki Rules on International Rivers 371, 374, 398, 404

Henderson, Arthur 40–1

Himachal Pradesh 17–18, 152

Himalayan glaciers 345–7, 352–3

Himalayas 5, 14, 17, 20, 344–7

Hindu Kush 344, 347

Huber, Max 127

Humanitarian Law 362, 401

Hussain, Sheikh Showkat 403

Hyderabad 20, 22, 32

I

Ibrahim, Pir Muhammad 97, 172

Idso, Craig D. 339

Ihlen Declaration 66

Iliff, William 81, 96, 125, 134, 140–3, 146, 149, 153, 156–7, 161, 172, 180–1, 183, 203, 210, 213

Immerzeel, Dr Walter 346, 348, 353

India 2–5, 8–11, 13–15, 17–20, 22, 24, 26–31, 33–4, 36, 39–46, 48–73, 75–8, 80–9, 92–3, 96–7, 100–8, 110–24, 129, 132–3, 137–41, 143–5, 147–68, 170, 172, 175, 178–82, 184, 186–94, 196–9, 202–43, 245–56, 258, 260, 264, 268–74, 276, 278–9, 281–307, 310–28, 330, 334, 346–8, 350–1, 354–61, 364, 366–96, 398–404, 406–7, 410–37: allegations of stealing Pakistan's water against 270–80; Reserve Bank 72

Indian: Central Water and Power Commission (CWPC) 110, 286, 289–90, 295; Constitution 369–70; Declaration on ICJ Jurisdiction (*see* World Court) 69, 144, 150, 201, 308, 310; Institute for Defence Studies and Analysis (IDSA) 359, 400; Irrigation Commission 29, 31; manifest destiny 424; National Commission for Integrated Water Resource Development (NCIWRD) 375; National Security Council Advisory Board 404; Policy Advisory Group 404; strategic worldview 423; Supreme Court 398; Water Resources Minister 240

Indira Gandhi Canal 376

Indus Basin Development Fund 158, 180, 182–3, 209

Indus: Delta 20–2; Indus I 356; Indus II 350, 356, 400; River 14–15, 17, 20, 33, 106, 113, 145, 149, 182, 184, 205, 251, 311–12, 368, 378, 403; Rivers 9, 13–14, 19–21, 27, 115–17, 120, 123, 131, 137, 139, 145, 188–9, 202–3, 206–8, 227, 231–2, 247, 253, 287, 289, 333–4, 347–8, 350, 417–18, 437

Indus Waters Treaty 3, 43, 125, 157–8, 170, 183, 187, 234, 246, 250, 267, 307, 358–9, 366–7, 369–70, 373, 378, 395, 399–401, 414–15, 427: Annexure D 217, 225; Annexure E 186, 211, 225–6, 286, 289, 299; John Briscoe on 12, 307, 377, 399, 410, 412–15, 417–18, 431–4, 436; Conflict Resolution Procedure and 99–202; emergency provisions of 187, Eugene Black on 3, 9, 81, 98, 102–3, 105, 108, 110–13, 120, 122, 133–4, 138, 141, 143–4, 146, 149, 152, 155, 157, 159–63, 168–9, 172, 179–80, 184, 213, 430; financial provisions of 187, 212; grounds for termination of 358–70; India's motive for revisiting 375–80; miscellaneous provisions of 207–10; and non-consumptive use 189, 222, 226, 249–50, 274; and Permanent Indus Commission 157, 193–4, 302, 304; and pollution 189–90, 399; ratification 68, 129, 208–9, 215, 370, 386–8; Raymond Lafitte on 210, 242, 265–7, 269, 319, 331; and run-of-river plant 4, 190, 196, 220, 242, 249, 284–5, 288–90, 294–5, 298–9, 305–6, 315; and safeguards for Pakistan and India 188–93; and storage works on the Western Rivers 180, 191, 225–6, 289–91, 298; territorial scope of 202–7; and Track II diplomacy 209, 354–5, 401; and transitional arrangements 118, 187, 190; uniqueness of 186–8

Integrated Water Resources Management (IWRM) 416–17, 419

Inter-Dominion Agreement (*see* Joint Statement) 2, 9, 71, 158

Intergovernmental Panel on Climate Change (IPCC) 334–40, 344–7, 350–1, 353, 437: and Synthesis Report (2007) 335, 337, 344; and Third Assessment Report (TAR) (2001) 335; and Working Group II 344, 347, 352, 437

International Centre for Integrated Mountain Development (ICIMOD) 347, 353

International Centre for Settlement of Investment Disputes (ICSID) 240–1, 275

International Commission on Large Dams (ICOLD) 250–1, 258, 265, 277, 313, 329

International Court of Justice (ICJ) (*see* World Court) 47, 68–9, 84, 95, 201, 270, 323

International Criminal Court (ICC) 360

International Environmental Law 286, 292

International Law: environmental customary 292; and non-state actors 361–3, 411; and oral declaration 66; and the principle of:
- Absolute Territorial Integrity 177
- effective control 363
- equitable apportionment 119, 126, 128, 130, 137, 176
- equitable utilization 373–4
- *jus cogens* 190, 313
- no appreciable harm 373
- optimal utilization 11, 394, 417
- *rebus sic stantibus* 368
- self-defence 360–1, 364
- *ut res magisvaleat quam pereat* 254

International Law Association 374

International Law Commission 66, 79, 245–6, 278, 361, 365

International River Law 125

International Treaty for the Renunciation of War (*see* Pact of Paris) 78

inundation canals 23–4, 29, 30, 35, 209

Islamabad Declaration 224–6, 360, 363, 420

Ismay, Lord 49, 52–3

Ispahani, Abul Hasan 167

Iyer, Ramaswamy 186, 192, 271, 307–8, 327, 356, 367, 370, 372, 374, 399, 401, 403–4, 413, 419, 434

J

Jammu and Kashmir 11, 16, 58–9, 68, 77, 92, 107, 123, 150–1, 153–5, 173–4, 194, 202–7, 213–15, 221, 223, 230, 232–3, 273, 278, 281, 294, 311, 325, 358–61, 363–4, 369–73, 400, 402, 404, 414, 418, 421, 426, 428, 431, 433, 435: Chief Minister 369–70; Legislative Assembly 369; Panthers Party 369

Jenkins, Sir Evan 48–50

Jhelum: Canal 17, 30, 223; Hydroelectric Plant 224; Lower Jhelum Canal 17; River 4, 14–17, 20, 116, 118, 123, 145, 148–9, 151, 174, 186, 206, 211, 221, 223, 225–6, 228–9, 281–3, 286–7, 289, 291, 294–5, 307, 325, 348–9, 370, 372–3, 377, 419; Upper Jhelum Canal 30, 223

Jinnah Barrage 15

Jinnah, Fatima 158

Jinnah, Muhammad Ali 9, 44–9, 53–5, 59, 87, 89, 92, 158: and the question of duress 9, 78, 82; Mountbatten on 45–6, 53; Beaumont on 48, 53, 91; Joint Statement (May 1948) 9, 71–87, 89, 97–8, 103, 106–7, 109–

13, 123, 131, 158, 366; Pakistan's denunciation of 82; validity of 71–87
Jullundur 49, 69, 88
Jumna: Canal 23–5; River 23, 75; Western Jumna Canal 23–5

K

Kabul 379–80, 420: River 15, 19–20, 368, 378, 404
Kailash Mountain 18–19
Kalpani River 20
Karachi 32, 76, 104, 115–16, 122, 151, 157, 175, 433
Karakoram 14–15, 346–7, 422
Kargil incident 224, 312, 400, 419
Kashmiris' right of self-determination 105
Kaufmann, K. 127
Kayani, Ashfaq Pervaiz 355, 400, 406, 411
Kellogg Briand Pact (see Pact of Paris) 78
Khan, Ayub 8, 10, 125, 146–7, 149, 152, 157–9, 161–2, 169, 179–80, 183, 307, 320: and charge of sale of Eastern Rivers 158–9, 161–2, 183; and Eugene Black 146, 149, 152, 159, 179–80; appraisal of 158–62; responsibility for accepting Indus Waters Treaty by 147, 159, 161
Khan, Liaquat Ali 47–9, 100, 169
Khan, Masood 237, 239
Khan, Palwasha 229, 412
Khan, Sir Zafrulla 57, 72–3, 81, 89–93, 100, 132–4, 174, 328
kharif 140, 227
Khosla, A. N. 75–6, 90, 97, 105, 110, 112, 121, 124, 171–3, 175
Ki-moon, Ban 1, 430
Kirpalani, Acharya 45
Kirthar Mountains 20–1
Kishenganga: case 155, 181, 187, 190, 196–7, 200–1, 203, 207, 257, 266,

269, 277, 282, 306–9, 320, 323, 327, 329–31; Dam 4, 10, 191, 281, 283, 322; Hydroelectric Project (KHEP) 228; Kishenganga/Neelum River 4, 16, 281–5, 287–91, 293–8, 300–2, 304–6
Kohala: Bridge 16; Hydroelectric Power Project 291–2
Korbel, Joseph 92
Kosi Agreement (see Nepal-India) 382–5
Kotri: Barrage (see Ghulam Muhammad Barrage) 16, 106, 184; Beghar Feeder 16
Kunhar River 16, 20
Kux, Dennis 125
Kyoto Protocol 312, 315

L

Ladakh 15, 311, 403, 422
Lafitte Raymond (see Indus Waters Treaty) 210, 242, 265–7, 269, 319, 331
Lahore 18, 22–3, 32, 42, 60, 64, 70, 72, 82, 102, 143, 160, 178, 183, 227, 279, 326, 329, 401, 419–20, 434
Lahore Declaration 358, 419
Lall, Arthur 124
Lamb, Alastair 58–9
Lashkar-e-Taiba (LeT) 11, 358–9, 363–4, 411, 419
Laylin, John 102, 104, 110, 167, 169, 173
League of Nations Covenant 78, 309, 328
Lilienthal, David E. 3, 9, 11, 80, 100–6, 108–14, 120, 122, 124, 142, 161, 163–6, 168–70, 173, 175, 366, 377, 417–18, 427: proposal on Indus Basin Settlement 100–3
Line of Control (LoC) 221, 224, 282, 298, 364, 420–2
Lipmann, Walter 100

Listowel, Lord 59, 87
Lloyd Barrage (*see* Sukkur Barrage) 16
Lok Sabha 182–3, 240
London 92, 145, 151–3, 196, 260, 358, 400
London Plan 145

M
Madhopur Headworks 2, 9, 39, 40–5, 56, 59, 72
Mahakali Treaty (*see* under Nepal-India) 186, 382, 386–8
Mahbub, S. I. 64, 94
Mailsi Canal 18–19
Majidulla, Kamal 314, 326, 329–30
Malé Declaration 435
Malik, Bashir 7
Malik, Ramiz A. 8, 42
Mandi Hydroelectric Scheme 227
Mangla Dam 8, 16, 148–9, 153, 190, 203, 213, 215–16, 226
Marhu Tunnel Proposal 146–7
Mathur, H. C. 182
McKinsey Report 375
Meghna River 388
Mehta, Ashok 182
Mehta, Jagat 43, 114–15, 167
Mehmood, Ashfaq 238, 259–64, 280, 326
Mekong River Basin 319
Menon, V. P. 52–3
Merala 17, 145, 186, 231, 270–1, 317, 348: Ravi Link Canal 17
Michel, Arthur Alloys 6–7, 24–5, 32, 36, 42–3, 115, 117, 120, 175, 184, 273, 373
Mirza, Asif Baig 305, 316
Mithankot 17, 20, 22
Miyar Dam 193, 316
Modi, Narendra 420, 422
Mosley, Leonard 44
Mountbatten, Edwina 92

Mountbatten, Lord 45–8, 50–3, 56–7, 59–60, 70, 87, 89–92, 94, 170, 233: Andrew Roberts on 87, 91; on Muhammad Ali Jinnah 45–6, 53; on Pakistan 46, 48, 51, 53, 89; on Radcliffe Award 45, 47–8, 50–3, 56–7, 59–60, 70, 87, 89, 91–2, 94; Philip Zeigler on 48, 50, 52
Mudie, Sir Francis 48–50
Mueenuddin 146, 151, 154–5, 157, 204
Muhammad, Din 40, 87
Muhammad, Ghulam 68, 70, 81
Mujib-ur-Rahman, Sheikh 390–2
Multan 22, 25
Mumbai terrorist attack 224, 360, 420–1, 431
Munir, Muhammad 39–40, 50, 60, 87
Murray Darling Basin 319
Musharraf, Pervez 235, 256, 278, 304, 315, 363–4, 400, 420–1
Muslim League 39–41, 44–8, 55–8, 60–1, 67–8, 87, 95
Muzaffarabad 16, 221, 281
Muzzafargarh Canal 15

N
Nangal Dam 19, 184
Naqvi, Saiyid Ali 8
Nariman, F. S. 82, 260–3
Nawab of Bhopal 60, 94
Nazimuddin, Khawaja 110
Neelum: Neelum-Jhelum Hydroelectric Project (NJHEP) 284, 325; Valley 290, 301–2, 306, 325
Nehru, Jawaharlal 44, 46–8, 52–4, 57, 59, 67, 69, 73–5, 77, 81, 83–7, 90, 92–3, 95, 98, 100, 103–7, 109–10, 121, 123, 125, 144, 147, 153, 157–8, 165–6, 168, 170–1, 179, 183, 210, 307, 404; comparison with Mahatma Gandhi 54; Joseph Korbel on 92; Kennedy on 93; Klaestad on 92;

Sheikh Abdullah on 95; Sir Owen Dixon on 92

Nehru, B. K. 105, 170

Nepal 186, 353, 367, 381–8, 392, 398, 407; United Marxist and Leninist Party (UML) 387

Nepal-India 386: Gandak Agreement 384–5; Kosi Agreement 382–4; Mahakali Treaty 386–8; Sarada Treaty 387; Tanakpur Agreement 385–6

Neutral Expert 4, 10, 157, 187, 195–200, 210, 220, 224, 232, 234–43, 245, 248, 250–65, 267–70, 272, 275–6, 278–80, 288, 293, 306–10, 315, 317–18, 323, 327, 399

Nile: Waters Agreement (1929) 114, 128; River 15, 114, 128, 176, 429

Nimoo Bazgo Dam 4, 10, 311–17, 319–20, 322, 330

Nizami, Majid 412

Noel-Baker, Philip 91

Non-Aligned Movement (NAM) 107, 166

non-state actors (*see* International Law) 361–3, 411

Noon, Firoz Khan 67, 159, 161, 174

Northwest Canal 16

O

O'Brien, John Lord 169–70

P

Pact of Paris 78

Pakistan 2–11, 13–20, 23, 27, 36, 39–40, 42–6, 48–65, 67–71, 73–84, 86, 88–9, 91–2, 94, 96, 100–25, 129, 131–4, 137–76, 179–82, 184–94, 196–8, 202–10, 213–29, 231–43, 245–73, 276–81, 283–94, 296–302, 305–7, 309–23, 325–8, 330–2, 346, 348, 350–1, 354–6, 358–61, 363–4, 366–72, 374, 376–81, 388–90, 395–6,

398–403, 406–7, 410–35, 437: and strategic depth 379; Attorney General 65, 236–7, 239, 318; Environmental Protection Agency (EPA) 313, 319; Executive Committee of the National Economic Council (ECNEC) 326; foreign office 237, 239, 315; Indus Commissioner 193, 209, 239, 321; Minister for Water and Power 255, 320; National Assembly 229, 315, 412; and Northern Areas 15, 206; Senate Committee on Water and Power 206; Senate Standing Committee on Cabinet Secretariat 330; and strategic depth 379, 406; Strategic Plan Division 412; Trans-Border Water Organization (PTWO) 330; Water and Power Development Authority (WAPDA) 7, 348

Pakpattan Canal 19

Pancheshwar Reservoir 386–7

Panjnad: Barrage 17; Canal 17; River 15, 17, 19–20, 23

Pannikar, Sardar 51–2

Partition: Council 47; Plan 60

Pat Feeder 16

Pathankot 39, 360, 420

Patterson, Anne W. 272

Pax Indica 426

peace process 4, 224, 235, 237, 406, 410, 419–23, 425, 428

Pearson 184

perennial: canals 23–4, 29–30, 32–3; irrigation 23–4, 34

Permanent Indus Commission 157, 193–4, 302, 304

Pinyari Canal 16

Poonch River 16, 221

princely states 59, 68, 95

Punjab 2, 6, 8–9, 15, 17–21, 23–37, 39–40, 42, 44, 46, 48–54, 56, 60–75, 77, 81–3, 87–8, 91–2, 94–5, 97,

100–1, 123, 131, 152, 174, 182, 223, 248, 358, 372, 376–7, 389, 403, 411, 427, 434: Boundary Commission 2, 9, 36, 39–41, 46–8, 50, 52, 55–6, 60–1, 87, 89, 94; breadbasket of northern India 28; Partition Committee 61–4, 69, 74, 82

Q

Qadirabad Balloki Link Canal 17
Qaimpur Canal 19

R

rabi 62, 140, 271
Radcliffe, Cyril 2, 6–7, 36, 39–45, 47–57, 60, 87–92, 267: Award 2, 6, 9, 36, 39, 41–5, 48, 50–7, 61, 63, 67, 88, 91–2; motives behind the Award 45–61
Rajasthan 123, 182, 372, 376, 403, 411: Canal Projects 114, 372, 376
Rangpur 396: Canal 17
Ranjit-Sagar Dam 19
Rao, K. L. 382
Raphael, Robin 387
Rasul: Barrage 17; Rasul Qadirabad Link Canal 17
Ratlé Dam 316–17
Rau, B. N. 34–5, 116, 130–1
Rau Commission 34–5, 116, 130–1
Ravi River 14–15, 17–20, 22–3, 25, 28, 39, 43, 76–7, 97, 110, 116, 118, 145, 188, 206, 372–3, 411
realpolitik 160, 413, 432
Rechna Doab 20–1
Responsibility of States for Internationally Wrongful Acts 359, 361
Rice, Condoleezza 436
Rice Canal 16
Roberts, Andrew 49, 87, 91
Rohtang Pass 18
Rohtas Dam 149, 179–80

Rose, Leo 424, 435
Roy, Dr 419

S

Saeed, Hafiz 411
Sain, Kanwar 51
Salal Dam 4, 10, 17, 200, 215–21, 231, 239, 271
Salman, Salman M. A. 258, 269, 323, 389, 393–4, 400
Sarada River 386–7
Sawalkot Dam 377
Sayeed, Mufti 371
Scott-Moncrieff, Sir Colin C. 29
Senge Ali Dam 368
Serageldin, Ismail 1
Seul River 18
Shah, Jamaat Ali 264, 313–17, 320, 322, 329–31
Shah, Syed Raghib Abbas 206
Shakargarh 39, 88
Shankardass 252, 258–60, 277
Shastri, Lal Bahadur 170, 233, 383
Shoaib, Mohammad 133–4
Shyok River 15
Siachen Glacier 311, 421–2, 426
Sidhnai: Barrage 18; Canal 28; Feeder Canal 18; Mailsi Link Canal 18
Simla Agreement 233–4, 358
Sindh Sagar Doab 20, 30
Sindh-Punjab Water Dispute 13, 26–36
Singer, Dr. S Fred 336, 339
Singh, Harshdev 369
Singh, Iqbal 182
Singh, Jaswant 425
Singh, Manmohan 10, 273, 316, 375, 378, 421–2
Singh, S. K. 358
Singh, Sarup 51, 61
Singh, Swaran 64, 69, 94
Singh, Teja 87
Sir Creek 422, 426, 435

Sirhind Canal 25–6
Siwalik Range 19, 407
Skardu Dam 356
Sommers, Davidson 110, 134, 140, 170, 185
South Asia 164–5, 344–5, 348, 387, 412, 414, 416, 424, 426, 432, 435
South Asia Network on Dams, Rivers and People 331
South Asian Association for Regional Cooperation (SAARC) Summit 239, 420, 435
Southeast Asia Treaty Organization (SEATO) 167–8
Soz, Saifuddin 255
Spain 24, 55, 323
Spens, Sir Patrick 64–5, 90
Srinagar 16, 221, 223, 227, 311, 431
Standstill Agreement 2, 9, 61–5, 71, 94
Suez Canal Compensation Agreement 170
Suhrawardy, Hussain Shaheed 143
Sukkur 20, 22, 32: Barrage 16, 33–4
Suleiman Range 20–1
Sutlej: River 8, 14–15, 17–20, 23, 26, 28, 31–2, 39–40, 43, 49, 76–7, 90, 94, 106, 110, 116, 118, 131, 184, 188, 368, 372–3, 376, 403, 411, 431; Valley Canals 3, 34, 76; Valley Project 31–2
Swat River 20

T
Tajwar, Imtiaz 314
Tarbela Dam 7–8, 16, 149, 179, 190, 215
Taunsa: Barrage 15; Panjnad Link Canal 15
Teesta: Barrage Project 396; River 396–8
Templar, Sir Gerald 89
terrorism 358–61, 369, 420, 426, 433, 435

Thal: Canal 15, 33; Doab (see Sindh Sagar Doab) 20–1
Thar Desert 20
Tharparker Desert 21
Tibet 14–15, 19, 368, 403
Tippetts-Abbett-McCarthy-Stratton (TAMS) 139
Tipton Report 137–8
Track II Diplomacy 209, 354–5, 401
travaux préparatoires 79, 205, 245, 285, 292, 296, 372
Trimmu: Trimmu-Sidhnai Link Canal 17; Trimmu-Sidhnai-Mailsi-Bhawal Canal 19; Canal 33
Triple Canals Project 28, 30, 223, 227
Truman, Harry S. 165
Tuglaq, Feroz Shah 23
Tulbul Navigational Project (see Wullar Barrage) 4, 10, 221–9, 423, 435
Tvedt, Terje 429, 437

U
United Kingdom (see Great Britain) 49, 91, 334
United Nations (UN) 1, 5, 11, 46, 54, 71, 77–8, 85, 92, 96, 100–1, 103, 105, 107, 153, 160, 166, 170, 176, 184, 194, 196, 233, 245, 274–5, 282, 309, 312, 328, 334, 336, 338, 353, 360–1, 364, 367, 390, 392, 394, 399, 402, 405, 425, 429, 430, 435. Charter 77–8, 85–6, 360, 364, 402; Commission for India and Pakistan (UNCIP) 402; Commission on International Trade Law (UNCITRAL) 275; Convention on Non-Navigational Uses of International Watercourses (1997) 274, 405; Economic Commission for Europe (UNECE) 176; ECE Convention on Environmental Impact Assessment (1991) 5, 189, 287, 293, 312–13, 315, 319; Environmental

Programme (UNEP) 336; Framework Convention on Climate Change (UNFCCC) 275, 312, 334; General Assembly 336; Representative on Kashmir 92; Resolutions 92, 105, 107, 233, 361, 402; Secretary-General 1; Security Council 70, 92, 100, 103, 105, 160, 161, 165, 166, 170, 184, 213, 233, 425; Trusteeship Council 328

United States (US) 95, 164–5, 167, 196, 246, 323, 362–3: Chief Justice 196, 212; Civil Nuclear Deal 425; Department of State 129; Mexico Convention 129, 176–7; National Oceanic and Atmospheric Administration 351; National Security Council 165; Secretary of Defence 381; Secretary of State 165, 173, 184, 436; Senate Committee on Foreign Relations 176, 405, 431; Supreme Court 130

Uprety, Kishore 389, 393–4

Uri Hydroelectric Plant 224

V

Vajpayee, A. B. 6, 224, 235, 358, 363, 419–20

Van Ypersele, Jean-Pascal 346

Verghese, B. G. 356–7

Vienna Convention on the Law of Treaties (1969) 66, 78, 95, 243–6, 248, 256, 276, 295–6, 329, 333–4, 350, 365, 368, 370: and definition 66
- Article 27: 370
- Article 31: 244–6, 253–4, 257, 295–6
- Article 32: 244–5, 254, 257
- Article 46: 370
- Article 52: 78–9, 97
- Article 53: 329
- Article 62: 333–4, 350, 365, 368
- Article 64: 329

W

Wajid, Hasina 398

Warsak Dam 20

Washington, DC 112, 122, 132, 141, 144, 152, 155, 400, 431, 436

Waslekar, Sandeep 400, 418–19

water terrorism 71, 80, 411–12

water wars 1, 367, 428–9

Wheeler, R. A. 108, 112–13, 120–2, 141–2, 172–3, 183

Wikileaks 272, 427

Wilson, Harold 170, 233, 401

Wirsing, Robert G. 91, 267

Wolf, Aaron T. 428–9, 437

World Bank 3–4, 7, 9, 14, 75, 81, 86, 100, 102, 104, 106, 158, 162, 169, 172, 187, 196, 202, 204–6, 208, 237, 254–5, 268, 350, 375, 377, 379, 426, 430: adjusted proposal 140–3, 146–8; aide memoire 140–2, 146, 148, 152, 161; good offices 70, 74, 98, 103, 106, 141, 144, 150, 170–1, 209; Good Samaritan 149, 163; honest broker 100–85; initiative on Indus Basin Dispute 103, 108–15, 120, 166; Operational Manual 207; proposal to break the deadlock 116–25
- Arthur Lall on 124
- Dennis Kux on 125
- Gulhati on 121, 124
- India's Reaction to 121
- Lilienthal on 122, 124
- Nehru on 121, 123, 125
- Pakistan's reaction to 122–3

World Commission for Water 1

World Court 55, 66, 69–70, 83–6, 92, 95, 98, 101, 105, 111, 159, 166, 212, 245–6, 280, 308–9, 312, 323, 328–9, 362, 364, 402: and *ex aequoet bono* provision 201, 267, 280; compulsory jurisdiction 96; Statute 84–6, 201, 212, 280, 308, 327, 360, 389

- Article 30: 327
- Article 38: 201, 280
- Article 59: 308
Wullar: Barrage 4, 10, 192, 221–2, 224–8, 418, 423, 435; Lake 16, 221–3, 229, 281–3, 419
World War I 328, 429
World War II 71

Z

Zada Gorge Barrage 368, 403
Zardari, Asif Ali 273, 421
Zaskar River 15
Zia, Khalida 398
Zira 39–40, 42, 48–9, 51, 53, 56, 88